Telecommunications

Marion Cole
DeVry Institute of Technology

Prentice Hall

Upper Saddle River, New Jersey   *Columbus, Ohio*

Dedicated to My Loving Wife, Mary

Library of Congress Cataloging-in-Publication Data

Cole, Marion.
    Telecommunications / Marion Cole
        p.   cm.
    Includes bibliographical references and index.
    ISBN 0-13-612129-2 (hardcover : alk. paper)
    1. Telecommunication.   I. Title.
TK5101.C635 1999   98-4274
621.382--DC21   CIP

Editor: Charles E. Stewart, Jr.
Assistant Editor: Kate Linsner
Production Editor: Alexandrina Benedicto Wolf
Cover Photo: © Unistock
Cover Designer: Lisa Stark
Design Coordinator: Karrie Converse
Production Manager: Deidra M. Schwartz
Marketing Manager: Ben Leonard
Editorial/Production Supervision: Custom Editorial Productions, Inc.

This book was set in Palatino by Custom Editorial Productions, Inc., and was printed and bound by R.R. Donnelley & Sons Company. The cover was printed by Phoenix Color Corp.

ISBN: 0-13-612129-2

Prentice-Hall International (UK) Limited, *London*
Prentice-Hall of Australia Pty. Limited, *Sydney*
Prentice-Hall of Canada, Inc., *Toronto*
Prentice-Hall Hispanoamericana, S. A., *Mexico*
Prentice-Hall of India Private Limited, *New Delhi*
Prentice-Hall of Japan, Inc., *Tokyo*
Pearson Education Asia Pte. Ltd., *Singapore*
Editora Prentice-Hall do Brasil, Ltda., *Rio de Janeiro*

Contents

Contents

CHAPTER 8 **STORED PROGRAM CONTROL DIGITAL MULTIPLEX SWITCHING SYSTEMS** **204**

Preface

The telecommunications industry has experienced tremendous changes during the last 25 years on both the legislative and technical fronts. This book was written to cover some of the most important changes that have occurred.

This book covers older technologies as briefly as possible. They will be covered in enough depth to provide the reader with an appreciation for how technology has changed and what benefits are offered by the new technologies. *Telecommunications* is written for a non-technical person. It will not provide you with the knowledge necessary to be a telecommunications technician, but it will provide you with an excellent overview of existing technology. Most people in the telecommunications field work in marketing, customer service, operator services, or other support roles. Many managers in these companies have limited knowledge of the technology used to provide telecommunications services, and many have never received technical training.

The main purpose of this book is to provide students with a comprehensive technical overview of the telecommunications technology in use today, but it can also serve to educate existing managers in the telecommunications industry about the technology used to provide the services they are marketing. A technical knowledge base will provide the tools necessary to make more informed decisions. It has been my experience that few people in the industry possess a *broad* technical knowledge base. The industry has been a forerunner in using division and specialization of labor disciplines to achieve high productivity gains. This specialization has led to a narrow, in-depth specialized knowledge base.

I was a telecommunications manager for 30 years, and it is not my intent to alienate existing managers in the telecommunications industry. They do an excellent job of providing our country with the best telecommunications system in the world, but the telecommunications industry is so large and varied it is impossible for anyone to be an expert in all areas of the industry. I think most managers will agree that they could use additional technical training in one or more areas. Talk to any manager or engineer at a switching manufacturer or operating telephone company and it is soon evident that they possess a great deal of knowledge about their specialty, but limited knowledge about areas outside their specialized field. The engineers designing networks for a switch have little knowledge about line or trunk circuit design and even less knowledge about outside plant design. This problem, brought about by specialization, also applies at the local telephone company level. The outside plant engineer usually has little understanding of how a switching system works.

This book can serve as one of the tools used by conscientious managers to broaden their technical knowledge base. Because most managers working for long-distance companies are involved in marketing services to business and residential telephone customers, they have business, marketing, or customer service backgrounds. This book can help these managers gain a better understanding of the telecommunications technology underlying the services they sell. A manager with telecommunications knowledge can deal more effectively with vendors to ensure the right mix of technology at the right price.

Telecommunications provides information about the various components which make up the Public Switched Telecommunications Network (PSTN) and the various technologies used for these components. The major emphasis of this book is placed on switching systems technology, transmission medium technologies, and multiplexing technologies, and how these components are used to form the PSTN.

The PSTN has evolved from a voice-only analog network to a digital network handling both voice and data. The on-ramp to the PSTN remains mostly an analog twisted copper wire pair. Technologies such as Integrated Services Digital Network (ISDN) and Asymmetric Digital Subscriber Line (ADSL) are being implemented to change this last remaining analog component to a digital-access medium. Many analog and digital technologies will be discussed in-depth, but the primary focus is on how the technology works, not how to repair systems or components. That level of detail can be found in an electronics engineering curriculum or in a manufacturing school.

Technicians must attend a manufacturer's school to learn a particular system and how to perform diagnostics on it. The technology behind a particular component in the PSTN is basically the same for all manufacturers of the component. Each manufacturer puts its own twist on a particular generic technology to turn the generic technology into a proprietary system with its brand name on it. This book will help you understand many telecommunications technologies in the generic sense. A technician with a basic understanding of technology, as provided by this book, will find manufacturer training easier to understand and will gain much more from the training.

College graduates who intend to work for one of the telephone companies and current telephone company managers can also benefit from this book. The knowledge gained will allow them to adequately manage the telecommunications technology used for voice and data networks. Prior to deregulation of telecommunications, many telephone companies hired college graduates with management degrees and placed them in a one- to two-year on-the-job training program. These individuals were called "management interns." They came to the company with a formal discipline in management technologies, and the company trained them on telecommunications technology. In a deregulated, competitive environment, companies are forced to reduce costs. One of the casualties of the efforts to reduce costs has been the management intern program. The industry still hires college graduates for most entry-level manager positions, but many of these graduates are placed immediately in supervisory positions without receiving technical training.

The local telecommunication companies have been very successful using non-technicians to manage the business. They leave the technical end of their business to their technicians. This has created a chasm between management and technicians. The technicians cannot rely on their managers for technical support and have lost some respect for management. It is my personal belief that better relations would exist between management

Preface

and technicians if managers had a better understanding of technology. Hopefully, this book will help managers replace the technology training lost with the demise of the management intern programs.

Please e-mail me at mcole@kc.devry.edu with any recommendations to improve future editions. Thanks for your interest in telecommunications.

ORGANIZATION OF THE CHAPTERS

Chapter 1

Introduction to Telecommunications Legislative History: Legislative changes which first shaped the telecommunications industry as a regulated monopoly are discussed. Legislation, which gradually moved the industry toward deregulation and a competitive environment, is also covered. The 1984 MFJ, the Carterphone Case, the MCI Decision, Computer Inquiry I, II, III, and the 1996 Telecommunications Reform Act are among the legislative rulings discussed.

Chapter 2

Telecommunications Technology: This chapter discusses changes in technology from the invention of the telephone and manual switchboard operations to automated switching systems.

Chapter 3

Basic Electricity Review: The transmission of voice and data is accomplished by using electrical signals over wire facilities or by using light signals over fiber optic facilities. This chapter provides a basic understanding of electricity because most telecommunication technologies are electronics-based. A manager needs to understand how electrical signals are impacted by their environment.

Chapter 4

Multiplexing: Frequency Division Multiplexing, Amplitude Modulation, Frequency Division Modulation, AT&Ts L Carrier systems, Pulse Code Modulation, Time Division Multiplexing, T1 Carrier systems, and SONET levels are discussed in detail.

Chapter 5

The Medium: This chapter provides an in-depth discussion on the twisted pair local loop and the design principles behind it. Subscriber Carrier technology and the fiber optic transmission medium are also covered.

Chapter 6

The Telephone Set: The components of a basic single line telephone are covered along with proprietary multiline telephones and key systems.

Chapter 7

Automated Local Exchange Switching: The evolution of switching systems technology is covered. Details are provided on how the switching systems convert dialed telephone numbers into switching instructions.

Chapter 8

Stored Program Control Digital Multiplex Switching Systems: In-depth details are provided on computer controlled switching systems and the digital switching matrix.

Chapter 9

Data Communications: Coverage includes ASCII Code, EIA-232 Interface, DTE, DCE, DSU, Modem, Serial vs. Parallel, Asynchronous vs. Synchronous, CRC, Baud, QAM, TCM, v.34 Modems, v.42bis compression, and Caller ID.

Chapter 10

Signaling: Supervisory, Addressing, Alerting, and Progress Signals are discussed. In-depth details are provided on the Signaling System 7 (SS7) Network and the Extended Super Frame Signaling technique used by T1 systems.

Chapter 11

ISDN and ADSL: This chapter covers Basic Rate ISDN in great detail. Topics covered are: 2B1Q line coding, AMI line coding, NT1, Reference Points, SPID, TEI, Channel D Signaling, LAPD, and ADSL.

Chapter 12

Private Switching Systems: How SPC technology is used to provide communications for a private business and features such as DNIS and DISA are covered.

Chapter 13

Automatic Call Distributors: In-depth coverage is provided on the marriage of an ACD system to personal computers on a Local Area Network (LAN).

Chapter 14

Voice Mail and Automated Attendants: Private and public voice mail systems are covered, along with voice synthesis, and Interactive Voice Response (IVR) technologies.

Chapter 15

Telecommunications Voice Network Design: The basic principles used to design a network, how to read and interpret information from a traffic study, and how to use traffic design tables are covered.

Chapter 16

Telecommunications Voice Network Design: Delays and Queues: This chapter continues the discussion of network design with emphasis on Erlang C Methodology.

Chapter 17

Personal Communication Systems: The evolution of mobile telephone systems from MTS, IMTS, AMPS, and DAMPS to PCS 1900 are covered in detail. Specific emphasis is placed on the design and operation of a PCS 1900 system.

ACKNOWLEDGMENTS

First, I must thank Charles Stewart for his encouragement to write this book. When Charles first approached me about writing a book, I was hesitant. Charles stated that he wanted a book on telecommunications written by me, because I could draw on the knowledge gained from 35 years in the industry. I also owe a huge thank you to Dr. Elizabeth Judd for correcting my grammatical errors. Thank you, Dr. Judd!

I drew upon my knowledge of telecommunications technology and managing telecommunications operations to write this book. However, I have never claimed to know everything. I simply know a little about most technology used in telecommunications. When I need more detail on a particular technology, I simply call an expert in that area or get a book on the subject. Therefore, I must thank every one of my past coworkers. I learned a lot from them. My knowledge of telecommunications was gained from reading books, attending company schools, attending manufacturer's schools, and on-the-job training. I have learned from all the authors of books I have read and all of the companies I have worked for. Nothing in my book is new or original. What I know, I learned from other people. I thank all of my past professors, teachers, and instructors.

The information contained in Chapters 15 and 16 was largely influenced by a fellow professor at DeVry, Karl Crum. Karl provided me with numerous handouts and Traffic Tables to teach TCM-410 (Systems Administration) at DeVry. Much of the information contained in those handouts has been incorporated, with Karl's permission, into Chapters 15 and 16. Thanks, Karl, for your assistance! I also wish to thank James Jewett, Jackie Shrago, and the Telco Research Corporation for their permission to use the 50% Table of Appendix E, the 100% Table of Appendix F, and the EQEEB Table of Appendix I. The methodology behind these tables is discussed in Chapter 16 with their permission. We are not sure who to credit for the Traffic Tables contained in Appendixes G and H. These tables have been modified from time to time by various authors. These tables are generated by computer outputs based on input data. We use a Traffic Design program at DeVry to generate these tables, and then modify them to suit our needs. I do not know who to credit for the original development of Traffic Tables outside of Mr. Poisson, Mr. Erlang, and Mr. Jacobsen. We also thank AT&T, because Mr. Jacobsen worked for them, and AT&T developed most of the engineering methodology used in designing telecommunications networks.

I would also like to thank the following reviewers for their invaluable feedback: Richard Anthony, Cuyahoga Community College; Robert Diffenderfer, DeVry Institute of Technology; and Jeffrey L. Rankinen, Pennsylvania College of Technology.

Finally, and most importantly, I thank my wife Mary for her encouragement and support in everything I do.

1

Introduction to Telecommunications Legislative History

Key Terms

Bell Operating Companies
(BOCs)
Carterphone Decision of 1968
Common Carrier
Communications Act of 1934
Computer Inquiry
Computer Inquiry II
Computer Inquiry III
Customer Provided
Equipment (CPE)
Deregulation
Divestiture
Equal Access
Feature Group
Federal Communications
Commission (FCC)
Graham Act of 1921

Interexchange Carriers (IECs
or IXCs)
Kingsbury Commitment
of 1913
Local Access Transport Area
(LATA)
Local Exchange Carrier (LEC)
MCI Decision of 1976
MCI Ruling of 1969
Modified Final Judgement
(MFJ)
Monopoly
Network
Preferred Interexchange
Carrier (PIC)
Point of Presence (POP)
Private-Line Services

Public Switched Telephone
Network (PSTN)
Public Utilities Commission
(PUC)
Regulated Monopoly
Revenue Sharing
Rural Electrification
Administration (REA)
Specialized Common Carrier
Decision of 1971
Station Equipment
Tariff
Telecommunications
Telecommunications Reform
Act of 1996

Telecommunications is the communications of voice or data over long distances using the *Public Switched Telephone Network* (PSTN) or privately owned *networks*. Many legislative and technological changes have affected the PSTN in the last two decades. Legislation governing the telecommunications industry has changed a great deal over the past 125 years. Laws covering the telephone industry did not exist when the industry was in its infancy, but as the *Bell Operating Companies (BOCs)* grew, such laws were enacted to direct growth in the industry. Initial legislation led to a regulated and monopolistic industry. Over the past 20 years, legislation has been aimed at deregulation of the industry to encourage competition.

1.1 MONOPOLY SERVICE PROVIDERS

Initial legislation provided for a regulated monopoly because a monopoly service provider could gain "economies of scale" to provide services at a lower cost. This is the basic economic principle underlying any *monopoly*. Legislators thought that by regulating the monopoly service providers, they could ensure that the monopoly would provide low rates. Legislators have done an about-face. They now believe a competitive environment will lead to lower rates. Some people question the wisdom of this argument. It is well known that a monopoly provider can provide service at a lower cost than any other form of business. The problem with a monopoly is its tendency to overcharge for services. If the government agencies charged with regulating the monopoly do their job, the cost savings achieved by the monopoly should be passed on to customers.

1.2 REGULATION OF MONOPOLY SERVICE PROVIDERS

Regulated monopolies are allowed to charge a rate for their services that provides them with a certain return on their investment. How much a monopoly can earn and thus the return on investment it is allowed to earn is set by the government regulatory agencies. The government agencies monitor rates and investments of the monopoly to ensure that rates are not inflated. If the government agencies do not do their job well, the monopoly service provider will find ways to inflate rates. Since the rate of return on an investment for a regulated monopoly is set, the only way for a regulated monopoly to raise what it earns is to increase what it invests.

The amount of investment allowed is the area that must be controlled to control rates. Some legislators believe that monopoly service providers have little incentive to improve productivity and lower costs. The higher their costs are, the more money they can earn. These legislators believe that rates are inflated because the monopoly service providers are not seeking to keep costs low. They believe costs will be lower in a competitive environment. This is tantamount to saying government regulators have not performed their job properly. If in fact these regulators have not been able to properly regulate what the monopolies have been doing, competition may result in lower rates. As we have seen with the deregulation of other industries, an initial reduction in rates can be achieved by sacrificing customer service.

1.3 INDEPENDENT TELEPHONE COMPANIES—1893

Between 1870 and 1875, several individuals were working on devices to convert sound waves into electrical energy that could be sent over telegraph lines and then to convert the electrical energy back into sound waves at the other end. Alexander Graham Bell was one of the successful inventors. He filed a patent for the telephone in 1876 and formed the Bell Telephone Company in 1877. The patent on his device ran out in 1893, after which others could use his technology to form their own telephone company. This led to the establishment of other telephone companies called "Independents." By 1894 there were about 90 Independents. In 1900 there were approximately 4000 Independents serving about 40% of all telephone customers. BOCs served the other 60%. Bell had created telephone companies in most large cities but did not consider rural communities profitable. Independent companies were formed all across America in large and small towns. Some Independents even competed directly against Bell. The largest cities had several telephone companies competing against each other and refusing to interconnect services. This meant a person needed phone service and a phone from each company. Bell bought out the Independents occupying prime territory. In 1882, Bell bought the Western Electric Company and in 1885 incorporated as American Telephone and Telegraph Company (AT&T). AT&T purchased large amounts of Western Union stock between 1900 and 1913 and bought out major competitors in their local telephone markets.

1.4 THE MANN-ELKINS ACT OF 1910

In 1910 Congress had passed the Mann-Elkins Act as the first step toward regulating the telecommunications industry. The Interstate Commerce Commission (ICC) was placed in charge of enforcing the regulations of the Mann-Elkins Act.

1.5 THE KINGSBURY COMMITMENT OF 1913

The Department of Justice was concerned that AT&T was monopolizing the telecommunications industry and was contemplating using the Sherman AntiTrust Act to bring charges against AT&T. The acquisitions made by AT&T had substantially lessened competition. In response to the concerns of the Justice Department, the vice president of AT&T, Nathan Kingsbury, made a commitment called the Kingsbury Commitment. The *Kingsbury Commitment of 1913* stated that AT&T would not buy any more Independents without Justice Department approval and would allow the connection of Independent companies to the AT&T network. This meant homes only needed one phone and one telephone company to provide service. The provider of their service could interconnect with other Independents and AT&T to complete calls anywhere in America. The Kingsbury Commitment also stated AT&T would sell its interest in Western Union. By entering into this agreement, AT&T successfully headed off action by the Justice Department and prevented sanctions by the Sherman AntiTrust Act.

1.6 THE GRAHAM ACT OF 1921

Between 1913 and 1921 AT&T lobbied the government to exempt telecommunications from the Sherman Antitrust Act. In 1921, Congress passed the Graham Act, which made the telecommunications industry (and AT&T) exempt from the Sherman Antitrust Act. As a result, companies were able to consolidate and become monopoly service providers. This measure prevented duplication of service providers and encouraged AT&T to build a nationwide network. Congress believed that a monopoly service provider under government control could provide phone service at the lowest cost to the public, but the *Graham Act of 1921* did not provide the controls needed to ensure low cost for customers. Congress recognized the error of this legislation and passed the Communications Act of 1934 to regulate the interstate telecommunications provided by AT&T.

1.7 THE COMMUNICATIONS ACT OF 1934

The *Communications Act of 1934* was designed to regulate broadcast radio and interstate telecommunications. This act also made it possible to pursue certain actions under the Sherman Antitrust Act. In 1910 Congress had passed the Mann-Elkins Act as the first step toward regulation. Under this act, enforcement was under the control of the Interstate Commerce Commission (ICC). The ICC was ineffective in its role as regulator of the telecommunications industry. The Communications Act of 1934 addressed this concern by establishing the *Federal Communications Commission (FCC)* to oversee interstate telecommunications regulation. The act has been modified over time but still regulates the telecommunications and broadcast industries. Congress left the regulation of telecommunications within a state (intrastate) to the jurisdiction of each state. States passed legislation and set up a Public Utilities Commission (PUC) in each state to govern intrastate telecommunications.

1.8 TARIFFS

Regulation required telephone companies to file a document detailing any new service and the proposed charge for the service. The document defining in detail any new service is called a *tariff*. For interstate services under the jurisdiction of the FCC, the tariff is filed with the FCC. For intrastate services the tariff is filed with the governing state's PUC. The tariff is a public document. Any company can view the tariff filed by a competitor and gain an understanding of the technology required as well as the costs and profit the competitor has anticipated. When AT&T was a monopoly service provider there was no competitor, and publication of tariffs served the public interest. With deregulation and competition, the filing of a tariff puts the filer at a competitive disadvantage. Therefore, the requirement to file tariffs has been relaxed as the industry has become deregulated. Local telephone services are still regulated and Tariffs must be filed with the PUCs. When the phone company desires to raise rates for local phone service, a new tariff must be filed and approved by the PUC. The regulation of rates by the government guaranteed that the monopoly service providers were giving fair rates to customers.

1.9 THE RURAL ELECTRIFICATION ACT OF 1936

During the first half of the 19th century, rural America was still being deprived of electricity and telephone service. It was more profitable to install equipment and wiring in large cities where the density of homes per square mile was high. The cost of equipment could be spread over a large customer base. In rural areas, wire had to be run many miles just to serve a few homes. To encourage investment in these rural areas, the Rural Electrification Act was passed in 1936 and created the *Rural Electrification Administration (REA)*. This act made low-cost, government-backed loans available to entrepreneurs willing to establish electrical service in rural areas. In 1949 the act was extended to cover low-cost loans for the establishment of telephone service in rural areas. The REA has been renamed and is part of the Rural Utilities Services of the U.S. Department of Agriculture.

1.10 STATE LEGISLATION: PUBLIC UTILITIES COMMISSION

State legislation guaranteed telephone companies a positive rate of return on their investment. A company could install electric or phone service in a community and provide the PUC with documentation on the amount of money invested to provide that service. The PUC would depreciate the investment over a definite time frame (5 to 25 years depending on the asset), add yearly maintenance expenses, and determine the rates a phone company could charge customers to receive as much as a 7% return on their investment. The investment could be borrowed from the REA at rates as low as 1%. A company could make 6% a year profit by providing the entrepreneurial resources and borrowing needed capital from the REA. The REA loan rates today are between 5% and 10%. The higher cost of money requires the PUC to authorize higher rates of return to the telecommunications companies. The PUC also grants a franchised territory to each company. Only one company could provide local telephone service and the provider of that service was regulated by the PUC. The regulations established by state legislatures eliminated multiple providers of phone service and the problems associated with multiple telephone companies in one town refusing to cooperate. Today, deregulation is beginning to impact the franchised territory principle. In 1996, Congress passed legislation leading to the authorization of more than one provider for local service in the same community.

1.11 AT&T CONSENT DECREE OF 1956

The Department of Justice (DOJ) sued AT&T in 1949 for violations of the Sherman Antitrust Act. The BOCs would only purchase equipment from Western Electric (both companies were part of AT&T). Absence of competition in the sales of equipment to Bell companies effectively prevented regulation of rates charged by these companies. Western Electric could charge high prices on sales of equipment to the BOCs. The BOCs included the cost of equipment in the tariffs filed for rate increases. High costs for equipment meant high rates for phone service. Western Electric was not a regulated company and could charge whatever price the market would pay. The BOCs were a captive market. They would pay whatever price Western Electric charged because they were guaranteed a return of the investment, plus a profit on the investment.

Independent telephone companies would buy their equipment from manufacturers offering the lowest price. The DOJ suspected that Western Electric was selling equipment to Independents at a much lower price than it sold to the BOCs. The DOJ thought Western Electric was undercutting the pricing of smaller equipment manufacturers. This type of anticompetitive practice could be very detrimental to the smaller equipment manufacturers. The DOJ was looking into allegations that AT&T was involved in the actions just described and determined that the results of its investigation warranted antitrust action.

To prevent the use of the Sherman Antitrust Act against it, AT&T settled the suit by reaching an agreement with the government. This agreement is known as the *Consent Decree of 1956* (it is also called the *final judgment of 1956*). AT&T was permitted to keep Western Electric but was ordered to limit its sales to the BOCs and to sell at competitive prices. The decree forced the Independents to buy switching equipment from companies such as Northern Telecom, Automatic Electric, Leich Electric, Stromberg Carlson, and Northern Electric. This ruling allowed Automatic Electric to become a major equipment provider to the Independents in the 1960s and 1970s. In the 1980s, Northern Telecom became a major switching systems manufacturer supplying computer-controlled switching systems to both Independents and the Regional Bell Operating Companies (RBOCs).

1.12 CARTERPHONE DECISION OF 1968 BY THE FEDERAL COMMUNICATIONS COMMISSION

In 1968, the FCC ruled that AT&T must allow customers to use a device called the *Carterphone*. This device would let ham radio operators attach a phone line to their radio system. The ham operator could use the ham radio to establish connections across the country with other ham radio operators. The ham radio operators would then use their phone line to call someone in their town and connect the called parties together with the Carterphone. Prior to the *Carterphone decision of 1968,* AT&T would disconnect phone lines where it found a Carterphone connected. Carter Electronics Corporation of Dallas sued AT&T and won. This is the decision that eventually let customers buy their telephones and business phone systems from someone other than the phone company. These alternate equipment providers were called *interconnect companies*.

1.13 STATION EQUIPMENT AND CUSTOMER-PROVIDED EQUIPMENT DEREGULATION: 1981 COMPUTER INQUIRY II

The telephone belongs to a larger classification of equipment known as station equipment. *Station equipment* is the equipment located at the telephone customer's residence or business location and is generally a telephone, modem, or personal computer terminal with internal modem. Station equipment was the first classification of telephone equipment to be deregulated. When customers provided their own equipment, it was called *customer-provided equipment (CPE)*. Now that telephone companies no longer provide station equipment as part of the services covered by the monthly service charge, all customers must provide their own station equipment and all station equipment is customer owned. Deregulation has reached out to include more than station equipment. Another area to be deregulated in the early 1980s was private telecommunications systems for businesses (keysystems and private switching systems called *private branch exchanges* or PBXs). If a business is large enough to require a keysystem or PBX, it must own or lease these systems from an interconnect company. The local telephone company could no longer provide these systems as of 1981 due to the Computer Inquiry II legislation. The Telecommunications Reform Act of 1996 allows local telephone companies to reenter this market by striking down the provisions of Computer Inquiry II.

All BOCs and Independents were forced by the deregulation rulings of the FCC (Computer Inquiry II of 1981) to set up separate subsidiaries as interconnect companies to handle the marketing and sales of keysystems and PBXs to business customers. The federal government felt a separate subsidiary was necessary to prevent the local telephone companies from using the revenues from local telephone service charges to subsidize discounts on sales of CPE. The existing business communications and marketing organizations of the local telephone companies were spun off in 1982 as separate subsidiaries that could not receive funds from their parent companies. The new subsidiary could sell CPE. The meaning of CPE was changed from "customer-provided equipment" to "customer-premise equipment." CPE includes all equipment at the customer's location that must be owned by someone other than the telephone company, according to the 1981 Computer Inquiry II legislation. In 1996, Congress passed legislation opening the way for competition in the local phone service markets. As part of this legislation, telephone companies are allowed to sell CPE and no longer need a separate subsidiary to handle CPE sales and service. Shortly after the passage of the Telecommunications Reform Act of 1996, many phone companies began merging the CPE subsidiary they spun off in 1982 back into the local phone company organization.

1.14 INTERCONNECT COMPANIES

AT&T fought deregulation of the telephone instrument on the grounds that interconnect companies would provide inferior equipment for attachment to the network and degrade its performance. The FCC initially tried to alleviate these concerns by ruling that an interconnect device would be required when using a piece of equipment not provided by the telephone company. The interconnect device was furnished by the local telephone company and was placed between the CPE and phone company equipment. This added unnecessary cost and inconvenience to purchasers of interconnect equipment. The manufacturers of telephones sold their telephones to telephone companies and to retail stores. If you bought the phone from a retail store, you needed a protective coupler from the telephone company. If you had the telephone company supply that same manufacturer's phone as part of your service, no protective coupler was needed. The FCC recognized the error of its earlier ruling on the need for protective devices and dropped the requirement for an interconnect device. In lieu of such a device, the FCC requires registration with it of all equipment intended for connection to the telephone network. Through the registration requirement, the FCC ensures that all telecommunications station equipment provided by manufacturers to interconnect companies, retail outlets, and local phone companies meets minimum established service standards.

1.15 MICROWAVE COMMUNICATIONS INC. 1969 RULING BY THE FEDERAL COMMUNICATIONS COMMISSION AND SPECIALIZED COMMON CARRIER RULING OF 1971

In 1969, the FCC ruled that AT&T could not prohibit customers from using local phone lines for access to Microwave Communications Inc. (MCI). The *MCI ruling of 1969* made it possible for businesses to use companies other than AT&T for private business communication networks. In the *Specialized Common Carrier Decision of 1971*, the FCC broadened the scope of the MCI decision. It allowed any qualified common carrier the right to carry *private-line* services. Southern Pacific Railroad was quick to join MCI in providing private line networks for business customers. In 1975, MCI begin providing long distance service to the general public. AT&T filed a complaint with the FCC asking it to stop MCI from providing regular long distance service. The FCC dutifully complied with the request and instructed MCI to cease providing this service. MCI sued, and the appeals court reversed the FCC ruling in 1976. As a result of the *MCI decision of 1976*, the FCC opened the long distance market to competition. The FCC ordered AT&T to provide access, via the BOCs, to MCI customers. In addition to providing private networks for businesses, MCI, Southern Pacific, and other companies could now provide long distance service to anyone.

1.16 DEREGULATION

Deregulation of the industry began in 1968. From the late 1960s until today, proponents of deregulation have steadfastly held that deregulation would provide consumers with more choice and stimulate competition within the telecommunications industry. Proponents stated this competition would lead to lower costs for telephone service. Opponents of deregulation have held that deregulation would lead to inferior service and higher costs. History has proved they were both wrong (but the opponents to deregulation were partially correct). Deregulation has not led to inferior service. In fact, service has improved. Long distance phone service was deregulated but local service is still regulated. Deregulation of long distance service reduced long distance charges but led to increased local service rates.

Prior to 1984, AT&T shared long distance revenues with the local phone companies. With deregulation in 1984, the local phone companies lost this important source of revenue. To ensure that they still received a guaranteed rate of return on their investment, these companies have been permitted to increase local rates and charge customers for access to long distance companies. Access charges were developed as a means of compensating the local phone companies for losing their share of toll revenue.

The long distance service providers and all telephone users pay access charges, which replaced *revenue sharing*. Access charges account for the largest expense for a long distance service provider. Local phone rates are also higher than they were with sharing of toll revenue. If long distance rates were a lot lower than they were prior to regulation, perhaps they would offset higher local service costs and access charges, but long distance rates have not fallen enough to offset the tremendous rise in local service rates. Business customers do have a significantly lower overall cost due to deregulation of long distance. Because of the volume of long distance calls they make, they can negotiate large discounts with the long distance service providers. The loss of long distance revenue to the local telephone companies had to be offset somehow. It has been offset by the higher rates paid by residential and business customers for local service.

In the past, business users complained that long distance and business charges to them were extraordinary high and they were subsidizing residential service. It is true that toll revenue and business rates subsidized local residential phone service. This made it possible to keep local residential service rates low and almost everyone could afford phone service. It is also true that a business customer uses the phone system much more than a residential user and should pay more. Now that the residential customer is paying rates that subsidize business rates, no one is complaining. In this respect deregulation has been a success. Business customers have a far wider range of choices for services, and competition results in lower costs for business. But you would have a hard time convincing opponents of deregulation that the residential customer has benefited. In 1980, residential service was about $10 a month and the phone was provided as part of the service. Today residential service is approximately $30 a month and the customer must purchase the phone.

1.17 HIGHER LOCAL SERVICE RATES AS A RESULT OF DEREGULATION

The local telephone companies have started filing tariffs to increase local rates again. They are citing the Telecommunications Reform Act of 1996 as one of the reasons they need to increase local service rates. They are stating that under the new law they must lower the access charges they are making to the long distance carriers (*interexchange carriers*, known as *IECs* or *IXCs*). Losing the IEC revenue must be offset by charging higher local rates. The PUCs have held hearings on this issue and have raised local residential rates again to offset the loss of revenue from access charges. It remains to be seen if deregulation of the local service arena will create competition that will lower local service rates. The local telephone companies (*local exchange carriers* or *LECs*) have reduced operating expenses by making capital investments in electronic switching systems, computerized record systems, and new outside plant. These systems require less maintenance and have led to dramatic productivity improvements. These improvements have allowed telephone companies to make drastic reductions in management personnel and the craft workforce. With deregulation, many people install their own CPE. This has reduced the workload of the LECs and has allowed the LECs to reduce personnel.

Although the LECs have dramatically reduced costs, these reductions have not been passed on to the consumer in the form of a rate reduction. The vast majority of all tariffs filed seek permission to increase rates. It is rare for a LEC to seek a reduction in rates. Because most PUCs approve most requests from LECs to raise rates, it appears as if the motto of many PUCs is: "If you file a tariff, we will approve it." The PUCs hold public hearings on the requested rate increase but schedule them at times and places that limit attendance by the public. Although they seem to approve all rate hike requests, the PUCs have ordered rate reductions in some instances. If the PUCs had not ordered these rate reductions, it is highly improbable they would have been made voluntarily by the LECs. If in fact local rates are artificially high, competition in the local services arena will lead to lower rates. Conversely, if competition does lead to lower rates, we know that our PUCs were allowing the LECs to overcharge us. If deregulation does not lead to lower rates, deregulation of local service is a failure from the consumer's point of view. In a deregulated environment, there is not much need for regulators; the competitive market serves as its own regulation device. Thus, elimination of regulated monopolies should lead to a reduction of personnel in the PUCs. It remains to be seen if our government agencies reduce their workforce in response to a reduced workload.

1.18 COMPUTER INQUIRY I OF 1971, COMPUTER INQUIRY II OF 1981, AND COMPUTER INQUIRY III OF 1986

Several rulings by the FCC during the 1970s and 1980s were aimed at defining whether regulation should apply to computer and data services; each ruling was known as a *Computer Inquiry*. The first was *Computer Inquiry I* in 1971. This ruling was issued after a study of the data industry by the FCC. The study found that the computer industry should not be subject to regulation and was beyond the FCC's control. In 1981 the *Computer Inquiry II* ruling was issued. As mentioned earlier, this ruling deregulated CPE and required telephone companies to set up separate subsidiaries to handle CPE. This ruling also allowed the BOCs to enter the nonregulated data processing business. In 1986, the *Computer Inquiry III* ruling was issued detailing the extent to which AT&T and the Bell companies could compete in the nonregulated enhanced services arena.

1.19 DIVISION OF TOLL REVENUE PRIOR TO 1984

The major opponents to deregulation were AT&T and the telephone companies. AT&T was the only provider of long distance service, and its BOCs furnished toll access to the Independents. Through toll revenue sharing, the Independents received a portion of the toll revenue their customers paid. If the customers of a local Independent phone company generated $100,000 a month in toll billing, the Independent would pass this amount on to Bell. The Independent would submit documents to Bell outlining what percentage of calls were toll calls and what percentage of their total investment was apportioned to provide both incoming and outgoing toll call completion. This process was called *separations and settlements*. Most Independents received more than half of the toll revenue they collected. Some Independents had Separations and Settlements Departments that provided documentation, which resulted in the Independent getting more money back than it gave Bell. In the example above, the Independent would probably receive $106,000 for the $100,000 paid to Bell. Without the Independent companies, Bell customers would not be able to make long distance calls to the Independents' customers. Bell kept all toll revenue generated by its own customers. Paying the Independents more toll revenue than their customers generated was basically a way to pay the Independents something to complete the toll calls of Bell's customers to the customers of the Independents. Deregulation of the telecommunications industry led to abolishing the separations and settlements process. The revenue lost from the demise of this process was replaced with access charges.

1.20 MODIFIED FINAL JUDGMENT OF 1984: DIVESTITURE AND DEREGULATION

Deregulation of the telephone industry has occurred mostly as a result of court rulings. Carter Electronics had to file suit to end anticompetitive practices by AT&T. It took a lawsuit filed by MCI against AT&T to open the doors to competition in the long distance market. The FCC was not in the forefront of establishing a competitive environment. Allegations of anticompetitive practices against AT&T led to involvement of the Justice Department in deregulation of the industry. The Justice Department begin to pursue antitrust action against AT&T and filed an antitrust lawsuit against AT&T on November 20, 1974. AT&T fought the lawsuit for eight years and then settled out of court, in 1982, with another "compromise." AT&T had come to regret the 1956 compromise agreement because it limited AT&T's business ventures to telecommunications. AT&T wanted to get into the computer industry but the 1956 agreement prevented it from entering the computer market. The 1974 lawsuit was settled on August 24, 1982, as a modification of the 1956 Consent Decree's Final Judgment and is known as the *Modified Final Judgment (MFJ)*. The MFJ took effect on January 1, 1984, replacing the 1956 Consent Decree.

The 1984 MFJ addressed *divestiture* and deregulation. AT&T agreed to divest itself of the BOCs but one wonders if this action really hurt AT&T. It appears that AT&T, not the DOJ, actually drafts these settlement agreements. The 1996 Telecommunications Reform Act allows AT&T to reenter the local telecommunications service market. AT&T will be using newer technology than it left behind with the BOCs. AT&T has aggressively entered several prime markets and is competing against the LECs and other IECs in these local markets to provide customers with local telephone service. Remember that local service was also defined as toll calls within a *local access transport area (LATA)*. All major IECs can compete against the LECs for this business and can use their existing networks to provide intra-LATA toll service.

1.21 RBOCS, LECS, IECS, AND COMMON CARRIERS

On divestiture, 21 BOCs were called *Baby Bells*. The Baby Bells merged and formed 7 RBOCs. These RBOCs are Atlantic Bell, Ameritec, Bell South, Nynex, Pacific Telesis, Southwestern Bell, and US West. These regional companies are made up of several of the original 21 BOCs. For example, Ameritec includes the former companies of Ohio

Bell, Indiana Bell, Illinois Bell, Michigan Bell, and Wisconsin Bell (Figure 1-1). These local BOCs and local Independent telephone companies are now called LECs. Long distance service providers (AT&T, MCI, SPRINT, and so on) are called IECs.

Companies offering telecommunications service as part of the PSTN are all referred to as *common carriers*. In addition to LECs and IECs, there are *satellite carriers* (such as RCA and American Satellite Company), *specialized common carriers* (SCCs) (such as Metropolitan Fiber), and *value added network carriers* (VANs) (such as GTE Telenet). Electronic Data Interexchange (EDI) networks are VANs. Some SCCs own their own networks; other SCCs use the networks of other carriers and act as resellers. SCCs such as Metropolitan Fiber offer business customers an alternative to using the LEC's local loop and central offices to access IECs. In large cities, Metropolitan Fiber has a cable network that rings the city and connects to IECs. A large business can buy IEC access from Metropolitan Fiber and bypass the LEC. Telenet is the oldest VAN in the United States and offers services such as electronic mail and access to services such as Prodigy and America Online (AOL) via Telenet's packet switched network. This packet switched network is made up of leased circuits from an IEC tied into packet assembler/disassembler (PADs) hardware owned by Telenet. Sprint also offers access to large, commercial Internet service providers (such as AOL and Prodigy) via its packet switched network called SprintNet. Common carriers provide service to the general public as part of the PSTN. They also provide services for private networks. Private networks are owned and operated by companies primarily for their own use but may lease services to other companies. Private networks may be constructed by leasing circuits from common carriers to form part or all of the private network.

FIGURE 1-1 A diagram of Regional Bell Operating Company territories.

1.22 EQUAL ACCESS TO LONG DISTANCE

In addition to forcing AT&T to divest itself of the BOCs, the MFJ contained language to deregulate the long distance market and required the BOCs to provide all IEC customers with *equal access* to their IECs. The MFJ ruling also required that local BOCs furnish space in their central offices for IECs to place their equipment. This space provides a point at which the LEC can connect customers to an IEC and is called the *Point of Presence (POP)*. Since the 1976 ruling opening the long distance market to competition, AT&T had an advantage in marketing its service because customers of AT&T only had to dial the digit 1 for access to AT&T long distance. Customers of MCI, Sprint, and other IECs had to dial a seven- or ten-digit phone number to access the IEC. After accessing the IEC, the customer had to dial a personal identification number (PIN) of four digits, then dial the area code and seven-digit phone number of the called party. AT&T customers dialed eleven digits to place a long distance call: 1 + area code + seven digit phone number. Other IEC customers dialed 25 digits to place a call: 1 + 800 + seven digits + four-digit PIN + area code + seven-digit phone number.

When AT&T handled all long distance calls, it could be reached by dialing the digit 1. The digit 1 was interpreted by the translator of the central office switch as a call that needed connection to an AT&T trunk circuit, and the call was connected to an AT&T toll trunk. This is referred to as *trunk side access*. Much information can be passed over trunk signaling circuits such as the originating caller's telephone number for billing purposes. Equipment called *Automatic Number Identification (ANI)* was added to each line circuit. On long distance calls, ANI automatically provided the caller's telephone number to AT&T. When other IECs came into being, they could not use the digit 1 for access because it had already been assigned to AT&T. Only one IEC could use the digit 1 for access to toll and since AT&T had already been given it, AT&T customers had 1+ access. It was necessary to use telephone numbers to reach other IECs. The use of telephone numbers for access is called *line side access*. The originating callers' telephone numbers (ANI) could not be passed through a line circuit; thus, line side access required callers to identify themselves to the IEC billing machine through the use of PINs. Line side access is called *Feature Group A* access (one of several types of *Feature Group* access).

1.23 FEATURE GROUP B ACCESS

The BOCs devised a way to provide trunk side access to non-AT&T IECs. This access was called *Feature Group B*. Feature Group B used the code 950-10*XX*. The last two digits of the code identified a unique IEC. This cut down on the number of digits required to place a long distance call using an IEC other than AT&T. The switching system could identify IECs from the translation of the seven-digit 950-10*XX* number and route the call to a trunk circuit instead of a line circuit. With trunk side access, no PIN was needed because the caller's telephone number was available via ANI over the trunk circuit to the IEC.

1.24 FEATURE GROUP D ACCESS

It was technically impossible for Bell to provide equal access in older electromechanical local exchanges. To provide all IECs with equal access via the digit 1 (*Feature Group D*) would require the replacement of these offices with computer-controlled switching systems. On dialing the digit 1, a computer-controlled system can look at the originating line's database to determine its *preferred interexchange carrier (PIC)* and route the call to that IEC. The Justice Department recognized that time would be required to comply with this MFJ directive and gave the BOCs until 1987 to meet the provisions of equal access. True equal access where all IECs can be accessed by the digit 1 is called Feature Group D. A customer can inform the LEC of a PIC, and the line for that customer is PICed to that IEC by making an entry in the switching equipment's database. Feature Group D also allows a caller to override a PIC and access any IEC by dialing the Feature Group B access code (10*XXX*). The access code for Feature Group B was changed from the seven-digit 950-10*XX* to the five-digit 10*XXX*. To use AT&T, callers dial 10288. To use Sprint, 10333 should be dialed. Other IECs also have five-digit Feature Group B access codes of the form 10*XXX*.

Deregulation required that the LECs assign IECs to pay stations in private business locations, based on the desires of the proprietor of the business where the pay phone is located. All pay phones located in public places must be assigned to an IEC by the public authorities or on an allocation basis by the LEC. The PIC used at pay stations might prevent people from using calling cards. If a pay station is picked to Sprint, it will not accept an AT&T calling card and vice versa. Feature Group B access allows callers to override the PIC. The 10*XXX* code allows pay phone users to override the PIC (the assigned IEC) and use the IEC of their choice. This allows them to use the phone card provided by their IEC.

1.25 UNITED TELECOMMUNICATIONS, INC. ADAPTS TO THE ERA OF DEREGULATION

Even though telephone companies originally opposed deregulation, some have benefited from this development. As any company should, they have been quick to adjust to the new rules of the game. A prime example would be United Telecommunications, Inc. United was an aggressively growing local exchange company in the 1970s. With deregulation of the long distance market, it quickly seized the opportunity to enter this market. It purchased some railroad right of way to lay fiber optic cable. It formed alliances with Southern Pacific Railroad and later with General Telephone. It subsequently bought out both companies and is the sole owner of Sprint. The company that grew up as United Telecommunications and formed a subsidiary called Sprint has seen the child grow up and take the dominant role. United Telecommunications companies are now subsidiaries of Sprint.

The major portion of Sprint's revenues come from its deregulated long distance services, not from the regulated local exchange companies (although most profit is still generated by the LECs). Company stockholders have benefited from deregulation due to the visionary leadership at United Telecommunications. Where others could only see deregulation as a problem, they saw deregulation as an opportunity. They have aggressively pursued this opportunity and are now forming alliances with telecommunications companies overseas to become as successful in providing intercontinental long distance. Sprint has also formed a division to provide local phone service in RBOC territories. Sprint will use personal communication systems (PCS) technology as well as alliances with cable TV companies to provide local telephone service.

1.26 LOCAL EXCHANGE CARRIERS

Before the deregulation movement got underway, there were about 14,000 Independent (non-Bell) local telephone companies. These Independent companies and the local BOCs provide local telephone service and are LECs. The LECs—briefly discussed earlier—are the local phone companies. They connect our telephones to a central office, from which we are able to call anywhere in the world over the *Public Switched Telephone Network (PSTN)*. When we pick up our telephone handset, we are automatically connected to switching equipment in the local central office. The switching equipment will receive dialed digits from our telephone and determine from those digits where the call should be switched to. Prior to deregulation, the LECs provided local service and AT&T was the sole provider of long distance service. Today there are several long distance service providers. As we have seen, companies offering long distance service are called IECs or IXCs.

Many of the Independent LECs have consolidated, and today there are about 1300 Independent LECs in addition to the 23 BOCs. The LEC is the carrier authorized to provide a physical connection from telephones to the telecommunications network. Recent legislation has approved competition for local telephone service in several areas of the United States. In these areas, the LEC must rent facilities to other competitors wishing to provide local exchange service using wire facilities. Some competitors will use PCS to provide local service. PCS employs signals similar to a mobile telephone and pager. PCS does not require a pair of wires to connect a phone to the central office switch. PCS providers will install their own switching systems and use radio waves to connect their switching system to their customers. The PCS switch will have trunks connecting it to the LEC switching center, and PCS customers will be able to call LEC customers.

1.27 LOCAL ACCESS TRANSPORT AREAS

The MFJ divided the country into 184 areas known as LATAs. A LATA is considered a local toll calling area, and the MFJ recognized toll calls within a LATA as calls that must be handled by an LEC and not an IEC. Many LATAs will serve the same basic geographic boundary covered by an area code. However, some LATAs serve more than one area code. LATA 524 has three POPs in Missouri (Kansas City, Moberly, and St. Joseph). The Kansas City, Missouri, POP serves customers making a toll call from both Kansas and Missouri. The area code for Kansas City, Kansas, is 913, and the area code for Kansas City, Missouri, is 816. Thus, this POP serves two area codes and also serves customers in two states (Figures 1-2 and 1-3).

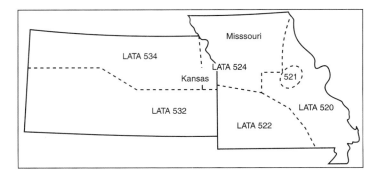

FIGURE 1-2 Kansas/Missouri LATAs.

1.28 POINT OF PRESENCE

In most LATAs the majority of intercity calls within the LATA (intra-LATA) are handled by the BOC for that LATA. Of the original 184 LATAs, 156 were served by Bell LECs and 28 by Independent LECs. The MFJ required calls between LATAs to be served by an IEC. The MFJ authorized AT&T and other IECs to provide service on inter-LATA calls (calls that cross from one LATA to another) and forbid the offering of this inter-LATA traffic by the RBOCs (or any LEC). Recent legislation by Congress has opened intra-LATA service to competition from the IECs. The RBOCs are petitioning for the right to provide inter-LATA (long distance) services.

Each LATA has established one or more access points—or POPs—for connection to the IECs for long distance services between LATAs. To comply with the 1984 provision, which mandated that the BOCs provide equal access to all IECs, the POPs were set up in the class 4 toll office and were called *equal access tandems*. Because class 5 electromechanical central offices have been replaced with computerized *Stored Program Control (SPC)* central offices, it is possible to place the POP at the class 5 office if desired. For a long distance call, the LEC will connect a customer to the POP where the call will be connected to that

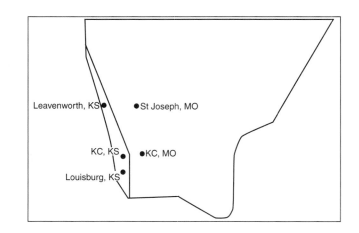

FIGURE 1-3 LATA 524.

customer's preferred IEC, and that IEC will handle the call. Telecommunications legislation of 1996 approved the entrance of RBOCs into the long distance market if they demonstrate compliance with the portion of the 1996 legislation providing for competition in the local service arena.

1.29 NETWORK HIERARCHY

When AT&T handled all long distance calls, it used a network hierarchy of five levels. The local exchange that telephones are wired to was designated as a class 5 office. There were about 20,000 class 5 offices across the United States. All calls originate at a class 5 office, and the class 5 office will connect to the network via a class 4 central office. When handling a long distance call, the local exchange would connect the call to a class 4 office when the caller dialed a 1 or 0. The class 4 office would serve as an access point to AT&T long distance. All class 5 offices within a small geographic area would connect to the class 4 office serving their area. The class 4 offices connected to a class 3 office. The United States was divided into different areas and an area code assigned to each area. Each area code would contain one class 3 office, which was known as a *Primary Center*. The class 3 office served all class 4 offices in the area code. Several Primary Centers (class 3 offices) would connect to a *Sectional Center* (class 2 office). Each *Sectional Center*, serving a geographic section of the country, would connect calls to a *Regional Center* (class 1 office). AT&T has 12 class 1 Regional Switching Centers in the United States (Figure 1-4).

If a long distance call was made between towns served by the same class 4 office, the caller's class 5 office would be connected to the class 4 office. The class 4 office would connect the call to the called class 5 office. Thus, the connection would be: originating telephone, originating class 5 office, class 4 toll office, terminating class 5 office, and terminating telephone. If the long distance call was from one region of the country to another, the connection would be: originating telephone, originating class 5 office, class 4 office, class 3 office, class 2 office, class 1 office serving the calling region, class 1 office in the region of the called number, class 2 office, the primary (class 3) office serving the area code for the called number, class 4 office, the class 5 office serving the called telephone, and the called telephone.

An analogy to airline travel may help clarify the different levels of switching and the trade-off that occurs between the number of transmission facilities and these levels of switching. The airline companies do not fly to every town in the country because the cost of having passenger terminals and routes to every town is prohibitive. If we wish to travel from a small town to a destination across the country, we will drive to a nearby airport and take a small commuter plane to a larger airport. At the larger airport, we catch a flight to a regional airport. Then we change planes and fly to the regional airport serving the region of our final destination. Here we board another commuter airline's plane for the

Five-Level Public Switched Telephone Network

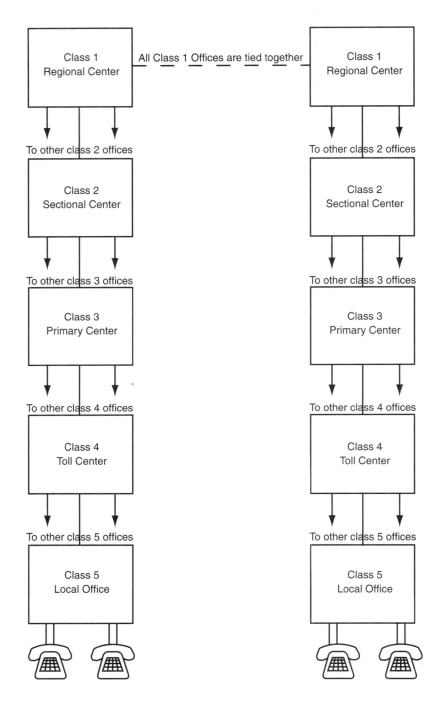

FIGURE 1-4 All telephones connect to a class 5 local central office.

final leg of our trip. For example, to fly from Dodge City, Kansas, to Pinehurst, North Carolina, we would need to change planes several times because no direct route exists between these two towns. We would drive from our small community (class 5 office) to an airport in Dodge City. Dodge City is the equivalent of our class 4 office. From Dodge City, we fly to Kansas City. Our transmission facility from Dodge City to Kansas City is a small airplane. There is no need to use a large airplane because the amount of traffic is small.

At Kansas City (our class 3 office), we board a larger plane (larger transmission facility) and fly to St. Louis (our class 2 office). Many passengers originating in various communities around the state of Kansas have joined us for our flight to St. Louis. At St. Louis, we leave some of our fellow passengers. Some have reached their final destination and will drive by car on the final leg of their journey to a local community in the St. Louis area. Others are going in a different direction and will board other planes heading for New York, Los Angeles, Denver, Dallas, and so on. We will board a plane to Chicago (our class 1 office), along with some passengers from Kansas City as well as other passengers who came from the St. Louis area or flew in from a different airport. The transmission facility is a large plane because a lot of people are flying to Chicago. At Chicago, we are switched again to another plane and board a plane heading for Atlanta (another class 1 office). In Atlanta, we board a plane bound for Charlotte, North Carolina (a class 2 switching center). In Charlotte, we switch to a plane flying into Raleigh (a class 3 office). In Raleigh, we are switched to a smaller plane flying to Fayetteville, North Carolina (a class 4 office), and we take a plane from Fayetteville to Pinehurst (our class 5 office and the end of our trip).

Of course, travelers were not happy with having to change planes so many times, and as the amount of air travel grew, additional transmission facilities (additional routes and planes) were added between class 3 centers to flatten the levels of the network. In telecommunications, it does not matter to us how many times our call is being switched or how many switching offices are used to connect our call. The switching is so fast that the number of offices we use to connect our call makes no difference. The PSTN network was flattened to reduce cost. The development of fiber optic transmission facilities has provided our industry with a transmission medium capable of handling 516,096 different conversations over one pair of glass fibers. This technology has made it possible to flatten the PSTN. Instead of needing thousands of smaller transmission mediums to connect offices, only one fiber optic facility is required. IECs have flattened the network to reduce the number of switching stages needed, thereby reducing capital expenditures for replacement offices and lowering maintenance expenses. Fewer offices require fewer maintenance people.

1.30 SPRINT'S FLAT NETWORK

This hierarchy has been flattened with the introduction of computer-controlled electronic switching systems, high levels of multiplexing using technology that allows thousands of conversations to be carried by one device, and fiber optic transmission systems. When AT&T set up the five-level hierarchy, it had to use electromechanical switching systems for the central offices and wire cable as the medium between switching systems. These older technologies and low orders of multiplexing limited the volume of calls that could be handled by one switching center. When Sprint and MCI set up their long distance networks using electronic switching systems, high levels of multiplexing, microwave technology, and fiber optic cables, they were able to combine the functions of the regional and sectional offices within one office. Sprint has approximately 45 sectional offices. These sectional offices are interconnected using fiber optic cable, and each office connects to many POPs (class 4 toll tandem offices and some class 5 offices). POPs were required because it was cheaper for the Bell LECs to furnish the switching technology necessary to provide equal access at the class 4 office rather than replace all class 5 electromechanical switching offices. These class 4 offices are called *Equal Access Tandems.*

As the class 5 electromechanical central exchanges were changed out to computerized switching systems, it became possible to provide equal access from that local office. Class 4 offices still serve as POPs to aggregate calls from small exchanges with low long distance calling volumes, but IECs also have direct connections to the larger class 5 computer-controlled central offices because these offices are capable of providing equal access. A caller using Sprint as the IEC of choice will connect from an originating class 5 office to the class 4 office serving as a POP (or directly from the local class 5 office to Sprint in some cases). The POP will connect to a Sprint sectional office. The originating caller's Sprint sectional office will determine which of the other 45 Sprint offices the call should be connected to, in order to connect to the POP serving the called number. The call will be connected to that office. This second Sprint office will then connect to the terminating POP. The terminating POP class 4 toll office connects to the terminating class 5 office for connection to the called telephone. In a small rural area, your IEC and the LEC use a POP in the class 4 toll exchange. In larger cities, the POP may be at the class 5 local exchange. (See Figure 1-5 for the Sprint Network.)

1.31 *TELECOMMUNICATIONS REFORM ACT OF 1996*

On March 30,1995, Senator Larry Lee Pressler of South Dakota introduced Senate Bill 652. This bill was amended many times over the next year and was signed into law on February 8, 1996, as Public Law 104-104. Telecommunications managers should be familiar with the *Telecommunications Reform Act of 1996* since it governs how we can conduct our business. A summary of this law is included in Appendix B. (The complete text of the bill can be downloaded over the Internet from the Library of Congress at http://thomas.loc.gov.) This law provided for further deregulation of both the telecommunications industry and the cable television industry. When Senate Bill 652 was under consideration, there was little debate on the portions of the Bill dealing with deregulation of telecommunications, but it received a tremendous amount of debate over its provisions to deregulate rates charged by cable television companies, to limit pornographic materials on the Internet, and to require manufacturers to include a programmable chip in television sets that parents could program to restrict what their children could watch.

The local BOCs and their parent RBOCs actively lobbied for support of the Telecommunications Reform Act of 1996 because it contained provisions that would remove restrictions placed on them by the 1984 MFJ. As applied to the telecommunications industry, Public Law 104-104 supersedes the 1984 MFJ three years from the date of becoming law. Therefore, the MFJ is effectively replaced by this law on February 8, 1999. This law removes the restrictions imposed on both AT&T and the RBOCs. It also eliminates restrictions on all IECs. They will be allowed to carry intra-LATA calls and can now compete in the local telephone service markets. LECs must cooperate with IECs and Competitive Access Providers (CAPs), such as Metropolitan Fiber Systems, that wish to provide local exchange services. Even before the president signed this legislation, the FCC and PUCs were authorizing multiple local access providers in the same locality. Under the Telecommunications Reform Act of 1996, the BOCs will be allowed to provide long distance (inter-LATA) service one year from the date the bill was signed into law, if they prove they have not hindered competition for local services in their local exchanges. AT&T will be allowed to reenter the local services market and the RBOCs will be allowed to manufacture telecommunications equipment.

One significant area this bill did not provide for was "universal service." Several amendments were offered to make enhanced services such as last number redial, caller I.D., and local area networks available in all local service areas, but these amendments could not gain support of the majority in either the Senate or the House. The bill asks that LECs strive to provide "universal service" at the same cost these services are provided at in large cities, but this bill does not mandate universal service. Therefore, many areas of rural America will not receive the enhanced services now available in larger cities. If the services are provided, they will cost more because the necessary investment cannot be spread over a large customer base.

1.32 SUMMARY

The telecommunications industry is a dynamic industry and is in a constant state of change. Legislation covering telecommunications and the technology available to provide various services are changing continually. As these changes occur, they create new market opportunities. The astute entrepreneur stays abreast of these changes and is quick to exploit new opportunities. The first to

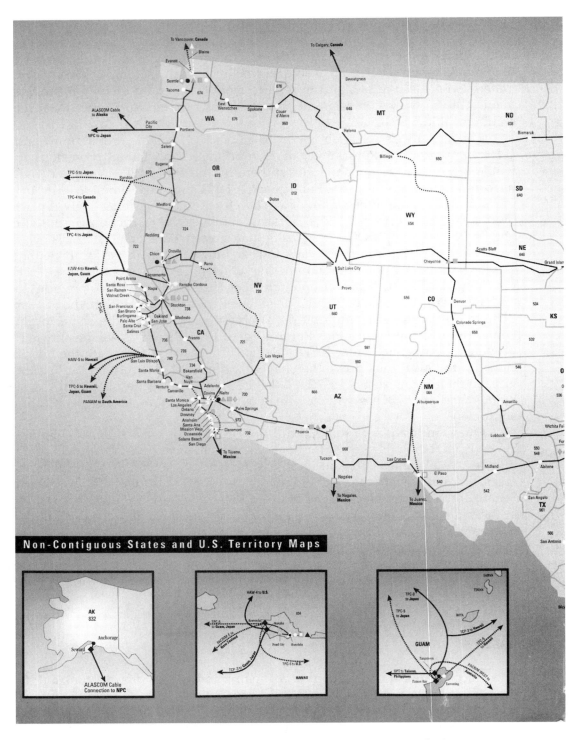

FIGURE 1-5 Sprint network. Used with permission from Sprint.

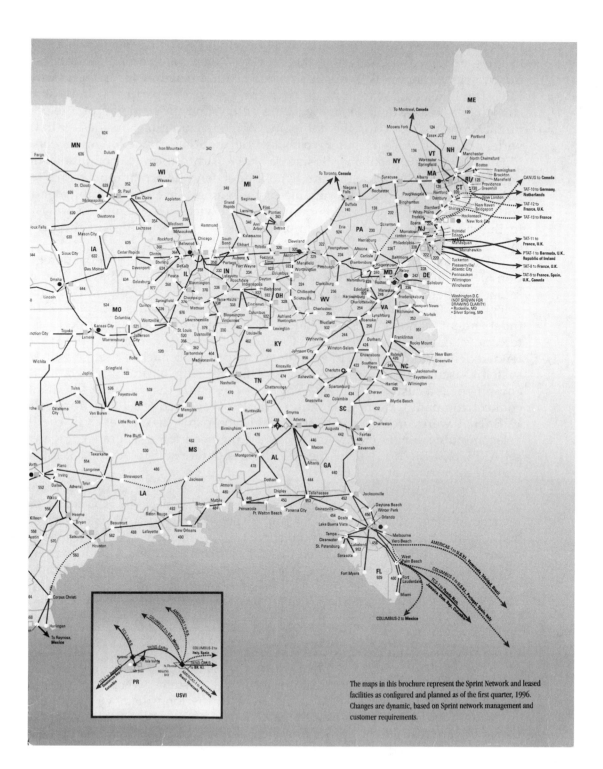

The maps in this brochure represent the Sprint Network and leased facilities as configured and planned as of the first quarter, 1996. Changes are dynamic, based on Sprint network management and customer requirements.

market reaps the greatest rewards, and you can only be first in a new market by keeping informed of changes in that market. When the telecommunications industry was a monopoly, a business owner had one source for all services, the LEC. The LEC had a marketing department, a service department, and a staff of consultants to assist customers with the design and provisioning of a telecommunications network for the company. With deregulation, businesses have a wider array of service providers to chose from, but they can no longer rely on the LEC to solve their problems. Businesses have had to form telecommunications departments and hire managers who understand telecommunications technology. Companies must be capable of designing, developing, implementing, and servicing their own telecommunications services. This requirement has led to the creation of many telecommunications job opportunities in the private business sector.

Deregulation of the telecommunications industry began with the deregulation of station equipment at the customer's premise. The event having the most impact on deregulation of station equipment was the Carterphone Decision of 1968 by the FCC. This decision allowed customers to purchase telephones from independent manufactures or retailers. These companies were referred to as interconnect companies. In 1981, the Computer Inquiry II decision of the FCC extended deregulation to cover keysystems and private branch exchange switching systems used by business customers. This decision basically stated that all equipment at a customer's premise must be owned by the customer (or an interconnect company). Computer Inquiry II also prevented the LECs from selling CPE. They were forced to establish separate companies to market CPE. The 1996 Telecommunications Reform Act abolished the need for a separate company to sell CPE, and the LECs have merged these separate subsidiaries back into the regular operations of the LEC.

Long distance telephone service was the second area of telecommunications to be deregulated. Deregulation of this service began with the 1969 FCC ruling often referred to as the MCI decision. This ruling allowed MCI to pursue its completion of a long distance network that could serve private-line networks of business customers. In 1971, the FCC broadened this decision to allow any common carrier the right to carry private-line services. This ruling is called the Specialized Common Carrier Decision of 1971. In 1975, MCI began offering long distance service to the general public. The FCC ordered MCI to cease providing service to the general public and to restrict its service to the private-line business. MCI sued. The appeals court reversed the FCC ruling. In 1976, the FCC opened the provisioning of long distance service to the general public to competition.

The Modified Final Judgment (MFJ) of 1984 addressed the deregulation of telecommunications and the breakup of the Bell System. AT&T was ordered to divest itself of the BOCs. AT&T was left with the long distance facilities and could not provide local service. The BOCs were left with the local service facilities and were restricted to providing local service. Local service was defined as calls that originated and terminated within a small geographic area called a LATA. The 1984 MFJ divided the country into 184 LATAs. According to this agreement, calls within a LATA must be carried by a LEC, even if they are toll calls, and calls between LATAs must be carried by an IEC. The 1996 Telecommunications Reform Act opened both areas to competition between the LECs, IECs, and other common carriers.

The 1984 MFJ also ordered the BOCs to provide Equal Access (1+ access) to all customers of an IEC by 1987. To meet this requirement the BOCs had to replace electromechanical toll offices with computer-controlled switching systems. They also had to establish locations for their facilities to meet the facilities of the IEC. These locations are called POP locations. The IECs were allowed to rent space in the BOC exchanges for the termination of their long distance network. Customers could specify who they wanted as their long distance service provider. The BOCs enter this information into the switching systems database. When the customer makes an inter-LATA call, the BOC automatically connects the caller to the preferred IEC.

Telecommunications Events and Legislation

Year	Title	Explanation
1876	Patent issued for telephone to Alexander Graham Bell	Recognized Alexander Graham Bell as inventor of telephone
1877	Bell Telephone Company formed	Investors needed to expand
1882	Bell buys Western Electric	Bell needed a manufacturer
1885	Bell incorporates as AT&T	To provide protection to company officers
1893	Bell's patent expires	Independents formed
1910	Mann-Elkins Act	Regulated using ICC
1913	Kingsbury Commitment	To settle antitrust action by DOJ
1921	Graham Act	Exempted AT&T from antitrust law
1934	Communications Act of 1934	Established FCC and state PUCs
1949	REA Act amended	To provide low-cost loans to Independents
1949	DOJ brings antitrust action	AT&T charged under Sherman Act
1956	1956 consent decree	1949 lawsuit settled by DOJ /AT&T
1968	FCC Carterphone decision	AT&T must permit use of CPE
1969	MCI ruling by FCC	AT&T must allow MCI access
1971	Specialized common carrier	Any common carrier can provide private line service
1971	Computer Inquiry I	Computer industry exempt from regulation
1974	DOJ files antitrust action	AT&T charged by DOJ
1975	MCI decision	Appeals court upheld MCI
1976	FCC orders AT&T to provide access to customers of SCCs	Any SCC can provide long distance service to the general public
1981	Computer Inquiry II	All CPE deregulated
1982	Modified Final Judgment (MFJ)	1974 lawsuit settled by consent
1984	MFJ takes effect	Replaced 1956 Final Judgment
1987	Equal access (1984 MFJ)	RBOCs must provide equal access
1996	1996 Telecommunications Reform Act	Replaced 1984 MFJ

1. Who invented the telephone and when was it patented?
2. When was the first BOC established?
3. When was the first Independent telephone company formed?
4. What was the purpose of the Kingsbury Commitment?
5. When was the Graham act repealed?
6. In telecommunications, what is the purpose of a tariff?
7. Which government agency regulates interstate telecommunications?
8. Which government agency regulates intrastate telecommunications?
9. How did the government encourage companies to build telecommunications systems as part of the PSTN in low-profit, rural areas?
10. What are companies called that provide telecommunications services to the general public?
11. What was the outcome of the antitrust lawsuit brought by the Department of Justice against AT&T in 1949?
12. What FCC ruling led to deregulation of station equipment?
13. What name is given to equipment on a customer's premises not owned by the LEC?
14. What is the name given to privately held companies selling equipment and providing services in the CPE market?
15. Which FCC ruling led the way for deregulation of long distance service?
16. Has deregulation of the telecommunications industry led to degradation of performance in the PSTN and lower local service rates?
17. Which ruling resulted in the BOCs buying equipment not only from Western Electric but from other manufacturers as well?
18. The MFJ accomplished two specific and separate actions. What are these two broad categories of actions called?
19. What was the toll revenue sharing procedure between the Independents and the BOCs called? What ruling killed this procedure?
20. Was the legislative process leading to deregulation and divestiture brought about by the FCC?
21. What is the name given to common carriers providing local phone service and connection to the PSTN?
22. What is the name given to common carriers providing long distance service?
23. What is the difference between the PSTN and private networks?
24. What did the Carterphone decision of 1968 accomplish?
25. What are the major provisions of the 1984 MFJ?
26. What is a PIC?

27. How can the user of a pay phone (or any phone) override the PIC?
28. What is a LATA? How many LATAs are there? Do LATAs follow the same boundaries established by area codes? Can LATAs cross state boundaries?
29. What is a POP? Where is it located?
30. What technologies have led to flatter network hierarchies?
31. As Public Law 104-104 applies to telecommunications, what were the significant provisions of the Telecommunications Reform Act of 1996?

GLOSSARY

Bell Operating Companies (BOCs) Often referred to after deregulation as the "Baby Bells." These were the local Bell Telephone companies. Before the 1984 Modified Final Judgment, 23 BOCs existed as subsidiaries of AT&T.

Carterphone Decision of 1968 A ruling by the Federal Communications Commission forcing AT&T to allow the attachment of a Carterphone to telephone lines at the customer's residence. This ruling was the beginning of deregulation of station equipment.

Common Carrier A company that offers telecommunications services to the general public as part of the Public Switched Telephone Network.

Communications Act of 1934 Legislation passed by Congress and signed into law by the president in 1934 to regulate interstate telecommunications and radio broadcasts. This law created the Federal Communications Commission to administer the act.

Computer Inquiry Any of a series of rulings by the Federal Communications Commission. Issued Computer Inquiry I in 1971; stated that the Federal Communications Commission would not regualte computer services.

Computer Inquiry II Issued in 1981; mandated that station equipment was to be deregulated and could not be provided by a local exchange carrier.

Computer Inquiry III In 1986, detailed the extent to which AT&T and the Bell Operating Companies could compete in the nonregulated enhanced services arena.

Customer-Provided Equipment Also called customer-premise equipment. This is station equipment, at a customer's premise, which according to Computer Inquiry II must be provided by someone other than the local exchange carrier. The 1996 Telecommunications Reform Act overrides Computer Inquiry II and allows LECs to reenter the CPE market.

Deregulation The change of the telecommunications industry from a regulated monopoly to a competitive nonregulated market. One of the major provisions of the 1984 Modified Final Judgment was to deregulate long distance services.

Divestiture A major provision of the 1984 Modified Final Judgment. This provision forced AT&T to divest itself of the 23 Bell Operating Companies.

Equal Access A key provision of the 1984 Modified Final Judgment. This provision forced the Bell Operating Companies to provide 1+ access to toll for all interexchange carriers.

Feature Group A series of services established by the Bell Operating Companies to provide access to interexchange carriers (IECs). Feature Group A gave customers access to IECs by having them dial a seven- or ten-digit telephone number plus a personal identification number. Feature Group B provided access via a seven-digit code of the form 950-10XX. This access code was later changed to 10XXX. Feature Group D is the 1+ access to IECs that was mandated by the 1984 Modified Final Judgment.

Federal Communications Commission (FCC) The federal agency created by the Communications Act of 1934 to oversee enforcement of the act by regulating interstate telecommunications and broadcast communications.

Graham Act of 1921 Legislation passed by Congress and signed by the president in 1921 to exempt telecommunications from antitrust legislation.

Interexchange Carriers (IECs or IXCs) Common carriers that provide long distance telephone service. Major IECs are AT&T, MCI, Sprint, LDI, and so on.

Kingsbury Commitment of 1913 An agreement signed by the vice president of AT&T in 1913. This agreement stated that AT&T would allow other telephone companies access to its network, would sell its Western Union stock, and would not buy any more Independent telephone companies without getting permission from the government.

Local Access Transport Area (LATA) A key provision of the 1984 Modified Final Judgment (MFJ). This provision established 184 geographic regions that conform to the Standard Metropolitan Statistical Index used by marketing organizations. The LATA was defined as an area within which calls must be carried by a local exchange carrier (LEC) and could not be carried by an interexchange carrier (IEC or IXC). These calls within the LATA are called intra-LATA

calls. The MFJ further stated that calls between LATAs (inter-LATA calls) must be carried by an IEC, not by a LEC. The 1996 Telecommunications Reform Act replaced the 1984 MFJ and opened the door to competition in this area. It allows either type of call to be carried by either a LEC or an IEC.

Local Exchange Carrier (LEC) The provider of local telephone services. Prior to the 1996 Telecommunications Reform Act, the LEC was your local telephone company. with the passage of that act, the local telephone company is now called the incumbent local exchange carrier (ILEC) and its competitors are called competitive local exchange carriers (CLECs).

MCI Decision of 1976 A ruling by the federal court in MCI's favor against AT&T; it allowed MCI and other common carriers to handle long distance services for the general public.

MCI Ruling of 1969 The ruling in 1969, by the Federal Communications Commission, that forced AT&T to allow private-line customers of Microwave Communications Inc. (MCI) to use local telephone lines for access to MCI's private-line network.

Modified Final Judgment (MFJ) The 1984 agreement—reached on August 24, 1982, between AT&T and the Department of Justice—to settle an antitrust suit brought against AT&T by the Justice Department, on November 20, 1974. This judgment modified the 1956 Final Judgment. The major provisions of the 1984 MFJ were deregulation of long distance services and divestiture of the Bell Operating Companies by AT&T.

Monopoly A form of market where one company is the sole provider of goods and/or services. There is no competition. Market demand does not set the price of goods or services. In a nonregulated monopoly, price is set by the monopoly service provider. In a regulated monopoly, price is set by the government agency charged with regulating the monopoly service provider.

Network An overused term. In telecommunications, the word network has many meanings, usually determined by the context. The Public Switched Telephone Network (PSTN) refers to the interconnection of switching systems in the PSTN. Switching network refers to the component inside a switching system that switches one circuit to another circuit. The network in a telephone refers to the hybrid network, which performs a two- wire to four-wire conversion process.

Preferred Interexchange Carrier (PIC) Determined as follows: A customer tells the local exchange carrier who it wants to use as its long distance service provider. The LEC makes an entry in the database of the switching system serving this customer. That database entry will inform the switch which IEC should be used on long distance calls placed by that customer. With the 1996 Telecommunications Reform act, customers must be able to choose a carrier for toll calls within a LATA as well as the same or a separate carrier for long distance service.

Point of Presence (POP) The point at which the local exchange carrier and interexchange carrier facilities meet each other.

Private-Line Services Services provided by a common carrier to a private organization to help that organization establish its own private network. A private network cannot be accessed or utilized by the public. A private network is exactly that—it is private and can only be accessed and used by the private organization.

Public Switched Telephone Network (PSTN) A network of switching systems connected together to allow anyone to call any telephone located in the United States. A public network is accessible to any members of the general public. The PSTN is composed of many nodes (switching systems) and many transmission links connecting these nodes. Stations are the end points of the network and connect to the PSTN via their local exchange.

Public Utilities Commission (PUC) The state government agency that regulates telecommunications within the state.

Regulated Monopoly A sole provider of goods or services; regulated by a government agency. A regulated monopoly must seek approval from the government agency for anthing it wishes to do. The government agency sets the price that a monopoly can charge for its goods and services.

Revenue Sharing The sharing of toll revenue between the Independent Telephone Companies and AT&T prior to the 1984 MFJ.

Rural Electrification Administration (REA) Department of the federal government established during the Depression under President Franklin D. Roosevelt by the Rural Electrification Act of 1936. This "Act" put people to work bringing electricity to rural America. The "Act" was amended in 1949 to bring telephone service to rural America. This department was renamed to the Rural Utility Services Department.

Specialized Common Carrier Ruling of 1971 An extension of the 1969 MCI Decision by the FCC. This ruling allowed any Common Carrier to handle Private Line Networks.

Station Equipment The largest segment of station equipment is the telephone but the term station equipment has been broadened to include anything a customer attaches to a telephone line. The most common piece of station equipment is the telephone. A modem, CSU/ DSU, personal computer, and keysystem are all referred to as station equipment.

Tariff The document filed by a Common Carrier which defines in detail any service proposed by the Carrier and the charge proposed for the service. For Interstate Service, the Tariff is filed with the FCC. For Intrastate services, the Tariff is filed with the PUC.

Telecommunications The communications of voice or data over long distances using the Public Switched Telephone Network or privately owned networks.

Telecommunications Reform Act of 1996 Effectively replaces the 1984 MFJ. This act is designed to accelerate competition for providing services in the local and long distance markets. It eliminates the franchised local teritory concept. The LEC is no longer the only service provider in town. Other common carriers can come into an incumbent local exchange carrier (ILEC) territory and provide local telephone service. These new local service providers are called competitive local exchange carriers (CLECs). Prior to this act, the older incumbent LECs (ILECs) were restricted to providing local services only. Under this act, if the ILEC can prove that it has not hindered competition in its local exchange territories, it can now offer long distance services.

2

Telecommunications Technology (1876–2000)

Key Terms

Alternating Current (AC)
Central Office
Communication
Direct Current (DC)
Frequency Division
  Multiplexing (FDM)
Full-Duplex Transmission
Half-Duplex Transmission

Internet
Interoffice Calls
Intraoffice Calls
Line Relay
Medium
Multiplexing
Receiver
Simplex Transmission

Sine Wave
Switchboard
Telecommunications
Time Division Multiplexing
  (TDM)
Transmitter

Telecommunications is the science or technology of communication by telephone or telegraph. *Tele* is a Greek word meaning "distant," and telecommunications is communications over a distance. *Communication* can take on any of several meanings. For our purposes it is the means of sending messages, orders, and responses by electronic, electrical, or electromagnetic means. The communication process allows information to pass between a sender (transmitter) and a receiver over some medium. The three parts of any communication system are:

—Transmitter
—Receiver
—Medium

The *transmitter* is the device responsible for sending information or a message in a form that the receiver and medium can handle.

The *receiver* is the device responsible for decoding or converting received information into an intelligible message. In a conversation between two people, one person will be transmitting information while the other is receiving the information.

The *medium* is the device or substance used to transport information or a message between the transmitter and receiver. If a conversation is face to face, the medium used to carry the message is air. Television and radio broadcasting also use air as a transmission medium. The signal from the station is connected to a transmitter. The transmitter is a very high powered electromagnetic wave generator located next to a high tower. The signal generated by the electromagnetic wave generator is sent to an antenna mounted on the high tower. The power of the signal is so high that the signal will leave the antenna and radiate out into the air. The signal is pulled out of the air using a receiver. The receiver includes a tuning device to allow tuning in, or selecting one station's signal out of the many different signals arriving at the antenna of the receiver. When air is used as a transmission medium the message is sent in all directions and is called a *broadcast* (scattered or spread over a wide area) signal. Broadcast signals can be picked up by anyone with a receiver tuned to the broadcaster's transmitting signal frequency. Using air as the transmission medium is the most economical way for a radio or television station to reach everyone.

2.1 SIMPLEX, HALF-DUPLEX, AND FULL-DUPLEX TRANSMISSION

The radio and television stations transmit signals and our television sets and radios receive them. We have no way of transmitting signals back to the radio or television station. Transmission in one direction only is called *simplex transmission*. When we have a conversation with someone, we transmit for a short time and the other person receives our message. We will then switch positions as transmitter and receiver several times during the conversation. This type of transmission, where transmitters on each end of a medium take turns sending over the same medium, is called *half-duplex transmission*. If a transmission system allows signals to be transmitted in both directions at the same time, the system is called a *full-duplex transmission* system. Many times a full-duplex system is actually made up of two simplex systems. On a highway we have traffic lanes going in opposite directions. Traffic can only go in one direction on one of the lanes (simplex). Used together, the two lanes allow traffic to flow in both directions at the same time (full-duplex).

2.2 TWISTED COPPER WIRE AS A MEDIUM

Copper wire can be used as a medium instead of using air. This would not be practicable for a radio or television station because they would need to run a wire to everyone's house. Cable television companies and local telephone companies do use wire as a medium. Using wire as a medium provides the transmitter with more control and security over the messages being transmitted. Local telephone companies use two wires to connect each phone to their switching office. Each set of two wires is referred to as a *pair*. The telephone company refers to wires serving a local telephone as the *local loop*.

2.3 INVENTION OF THE TELEPHONE—1876

The telephone was invented by Alexander Graham Bell in 1876. He used a pair of wires to connect two telephones to each other and used one device for both the transmitter and receiver (referred to as a *transceiver*). This device required no external power and could convert sound waves into low-current electrical waves. It could also convert the electrical energy received from the distant phone back into sound waves at the receiving phone.

2.4 INVENTION OF THE CARBON TRANSMITTER

One year later, in 1877, Thomas Edison invented the carbon microphone. This transmitter required a battery for the supply of electrical energy. Edison's transmitter did not produce electrical energy. His transmitter consisted of carbon granules enclosed in a capsule containing a diaphragm. As sound waves struck the diaphragm, the diaphragm would compress and decompress the carbon granules, causing the resistance of the carbon granules to vary. As the resistance varied, it caused the electrical current flowing from the battery to increase and decrease in unison with the speaker's voice. This varying electrical current was sent over the phone line to the distant phone, and the receiver converted the varying electrical current into sound waves. Edison's transmitter and battery supply generated a much stronger signal than the Bell transceiver. This type of carbon microphone is still used today as the transmitter in many telephones. The original transmitter/receiver device invented by Bell now serves only as the receiver of the telephone. Transmitters and receivers fall into a category of equipment called transducers. A *transducer* is a device that converts one form of energy into a different form. The transmitter coverts sound-wave energy into electrical energy. The receiver converts electrical energy into sound waves (Figures 2-1 and 2-2).

2.5 LOCAL POWER SUPPLY AT EACH TELEPHONE

As telephones were introduced to the public, each phone had its own battery. The battery that powered the telephone was composed of two large $1^1/_2$-V DC cells connected in series to make up a 3-V DC battery. A *direct current (DC)* battery supply maintains a constant voltage that gradually gets lower as the battery discharges. Each phone was equipped with a ringer and a hand crank connected to a magneto. A *magneto* is a device containing a coil of wire inside a magnetic field. When the wire coil is

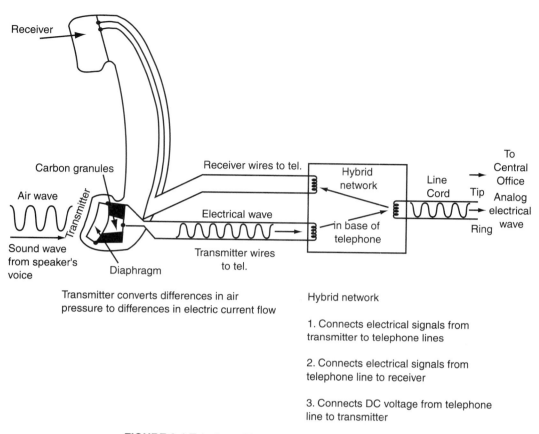

FIGURE 2-1 Telephone Transmitter and hybrid network.

turned so that the wire of the coil cuts through a magnetic field, electric energy is created in the wire. This is the same principle used by commercial electric companies to generate electricity for our homes. This type of electricity is referred to as ***alternating current (AC)*** electricity. Operating the hand crank on a phone would cause a 90-V AC signal to go out over the telephone line to the distant phone. The AC signal would cause the ringer to operate on the receiving telephone. An AC voltage supply generates a voltage that is constantly changing and reversing polarity. The voltage rises from zero volts to a positive value, then gradually becomes zero, then drops to a negative value, and finally returns to zero. This trip from zero to maximum positive, to zero, to maximum negative, and back to zero volts is called a *cycle*. The number of cycles an AC signal goes through in one second is called the *frequency* of the signal and is measured in cycles per second (cps). Cycles per second are often called Hertz as a tribute to Heinrich Hertz, who discovered electromagnetic radiation (the birth of radio waves) in 1886. The frequency of the signal generated by a magneto depends on how fast the crank could be turned. Frequency pertains to a process that repeats itself; the term denotes how often the process repeats. An event that occurs at noon every day can be said to have a frequency of once a day, or a frequency of 30 times a month

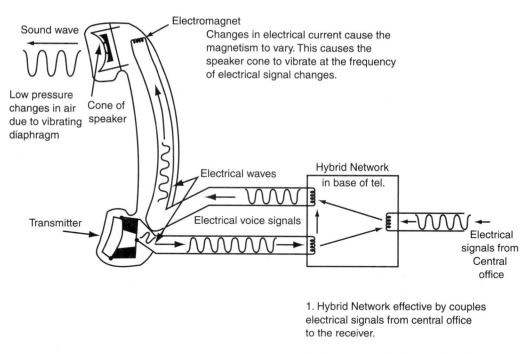

Sound wave

Low pressure changes in air due to vibrating diaphragm

Cone of speaker

Electromagnet
Changes in electrical current cause the magnetism to vary. This causes the speaker cone to vibrate at the frequency of electrical signal changes.

Electrical waves

Hybrid Network in base of tel.

Transmitter

Electrical voice signals

Electrical signals from Central office

1. Hybrid Network effective by couples electrical signals from central office to the receiver.

2. A small amount of the electrical signal from transmitter is coupled to the receiver.

FiGURE 2-2 Telephone receiver and hybrid network.

(sometimes 28, 29, and 31 times a month), or a frequency of 365 times a year. The time frame used in electrical frequency measurements is the second. The more cycles completed in one second, the higher the frequency. Figure 2-3 shows two AC (sine) waves. The *sine wave* in (a) completes two full cycles in one second. The sine wave in (b) completes four full cycles in one second.

Voice signals are composed of several different sine-wave frequencies that are mostly between 300 and 4000 cps mixed together. If we hold our voice signal at a steady output (such as when we whistle), our voice approaches a pure sine wave. As we speak, the sounds we make are the result of continuously varying the frequency of the signal. Therefore, speech is not a pure sine wave but is a variable signal composed of several sine waves mixed together. A speech signal looks like the signal in Figure 2-4.

The original telephone unit was self-contained. Each phone had its own power supply furnished by batteries similar to the batteries that power a large camping lantern, and each phone could generate a high-voltage ringing signal to alert the called party. Telephones were connected to each other using iron wire strung on poles between the two phones. A separate pair of wires and phone was needed for each different location you wished to call. This cumbersome and expensive situation led to the development of the central wire center and manual switching system commonly referred to as central.

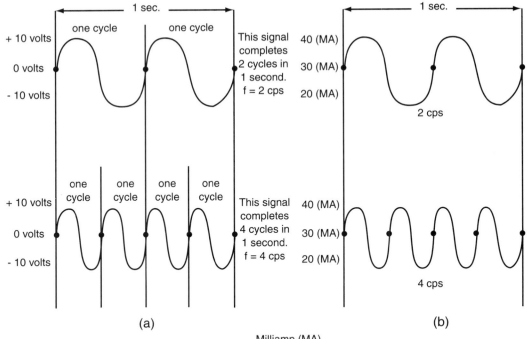

FIGURE 2-3 Analog electrical signals.

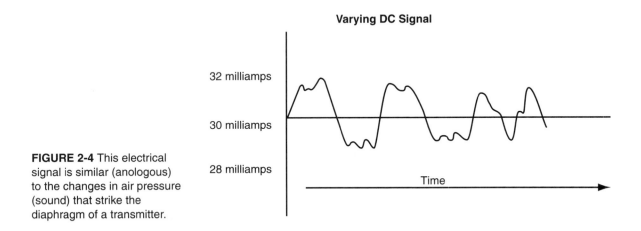

FIGURE 2-4 This electrical signal is similar (anologous) to the changes in air pressure (sound) that strike the diaphragm of a transmitter.

In Figure 2-5a one pair of wires connects one telephone directly to another telephone. In Figure 2-5b, all telephones are wired to the "central" wire center. At the central wire center, all lines are connected to an operator's switchboard. The operator can connect any line to any other line. This manual switching was the birth of switching systems.

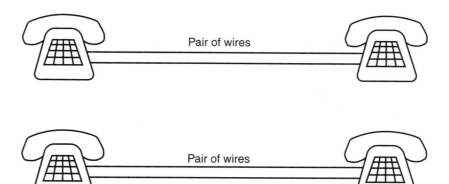

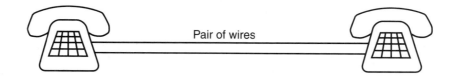

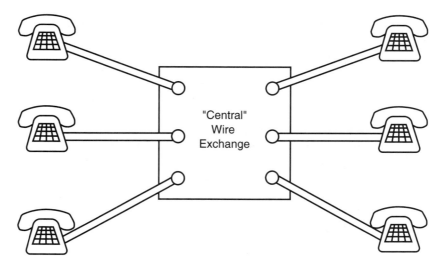

FIGURE 2-5 (a) One pair of wires connects one phone directly to another phone. (b) Each phone is connected by a pair of wires to the wire center. An operator can connect any phone line to any other phone line.

One serious drawback to the early phone system was the need for the transmitting party to have a separate (individual) phone for everyone they wanted to call along with a different pair of wires for each phone. If you wanted the capability to call ten different people you needed ten phones connected to ten different pairs of wires. The other end of these pairs of wires would go to phones at the ten different locations you wanted to call. A large business would have several phones in place and still have limited calling capabilities. The first step toward solving this problem was to wire all telephones through a central location. The next step was to install a manual switchboard and wire all lines coming into the central location to this switchboard. Manual switching systems were introduced to allow the connection of one phone to any other phone connected to the system.

2.6 CENTRAL EXCHANGE—1878

The first telephone exchange was installed in 1878, two years after Bell invented the telephone. The Bell Company recognized the need for a method to connect one telephone to any other telephone. It also recognized the cost savings to be achieved by wiring all phones through a central location. The central location is referred to as *central office, central, central wire center, central exchange,* or *exchange.* To provide everyone who had a phone with the capability to call any phone connected to the central office, an operator switchboard (patch panel) was added to the system. All wire pairs coming into the central office were connected to individual jacks, signal relays, and lights on the patch panel. When someone wished to place a call, the person would use the hand crank at the phone to send a ringing signal to the operator. The ringing signal would operate the relay and lamp on the patch panel associated with this particular telephone line. The operator would answer the call through a plug-in and jack arrangement that would connect a pair of wires from her telephone to the incoming line.

The wires of the operator's phone were connected to a plug. The wires coming in from phone lines were connected to jacks. Each phone line could have telephones from as many as ten residences connected to the line. A private line had telephones at only one location (a residence or business) attached to the line. If a town had 200 telephone lines, the patch panel at the operator's switchboard had 200 jacks, relays, and lamps. The operator could answer any one of these by plugging her phone into the jack of the calling line. The operator answered calls using the phrase "central." On being told who the caller wished to call, the operator would plug her phone into the jack of the person being called and ring their line using the crank on her phone. When the called person answered the phone, the operator would remove her phone from the jack and connect the jack of the calling party to the jack of the called party using a wiring cord that had a plug on both ends of the cord. This cord was called a *patch cord.*

The phone lines were wired to jacks very similar to the headset jack on a portable AM/FM cassette tape player. The plug on the operator's phone and the plugs on the patch cord are similar to the plug on the end of the wires connected to a cassette player's headset. If you look closely at the plug on the headset of a portable cassette player, you can see a copper metal tip on the plug, a band of plastic around the plug behind the tip, then a narrow copper ring followed by another band of plastic. In the telephone industry the wire that connected to the tip of the plug was called the *Tip lead* and the wire that connected to the copper ring was called the *Ring lead*. The wires on the jacks were also called *Tip* and *Ring* according to which part of the plug they made contact with (Figure 2-6).

2.7 OPERATOR SWITCHBOARD

When the operator connected two lines using a patch cord, there was no indication when the call was completed. To alleviate this condition an operator *switchboard* was invented. The switchboard consisted of patch cords, patch panels, and switches (see Figure 2-8). The switches are similar to a light switch but have more contacts. A light switch has one set of contacts. When the switch is <u>off</u> the contacts are <u>open</u>. When the switch is thrown <u>on</u>, the contacts are <u>closed</u>. Operation of the light switch connects the AC power line to a light socket. The switches used on the switchboard had more than one set of contacts. Some contacts opened when the switch was thrown on and some contacts would close when the switch was thrown on.

The patch cords used on the switchboard were made with a plug on one end only. The other end of the patch cord was wired to one of the switches on the switchboard. The patch cords were arranged in pairs. One cord was placed in front of the other; both cords sat directly behind the switch they were wired to. Two patch cords (a pair of patch cords) were wired to one switch. A pair of patch cords was used to handle one call. The switchboard was equipped with 20 patch cords (10 pairs of patch cords) wired to 10 switches. The patch cords and switches were placed in a desktop, and the patch panel containing jacks connecting to phone lines were placed in a vertical position, behind the desktop, in front of the operator. The operator's phone was replaced by a headset and a steady ringing source was available at the switchboard position. The operator's headset was connected through a switch called the *monitor switch* to the contacts on the patch cord switches. Every set of patch cords was equipped with a monitor switch in addition to the patch cord switch. Each of the ten monitor switches was placed directly in front of each patch cord switch.

The monitor switch allowed the operator the option of connecting both the transmitter and receiver of the headset to a patch cord or connecting only the receiver to the patch cord. The monitor key made it possible for the operator to talk

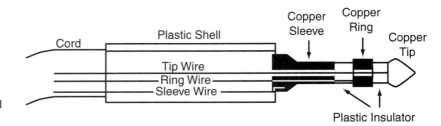

FIGURE 2-6 Switchboard cord at operator's switchboard.

to either person on the patch cords and to monitor a call without being heard. Operators would monitor a call once every few minutes—long enough to determine if the call was over. On determining that a call was over, the operator would take down the connection by removing the patch cords from the jacks.

The patch cord switch was a multiposition switch. In one position the front patch cord of a pair was connected to the rear patch cord. The switch was placed in this position to connect the calling party to the called party. The switch could be placed in a second position to allow the operator to talk with the calling party without the called party hearing the operator's conversation. The switch could also be placed in a third position allowing the operator to talk in private with the called party. When the patch cord switch was in either the second or third position, the operator could operate a ring key on the position. The operation of the ring key would open the headset from the call and attach a 90-V AC ringing signal to the patch cord. This allowed the operator to ring over either patch cord when placing a call or to ring someone if the person hung up before the other party was through with the call.

An operator could handle ten different calls at one time. If more than ten calls needed to be handled at one time, a second operator switchboard position was added and all incoming jacks were cabled to both switchboards. The operator switchboard had a total of 21 switches and 20 cords. The 20 cords were arranged in 10 pairs, and 2 switches were associated with each pair. The 2 switches were an operator talk/monitor switch and a patch cord switch to either separate or tie the pair of cords to each other. The switchboard had one switch to connect either the operator's headset or the ringing signal to the patch cord keys. This switchboard was known as a *manual switchboard* because the operator manually switched incoming calls to their calling destination (Figures 2-7 and 2-8).

Refer to Figure 2-8. An incoming call was identified by a lit light above a jack. When a caller wanted to get the operator, it was necessary to turn the hand crank on the telephone or lift the receiver off the hook. This action would operate a line relay in the central office and cause the light above the individual's jack to light. The operator would take a rear patch cord and plug it into the jack below the light (which would cause the light to go out) and find out who the caller wished to call. On being told who to call, the operator would take the front cord, directly in front of the cord used to answer the call, and plug it into the jack of the desired line. She would then operate the ring key to ring the called telephone. The lamps behind the cords on the desktop are called *supervisory lamps*. The rear lamp would light if the calling party hung up. The lamp just in front of the rear lamp would light if the called party hung up. The operator used the key in front of the patch cords to select which party she wanted to talk to, when a private conversation was desired. This key was also used to select which cord the ringing signal would be sent over, when the ring key was operated.

2.8 LONG DISTANCE OPERATOR SWITCHBOARDS

In addition to the jacks at the manual switchboard that connected to wires going to local telephones, there were also jacks connecting to wires that ran to switchboards located in other towns and cities. The local operator could plug a patch cord into one of these jacks

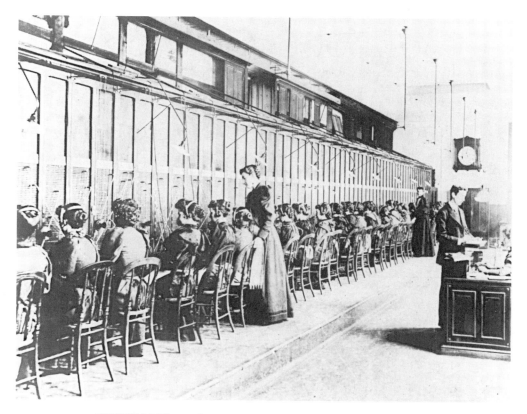

FIGURE 2-7 The early manual telephone switchboard was installed in New York City in 1888. (Courtesy of AT&T Bell Laboratories)

and ring the distant operator. AT&T and the Bell companies installed operator switchboards in several large cities in each state for the purpose of handling only long distance calls. The long distance centers were wired to each other. The wires going between them terminated on jacks at the long distance operator's switchboard position. Each town in a state had several jacks on its switchboard wired to the nearest long distance switchboard. The local operator would place all long distance calls to the Bell long distance operator for completion over the AT&T network. On some long distance calls, several long distance operators would be used to connect from one long distance center to another long distance center before reaching the called party's local operator.

2.9 THE DIVISION OF LARGE CITIES INTO SEVERAL LOCAL EXCHANGE AREAS

In large cities such as New York and Chicago, the local exchange would be split up into several exchanges scattered around the city. This reduced the congestion of wires coming into each exchange. Instead of one local exchange with thousands of wires

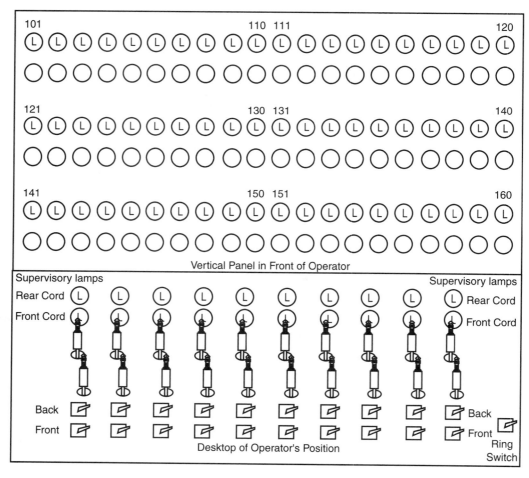

FIGURE 2-8 Manual Switchboard

coming to it, each of the local exchanges had a few hundred wires coming in. The operator at each exchange had access to jacks that were wired to the other local exchanges around the city for *interoffice calls*. During this era, few problems were encountered using this arrangement, because most local telephone calls were made within one office. Few lines were needed between offices because there were few interoffice calls. As the number of interoffice lines grew it became necessary to distinguish local phone lines from the lines going to other offices. The lines going to other offices were called *interoffice trunks* or just *trunks*. This change in terminology helped to avoid confusion when describing trouble to a repair person. If a line was out of service, the repair person knew only telephones attached to that line were out of service. When told a trunk was in trouble, the repair person knew that this outage was impacting all interoffice calling going over that trunk to another exchange (Figure 2-9).

To differentiate telephone numbers in one local exchange from the numbers in another, each local exchange was given the name of the local area. Some names

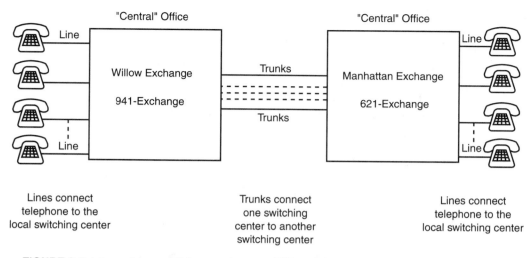

FIGURE 2-9 Interexchange call from customer at Willow to customer at Manhattan or vice versa.

were Manhattan, Downtown, Bronx, Willow, Riverside, Little Chicago, and so on. If a caller were placing a call to telephone number 4567 in the Manhattan exchange, the caller would tell the operator they wanted Manhattan 4567. The operator knew which local exchange to place the call to from the name of the exchange given. If the caller's phone was connected to the Willow exchange, the operator at the Willow exchange would place a patch cord into a trunk jack wired to the Manhattan exchange switchboard. The Willow operator would tell the Manhattan operator to connect the call to number 4567. The number 456 would tell the Manhattan operator to plug into line jack number 456. The last digit would tell the operator which party on the line to ring. In our example, party number 7 on the 456 line would be rung by the Manhattan operator.

Even when offices were distributed around the city, some of the local exchanges would be so large that hundreds of switchboards and operators were needed. Each switchboard would have specific lines cabled to its jacks; incoming calls could not be accessed from all switchboard positions. Therefore, it was necessary to add interposition jacks to the switchboards. Switchboard 1 had several jacks wired to jacks at switchboard 2, 3, 4, and so on. Every switchboard position had jacks wired to jacks at other positions. This allowed the operator to place calls to lines not appearing at her position. On interposition calls, the operator plugs into a jack associated with the position where the called line is terminated and ask that operator to ring the desired line.

The room in which all operator switchboards were installed was called a *switchroom*. The BOC tried to keep the size of the switchroom to 50 positions or less by using more than one exchange in large cities. In larger cities it was impossible to keep the downtown exchange small. Each city also had only one long distance operator center. In a large city the switchboards in these long distance centers were so long that management personnel sometimes used roller skates to move around the room in providing assistance to operators.

2.9   The Division of Large Cities into Several Local Exchange Areas

2.10 LINE AND TELEPHONE NUMBERING

Operators remembered which line each person was on in small towns. A caller would just ask the operator to ring someone by name and the operator knew which line to ring. As more people were added to the phone system, it became more difficult for the operator to remember who was on which line. To make calling easier in large cities, it was necessary to give the operator line numbers (telephone numbers) instead of a name. By the late 1950s, the growth of telecommunications led to the development of the North American Telephone Numbering Plan. This plan assigned three-digit area codes to each area of North America and assigned a unique three-digit exchange code to each central exchange within an area code. With this ten-digit numbering plan, every telephone in North America had its own unique identifier.

2.11 INTRODUCTION OF CENTRAL POWER AND LINE EQUIPMENT

One of the first improvements made in the early telecommunications system was to eliminate the batteries and hand crank at the telephone. The hand crank was used initially to generate a signal that would cause the ringer of the distant phone to operate. The batteries were used to provide electrical energy for the transmitter of the telephone. The hookswitch of the phone would connect the battery to the transmitter and telephone line when the receiver was lifted from the hookswitch. When the receiver was hung up on the hookswitch, the contacts of the switch would open the circuit to the battery and telephone line. Thus, the battery was only connected when the phone was in use to conserve the battery and lengthen its life.

With the introduction of central exchanges and operators, a phone was now rung by the switchboard operator. Engineers came up with a new way to signal the operator instead of using the hand crank. They equipped additional contacts on the hookswitch that would operate when the receiver was lifted and connect the transmitter and receiver directly to the telephone line. The engineers installed a 24-V DC battery in the local exchange office. A 24-V battery is the equivalent of two 12-V car batteries connected together in series. The negative lead of the 24-V battery was wired to one side of a device called a *line relay*. The other side of the line relay was wired to the ring wire of a local phone line, at the jack on the operator switchboard. There was one relay for every line connected to the switchboard. The positive side of the battery was connected through a second winding of wire in the line relay to the Tip wire of the phone line.

When the receiver at a telephone is lifted, the hookswitch closed the Ring wire to the transmitter and receiver. The other side of the transmitter and receiver was closed by a separate set of contacts on the hookswitch to the Tip wire. When the hookswitch was operated by a phone going off hook, an electrical circuit was completed. Electrical current then flowed from the negative terminal of the 24-V battery through the wiring of the line relay, out over the Ring wire to the telephone, through the hookswitch contacts (which were closed), through the transmitter, back to the other closed hookswitch contacts, back over the Tip wire, and returned through the second winding of wire in the line relay to the positive terminal of the 24-V battery.

Electrical current flowing through the wiring of the line relay attached to the line will cause this relay to close a set of contacts together. The operation of the contacts on this relay caused a miniature light to turn on above the jack on the operator's switchboard and sound a buzzer. The operator heard the sound and saw which lamp was lit. The operator answered the incoming call by plugging into the jack below the lit lamp. When the operator plugged into the jack, contacts on the jack opened, causing the lamp to go out and the buzzer to quit sounding (Figure 2-10).

The 24-V battery was still connected to the phone and would supply the energy necessary to power the transmitter. Consequently, the batteries at the phone were no longer needed. When the receiver was hung up the hookswitch returned to normal. This would disconnect the transmitter and receiver from the ring lead and connect the ringer of the phone to the ring lead. The ringer of a phone will only respond to an AC signal and looks like an open circuit to a DC voltage. When the receiver is hung up and the hookswitch is normal, the path for DC current flow is opened. With no current flowing through the relay attached to this line at the switchboard, the relay will release and turn off the switch contacts that lit a lamp and caused the buzzer to sound. With the ringer at the phone now attached to the line, the operator can ring the phone when necessary. Because all telephones attached to a party line had the same ringers in them, every phone on the line would ring when the operator sent a ringing signal on the line. All parties on a party line were assigned a certain code for their ring. One party could be assigned a ring of two long rings followed by one short ring, another party's ring would be three short rings, and so on. The operator would use these codes and people would only answer when they heard their code.

Later advances in operator switchboard design added a third lead to the plug of the patch cord and to the jacks. This third lead was called the *sleeve lead*. From the jack, this third lead was wired to a contact on the line relay that was closed to ground when

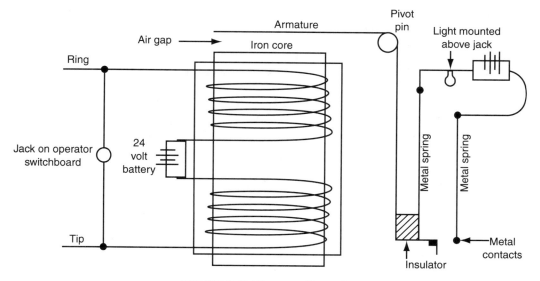

FIGURE 2-10 Diagram of line relay.

the line relay was released. The sleeve of the patch cord would make contact with the sleeve of the jack when the operator inserted the plug into the jack. The other end of the patch cord had the sleeve lead wired to a lamp next to the switch. The other side of the lamp was connected to battery. Now as long as the customer had the phone off hook and the line relay was operated, the lamp would stay dark. When the customer hung up the phone, the relay released. This would place a ground on the sleeve lead and the lamp would light. Operators no longer needed to monitor the call but could watch the lamps (called *monitor lights* or *supervisory lamps*) to see when a call was over.

Manual operator switchboards were also used at business locations. When installed at one business location to handle all incoming and outgoing calls, the system was called a *Private Branch Exchange* or *PBX*. The operation of a PBX was much the same as the operation of the switchboard in a central exchange. If you wished to call a large business, you would ask the local operator to connect you to the business. The PBX operator would answer the call and patch you through to the telephone of the person you wanted to talk with.

2.12 AUTOMATED SWITCHING

In 1892, Almond B. Strowger invented the first automated switching system. This system was designed so the telephone customer could provide directions to the switch instead of an operator. The instructions to the switch were provided by a dial. A dial was added to the telephone so that it could be connected to the automated switching system. This automated switch had stepping relays that followed the dial pulses. At the completion of a digit, the stepping relay connected the caller to another stepping relay. The completion of a call was done by using several stepping relays, and the system was called a *step-by-step switching system* or *Strowger switch*. Although the Strowger switch automated the switching process, these switches were not deployed in any great numbers until about 1925. Many rural areas did not replace manual switchboards with automated systems until after 1950. In the early 1960s, I installed many Strowger switches in Wisconsin to replace manual switchboards. The invention of automated switching systems has been one of the greatest developments in this century. If we were not using automated switching systems today, every man, woman, and child in America would have to work as an operator, to handle the volume of telephone calls handled by today's switching systems.

The Strowger switching systems have been replaced by computerized switching systems. All switching systems manufactured for use as public switching systems now use computers and software programming to control the switching of calls. These switching systems are called *Stored Program Control (SPC)* switching systems. The SPC system makes it possible to offer the enhanced services customers want such as three-way calling, call waiting, caller identification, last number redial, and so forth. These systems also make it possible for the LEC to provide Feature Group D (equal access). The SPC system requires almost no maintenance and takes much less floor space than its electromechanical predecessors. When Strowger switching systems were used in a private company to replace manual PBXs, they were known as *Private Automated Branch Exchanges (PABXs)*. SPC switching systems have now replaced Strowger Switches in private branch exchanges. SPC systems designed for use as a PABX are called *Computerized Branch Exchanges (CBX)*. Today, the acronym PBX no longer refers to a manual private switchboard, because these switchboards no longer exist. All PBXs manufactured today are actually computerized PABXs (CBXs). It is understood that PBX now means the same thing as PABX and CBX. The only major differences between the SPC used for a private switching system (PBX) and the SPC used for a public switch is the speed of the central processor, the amount of memory, the size of secondary storage devices, and the number of peripheral devices attached to the system. The SPC switching system used for a PBX will be quite small and usually occupies less space than a desk or file cabinet. The SPC switching system used in a public central office usually takes about 400 square feet for a large office (10,000 lines) and about 100 square feet for a small office (1000 lines).

2.13 ADVANCES IN OUTSIDE PLANT DESIGN

Technology has also improved tremendously in the area of outside plant. Outside plant at the turn of the century was nothing more than wire on a fence post. As more people started using telephones, the outside plant used the same wire and poles that were used for telegraph wire and telegraph poles. This type of wiring for telephones was called *open wire*. Iron wire was used and was attached to the telephone poles with glass insulators. This wire was exposed to weather. The wire would rust, deteriorate, and break in two, when wind speed was high. With the development of paper insulation, cables were developed with many wires inside the cable. Each wire was insulated from other wires by coating the wire with shellac and wrapping the wire in paper. The invention of plastic led to the use of a plastic covering on each wire as an insulator. These cables were called *plastic insulated cable (PIC)*. These cables could be buried, which eliminated damage from bad weather.

2.14 ADVANCES IN MULTIPLEXING TECHNOLOGY

A third area where technology has improved tremendously is multiplexing technology. *Multiplexing* is the placement of more than one telephone call over the same facility. Several multiplexing techniques were used, but the most favored technique until the 1970s was frequency division technology. As AT&T developed and continuously improved this technology, it was able to produce a *frequency division multiplexer (FDM),* which could place 10,200 telephone calls over one coaxial cable pair. In the 1970s, another multiplexing technology took hold and is now favored over FDM. This new multiplexing technology is called *Time Division Multiplexing (TDM).* It is really not a new technology; it was developed when vacuum tubes were state-of-the-art technology. Vacuum tubes could not handle TDM efficiently. The development of transistors and their evolution into integrated circuits has made efficient use of TDM possible. Basically, TDM was a technology that emerged ahead of its time, but today it is the technology underlying switching and transmission systems.

TDM technology is being advanced in the local telephone services arena. Integrated Services Digital Network (ISDN) is a TDM technology. The facility between a telephone and a central exchange is predominantly copper wire. Improvements are being made to TDM technology that uses copper wire as a medium. TDM technology will allow the local phone line to handle more than a telephone call. One TDM facility to the house will handle telephone calls, Internet access, alarm systems, and control systems. Copper wire can handle these needs easily. If coaxial cable or fiber optic cable is used as the facility to the central exchange, it is also possible to carry television signals on the TDM system. The TDM technology will continue to improve, and it will not be long before copper wire can also be used to carry television signals to our homes as part of a TDM system.

2.15 THE AGE OF COMMUNICATIONS

All the technologies mentioned above have improved tremendously over the past 20 years and will continue to improve in coming years. These technologies provide a vehicle allowing us to reach out and access any information we need. In this information age, we can get on the *Internet* highway and access many libraries. As we move into the future, the developments in deregulation, personal computer hardware, software, and telecommunications technology will allow us to make even greater accomplishments. LECs, IECs, and other common carriers will strive to become our one-stop communications center. They will offer us local telephone service, long distance service, television service, and Internet access service. Since they will now be able to provide all these services, they will seek to provide bundled services at discount prices. Can we do without these services? Of course we can. We do not need television. We can drive to the library and get books. We can call our stockbroker to get financial information. We can also do without a car. We can walk everywhere we want to go. Just as a car allows us to be more productive, the enhanced services that are available via telecommunications allow us to be more productive.

President Bill Clinton has been a strong advocate of the Internet Information Highway and has proposed that Internet access be available in all public schools by 2000. Access to information will be critical in the twenty-first century. This access must be available in our homes. Many children will be denied equal educational opportunities because their family can not afford a personal computer and Internet access. Still other children (and adults) will not take advantage of these technologies. I am not advocating that everyone should become an Internet junkie. I am merely saying that a person can take advantage of the good Internet sites to broaden their horizons and expand their intellectual database. In the 1940s, everyone did not have a car, nor did everyone have a radio. In the 1950s, not everyone had a television set, nor did everyone have a telephone. Internet access will be the next service to fall in to this category. Within the next ten years, more and more people will take advantage of the emerging telecommunications technologies. As growing numbers of people install these technologies, think of the new business opportunities that will be created. I see great applications in the educational field. The data capacity of a CD-ROM provides a great vehicle for interactive education on any subject matter, at any age level, from prekindergarten to master's programs.

Having many libraries and live meeting room forums available over the Internet can also enhance the learning process. Many meeting rooms are used today for people with mutual interests. Some companies now set up meeting rooms over Internet or intranet facilities. Many companies make use of customer knowledge by providing technical rooms. A person can access these rooms via Internet and ask if anyone present can help solve a problem. Since many people come and go in these rooms, the problem can be posted for access by later attendees. The company has technicians assigned to monitor the bulletin board of the meeting room. If the technician has an answer to posted problems, they will provide a reply. Many times, one customer's problem can be solved by another customer that has experienced the same problem. Establishing meeting rooms can provide a company with many intellectual resources. Many software companies are now charging a flat fee of $25 to provide support on their products. They would enjoy better customer relations if they provided an address on the Internet for a technical room. I also see opportunities for an entrepreneur to set up pay-per-question sites as the quest for knowledge becomes greater. You could establish sites which require a user identification number. The user pays a flat fee for enrollment. They could then access the site for information over a certain time frame and ask unlimited questions. I could offer a course in telecommunications! Everyone would not have to start school on the same date. They could progress through modules at their own pace and time.

It takes some time for technologies to be accepted. We still use checks and the mail to pay our bills. Many companies have set up direct deposit of checks and many employees use direct deposit. Most of us have a problem setting up direct payment from our checking to pay bills. We are hesitant to allow companies access to our accounts, so we do not do direct payments. I do not have any direct payment accounts set up, but I would be receptive to paying by phone or Internet if companies would provide such a setup. Companies are reluctant to accept this method of payment due to possibilities of fraud. In the near future, I expect that someone will provide a secure method for making transactions via telephone or Internet access. Instead of a bankbook, we will have a software program and ledger. We will be able to use dial-up access to check on our account balance at the bank and view all deposits and withdrawals. Many banks provide this service over voice mail. It will not be that difficult to provide the same information as a file transfer. This information could be downloaded directly into a software program at our personal computer. We can enter bills we wish to pay and the amount for each. The software program can automatically launch calls over the telephone network or Internet to provide an amount, account number, and authorize transfer of funds for payment of bills.

There are many ways to use the technologies coming on line. The extent to which they will be used is limited more by our acceptance of technology than by the lack of technology. ISDN has been around for ten years with almost no acceptance by the average consumer. The future of the information age, and how much we benefit from emerging technologies, will only be limited by our vision and active involvement in the use of technologies developed. When personal computers were first introduced, many of us did not accept the technology because you had to be a technical expert to use a computer. You had to build your own system and write your own software. The development of standards brought many manufactures into the business. Some people started software companies and developed operating system software and application software. People who knew nothing about hardware or software were now able to use the technology developed by others and many people became computer users. I was slow to accept Internet, because it had limited capabilities for someone who cannot remember site addresses. The development of browser software and fast modems has won me over. I am now looking forward to some company providing me with the speed of a cable access modem.

2.16 LOCAL EXCHANGE LINE

The local exchange line that connects our telephone to the central office will undergo major changes in the next few years. The Telecommunications Reform Act of 1996 allows almost any common carrier to provide local telephone service. Many IECs and cable television companies have formed strategic alliances to provide phone service. Many IECs are also using PCS technology to provide local telephone service. AT&T bought McCaw Cellular Telephone Company and will use this technology as a vehicle to provide local telephone service. AT&T has announced that this system has 70 million potential customers. The LECs must do something to protect their business. Most competition for local phone service will be in the major metropolitan areas. The LECs will accelerate deployment of digital services in these areas to provide more enhanced services in the competitive environment. I have seen several news releases that state that LECs are deploying fiber and coaxial cable facilities to the customer's doorstep. The days of a twisted pair of wires providing analog service are numbered. The services to our homes and business will soon be furnished by a digital subscriber line facility with capacity to furnish much more than local phone service. Some business owners are already using ISDN as a vehicle to carry the signal from a surveillance video camera to a remote monitor. The remote monitor may be at the home of the business owner, or any other location they desire.

2.17 STANDARDS ORGANIZATIONS

Standards for the telecommunications industry are generally established by the American National Standards Institute (ANSI), the Electronic Industry Association (EIA), and the Consultative Committee on International Telephone and Telegraph (CCITT). CCITT is part of the International Telecommunications Union (ITU). The ITU is chartered by the United Nations. CCITT standards are now being changed to the name of the parent organization and are called ITU standards. ANSI established most formal standards used by the telecommunications industry in the U.S. ANSI does not actually create and write standards. ANSI coordinates and sanctions the activities and standards written by other organizations. The Exchange Carriers' Standards Association, which is recognized by ANSI, produces some of the standards for telecommunications in the U.S. After the 1984 Divestiture, the RBOCs formed the Bell Communications Research Company (Bellcore) to do joint research and development activities. AT&T Labs and Bellcore participate in ANSI and CCITT standards activity. There are many informal or de facto standards that were set by AT&T's Bell Labs.

There are no set standards for telephones, central office switches, PBX switches, and key telephone systems. Although there are no set standards, telephones must be able to work on any switching system. This can only be accomplished if the telephone set manufacturer designs the telephone to meet established interface standards. Bell Labs published a document known as the *Local Switching Systems Generic Requirement* (LSSGR). This document outlines the basic interface design criteria for local (class 5) central offices. Knowing the interface requirements allows independent manufacturers to design to that interface. There is no standard established for the telephone, but it must be designed to work with the central office.

When party lines existed, many different types of ringers were available in phones, which caused problems when trying to buy a phone that would ring properly. The FCC established a ringer equivalency code for different types or ringers. This ringer equivalency code is also called the *ringer equivalency number (REN)*. When buying a telephone, people first had to check with their LEC to find out what the ringer equivalency code was for their telephone number. When they purchased the telephone, they would look at the REN number on the bottom of the telephone for a match. The elimination of party lines has led to the elimination of the requirement for the central office to be able to send many different ringing signals. Telephones manufactured today do not use a mechanical bell for the ringer. They use an electronic circuit that detects the presence of *any* ringing signal. Quite appropriately, this circuit is called a *ring detect circuit*. When this circuit detects a ringing signal, it will send an audible warble tone out through a speaker. This ability to ring on any frequency of ringing signal has led to the elimination of the need for ringer equivalency codes.

The formulation of standards is necessary for equipment made by different manufactures to work together. A standard such as EIA 232 allows a manufacturer to design equipment intended to work with a personal computer's serial port simply by knowing the EIA 232 standard. The personal computer manufacturers design their computers to send and receive data over the serial port in conformance with the EIA 232 standard. Open standards allow many manufacturers to make equipment. Some large companies try to establish their own proprietary standards, but this stifles growth in the industry. When ISDN was first introduced several proprietary standards were in use. Northern Telecom used a different standard than AT&T. An ISDN phone purchased for use on an AT&T switching system could not be used at a later date on a Northern Telecom switch. With the adoption of an international standard, the phones are now designed to one standard and will work on both AT&T and other manufacturers' central offices. Prior to deregulation, the LEC provided all station equipment, and it was the LEC's responsibility to ensure compatibility between devices. The LEC is no longer responsible for end-to-end service, and the interconnection of equipment from many manufacturers and carriers is only possible by having published standards.

IBM used proprietary standards on its first personal computers and the systems did not sell well. Only after making proprietary standards public and adopting existing standards did the personal computer business soar. Now if you buy an IBM PC, you do not have to buy all your hardware and software from IBM. If you want to add a disk drive or new software program, you have many competing manufacturers to choose from. With published interface standards in the telecommunications industry, many manufacturers can supply telecommunications equipment. The equipment from different manufacturers can be used together because the interface between them meets the same standard. While there is no standard for a telephone, and it can be designed as a manufacturer sees fit, it must be designed to meet the interface standards for a local central office exchange or it will not work. When Integrated Services Digital Network (ISDN) was first introduced, it was introduced with each manufacturer using proprietary standards. Therefore, equipment from different manufacturers would not work together. You could not use an ISDN station device manufactured for use on an AT&T switch on a Northern Telecom switch. CCITT published ISDN standards in 1984 and an update was published in 1988. American companies were slow to adopt these standards and this delayed acceptance of ISDN.

2.18 SUMMARY

The implementation of central exchanges with switching capabilities was the single biggest factor in the rapid growth of telecommunications at the turn of the century. Most manual switchboards were staffed by young boys when they were introduced. The service provided by young boys was less than professional and they were quickly replaced by female operators. "Central" was a phrase heard by many people for the first half of the 20th century. Manual switchboards were not replaced in many rural areas until after World War II. The introduction of a central power supply and line equipment was the first step toward automating the switching process. This automation of switching would lead to the loss of jobs for local exchange operators but would be a necessary step in the growth of telecommunications. If manual switching were still used in the local exchange, every person in the United States would have to work as an operator in order to handle the volume of local calls being placed today. The next major development in telecommunications was the ability to put many conversations over one facility. This multiplexing technology was first deployed in transmission facilities connecting central offices together. It was later deployed as part of the central office and PABX switching system in computer controlled switching systems. Accompanying the development of multiplexing technologies was the development of mediums such as coaxial cable and cables containing glass fibers to handle high levels of multiplexing. This multiplexing technology has grown to a point which allows 516,096 conversations to be placed over one pair of glass fibers. Multiplexing technology is now being extended from the central office to the residence.

REVIEW QUESTIONS

1. What are the three parts to a communications system?
2. What are the functions a receiver must perform?

3. What is the difference between half-duplex and full-duplex?
4. Why does the LEC use wire as a transmission medium?
5. Who invented the carbon transmitter?
6. What type of electricity is used to provide power for the transmitter of the telephone?
7. What type of electricity is used to provide power for ringing the telephone?
8. What is the central wiring center called?
9. When was the first central exchange installed?
10. How many wires connect the telephone to the central exchange?
11. What are the names given to the wires connected to the phone?
12. How many calls can a switchboard handle at one time?
13. What is a line?
14. What is a trunk?
15. Why were exchange names needed?
16. In large cities with thousands of lines on a switchboard, how were calls placed to lines not appearing on the position of the originating line?
17. What were the advantages of a central exchange?
18. What standards organization provides most standards for telecommunications in the United States?
19. What standards organization provides international standards?
20. What company is responsible for most de facto standards in telecommunications?
21. What factor is the major impediment to the deployment of new technologies?
22. When was the first automated switching system invented?
23. What are computer-controlled switching systems called?
24. What major factor delayed implementation of TDM technology?
25. What is plastic insulated cable called?

GLOSSARY

Alternating Current (AC) A signal that is continually alternating the direction of current flow due to changes in the polarity of the voltage of the signal. In other words, it is a signal that starts at zero voltage, rises to a maximum positive potential, declines to zero, and continues the decline until it reaches a maximum negative potential, then returns to zero to complete one cycle of a signal. The number of cycles the signal completes in one second is the frequency of the signal. This is the type of electricity supplied by the local power company to our homes and businesses.

Central Office The central wire center or central exchange. All telephones in a small geographic area are wired to a central exchange, which serves all telephones in that area.

Communication A process that allows information to pass between a sender (transmitter) and a receiver over some medium.

Direct Current (DC) The electric current flow is at a constant rate and flows continuously in one direction. The direction of electric current flow (outside the battery) is from the negative terminal of the battery, through a device attached to the battery, to the positive terminal of the battery. The amount of electric current flowing is measured in amperes.

Frequency Division Multiplexing (FDM) The process of converting each speech path to different frequency signals and then combining the different frequencies so they may be sent over one transmission medium.

Full-Duplex Transmission If a transmission system allows signals to be transmitted in both directions at the same time, the system is called a full duplex transmission system.

Half-Duplex Transmission A type of transmission where transmitters on each end of a medium take turns sending over the same medium.

Internet A network of computers that can be accessed by other members of the Internet community. Most individuals become members of the community by purchasing access to the Internet from a local Internet Service Provider or Commercial on-line Information Access Provider such as Prodigy or America-On-Line (AOL).

Interoffice Calls Calls completed between telephones served by two different central exchanges.

Intraoffice Calls Calls completed between telephones served by the same central exchange.

Line Relay An electromechanical device attached to a telephone line at the central exchange. When a subscriber took the handset of the telephone off-hook, electric current flowing in the relay caused it to operate and signal the operator.

Medium The device used to transport information or a message between the transmitter and receiver.

Multiplexing In telecommunications, a process combining many individual signals (voice or data) so they can be sent over one transmission medium.

Receiver The device responsible for decoding or converting received information into an intelligible message.

Simplex Transmission Transmission of signals in one direction only is called simplex communications.

Sine Wave A graphical representation of an AC signal along a time line.

Switchboard The manual switchboard was used by operators to establish calls between two telephones. The manually switchboard was replaced by an automated switch.

Telecommunications The communications of voice or data over long distances using the Public Switched Telephone Network or privately owned networks.

Time Division Multiplexing (TDM) The process of converting each speech path into samples, then combining the different samples by transmitting each sample at a different time, so they may be sent over one transmission medium.

Transmitter The device responsible for sending information or a message in a form that the receiver and medium can handle. The device in a telephone that converts the air pressure of a voice signal into an electrical signal that represents the voice.

3

Basic Electricity Review

Key Terms

Alternating Current (AC)
Ampere-Hours
Battery
Cell
Conductor
Direct Current (DC)

DC Voltage
Decibel (dB)
Decibels Referenced to 1 MW
  (dBm)
Dry Cell
Kirchhoff's Current Law

Kirchhoff's Voltage Law
Ohm's Law
Resistance
Transducer
Voltage
Wet Cell

The preceding chapter referred to DC voltage, AC voltage, current, and relays. This book is not intended to be a basic electricity or basic electronics book. For an in-depth understanding of these subjects, you should consult a book covering these subjects in detail. This chapter is intended to acquaint you with how electricity acts in telecommunications circuits. Electronic devices, integrated electronic circuits, and electromagnetic relays are used extensively in telecommunications. A telecommunications manager should have a basic appreciation of how these devices work and how electricity is used in telecommunications. The telecommunications system carries messages from a sender to a receiver. The system uses electrical signals to convey these messages. The messages have to be converted into electrical signals before the system can carry them. Therefore, a brief overview of electricity forms the core of this chapter. This overview should leave you with a basic understanding of how electrical signals are affected by the wire media used to carry the signals, and some idea of the immediate environment surrounding those media.

    Electricity used in telecommunications takes two forms: AC electricity supplied by the local power company and DC electricity supplied by batteries installed at the telephone exchange. Central office exchanges use batteries to supply DC voltage for powering both the telephone and central office equipment. They also use AC voltage supplied by the local power company. The AC power is connected to a battery charger and keeps the batteries charged. With this arrangement, power for the switching system is always supplied from the battery, and the system will still work when a commercial AC power outage occurs.

3.1 DC VOLTAGE / BATTERY

DC voltage is the voltage produced by a chemical reaction. Devices that produce electricity via a chemical reaction are called *cells* or *batteries*. Batteries produce a voltage that results in a steady flow of electric current. This steady electric current flow is called *direct current (DC)*. Direct current is electric current that flows from the negative terminal of the battery out over a wire connecting some device to the battery, through the device and returning over a second wire from the device, to the positive terminal of the battery. The electric current flow is at a constant rate and moves continuously in one direction. The direction (outside the battery) is from the negative to the positive terminal of the battery. The amount of current flowing is measured in amperes (amps) and is directly proportional to the voltage produced by the battery.

Voltage is a measure of how much potential difference or electrical pressure exists between the two terminals of a battery (or some other device that generates electricity). This electrical pressure is called *electromotive force (EMF)* and is measured in volts. The difference in pressure is what causes electric current flow. In our universe, many different occurrences are the result of things moving from an area of high pressure to an area of low pressure, or from an area of high concentration to an area of low concentration. Flows of this nature are responsible for weather changes and for the osmosis process by which plants take in food and water for survival. The movement of electric current is the movement of electrons from an area of high concentration to an area of low concentration. The higher the difference between the area with a high concentration of electrons and the area with a low concentration of electrons, the higher the potential difference between the two and the higher the voltage developed between the two areas.

3.2 MATERIALS, ELEMENTS, MOLECULES, AND ATOMS

All materials in the universe are composed of substances called *elements*. There are over 100 elements in the universe. Some materials are made by combining two or more elements. The smallest unit that these *materials* can be broken down into and still retain the characteristics of the material is a *molecule*. For example, water can be broken down or subdivided into one molecule and still be water. If the water is broken down any further, it will no longer be water but will be atoms of the elements that make up water. It takes two elements to make water (hydrogen and oxygen).

The smallest piece of any element that will still retain the characteristics of that element is an *atom*. Molecules are made up of atoms. A water molecule is made up of two atoms of hydrogen (the designation for two atoms of hydrogen is H_2) and one atom of oxygen (O). The symbol for a molecule of water is thus H_2O. When a material is made strictly from one element, it is called a *pure material*. Copper is a pure material. Copper is composed solely of atoms of the copper element. Copper can be broken down into a molecule of copper, which is composed of one atom of the element copper.

3.3 ELECTRONS

Atoms are made up of protons, neutrons, and electrons. Electrons exist in the outer part of an atom (Figure 3-1). Materials or elements that conduct electricity have electrons that are held loosely in place. These materials are called **conductors**. Copper is a material (an element) that conducts electricity well because it has one electron held loosely in place (Figure 3-1). If a voltage source is attached to these materials, the voltage source will dislodge the electrons and force them to move down the conductor. Electrons flow out from the voltage source (that is, battery) into the conductor to replace the electrons that have moved down the conductor away from the negative battery terminal, and the process continues.

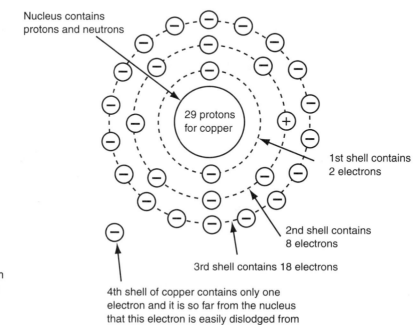

One Atom of Copper (CU)

Nucleus contains protons and neutrons

29 protons for copper

1st shell contains 2 electrons

2nd shell contains 8 electrons

3rd shell contains 18 electrons

4th shell of copper contains only one electron and it is so far from the nucleus that this electron is easily dislodged from the atom

FIGURE 3-1 An atom of copper. Positive protons in the nucleus attract negatively charged electrons to keep them from flying away. The electrons circle the nucleus at high speed, much like planets circling the sun.

The electron flow process is similar to what would happen if you had a tube filled with tennis balls (or baseballs or billiard balls) and added a ball to one end of the tube. Assume the tube represents the conductor and the balls represent electrons. If you were to force a ball into one end of the tube, all the balls move down the tube in a chain reaction until the ball at the far end will pop out of the tube. This is how electricity flows. The electrons bump each other down the conductor at the speed of light. If you had a tube from New York to California filled with balls and forced a ball into the tube in New York, a ball would pop out the California end almost immediately.

3.4 ELECTRON FLOW, CONDUCTORS, AND INSULATORS

The chain reaction flow of electrons in a conductor is called electric current. *Electric current is the flow of electrons.* How fast the current flows depends on the pressure (called electromotive force) or voltage applied to the conductor and how much resistance the conductor has to the current flow. *All conductors have resistance or opposition to current flow.* The amount of resistance depends on how easy it is to dislodge electrons from their atoms. *Different elements have different **resistance**.* Some elements have their electrons held so tightly to the atom that it is hard to dislodge them, and they have great opposition or resistance to electric current flow. The elements or materials that have a high resistance to electric current flow are called insulators. Materials having a low opposition to electric current (electron flow) are called conductors. *EMF is the pressure that forces electrons to flow and is measured in volts.*

Electricity will only flow if a complete loop of conducting materials between the negative battery terminal and the positive battery terminal exists. If the loop is opened, electron flow ceases. The purpose of a switch in a circuit is to provide a means whereby the loop can be opened or closed. When the switch is on, the loop is closed and current flows. When the switch is off, the loop is opened and current flow ceases (Figure 3-2).

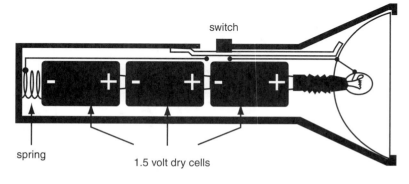

FIGURE 3-2 A 4.5-V battery composed of three 1.5-V cells.

Three dry cells connected in series equals a 4.5-volt battery. When switch is moved forward, the negative end of battery is connected to light bulb, and it will light.

3.5 BATTERIES VERSUS CELLS

The battery used in Figure 3-2 is composed of three cells. Figure 3-3 is the electrical schematic diagram for Figure 3-2. Many people refer to these cells as flashlight batteries, but this is an incorrect use of the word *battery*. A battery is a DC voltage source consisting of two or more cells in series. Earlier we referred to a battery as the device that generates electricity via chemical means. This is also incorrect usage. A *cell* is a device that generates electricity by a chemical reaction. When two different materials are immersed in an acid solution or acid paste, the acid causes a chemical reaction that removes electrons from atoms of one material and deposits them on the other material. This chemical action results in an excess supply of electrons on one of the materials and a shortage of electrons on the second material. Because electrons have a negative charge, removing them from one material leaves that material positively charged. The material that gains extra electrons becomes negatively charged.

The positively charged material has a deficiency of electrons and is an area of low electron concentration. The negatively charged material has many excess electrons and is an area of high electron concentration. The measure of the differences between these two different areas of concentration is expressed in volts. If an external device is connected between the material having an excess of electrons (the negative terminal) and the area with a deficiency of electrons (the positive terminal), the electrons will flow through the external device as they move from the area of high concentration (high-pressure area) to the area of low concentration (low-pressure area). Electron flow will continue until most of the excess electrons on the negatively charged material have flowed through the external circuit and reached the positive material.

As electrons leave the negatively charged material, it slowly loses its negative charge. As electrons flow into the positively charged material, it slowly loses its positive charge. Both materials would slowly lose their electrical charge as current flows from the negative terminal to the positive terminal, if the chemical reaction in the cell stopped. The chemical reaction within the cell continues to maintain the electrical charge on each material. Eventually, the chemical reaction will stop as materials are changed by the chemical reaction. The voltage difference between the two terminals is slowly reduced until the difference in potential is so low that current will not flow. At this point, we say the battery is discharged.

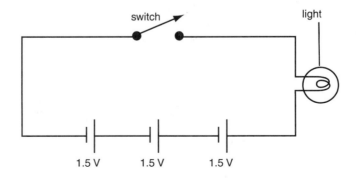

FIGURE 3-3 Schematic diagram for Figure 3-2.

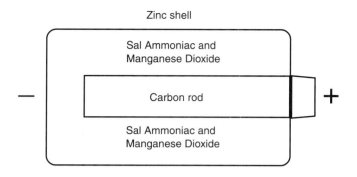

Zinc shell

Sal Ammoniac and
Manganese Dioxide

Carbon rod

Sal Ammoniac and
Manganese Dioxide

FIGURE 3-4 Typical dry cell.   Develops 1.5-volt "dry cell"

The cell for a flashlight uses a carbon rod as one of its active materials. The carbon rod is located in the center of the cell. The carbon rod is surrounded and enclosed with an acid paste and an outer shell made of zinc (Figure 3-4). The acid paste chemically removes electrons from atoms of carbon and deposits them on the zinc material. The carbon rod is positive because it has lost many electrons. With many additional negative electrons added to it, the zinc shell has a negative charge. The chemical reaction continues until the pressure difference created between the carbon and zinc reaches a point where the chemical action stops. At this point the zinc material has accepted all the extra electrons it can store and will repeal any attempt to add more electrons to it. The cell is then said to be fully charged. The point at which a cell composed of carbon and zinc becomes fully charged results in a pressure difference (EMF) of 1.5-V per cell. If two of these cells are connected in series, they make a 3-V battery. If three of these cells are connected in series, they make a 4.5-V battery (Figures 3-4 and 3-5). This type of cell is called a *dry cell* because no liquid acid is used. The acid is in a paste.

3.6 RECHARGEABLE CELLS

Another type of chemical cell is referred to as a *wet cell* because it uses a liquid acid instead of a paste. The battery used in an automobile is composed of six wet cells connected in series. Wet cells can be recharged by forcing electricity to flow backward through the cell. The most common wet cell uses lead and lead peroxide as the two active materials. These two materials are immersed in a sulfuric acid solution. Figure 3-6

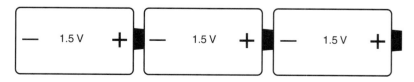

FIGURE 3-5 A 4.5-V battery.

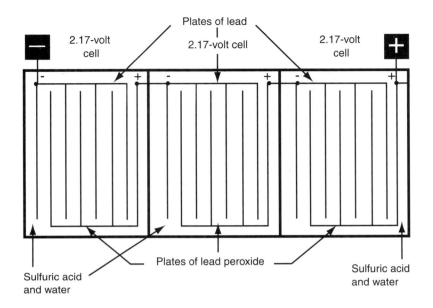

Plates of lead

2.17-volt cell   2.17-volt cell   2.17-volt cell

Plates of lead peroxide

Sulfuric acid and water   Sulfuric acid and water

FIGURE 3-6 A 6-V battery composed of three 2.17-V wet cells.

The chemical action in a "wet cell" produces 2.17 V when the active materials are lead, lead peroxide, and sulfuric acid. This device has three "wet" cells connected in series. Thus, this is a 6-V battery.

is a diagram of a 6-V battery containing three wet cells. The sulfuric acid causes a chemical action to take place that results in the lead plate becoming negative and the lead peroxide plate becoming positive. This type of cell develops an EMF of 2.17 V per cell. A car battery has six cells connected in series and results in 13.02 V output for the battery ($6 \times 2.17 = 13.02$ V). These batteries are commonly called 12-V batteries because most people use 2-V instead of 2.17 V per cell when making a rough approximation of the voltage.

The size of a wet cell has no bearing on the EMF developed between the positive and negative plates. All wet cells composed of lead and lead peroxide develop approximately 2.17 V per cell regardless of size. The size of the cell determines how long a cell can deliver electricity before it discharges. The larger the cell, the more active material it contains, and the longer it will supply electricity without being recharged. For this reason, the batteries in a central office are made using large wet cells. Each cell is about 2 ft high, 1 ft wide, and 1 ft long. An engineer determines the size of a cell to use by calculating how much electric current is needed for eight hours. Electric current is measured in amperes. If a central office uses an average of 200 amps each hour, then in 8 hr it will use 1600 amps. Cells are rated in *ampere-hours*. This is a rating that states how many amps can be delivered in a stated time frame. A 1600 ampere-hour cell can deliver 1600 amps for 1 hr, or 200 amps for 8 hr, or 100 amps for 16 hr, and so on.

The engineer designing the battery for a central office will select cells large enough to provide the office with electricity for 8 hr. If a commercial AC power failure occurs, the battery chargers will stop working and the batteries will be large enough to supply electricity for 8 hr. This is sufficient time to get an emergency power generator at the site to provide alternating current for the battery chargers. A central office is equipped with 24 cells connected in series. This provides approximately 52 V to the equipment (24 × 2.17 = 52.08 V). Most people refer to the central office battery as a 48-V battery. They use 2 V per cell in their calculations (24 × 2 = 48 V). Because some people use 2 V per cell in their calculations and others use 2.17 V per cell, the central office battery may be called a 48-V supply, a 50-V supply, or a 52-V supply. In the older electromechanical switching systems, the battery consisted of 23 cells and was called a 48-V battery even though it was actually producing 49.9 V (23 × 2.17 = 49.91 V). Remember that the cell's size has nothing to do with its voltage. A lead-acid wet cell produces 2.17 V per cell regardless of how large the cell is. What determines the voltage of a battery is how many cells we connect in series.

3.7 VOLTAGE / CURRENT / RESISTANCE RELATIONSHIP

The more voltage supplied to a device, the more current will be flowing through the device. The more opposition or resistance the device offers to current flow, the less current there will be flowing through the device. Voltage and current are directly related. Double the voltage applied to a device and the current flow will be doubled. Resistance and current are inversely related. If resistance goes up, current goes down. If resistance is doubled, the current flow will be cut in half. These relationships between voltage, current, and resistance are stated by a formula called *Ohm's law*:

Ohm's Law

$$I = V / R$$

In other words, current = voltage/resistance. This formula can be rearranged to find the voltage that will be dropped across a device with current flowing through it: voltage dropped = current × resistance.

$$V = I \times R$$

The formula can also be rearranged to find the resistance of a device when we know the voltage dropped across it and the current flowing through it: resistance = voltage/current.

$$R = V / I$$

Voltage is represented by the symbol V (some people use the symbol E, which stands for EMF). *Current is measured in amperes* and amperes are represented by the symbol A (amps) or I (French word for intensity of current). *Resistance is measured in ohms (Ω).* The formulas above are usually expressed in units of measure such as $V = I \times R$; $I = V/R$, and $R = V/I$. Ohm's law formulas are used to determine how much voltage, resistance, and current we have flowing in any device or circuit.

3.8 THE COMPLETE SYSTEM

A battery can be thought of as being similar to a water tower. A water tower is filled with water and the weight of the water causes it to exert pressure downward. When a pipe connected to the water tower is opened, the gravitational pull results in water flowing out of the pipe. In today's water distribution system, additional pumps are placed at intervals along the main water distribution pipe to increase the pressure behind the water in the pipe. When we open the faucet at our sink, the pressure behind the water causes water to flow out of the pipe. If there was no way to get additional water back into the water tower, we would soon run the tower dry. But water coming out of our faucets finds its way back into streams and rivers and is pumped back up into the water tower. Our water system has a way to supply us with water and also has a way to return water to the tower.

Batteries are like water towers because a supply of electrons is stored in the battery under pressure. The battery supplies electrons to our devices and circuits when a circuit is completed between the negative and positive terminals. These circuits must have a return path to the positive side of the battery to return the electrons we do not use to the battery. If a device is connected only to the negative terminal of a battery, electric current will not flow. The excess electrons on the negatively charged plates are trying to get back to the positive terminal. If the positive terminal is not attached to the circuit, the electrons refuse to flow. When both terminals are hooked to a device, electrons will leave the negative terminal and flow through the device in their journey back to the positive terminal.

Resistance can be thought of in terms of water pipes. Small water pipes have a higher resistance to the flow of water than larger pipes. The higher the resistance, the less water current we will have. The same is true in electrical circuits. Higher resistance in the circuit results in less electric current.

Now that we can generate electricity using a battery and we know current will flow through any device attached between the negative and positive battery terminals, we can put switches in the circuit to turn the device on and off. In Figure 3-3, the flashlight is turned on and off by operating the switch. When the switch is closed, the flashlight is turned on. Electrons will flow from the negative terminal, through the spring at the bottom of the cells, through the metal case to the switch, through the closed switch, to the lightbulb, through the lightbulb, and return to the positive terminal. This flashlight has three cells connected in series. The negative terminal of one cell connects to the positive terminal of another cell. The overall voltage of any battery is equal to the voltage of each cell, connected in series, added together. Since each cell produces 1.5 V, the voltage for this battery (three cells in a series connected arrangement) is 4.5 V. If a water main had three pumps in series along its route, the total pressure is found by adding the pressure developed by each pump. Each cell in a battery is an electron pump and when they are connected in a series arrangement, the total voltage equals the sum of the individual cell voltages. Voltage can be measured by a device called a *voltmeter*. Many times the voltmeter is part of a device called a *multimeter*. This multimeter can be used as a voltmeter to read voltage, as an ammeter to read current, or as an ohmmeter to read resistance.

3.9 SERIES RESISTANCE IN A CIRCUIT

When components in a circuit are connected in series, the total resistance for the complete circuit is arrived at by adding the value of all the resistances. A series circuit is a circuit in which the output of one resistor connects to the input of the next resistor. In a series circuit the electron current flow will flow through one resistor, then the next, then the next, and so on. You can tell when a circuit is a series circuit because the electron flow can only take one path (Figure 3-7).

We can tell that the resistors of Figure 3-7 are wired in a series arrangement because the current flow from the negative terminal of V_s will first flow through R_1, then R_2, then R_3, then R_4, then R_5, before returning to the V_s positive terminal. Also note that the output of the negative terminal of V_s is wired to the input lead of R_1, the output of R_1 is wired to the input of R_2, the output of R_2 is wired to the input of R_3, and so on. The total resistance (RT) of a series circuit is equal to the sum of the individual resistance:

$$RT = R_1 + R_2 + R_3 + R_4 + \text{ and so on.}$$
$$\text{In Figure 3-7, } R \text{ total} = R_1 + R_2 + R_3 + R_4 + R_5$$
$$= 50 + 100 + 25 + 10 + 15 = 200 \ \Omega$$

The first step in trying to determine how much current flows in a circuit is to determine the total voltage applied to the circuit as well as the total resistance of the circuit. Ohm's law can then be applied to determine the current flow. In Figure 3-7, let us assume that the supply voltage (V_s) is 50 V. Ohm's law states that current in amperes (I) = EMF in volts (V) divided by resistance in ohms (R). For Figure 3-7, $I = V/R = 50/200 = 0.25$ amp. In a series circuit the current flow is the same at any point in the circuit. The amount of current leaving the battery is 0.25 amp and the amount of current returning at the positive terminal of the battery is also 0.25 amp. The current through each of the resistors is 0.25 amp as well.

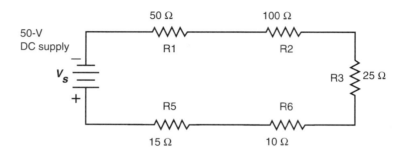

FIGURE 3-7 Series circuit.

Many people find this point counterintuitive. They think each resistor will eat up or use current and less is sent to the next component in line. Each component will use up energy, but the energy it uses is what creates the resistance of the component. The difference in resistance has compensated for each component's use of electricity. This is much the same as trying to pump water through several pipes in series with each pipe having a different size. The smallest pipe (largest resistance) is the major component limiting current flow, but the resistance of all pipes (total resistance) limits the overall water flow. The same amount of water flows through each pipe and the current flow is the same everywhere in the pipeline. In an electrical circuit, the current flow is limited by total resistance, and the current flow is the same at any point in the circuit.

Another factor to consider in a series circuit is the voltage supplied to each resistor. This can be determined by using Ohm's law, which states that voltage (V) = current (I) × resistance (R). In Figure 3-7,

$$\text{Voltage across } R_1 = 0.25 \text{ amp} \times 50 \ \Omega = 12.5 \ V$$
$$\text{Voltage across } R_2 = 0.25 \text{ amp} \times 100 \ \Omega = 25 \ V$$
$$\text{Voltage across } R_3 = 0.25 \text{ amp} \times 25 \ \Omega = 6.25 \ V$$
$$\text{Voltage across } R_4 = 0.25 \text{ amp} \times 10 \ \Omega = 2.5 \ V$$
$$\text{Voltage across } R_5 = 0.25 \text{ amp} \times 15 \ \Omega = 3.75 \ V$$
$$\text{total} = 50 \ V$$

Notice how the sum of the voltages across each resistor is equal to the applied voltage. The voltages across each resistor are called *voltage drops*. The person who first observed this was the 19th-century German physicist Gustav Robert Kirchhoff, and this observation is known as **Kirchhoff's voltage law.** This law states that the sum of all the voltage drops around a single closed loop in a circuit is equal to the total source voltage in that loop. Now you can see that it is not current that drops in a circuit, it is voltage. The current is the same everywhere in a series circuit but the voltage is not. We start out with a supply voltage of 50 V. If we use a voltmeter to measure between the + and – terminals of the battery in Figure 3-7, we read 50 V. If we measure between the + terminal of the battery and the junction between R_1 and R_2, we will read 37.5 V because 12.5 V was dropped across R_1. Each resistor will cause the voltage to drop until at the output of R_5, the voltage is zero.

Kirchhoff's Voltage Law

$$\text{Total voltage} = VR_1 + VR_2 + VR_3 + \text{and so on.}$$

3.10 PARALLEL RESISTANCE IN A CIRCUIT

When components are connected such that all their inputs are connected together and all the outputs of each are connected together, they are wired in parallel. This is similar to a multilane highway. The lanes are in parallel. Traffic can flow over any lane. Figure 3-8 is a parallel circuit. Notice how point A is connected to one

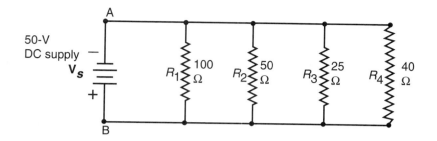

FIGURE 3-8 Parallel circuit.

end of all resistors and point B is connected to the other end of all resistors. Each resistor has its input lead connected to point A and output lead connected to point B. If current is flowing from point A, when it gets to resistor R_1 it has two paths to choose from. Some of the current will flow through R_1 and some of the current will flow toward R_2 and R_3. Thus, in a parallel circuit current *is not* the same at every point in the circuit. If a 50-V battery were connected across points A and B, the 50 V would appear across each and every resistor in parallel. Thus the voltage across R_1 is 50 V, the voltage across R_2 is 50 V, and the voltage across R_3 is also 50 V.

Because a parallel path provides multiple paths for current flow, when resistors are placed in parallel, the total circuit resistance decreases. The formula used for calculating total resistance of a parallel circuit is:

Parallel Resistance Formula

$$1 \,/\, RT = 1/R_1 + 1/R_2 + 1/R_3 + 1/R_4 + \text{ and so on.}$$

This formula can also be stated as follows: the reciprocal of the total resistance is equal to the sum of the reciprocals of each resistance. In Figure 3-8, the reciprocal of $R_1 = 1/R_1 = 1/100 = 0.01\ \Omega$. The reciprocal of $R_2 = 1/R_2 = 1/\ 50 = 0.02$, and the reciprocal of $R_3 = 1/R_3 = 1/\ 25 = 0.04$. The reciprocal of $R_4 = 1/R_4 = 1/40 = 0.025$. The next step is to get the sum of these reciprocals: $0.01 + 0.02 + 0.04 + 0.025 = 0.095$. The reciprocal of the total resistance equals the sum of the reciprocals: $1/RT = $ sum of the reciprocals. In Figure 3-8, $1/RT = 0.095$. Therefore, to get total resistance we take the reciprocal of both sides of the equation. The reciprocal of $1/RT = 1/1/RT = RT$. The reciprocal of the sum of the reciprocals: $1/0.095 = 10.526316\ \Omega$. Therefore, $RT = 10.526316\ \Omega$. In a parallel circuit RT will always be less than the lowest resistance, and we meet this test (10.526316 is less than 15). Now that we have RT, we can use Ohm's law to find total current.

Total current (*I* total) = total voltage (50 V) / total resistance (10.526316) = 4.75 amps total current. The current in each resistor can also be found using Ohm's law. By this method, IR_1 (current in R_1) = VR_1 (voltage across R_1) / R_1 (resistance of R_1) = 50/100 = 0.5 amp. Likewise, $IR_2 = VR_2 / R_2 = 50/50 = 1$ amp and $IR_3 = 50/25 = 2$ amps. $IR_4 = 50/40 = 1.25$ amps. The total current for all four resistors equals $IR_1 + IR_2 + IR_3 + IR_4 = 0.5 + 1 + 2 + 1.25 = 4.75$ amps. This equals the

Basic Electricity Review Chap. 3

I total found above. This proves another theorem of Kirchhoff. The second law of Kirchhoff is called **Kirchhoff's current law**:

Kirchhoff's Current Law

Total current approaching a point = Total current leaving a point

The total amount of current approaching any point in a circuit is equal to the total current leaving that point. The total current at point A = 4.75 amps; 0.5 amp of that current flows through R_1, 1 amp of that current flows through R_2, 2 amps of that total current flow through R_3, and 1.25 amps of that current flows through R_4. These individual currents flowing through R_1, R_2, R_3, and R_4 will combine on the output side of the resistors, and a total current of 4.75 amps will flow back into the + terminal of the battery connected to point B.

No matter how many devices are connected in parallel, the voltage across each is equal to the voltage applied and the voltage across every resistor will be the same. Calculators make it easy to find total resistance for a parallel circuit. First enter the resistance value for R_1, then press the $1/x$ key, press the + key, enter the value of R_2, then press the $1/x$ key, press the + key, enter the value of R_3, press the $1/x$ key, and continue this process for all resistors in parallel. When the last resistor value has been entered and the $1/x$ key pressed, press the = key, then press the $1/x$ key. Voilà, we have the total resistance for the parallel circuit.

3.11 POWER IN TELECOMMUNICATIONS AND DECIBELS

We are all familiar with power as used in our homes. This power is stated in watts. One watt of electricity is equal to one ampere of electricity flowing through one ohm of resistance. If the current is doubled, the power goes up by a factor of 4. The formula is as follows:

$$\text{Power (in watts)} = I^2 R$$

Power can also be restated in the following way: $P = IE$. Remember that according to Ohm's law, $R = E/I$. If we substitute in the original formula ($P = I^2 R$) for R, we get $P = I^2 E/I$. This becomes $P = EI$. One watt = $1V \times 1$ amp. The current will be doubled through a stated resistance if the voltage is doubled. Thus, doubling the voltage doubles the current and $P = 2 \times 2 = 4$ W. The current was doubled and the power went up by a factor of 4. Power can also be stated as follows: $P = E^2 / R$. In this case we use Ohm's law ($I = E/R$) to substitute for I in the equation ($P = EI$) to get $P = E (E/R) = E^2 / R$. In telecommunications, we deal with extremely low power levels such as one-thousandth of a watt (1 mW). The milliwatt is the unit of measure used for power in telecommunications circuits.

In telecommunications, we are often concerned with the comparison of one power level to another. The unit of measure used to compare two power levels is the *decibel (dB)*. A decibel is not an *absolute* measurement. *It is a relative* measurement. The decibel level indicates the relationship of one power level to another. The relevant formula is:

Decibel Formula

$$dB = 10 \log P_2 / P_1$$

If we have an output signal of 2 mW at an amplifier and the input to the amplifier is 1 mW, the relative power of output to input will tell us the gain of the amplifier in decibels:

$$dB = 10 \log \text{(power ratio)}$$
$$dB = 10 \log \text{(output power / input power)}$$
$$dB = 10 \log (2/1)$$
$$dB = 10 \log \text{ of } 2$$

The log of 2 is .30103; therefore,

$$dB = 10 \times 0.3 = 3$$

Suppose the output of an amplifier is 2000 W and the input is 1000 W. You can see that this provides a ratio of 2 and the gain of the amplifier is also 3 dB. The first amplifier increased the power by 1 mW, and the second amplifier had to increase the power by 1000 W to result in a 3-dB gain. That is why decibels are relative measurements and not absolutes. A 3-dB gain in one circuit is quite different from a 3-dB gain in another, but they are similar in one respect: when comparing two power levels, if one power is twice another, the difference between the two is 3 dB. No matter what power levels we deal with, if the power is doubled, we have a 3-dB gain. Also notice what happens when we reverse our comparison. If we compare the ratio of input to output, the ratio is 1/2 or 0.5. The log of 0.5 is –0.30103.

$$dB = 10 \log \text{(power input / power output)}$$
$$dB = 10 \log (1/2)$$
$$dB = 10 \times -0.30103$$
$$dB = -3$$

When the power is cut in half, it is –3 dB. When the power is doubled, it is +3 dB. Let's check a few more power levels. If the power ratio is 10, the log of 10 is 1 and we find that for any power change by a factor of 10, it is a 10-dB change. For example, a change in power level from 1 to 10 mW is a gain of 10 dB. A change from 10 to 100 W is also a gain of 10 dB. A change of 10 dB means the power level changed by a factor of 10.

$$dB = 10 \log (P_2 / P_1)$$
$$dB = 10 \log (10/1)$$
$$dB = 10 \log 10$$
$$dB = 10 \times 1$$
$$dB = 10$$

If the power changes by a factor of 100, it is a 20-dB change, and a power ratio of 1000 is a 30-dB change, because the log of 100 is 2 and the log of 1000 is 3. Thus, a comparison of 1 W (1000 mW) to 10 mW, and 1 W to 1 mW, results in the following:

1 W to 10 mW	**1 W to 1 mW**
dB = 10 log (P_2 / P_1)	dB = 10 log (P_2 / P_1)
dB = 10 log (1000 / 10)	dB = 10 log (1000 / 1)
dB = 10 log 100	dB = 10 log 1000
dB = 10 × 2	dB = 10 × 3
dB = 20	dB = 30

Similarly, if the power ratios are turned around and we state that one power is a tenth of another power, the log of 1/10 (0.1) is –1 and 10 times –1 is –10. Thus, we state this as –10 dB. When one power is 1/100th of another, it is –20 dB, and 1/1000th power ratio is –30 dB. We usually state overall circuit loss in a circuit in one of two ways. In terms of gain, we can say that the gain of the circuit is –10 dB and so on, or we can state it as a loss, indicating that the loss is 10 dB and so forth. The use of decibels is a convenient method of determining overall circuit gains and losses, because decibels are added together (Figure 3-9).

In Figure 3-9, the total gain or loss for the circuit is found by adding the decibel gain and loss of each section of the circuit (–5, +3, –4, +4, –6, +3, –5) = –10-dB overall circuit loss. What does this mean? The overall loss of –10 dB means we receive 1/10th of the transmitted power level. The power level of 0 dBm is 1 mW. Therefore, the received power level is 0.1 mW. What if the overall loss had been –13 dB? We can come close to calculating the total power loss in our head by calculating the losses independently. We know that –10 dB = 1/10th of the original power and a –3-dB loss would be 1/2 the power it received. So if the overall loss was –13 dB, we would have 1/2 of 1/10, or 1/20 (0.5 × 0.1 = 1/20th). Thus, we would have approximately 1/20th of the original power from a transmitter available at the receiver, if the overall loss was 13 dB. For –10 dB, we have 1/10th of the original power. For –13 dB, we have 1/20th of the original power.

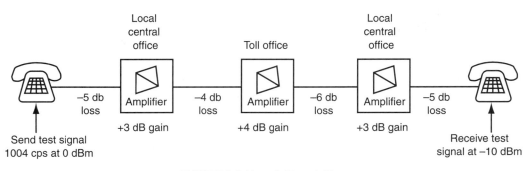

FIGURE 3-9 Use of dB and dBm.

If we wish to provide a precise calculation, we can do so by rearranging the dB formula to find the power ratio:

$$dB = 10 \log (P_2/P_1) \text{ Therefore:}$$
$$dB/10 = \log (P_2/P_1)$$
$$\text{Antilog } (dB/10) = \text{antilog } (\log (P_2/P_1))$$
$$\text{Antilog } (dB/10) = P_2/P_1$$

Formula to Convert Decibels to Power Ratio

$$\text{Antilog } (dB / 10) = P_2/P_1 \text{ (the power ratio)}$$

Substituting the values from Figure 3-9 (that is, we want to find the power ratio of −10 dB), we arrive at the following:

$$\text{Antilog } (-10/10) = P_2/P_1$$
$$\text{Antilog } 1 = P_2/P_1$$

The antilog of 1 = 10; therefore,

$$10 = P_2/P_1$$

We can also find the overall power loss of a 13-dB loss (−13-dB gain) circuit. Since −10 dB = 1/10th (0.1), and −3 dB = 0.5, −13 dB = 0.1 × 0.5 or approximately 0.05, which is 1/20 (1/20th = 0.05). Thus, we can state that the −13-dB circuit results in an overall power loss such that 1/20th of the power at the transmitter reaches the receiver. I think this exercise shows you the advantage of using decibels. It is much easier to add decibels, than to try calculating how much power is lost over each part of the circuit.

3.12 DECIBELS REFERENCED TO 1 mW (dB)

In Figure 3-9, a test signal generator is attached to one end of the circuit and sends a test tone at 0 dBm. Where did the *m* in *dBm* come from? dBm means a comparison of the measured power level to a 1-mW power level. By comparing every signal to a power level of 1 mW, we achieve an absolute rather than a relative measurement. In actuality, we are comparing all signals relative to one signal level, and that one signal level is 1 mW. The use of one level to compare all other signals against results in an absolute measurement. When signals are measured, they are measured relative to a 1-mW signal and are stated as *decibels referenced to 1 mW* **(dBm)**. But dBm is only used to state a measured signal or a transmitted signal from a test set. The gain (or loss) of each circuit path is stated in dB, not dBm. Let's return to Figure 3-9. If we were to measure the test signal where it enters the first central office, the measurement would be –5 dBm. The signal is sent by the test tone generator at 0 dBm. It travels over wire that has a loss of 5 dB. Thus, it will measure 5 dB down in level from the transmitted level of 0 dBm and will be measured as a –5-dBm signal level.

What is the power level of a 0-dBm test tone?

$$0 \text{ dBm} = 10 \log (P_2 / 1 \text{ mW})$$

because the reference signal (P_1) is 1 mW. Remember that the formula used to find our power ratio is the following: Antilog of dBm/10 = power ratio. If dBm/10 = 0, then the antilog of 0 equals the power ratio. The antilog of 0 is 1. Therefore, $P_2 / P_1 = 1$. In dBm, P_1 is 1 mW. For P_2 / P_1 to be 1, P_2 must also be 1 mW. Thus, 0 dBm = 1 mW.

Power Level of 0 dBm

$$0 \text{ dBm} = 1 \text{ mW}$$

If we measure all power levels relative to 1 mW, and a doubling of the power = 3 dB, then a measured power level of 3 dBm will be double the power at 0 dBm. Since the power of 0 dBm is 1 mW, the power at 3 dBm must be 2 mW. Another doubling of power would result in a reading of 6 dBm and would be 4 mW. An 8-mW power level would give a reading of 9 dBm. Let's double-check this:

$$dBm = 10 \log (P_2/1 \text{ mW})$$
$$dBm = 10 \log (8 \text{ mW} /1 \text{ mW})$$
$$dBm = 10 \log 8$$

The log of 8 is 0.90309; therefore,

$$dBm = 10 \times .90309$$
$$dBm = 9.0309$$

Now let's look at the overall circuit loss of 13 dB. If we transmit at 0 dBm and have a circuit loss of 13 dB, the signal will measure as a –13-dBm signal. What is the actual received power? Earlier, we stated that it would be about 1/20th of the transmitted level. If the transmitted level is 1 mW (0 dBm), the received level is about 1/20th of a milliwatt. We can double-check this assumption by using the following formula:

$$-13 \text{ dBm} = 10 \log (P_2 / 1 \text{ mW})$$
$$-13 / 10 = \log (P_2 / 1 \text{ mW})$$
$$-1.3 = \log (P_2 / 1 \text{ mW})$$
$$\text{Antilog} -1.3 = \text{antilog} (\log (P_2 / 1 \text{ mW}))$$
$$\text{Antilog} -1.3 = P_2 / 1 \text{ mW}$$
$$0.0501187 = P_2 / 1 \text{ mW}$$
$$1 \text{ mW} \times 0.0501187 = P_2$$
$$0.0501187 \text{ mW} = P_2 = \text{the power level at the receiver}$$

This is approximately 1/20th of a milliwatt and agrees with our earlier findings for Figure 3-9.

When using decibels to describe the relationship in sound volume changes, a decibel is a change in power that the human ear can barely detect. A change in power level that sounds twice as loud is a change of 3 dB. If one sound is 10 times as loud as another, it is 10 dB higher, and so on. The starting point for measuring sound is one thousand-trillionth of a watt (0.000000000000001 W) or 0.001 pW. A 1000-W speaker produces a sound power of 1 mW (0.001 W or 1/1000 of a watt). Anything louder than this will probably break your eardrum. Levels between 100 and 1000 W will cause some damage to the eardrum. In telecommunications, we did not choose to use 0.001 pW as a starting reference point because most power levels used in telecommunications are in milliwatts. This is the reason 1 mW was chosen as the starting reference point. Sound decibels and electrical decibels do not have direct correlation because of different starting reference points. The ratios do have the same meaning in either case. A double in power is 3 dB, and so forth.

3.13 NOISE MEASUREMENTS

Decibels referenced to 1 mW provides a convenient measure for measuring the signal levels in telecommunications because most signal levels in the telecommunications system fall within +10 and –10 dBm. The level of noise present on a circuit should be well below the level of electrical voice signals. *To achieve a convenient starting reference point for noise measurements, –90 dBm was chosen* for the starting reference point. The power level of –90 dBm equals 1 picowatt (pw) of power. Let's double check this with our formula for finding power levels referenced to 1 mW:

$$dBm = 10 \log (P_2 / P_1)$$
$$dBm/10 = \log (P_2 / P_1)$$
$$\text{Antilog } (dBm/10) = P_2 / P_1$$
$$\text{Antilog } (dBm/10) \times P_1 = P_2$$

Since $P_1 = 1mW$,

$$\text{Antilog } (dBm/10) \times 1 \text{ mW} = P_2$$
$$\text{Antilog } (-90/10) \times 1 \text{ mW} = P_2$$
$$P_2 \text{ in milliwatts} = \text{antilog } (-90/10)$$
$$P_2 \text{ in milliwatts} = \text{antilog } -9 \text{ and the antilog of } -9 = 0.000000001$$
$$P_2 \text{ in milliwatts} = 0.00000001 \text{ mW}$$

this can be converted into watts by moving the decimal point three places to the left:

$$0.00000001 \text{ mW} = 0.000000000001 \text{ W}$$

Watts can be converted into picowatts by moving the decimal point 12 places to the right:

$$0.000000000001 \text{ W} = 1 \text{ pW}$$

The results of our formula inform us that a signal 90 dB below 0 dBm is in fact a power level of 1 pW. This signal is called –90 dBm. We choose –90 dBm as the reference starting point for measuring noise but give it a new name. *We call this noise reference point 0 dBrn* (0 dB reference for noise). Another factor enters into measuring noise in telecommunications. We only care about noise that will be passed by our telecommunications system. Since the telecommunications system is designed to only pass voice frequencies between 300 and 3400 Hz, we are only concerned with noise that falls between these two frequencies. To measure only the noise signals between 300 and 3400 Hz, a filter is placed in the transmission measuring test set. This filter is called a *C message* filter. This filter will only pass signals between 300 and 3400 Hz to the test set. A transmission measuring test set is called a *TMS*. The C message filter causes a loss in the signals it passes to the TMS. This loss is approximately 1.5 dB. Thus, if the actual noise on a circuit is –90 dBm (0 dBrn), the TMS will measure –91.5 dBm (1.5 dBrn). To adjust for the 1.5 dB of loss in a C message filter, the TMS meter display is adjusted such that –91.5 dBm (1.5 dBrn) will read 0 on a scale called *dBrnC0* (dB referenced to noise through a C message filter at a 0 level test point). Thus, the scale is calibrated to compensate for the loss of the filter. The actual noise prior to going through the filter is –90 dBm and the meter displays this as 0 dBrnC0. This scale is called a C message weighted scale. One point to note is that noise readings in dBrnC0 can be converted to dBm simply by subtracting 90 from the dBrnC0 number. For example, 0 dBrnC0 = –90 dBm, and 30 dBrnC0 = –60 dBm. A conversion from dBm scale to dBrnC0 scale can be done by adding 90 (0 dBm = 90 dBrnC0). An advantage of using the dBrnC0 scale is that all measurements are usually positive numbers.

Noise Measurements in dBrnC0

$$0 \text{ dBrnC0} = -90 \text{ dBm}$$

3.14 SIGNAL-TO-NOISE RATIO

The term *signal-to-noise* ratio is used to express in decibels how much higher in level a signal is to the noise on the circuit. Suppose, at some test point, we measure the voice signal a facility is carrying and measure the noise at the same point. We get the following readings: voice signal = –5 dBm and noise = 50 dBrnC0. The two signals can be compared if the noise signal is converted to dBm (dBm = dBrnC0 –90). Thus, dBm = 50 dBrnC0 –90 = –40 dBm. To compare the signal-to-noise ratio, the noise level (in dBm) is subtracted from the voice level (in dBm):

Signal-to-Noise Ratio

$$S/N \text{ ratio} = \text{signal-level dBm} - \text{noise level dBm}$$
$$S/N \text{ ratio} = S \text{ dBm} - N \text{ dBm}$$

In our example,

$$S/N \text{ ratio} = -5 - (-40)$$
$$S/N \text{ ratio} = 35 \text{ dB}$$

In telecommunications, a signal-to-noise ratio of 30 dB or more is satisfactory.

3.15 AC VOLTAGE GENERATION

AC voltage is voltage that causes an *alternating current* to flow in a circuit. AC voltage is what we have in our homes and business to light our homes and supply electricity at the wall outlets. This type of voltage is generated by mechanical means. The device is called quite appropriately a generator. A generator consists of a device that has two large magnets facing each other. One magnet is a south pole and the other is a north pole. Lines of magnetic flux extend from one pole to the other. Between these two magnets is a coil of wire. When wire moves in a magnetic field a voltage is induced in the wire. The generator uses a steam turbine to turn the coil of wire placed between the magnets. As the wire turns, a voltage is induced in the wire. The way the generator is constructed, the wire is being spun in the magnetic field. As the wire spins, it moves down through the magnetic field and then comes back up through the field. On the return path, the motion of the wire is in an opposite direction and the voltage induced will also be opposite. The lines of magnetic flux are much stronger in the middle of the two poles and more voltage will be induced when the wire is in the center of the magnetic flux field than when it is moving in the outer regions of the flux. Because the voltage polarities are reversed when the wire moves back up through the magnetic field relative to when the wire was moving down through the field, the voltage across the wire is continually changing polarity, which will cause current in the external circuits attached to the output of a generator to continually change direction. The amount of current flow depends on the amount of voltage generated. The voltage varies depending on how many lines of magnetic flux are being cut by the moving wire. Voltage and current are maximum when the wire is moving in the center of the magnetic field. Voltage and current are zero when the wire is moving in the outer region of the magnetic field where it is moving parallel to the lines of magnetic flux and does not cut across the lines of flux (Figure 3-10).

3.16 *ROOT MEAN SQUARE VOLTAGE*

AC signals, voltages, and currents can be evaluated in terms of how they compare to a DC signal, voltage, or current. We use the term *root mean square (RMS)* to correlate alternating to direct current. A 120-V RMS AC signal will do the same work as a 120-V DC volt signal. The formula for converting an AC signal to RMS is:

RMS Formula

$$RMS = 0.707 \times 1/2 \text{ the peak-to-peak signal}$$

The power supplied to our homes has an AC voltage with a positive peak of +170 V and a negative peak of –170 V. The peak to peak is 340 V. Thus, RMS = 0.707 × 1/2 (340) = 0.707 × 170 = 120 V. We state that the voltage at our electrical outlets is 120 V. It is 120 V RMS. When AC is stated in the RMS values, we can use the RMS voltage and RMS current values in Ohm's law formulas and in power formulas.

An AC voltage signal is voltage that continuously changes polarity. The voltage will start at 0 V and go to a maximum positive voltage. From a maximum positive voltage, the signal gradually decreases in voltage until it is 0 again. The signal then goes to maximum negative voltage, and then back to 0. This trip from 0 to maximum positive, back to 0, to maximum negative, and then back to 0 V is called a *cycle*. The number of cycles that an AC signal completes in one second is called the *frequency of the signal*. The AC voltage to our house is a 60-cycle-per-second (60 cps) signal. The signal goes from 0 V, to +170 V, to 0 V, to –170 V, and back to zero in 1/60th of a second. Thus the signal can complete 60 of these cycles in one second. The number of cycles completed in one second is the frequency of

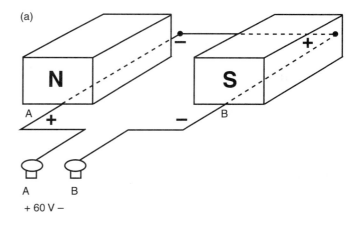

(a)

As wire rotates up through magnetic field it creates voltage in one direction. As wire rotates down through magnetic field it creates voltage in the opposite direction. Thus, if voltage induced in wire A is 30 V and voltage in wire B is 30 V, the total voltage produced = 60 V.

FIGURE 3-10 AC voltage generator.

(b)

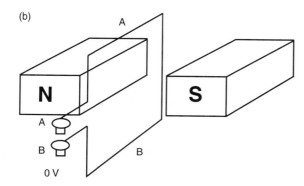

When wire has rotated to a vertical position it is travelling parallel to the magnetic liner of flux between N and S. Since it cuts no magnetic flow, 0 V are produced.

(c)

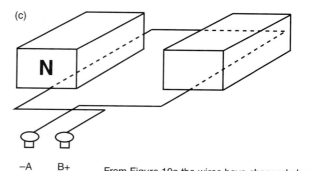

−A   B+

− 60 V +

From Figure 10a the wires have changed places, which causes a reversal of voltage output.

(d)

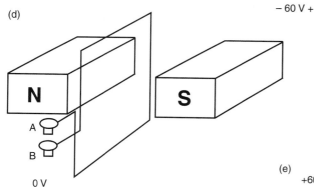

Wires are outside of magnetic field and moving parallel to it, therefore 0 V is produced.

(e)

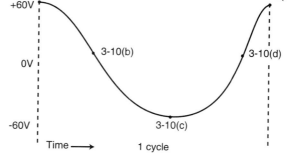

Voltage of A relative to B graphed over the time for 1 cycle.

FIGURE 3-10 (Continued).

Basic Electricity Review   **Chap. 3**

the signal and can be stated as cycles per second (cps). Cycles per second are often called *Hertz (Hz)* as a tribute to German physicist Heinrich Hertz, who discovered electromagnetic radiation (the birth of radio waves) in 1886.

3.17 INDUCTIVE REACTANCE

AC signals do not behave like DC signals. With a DC signal, we are only concerned with voltage, current, and resistance. An AC signal entails two additional factors: inductive reactance and capacitive reactance. The combination of resistance, inductive reactance, and capacitive reactance is called *impedance*. All three components of impedance are measured in ohms; therefore, impedance is stated in ohms. Impedance is the opposition to flow of an alternating current. Inductive reactance exists in all conductors due to a property called *inductance*, which is present in all conductors. Inductive reactance is very high when changes occur in a signal and is zero when no change is occurring. DC circuits have a steady current flow. There is no change in the flow of current and no inductive reactance is present in a DC circuit. However, an AC signal is constantly changing, and inductive reactance comes into play in an AC circuit. The faster a change occurs, the higher the inductive reactance. How fast a signal changes depends on the frequency of the signal. The higher the frequency, the more times the signal changes, and the higher the inductive reactance.

Inductive reactance is directly related to the frequency of the signal and the inductance in the circuit. The higher the inductance, the higher the inductive reactance. The inductance for a 24-gauge copper wire, 1 mi long, is approximately 0.5 millihenry. Since the local loop is composed of two wires (Tip and Ring), the inductance per loop mile is approximately 1 millihenry (0.001 henry). Inductive reactance is represented by the symbols XL. The formula for finding inductive reactance in a circuit is: $XL = 2 \times 3.14159 \times F$ (frequency) $\times$ inductance (L) in henries.

Inductive Reactance Formula

$$XL = 2 \times 3.14159 \times F \times L$$

Inductive reactance of a 100-cps signal on a 24-gauge cable pair is:

$$XL = 2 \times 3.14159 \times 100 \times 0.001$$
$$XL = 0.628318 \ \Omega$$

Because a direct correlation exists between inductive reactance and frequency, a 500-cps signal will have 5 times the inductive reactance of a 100-cps signal and a 1000-cps signal has 10 times the inductive reactance of a 100-cps signal (6.28 Ω on a local loop). Note that the inductive reactance for a 5000-cps signal on a local loop is as follows: $XL = 2 \times 3.14159 \times 5000 \times 0.001 = 31.4159$ ohms. Since a 5000-cps signal is equal to 50 times a 100-cps signal, the XL for a 5000-cps signal should be 50 times the XL of a 100-cps signal. Thus, $50 \times 100 = 5000$ and $50 \times 0.628318 = 31.4159 \ \Omega$.

3.18 CAPACITIVE REACTANCE

Capacitive reactance occurs in all AC circuits and is due to the capacitance existing in a circuit. Capacitance exists when two conducting materials are separated by an insulator. Every cable pair in the local loop has capacitance between the two conductors because the conductors are separated from each other by the insulation around each wire. Capacitance per loop mile of a 24-gauge cable pair is approximately 0.083 microfarad. Capacitive reactance (XC) is a measure of how the capacitance of a circuit is affecting the circuit. It is inversely related to the frequency of the AC signal and the value of the capacitance. The higher the signal or the higher the capacitance, the lower the capacitive reactance will be. The formula for finding the capacitive reactance of a circuit is: $XC = 1/(2 \times 3.14159 \times F \times C)$ (where C = capacitance in farads).

Capacitive Reactance Formula

$$XC = 1 / (2 \times 3.14159 \times F \times C)$$

For a 1000-cycle signal on a mile of local loop, the capacitive reactance is:

$$XC = 1/(2 \times 3.14159 \times 1000 \times 0.000000083) = 1917.5 \ \Omega.$$

Capacitance exists down the length of the cable pair and is said to be distributed along the cable pair. The distributed capacitance is 0.083 microfarad per mile. The capacitance for 4 mi of cable is four times the capacitance of 1 mi. Since capacitive reactance is inversely proportional to capacitance, the capacitive reactance for 4 mi of cable is one fourth the capacitive reactance at 1 mi. Thus, the capacitive reactance for 4 mi is 1917 / 4 = 479 Ω at 1000 cycles. Since capacitive reactance is also inversely proportional to frequency, the capacitive reactance for 4 mi and a 5000-cycle signal is 479 / 5 = 96 Ω. Remember that the capacitive reactance exists between the pairs. This 96 Ω of capacitive reactance between the pairs will effectively short the signal across the pair, and very little of the signal placed on one end of the loop reaches the other end of the circuit if we do not do something to correct the problem caused by distributed capacitance. To correct this problem, inductance is purposely placed in the loop. Capacitance and inductance are affected in opposite directions by frequency. The higher a frequency, the higher the inductive reactance and the lower the capacitive reactance. Because inductive reactance tends to offset capacitive reactance, we can use inductance to offset the effect of capacitance.

3.19 TUNED CIRCUITS

Because inductance and capacitance have opposite effects on an AC signal, we can design circuits to take advantage of this condition. When an AC signal is placed on an inductance, the inductive reactance (XL), or opposition to current flow, becomes greater as the frequency of a signal goes higher. When an AC signal is placed on a capacitor, the capacitive reactance (XC), or opposition to current flow, diminishes as the frequency of a signal increases. At very high frequencies, inductors have so much opposition to current flow that very little current will flow. At very high frequencies, capacitors have very little opposition to current flow. The reverse is true for very low frequencies. At very low frequencies, inductors have very little XL and capacitors have very high XC. A DC circuit has no frequency; the frequency is 0. The formula for XL is:

$$XL = 2 \times 3.14 \times F \times L$$

If F (frequency) = 0, then XL is 0 because anything multiplied by 0 is 0. The formula for XC is:

$$XC = 1 / (2 \times 3.14 \times F \times C)$$

If $F = 0$, this formula becomes 1/0.

Anything divided by 0 is an extremely high number that cannot be defined. Thus, if the frequency is 0, the XC is so high that no current will flow. This is why a capacitor blocks the flow of current when placed in a DC circuit.

Let's look at the XL and XC at a particular frequency for a particular inductor and capacitor. We will use a frequency of 10,000 cps, an inductor of 1 millihenry, and a capacitor of 253.3 nanofarads. Let's calculate the XL and XC for these two components:

$XL = 2 \times 3.14159 \times F \times L$
$XL = 6.28318 \times 10,000 \times 0.001$ (inductance must be stated in henries)
$XL = 62.83\ \Omega$
$XC = 1/(2 \times 3.14159 \times F \times C)$ (capacitance must be stated in farads)
$XC = 1/(6.28318 \times 10,000 \times 0.0000002533)$
$XC = 1 / 0.015915308$
$XC = 62.83\ \Omega$

Notice how $XL = 62.83\ \Omega$ and $XC = 62.83\ \Omega$. I purposely selected an inductor and capacitor that would provide the same impedance to the flow of electricity when the frequency of the AC signal is 10,000 Hz. When $XL = XC$, we say the circuit is a *tuned circuit*, and the frequency of the signal is called the *tuned frequency*. The total impedance of a circuit is composed of total resistance, total inductive reactance, and total capacitive reactance. Because inductive reactance and capacitive reactance have opposite effects on current, they tend to offset each other in a series circuit. The formula for total impedance in a series circuit is: Z (total impedance) = $(R^2 + (XL - XC)$.

Total Impedance Formula

$$Z = (R^2 + (XL - XC_2))$$

If $XL = XC$, total impedance is equal to the resistive component only because when XL and XC are equal, $XL - XC = 0$. But what if they are not equal? Let's look at a frequency that is not the tuned frequency. Connect the capacitor and inductor in series and pass a 12,000-Hz signal through the circuit.

$$XL = 2 \times 3.14159 \times 12,000 \times 0.001$$
$$XL = 75.398224\Omega$$
$$XC = 1/(2 \times 3.14159 \times 12,000 \times 0.0000002533)$$
$$XC = 52.36\Omega$$
$$Z = \sqrt{R^2 + (XL - XC)^2}$$
$$Z = \sqrt{R^2 + (75.398 - 52.36)^2}$$
$$Z = \sqrt{R^2 + 23.038^2}$$

You can see from the example above that when XL does not equal XC, the circuit will have a higher impedance. Also notice the same happens if XC is larger than XL. When XC is larger than XL, we get a negative number by subtracting XC from XL. When we square this negative number, we have a positive value. Thus, it does not matter whether XL or XC is larger. If they are not the same, the series circuit will have greater opposition (impedance) to current flow when they are not the same value than when they are the same value. Let's look at the circuit when an 8,000-Hz signal is passed through it:

$$XL = 2 \times 3.14159 \times 8,000 \times 0.001$$
$$XL = 50.26 \ \Omega$$

This agrees with the statement made earlier: the lower the frequency, the lower the XL. Now let's look at XC:

$$XC = 1 / (2 \times 3.14159 \times 8,000 \times 0.0000002533)$$
$$XC = 78.54 \ \Omega$$

This also agrees with an earlier statement: the lower the frequency, the higher the XC. Because XL does not equal XC at 8000 Hz, the circuit has more impedance than it has at the tuned frequency of 10,000 Hz. There is only one frequency for a series circuit at which impedance will be its lowest. This tuned frequency (resonant frequency) is found by the following formula: $Fr = 1 / (2 \times 3.14159 \ \sqrt{LC})$.

Tuned Circuit Formula to Find Resonant Frequency
$$Fr = 1 / 2 \times 3.14159 \ \sqrt{LC}$$

This formula was arrived at by using our knowledge of the fact that at the tuned frequency for the tuned circuit $XL = XC$:

$$XL = 2 \times 3.14159 \times F \times L \text{ and } XC = 1 / (2 \times 3.14159 \ FC)$$

Therefore, at the tuned frequency where XL must equal XC:

$$XL = XC$$
$$2 \times 3.14159 \, FL = 1 \,/\, (\, 2 \times 3.14159 \, FC)$$
$$(2 \times 3.14159 \, FL) \, (\, 2 \times 3.14159 \, FC) = 1$$
$$(6.28318 \, FL) \, (6.28318 \, FC) = 1$$
$$FF \times 6.28318 \times 6.28318 \, LC = 1$$
$$FF = 1 \,/\, (6.28318 \times 6.28318 \, LC)$$

Now take the square root of each side of the equation:

$$F = 1 \,/\, 6.28318 \, \sqrt{LC}$$

This formula can be used to find the tuned frequency (called the *resonant frequency*) for any value of inductance and capacitance. In telecommunications, we use inductors and capacitors to tune devices to a particular frequency. This is what tuners in a radio do. The tuner has one device that is variable. In older radios the tuning knob was attached to a variable capacitor. Turning the knob changed the value of the capacitor. By changing the value of the capacitor, we changed the frequency the radio was tuned to.

3.20 AC TO DC CONVERTERS

It was stated earlier that an AC voltage can be converted into DC voltage by a device called a battery charger. The battery charger in a central office is large enough to keep the batteries charged and also supplies DC voltage and current to the central office equipment. Just as your car does not run off its battery (it runs off of the generator/alternator), a central office does not run off of the batteries. The battery in a car is only used to supply power when the engine is not running and the alternator is not working. The battery in a central office is only used to supply power when the AC powered battery charger stops working. The 52-V DC supply not only supplies voltage for central office equipment, it also supplies power for the telephone. This supply must be free of noise. In addition to providing a standby emergency supply, batteries also act as a noise filter. If a battery charger were used to supply the energy by itself, the supply would be noisy. The battery attached across the output of a battery charger absorbs the noise and it does not reach our phone. In computer-controlled central offices, additional noise filters are placed throughout the office on the battery supply that is going to be sent out to telephones.

Just as there are devices to convert AC electricity to DC electricity, there are also devices to convert DC to AC The ringing signal used to ring our telephones is generated by a solid-state, transistorized device that converts 52-V DC to an AC ringing signal. This device is called a ringing machine. Today, most ringing machines have an output of 20 cps, 90 V RMS.

3.21 RINGING GENERATORS

When the first solid state ringing machines were introduced in 1960, there were ringing machines that generated signals between 16 and 66 cycles AC at voltages between 90 and 150 V AC. These machines were used in rural areas for ringing party-line phones and will be phased out over time as older phones with ringers other than 20-cycle ringers are replaced by ringers that will operate on 20-cycle AC. Rural areas had many different ringers because of party lines. It will be some time before these ringers are replaced. Therefore, rural areas will continue to have five different ringing machines to generate the five different ringing signals used in those areas. Even though almost all rural areas have converted most party lines to private lines, the telephone company left the old phone with the existing ringer in place when the line was converted to a private line.

Some rural areas had a ringing scheme called *decimonic ringing*. The ringing machines produced ringing frequencies of 20, 30, 40, 50, and 60 cps. Other rural areas used a ringing scheme called *harmonic ringing*. The ringing machines for these offices produced ringing signals with the following frequencies: 16 2/3, 25, 33 1/3, 50, and 66 2/3 cps. Whichever scheme was used would provide five different ringing signals. On a five-party line, each phone would have a different ringing signal assigned to it. The phone had a specially designed ringer called a *tuned ringer*. A tuned ringer will only respond to one frequency of AC signal. A ringer tuned to ring only when 20 cps AC is present is called a 20-cycle ringer. Each phone on a party line had a tuned ringer to match the ringing frequency assigned to it. Therefore, only one phone on the line would ring when a call arrived for one number on the party line. This was superior to early party lines without tuned ringers. Prior to tuned ringers, *coded ringing* was used. With coded ringing every phone on a party line would ring. Since all phones would ring, a person could tell if the call was for them by the code of the ring. Codes were: one long ring followed by two short rings, two short rings followed by two long rings, three long rings and one short ring, and so on.

For party lines with more than five parties, all telephones had one side of the ringer connected to earth ground at the residence. The other side of the ringer was attached to either the Tip lead or the Ring lead of the pair of wires from the central office. Five parties could have the ringer attached to the Tip lead and five could have the ringer attached to the Ring lead. By using five different tuned ringers for Tip party and Ring party phones, it was possible to ring any one of the ten phones without other parties on the line hearing a ring. In rural areas, for Tip party phones, the last digit of the phone number was 1 to 5 and ring frequency 1 to 5 was applied to the Tip lead at the central office corresponding to what the last digit of the phone number was. This signal would flow over the Tip lead to all Tip party phones. The signal would only pass through the tuned ringer matching the frequency of AC Ring signal being sent by the central office and return to the central office via earth ground. For parties 6 through 10 (Ring lead parties), the last digit of the phone number was 6 to 10 and one of the five ringing machine signals (1 to 5) would be sent over the Ring lead. Since parties 6 to 10 have their ringers wired to the Ring lead, only these five phones received the ringing signal and of these five, only the tuned ringer matching the signal sent would ring.

In metropolitan areas all private lines are equipped with phones that have a non-tuned ringer or a transistorized device called a *ring detector* in them. Both of these will operate on any ringing frequency. Most central offices in metropolitan areas use one ringing frequency. They use a 20-cps ringing signal at 90 V RMS AC. Rural areas have been forced to keep ringing machines for five different frequencies even though party lines have been eliminated. Though a person may live in a rural area and have a private line, that phone may be equipped with one of the old tuned ringers from its party-line days. Today, when new phone service is installed only 20-cycle ringing is assigned. The local phone company keeps track of ringing frequencies assigned. The number of ringers tuned to other frequencies will decrease through attrition (as existing customers move and have their service disconnected). At some point in time, the number of older ringers will diminish enough that the LEC can inform customers that the LEC will no longer support tuned ringing and will only use 20-cycle ringing to ring telephones. These customers will then need to purchase at least one straight-line ringer for their home. Their existing tuned ringer phones can still be used to place calls; they just will not ring on incoming calls. Of course, customers can change the ringer in existing phones to a straight-line ringer and it will work fine. With the Telecommunications Reform Act of 1996, LECs can once again sell station equipment. This will allow rural LECs the opportunity to change all ringing to 20-cps ringing by offering to change the phone's ringer at a nominal charge (Figure 3-11).

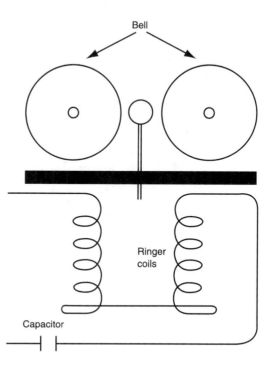

FIGURE 3-11 Tuned ringer.

The value of Inductance used for the ringer coils and the value selected for the capacitor tune the ringer to one frequency.

3.22 PUTTING ELECTRICITY TO WORK: THE TELEPHONE SET

When Alexander Graham Bell developed his first telephone device, he used one device for both the transmitter and receiver. This device was called a *transceiver*. This device is constructed of a wire wound many times around an iron magnet. A thin piece of flexible metal (called a *diaphragm*) is placed a short distance from the magnet. When vibrating air waves from a voice hit this diaphragm, it vibrates. As it vibrates it will move closer and then farther away from the ends of the magnet. When it is closer to the ends of the magnet, it creates a low reluctance path between the north and south poles of the magnet. This increases the strength of the magnetic field. When it moves away from the ends of the magnet, it decreases the strength of the magnetic field. A varying magnetic field causes an AC voltage to be developed in the wire wrapped around the magnet. This varying AC voltage signal is coupled by wires to another receiver where the process is reversed. A varying voltage in wire wrapped around a magnet will cause additions and subtractions to the magnetic field. Thus, a signal in the wire causes the magnetic field to vary in unison with the signal. As the magnetic field varies, it causes the diaphragm to vibrate in unison with the signal. The vibrating diaphragm causes air movement at the frequency of the signal. If your ear is placed close to the receiver, you can hear the low changes in air movement. The frequency of air movement recreates the speaker's voice at the transmitter. Using the same device for a transmitter and a receiver had one major drawback. The signal created on the transmitting end was a very low voltage signal. The transceiver was replaced by two devices: a transmitter and a receiver. These two devices belong to a classification of equipment known as *transducers*. A transducer is a device that converts energy from one form into another. The transmitter converts sound energy into electrical

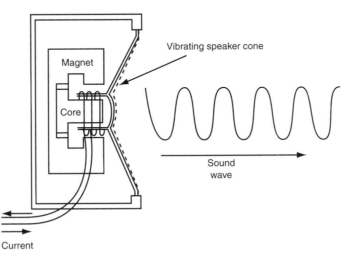

FIGURE 3-12 In the telephone receiver, the electrical current activates the voice coil, causing the speaker cone to move. The vibrating speaker cone creates air pressure waves that are heard as sound by the ear.

energy and the receiver converts electrical energy into sound energy. The original transceiver now serves only as a receiver (Figure 3-12). Another device was invented to serve as a transmitter.

Bell experimented to find a better device to use as a transmitter. He found that by using a battery and variable resistance transmitter, he could obtain a much higher signal voltage. Thus, the variable resistance transmitter was born. Bell used a transmitter much like one proposed by Elisha Gray. Elisha Gray also filed for a patent on this device, but the patent was awarded to Bell because he filed five hours ahead of Gray. The transmitting device proposed by Gray used a liquid acid solution as a variable transmitter. Bell also used this type of transmitter for his invention. This worked great in the lab but would not lend itself to commercial use. Thomas Edison picked up on a carbon granule type of transmitter invented in Great Britain by Henry Hummings. Edison made some improvements and patented the variable resistance carbon granule transmitter used in many telephones. A carbon granule transmitter cannot create an electrical voltage like the early transceiver because it does not use a varying magnetic field to produce a signal. A carbon granule transmitter is a device that will vary in resistance when a voice signal strikes the diaphragm (Figure 3-13).

To make this transmitter work, an external voltage source is required. The early telephones used a 3-V battery at the telephone to power the transmitter. When the telephone was in use, the battery was connected to the transmitter. When a voice signal struck the diaphragm of the transmitter, it causes the resistance to vary and this caused the current of the transmitter circuit to vary (Figure 3-14). This varying current had the same frequency as the voice signal. This varying current was much higher than the varying current of the earlier transceiver device. The varying current was connected to a pair of wires attached to another telephone. The receiver at this telephone converted the varying current back into sound waves.

3.23 PUTTING ELECTRICITY TO WORK: ELECTROMECHANICAL SWITCHING

Electromechanical switching uses relays to perform switching functions. A relay is constructed by wrapping wires around a soft iron core. Soft iron will not retain magnetism very long but is quickly magnetized when a magnetic field is brought close to it. When electrical current passes through a wire, a magnetic field is developed around the wire. By wrapping many tuns of wire around a soft iron core, the magnetic field can be concentrated when current is present in the wire. This concentration increases the magnetic field. When current is present in the wire the soft iron core becomes a magnet. A magnet created by passing electric current through a wire wrapped around iron is called an electromagnet. When electromagnets activate switching contacts, they are called *relays*. Electromagnets are used to make electromechanical switching systems. Each electromagnet has a device called an *armature* placed at one end of the magnet. The armature is about a quarter of an inch away from the magnet. When no electric

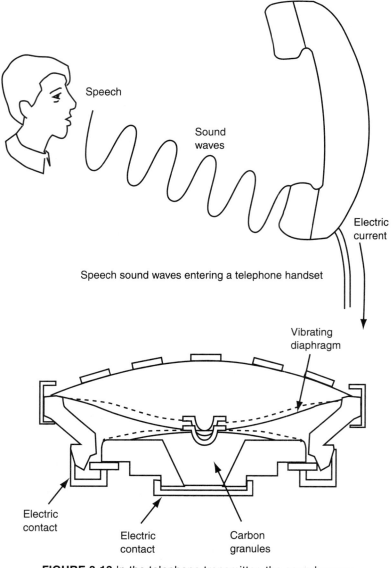

Speech

Sound
waves

Electric
current

Speech sound waves entering a telephone handset

Vibrating
diaphragm

Electric
contact

Electric
contact

Carbon
granules

FIGURE 3-13 In the telephone transmitter, the sound waves
of the voice cause the diaphragm to vibrate. The vibration puts
varying amounts of pressure on the carbon granules, causing
more and less electrical current to flow.

current is flowing in the wire around the core, there is no magnetic field and a
one-quarter inch air gap remains between the magnet and the armature. The
magnet is energized by current flow in the wire around the iron core. When
current flows, the magnetic field developed in the core attracts the armature. The
armature moves a quarter inch and will be against the core. Movement of the

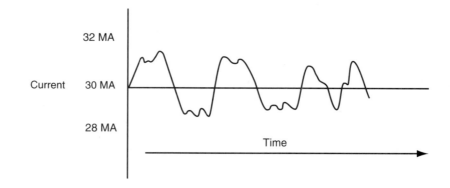

FIGURE 3-14 Wave of a typical voice signal.

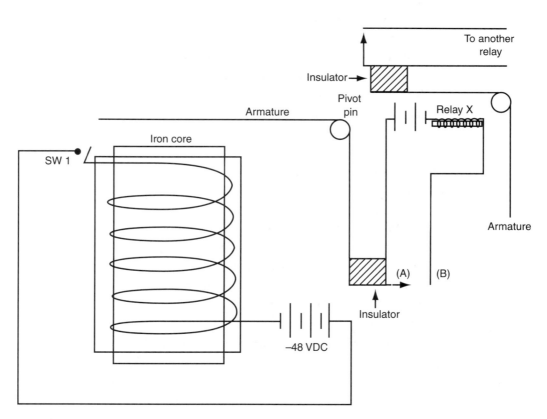

FIGURE 3-15 Electromechanical switching system. When switch 1 is closed, electric current will flow from the battery through the windings of wire around an iron core. The flow of electric current establishes a magnetic field that attracts the armature. When the armature is pulled up against the iron core, it will cause contacts A and B to make contact. Closing A to B causes relay X to operate. Each relay would contain from one to ten sets of contacts. Only one set of contacts is shown for clarity.

armature activates switching contacts. The electromagnet, armature, and switched contacts comprise one relay. The operation of one relay will cause other relays to operate (Figure 3-15). This is the principle behind all electromechanical switching systems. The systems are designed such that a particular action causes certain specific relays to operate. The operation of relays can be used to make each call take a specific and defined path through a switching system. Using their knowledge of electricity and relays, engineers developed several different types of electromechanical switches. These electromechanical switching systems gradually replaced manual switching systems.

REVIEW QUESTIONS

1. What type of voltage was used for power to the telephone?
2. What type of voltage was used to ring a telephone?
3. How is DC voltage produced?
4. What is the difference between a cell and a battery?
5. How much voltage did the battery of the first central exchange produce?
6. What is a molecule?
7. What is an element?
8. What is an atom?
9. What is an electron?
10. What is electric current flow?
11. What is resistance?
12. What is voltage?
13. What determines the amount of electric current flow in a circuit?
14. How much voltage is developed by a dry cell?
15. How much voltage is developed by a lead acid wet cell?
16. How much voltage is supplied by the central office battery used today?
17. Why is the physical size of a battery important?
18. What type of electricity is supplied by the power company?
19. How is AC voltage different from DC voltage?
20. Why did the telephone engineers select DC voltage as a power supply?
21. What is the value of expressing AC voltage in RMS values?
22. How many different ringing signals were used in rural exchanges?
23. What is a tuned ringer?
24. What is a straight-line ringer?
25. If four 100-Ω resistors are connected in series, what is the total resistance they offer?
26. If four 100-Ω resistors are connected in parallel, what is their total resistance?

27. What measurement is used for electrical voice signals?
28. What measurement is used for noise?
29. If noise measures 30 dBrnC0, what is this in dBm?
30. How are *XL* and *XC* related to one another at the tuned frequency?

GLOSSARY

Alternating Current (AC) A signal that is continually alternating the direction of current flow due to changes in the polarity of the voltage of the signal. In other words, it is a signal that starts at zero voltage, rises to a maximum positive potential, declines to zero, and continues the decline until it reaches a maximum negative potential, then returns to zero to complete one cycle of a signal. The number of cycles the signal completes in one second is the frequency of the signal. This is the type of electricity supplied by the local power company to our homes and businesses.

Ampere-Hours Units in which a cell or battery is rated, indicating how much electrical storage it has. A 1600 ampere-hour battery will deliver 1600 amps for 1hr, 1 amp for 1600 hr, 400 amps for 4 hr, 200 amps for 8hr, and so on.

Battery The connection of two or more cells in series to form a power supply. A typical car battery has six 2.17-V cells connected in series to form what is commonly referred to as a 12-V battery. In reality it is a 13.02-V battery.

Cell A device that generates electricity by a chemical reaction

Conductor A device or transmisson medium that readily conducts or carries an electrical signal. Copper wire is the most commonly used transmission medium in telecommunications and is a good conductor because it has a low resistance to electric current flow.

Direct Current (DC) The electric current flow is at a constant rate and flows continuously in one direction. The direction of electric current flow (outside the battery) is from the negative terminal of the battery, through a device attached to the battery, to the positive terminal of the battery. The amount of electric current flowing is measured in amperes.

DC Voltage The voltage produced by a chemical reaction. DC stands for "direct current". Devices that produce electricity via a chemical reaction are called batteries or cells.

Decibel (dB) The unit of measure used to compare two power levels is the decibel. A decibel is not an absolute measurement; it is a relative measurement. The decibel tells the relationship of one power level to another power level. The formula for the decibel is as follows: dB = 10 log (power ratio). Also see dBm.

Dry Cell A chemical cell that generates electricity via chemical means between an electrolytic solution and two dissimilar metals. The electrolytic solution is in the form of a paste.

Kirchhoff's Current Law States that at any given point in a circuit, the total current leaving a point equals the total current entering the point.

Kirchhoff's Voltage Law States that the sum of all voltage drops in an enclosed loop equals the total voltage supplied to the loop.

Ohm's Law A formula showing the relationship between the voltage applied to a circuit, the resistance in the circuit, and the resulting current flow in the circuit. I = V/R.

Resistance The amount of opposition to electric current flow that a device has.

Transducer A device that converts energy from one form into another. The transmitter converts sound energy into electrical energy, and the receiver converts electrical energy into sound energy.

Voltage A measure of how much potential difference or electrical pressure exists between the two terminals of a battery (or some other device that generates electricity). This electrical pressure is called electromotive force (EMF) and is measure in volts.

Wet Cell A chemical cell that generates electricity via chemical means between an electrolytic solution and two dissimilar metals. The electrolytic solution is a liquid. The battery in a Stored Program Control central exchange is composed of 24 wet cells.

4

Multiplexing

Key Terms

Amplitude Modulation (AM)
Bandwidth
Channel Unit
Frequency Division
  Multiplexing (FDM)
Mixer (Modulator)

Multiplexing
Pulse Amplitude Modulation
  (PAM)
Pulse Code Modulation (PCM)
Sidebands
Space Division Multiplexing

Time Division Multiplexing
  (TDM)
Wave Division Multiplexer
Wave Division Multiplexing
  (WDM)

The term *Multiplexing* is derived from the word *multiple*. Multiple means consisting of, or having, many individual parts. In telecommunications, multiplexing means to combine many individual signals (voice or data) so they can be sent over one transmission medium. The medium used is usually a cable pair, a fiber optic strand, or a microwave link. A multiplexer actually contains the equipment to do multiplexing and demultiplexing. A multiplexer/demultiplexer is attached to each end of a transmission medium. The equipment on each end is simply called a *multiplexer*. It is understood that a multiplexer also contains a *demultiplexer*. The multiplexer equipment on each end of the medium performs both functions.

It is important to note that many people include space division multiplexing as a form of multiplexing. In space division multiplexing, multiple communications are placed over many different wire pairs inside one cable. If the cable is considered to be the medium, multiplexing has occurred, when many conversations are carried by one cable. If a cable has 100 cable pairs in it, the cable can handle 100 calls and each call occupies its own pair (or space) within the cable. Cables connect the switching stages of a space division switching system, and each call will occupy its own set of wires between and within switching stages.

Space division multiplexing requires that each communication channel or voice path occupy its own set of wires for transmission of signals. If we consider the wire to be a medium instead of the cable, we cannot use space division multiplexing. By definition, space division multiplexing requires that each signal have its own transmission medium within the cable. To multiplex many conversations over *one* wire, we must use a different multiplexing technique. We can use a process called *frequency division multiplexing (FDM)* or another process called *time division multiplexing (TDM)* to multiplex many conversations over a single wire.

FDM is a process whereby each communication channel is made to occupy a certain band of frequencies and each channel is forced into different frequency bands. The human voice band is basically in the frequency band 0 to 4000 Hz. Through the use of a technique called *modulation*, we can force each communication channel to occupy a different band of frequencies. We cannot combine voice signals that are all 0 to 4000 Hz. We must change these signals to different frequencies for each communication channel before they can be multiplexed. For example, we can force one communications channel to be in the band from 60,000 to 64,000; a second channel is forced to be in the 64,000-to-68,000 band; a third channel is forced to occupy the 68,000-to-72,000 band; and so on. If the transmission medium we choose to use for the multiplexed signal is capable of carrying signals between 60,000 and 108,000 Hz, we could multiplex 12 channels of conversation over the medium using FDM. The band of frequencies from 60,000 Hz to 108,000 Hz provides a 48,000 Hz bandwidth. This 48,000-Hz bandwidth contains 12 channels and each channel has the 4000-Hz bandwidth needed by each voice channel (Figure 4-1).

Bandwidth refers to the width of a signal. The width of a signal is determined by subtracting the highest frequency of the signal from its lowest frequency. A voice signal is usually thought of as a signal between 0 and 4000 Hz. The bandwidth of this signal is 4000 Hz, and we say the voice signal has a nominal bandwidth of 4000 Hz.

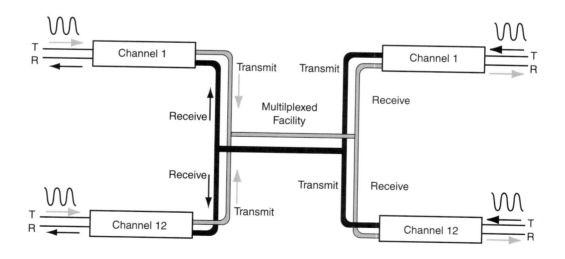

FIGURE 4-1 Twelve channel frequency division multiplexing

In the United States, AT&T designed its FDM systems to handle the band of signals between 200 and 3400 Hz. Thus, the FDM system used in the United States provides a bandwidth of 3200 Hz for voice (or data) signals. Voice signals below 200 Hz and above 3400 Hz will not be carried by the FDM system. We lose these signals, but this loss does not affect voice since most of the intelligence in a voice signal falls within the frequencies passed by FDM. The frequencies passed by a device or system is called the *passband*. The passband for FDM is 200 to 3400 Hz.

4.1 THE CHANNEL UNIT OF A FREQUENCY DIVISION MULTIPLEXING SYSTEM

Each signal fed to a FDM system interfaces to the multiplexer through a device called a *channel unit*. If a multiplexer can combine 12 different signals, it will have 12 channel units. The channel unit performs the function of interfacing the multiplexer to a specific type of input signal. The channel unit makes changes to the input signal so it can be combined (multiplexed) with other signals for transmission over the transmission medium. All voice signals arriving at a channel unit are between 0 and 4000 Hz. The channel unit contains a sharp cutoff bandpass filter to pass only the frequencies between 200 and 3400 Hz to a device called the **mixer** (or **modulator**). The modulator takes the 200-to-3400 Hz input signal and modulates a higher-frequency signal. Signals of the same frequency cannot be multiplexed together. The output of each modulator is a different frequency. This allows the signals to be multiplexed.

After the channel unit changes the frequency of the input signal to a higher frequency, the signals are combined. The 12 different output signals of the channel units are multiplexed and the composite signal is sent over the transmission medium to the multiplexer on the other end of the medium for demultiplexing. As the signal is received at the distant end, it is demultiplexed into 12 different signals and assigned to the same channel number used at the originating end. If an input signal comes in over channel 1 on the transmitting end, it will be received by channel 1 on the receiving end (Figure 4-1).

4.2 TRANSMIT AND RECEIVE PATHS: THE FOUR-WIRE SYSTEM

If multiplexers are placed on a medium between two cities such as New York and Chicago, the signal sent from New York will be placed over a transmit medium to Chicago. Since Chicago is receiving the signal, the transmit medium from New York is attached to the receive terminal of the multiplexer at Chicago for demultiplexing. This allows the Chicago receiver to hear the New York transmitter. For New York to hear Chicago, it is necessary to attach the transmit terminal of the Chicago multiplexer to a transmit medium to New York. This transmit medium from Chicago is attached to the receive terminal of the New York multiplexer for demultiplexing. Thus, multiplexing involves attaching multiplexers to two transmission mediums. When cable is used as the transmission medium, we have one pair of wires for each transmit/receive direction. Each of these two paths is usually placed in separate binder groups in the cable to reduce the possibility of crosstalk between the two pairs.

The first multiplexers used wire for the transmission medium. One pair of wires was used for the signal going in one direction (that is, east to west) and a separate pair of wires for the signal going in the opposite direction (west to east). Thus, four wires were needed to connect one multiplexer to another. When multiplexers are used on fiber optic cable strands, two strands are needed (one strand for each direction). When microwave is used as the transmission medium, multiplexers are assigned to a different transmitting frequency in each direction.

4.3 WHY USE MULTIPLEXING?

Multiplexing was first used to reduce the number of transmission mediums needed between cities or towns. If there were 24 trunks between two cities, 24 cable pairs were needed. By using two 12-channel multiplex systems, only four cable pairs were needed (two pairs for each system). Multiplex systems significantly reduced the number of mediums used between cities. This resulted in significantly reduced costs for trunk circuits. If multiplexers did not exist, we could not use fiber optic cable. Fiber optic cable allows the multiplexer to combine as many as 6 million signals in one direction on one fiber strand. The multiplexer on fiber optic cable combines many signals and transmits the signals as pulses of light. The receiving multiplexer uses a photodiode to detect the individual pulses of light.

The different sounds of our language such as *a, e, i, o, u, by, my, car, far, tar,* and so on, are due to the different frequency of the sounds. Spoken words are nothing more than a string of signals that continually vary in frequency. When we speak into a transmitter or microphone, these devices convert the varying sound frequencies into varying electrical currents that have the same frequency as the sound. In FDM, these varying electrical signals are input to a device in the channel unit called a mixer (or modulator). Another signal called the carrier signal is also input to the mixer. A *carrier signal* is a signal having a much higher frequency than the voice input signal. In **amplitude modulation (AM)**, the *amplitude* of the *carrier* frequency coming out of the mixer will vary according to the changing *frequency of the input* voice signal (Figure 4-2).

4.4 AMPLITUDE MODULATION

Amplitude modulation is used by AM broadcast stations. These stations transmit on carrier frequencies between 540 kilohertz (540,000 Hz) and 1.65 megahertz (1,650,000 Hz). How far these signals can be transmitted depends on the power of the transmitter. Most AM transmitters used by commercial broadcast stations have sufficient power to send a signal 40 mi. This allows the FCC to reuse a frequency in many parts of the United States, being careful not to assign the same frequency to

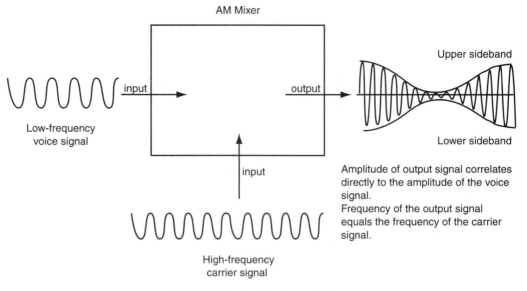

FIGURE 4-2 Amplitude modulation.

radio stations within 100 mi of each other. Many stations will exist in each specific geographic area of the country, but each station will be transmitting using different carrier frequencies. All these carrier frequencies arrive at the antenna of your radio. The radio has a tuning device that allows it to tune in and pick out the carrier signal desired. The radio will then demodulate the carrier signal we have tuned to. The demodulation circuitry in the radio will recognize the changes in amplitude of the carrier signal and convert the changing amplitudes into an electric current that changes in frequency. This signal is coupled to a speaker where the frequency of electrical energy is converted to vibrating sound waves of the same frequency. The radio listener hears the sound from the speaker, and it is the same as the sounds spoken into the microphone at the transmitter.

The signal emanating from a radio tower is possible due to an effect called *electromagnetic radiation*. When a high-frequency signal is attached to an antenna, the signal will radiate off the antenna. The higher the frequency and power of the signal applied to the antenna, the more easily it will radiate into the air. Broadcast stations must use high-frequency signals, and high power, to force the signal off the transmitting antenna and out into the air. By using a modulation technique, we can take the intelligence from a low-frequency signal and insert the intelligence into a higher-frequency signal easily radiated into space.

In telecommunications, we use wire as a transmission medium instead of air. We do not want our signals broadcast to many receivers. We only want one receiver to receive our transmitted signal. The use of wire to connect the transmitter and receiver allows us to establish a point-to-point communications link. Since we do not want the signal to radiate off the wire, we use low-powered transmitters. We use amplitude modulation in telecommunications to gain many different voice channels. Every voice signal is 0 to 4000 Hz. If we are going to combine many voice channels over one transmission medium, each signal must be different. We can convert the voice signals into different signals by using amplitude modulation. This conversion process takes place in the channel unit of a multiplexer system. With 12 input voice signals, we need 12 channel units, where each channel unit has a different frequency of carrier signal input to its mixer. The output of each channel unit will be the higher carrier frequency, modulated by the voice input signal. We will have 12 different output signals, and each signal will contain the intelligence of the original voice signal fed to the input of the channel units.

4.5 FREQUENCY DIVISION MULTIPLEXING

Using FDM, many telephone calls can be multiplexed over two pairs of wires. If we wish to multiplex 12 calls over two pairs of wires between cities A and B, we hook up a multiplexer containing 12 transmitters to the wire pair in city A. Each of these transmitters uses a different carrier frequency. At the other end of the cable pair, we attach 12 receivers in city B each tuned to receive only one of the 12 transmitted carrier signals, since this is a simplex circuit from A to B. We would also use 12 transmitters in city B, hook them up to another pair of wires between A and B, and place receivers in city A. This gives us a simplex circuit from B to A. The two simplex circuits together make up a full-duplex circuit between A and B. As we have seen, the use of multiple carrier frequencies to provide multiple conversations over two pairs of wires is called FDM (Figure 4-1).

4.6 THE HYBRID NETWORK

The inputs to the channel units of a multiplexer usually come from a local telephone circuit. The local telephone circuit is a pair of wires that carries voice signals in both directions. The circuit from a telephone arrives at the input to the channel unit of a multiplexer as a two-wire circuit. The channel unit contains a device (hybrid network) that interfaces the two-wire input to a four-wire transmit/receive path. The transmit path connects to the transmitter for this channel, and the receive path connects to the receiver for this channel. The hybrid network connects the one-directional, transmit, and receive circuits of the multiplexer to the two-directional telephone circuit. The channel units contain the hybrid network, modulator, transmitter, receiver, and demodulator (Figure 4-3).

4.7 AMPLITUDE MODULATION TECHNOLOGY

The first multiplexer to combine as many as 12 different signals into one signal for transmission over one medium in each direction used amplitude modulation technology. Amplitude modulation is a technique in which the amplitude of a high-frequency signal changes when the high-frequency signal is mixed with a low-frequency

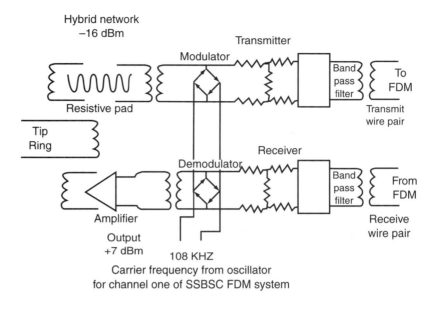

Channel Unit

FIGURE 4-3 Channel unit.

Carrier frequency from oscillator
for channel one of SSBSC FDM system

signal in a device called a mixer or modulator. The high-frequency signal is called the *carrier frequency*; the low-frequency signal is called the *modulating signal*. When the carrier frequency and the modulating signal are mixed together, the output of the mixer will be four signals. One signal will be at the frequency of the carrier frequency and one will be at the frequency of the modulating frequency. The third signal output by the mixer has a frequency representing the sum of the carrier and modulating frequencies. The fourth signal will have a frequency that is the difference between the carrier frequency and the modulating frequency. For example, if a 64,000-Hz signal is modulated by a 4000-Hz signal, the outputs of the modulator will be 64,000 Hz; 4000 Hz; 68,000 Hz (64,000 + 4000); and 60,000 Hz (64,000 − 4000). The two additional signals generated by the mixing process are called *sidebands*. The higher signal is called the *upper sideband*. In our example the frequency of the upper sideband is between 64 and 68 KHz. The lower signal is called the *lower sideband*; in our example it is between 60 and 64 KHz. Both these sidebands are replicas of the original input signal but at a higher frequency (Figure 4-2).

4.8 BANDWIDTH AND SINGLE-SIDEBAND SUPPRESSED CARRIER

The **bandwidth** of an AM signal is the difference between the highest frequency and the lowest frequency. In the example used, the bandwidth is 8000 Hz (68,000 − 60,000). The bandwidth of an AM signal is twice the modulating frequency. When a low-frequency signal is present as an input to the mixer, the amplitude of both sideband signals changes as the frequency of the input signal changes. We say the sidebands have been modulated (changed) by the input signal. Since both sidebands have been modulated, they both contain the frequency changes of the modulating signal. We only need to demodulate one of these sidebands to recapture the intelligence of the modulating signal. Since only one sideband is needed to demodulate the modulating signal, some systems will transmit only one sideband with the carrier frequency. These systems are called single-sideband (SSB) systems. Systems that do not transmit the main carrier frequency, but transmit the sideband only, are called *single-sideband suppressed carrier (SSB-SC)* systems.

4.9 CCITT STANDARDS FOR FREQUENCY DIVISION MULTIPLEXING

Telecommunications standards are set by several organizations. The predominant organizations are the American National Standards Institute (ANSI), which sets standards for North America, and the Consultative Committee on International Telegraphy and Telecommunications (CCITT) organization within the International Telecommunications Union (ITU), which sets worldwide standards. Figure 4-4 provides a description of the SSB-SC system specifications per CCITT Recommendation G.232.

4.10 GROUPS AND SUPERGROUPS

The basic building block for any multiplexing system is the **channel unit**. One channel unit is needed for each input voice signal. Each channel unit is designed with a 3200-Hz bandwidth to pass the voice frequency signals between 200 and 3400 Hz, (Figure 4-4). In the SSB-SC system specifications per CCITT Recommendation G.232, channel 1 will modulate a 108-kHz carrier signal. By using only the lower sideband, channel 1 occupies frequencies 104 to 108 KHz. Channel 2 will modulate a 104-kHz channel and occupy the 100 to 104-kHz band. Each channel will have a 4-kHz bandwidth assigned to provide guardbands on each side of the voice signal. Channels 3 through 12 each occupy the next lower 4-kHz band. Channel 12 modulates a 64-KHz carrier and occupies the 60- to 64-KHz band. The total bandwidth for a 12-channel Group is from 108 KHz (the top frequency of the lower sideband for channel 1) down to 60-KHz (the bottom of the lower sideband of channel 12). This is a bandwidth of 48 KHz (108 − 60 = 48). The bandwidth for 12 channels needs to be 48 KHz (12 channels × 4 KHz per sideband = 48 KHz). Twelve channels grouped together as above form a Group. Five groups, of 12 channels each, are combined into one 60-channel group called a Supergroup.

To build a Supergroup, we must change the frequencies of the Groups before they can be combined (multiplexed together). The frequency band of each group is from 60. to 108 KHz. *Each* of the five Groups in a Supergroup consists of 12 channels multiplexed together, as stated above. The five Groups are identical. To change the frequency of each group, we do a second modulation step. The first Group of 12 channels, is input to a second modulator where they modulate a 420-kHz carrier. The second Group of twelve channels modulate a 468-KHz carrier. The third Group will modulate a 516-KHz carrier, the fourth Group will modulate a 564-KHz carrier, and

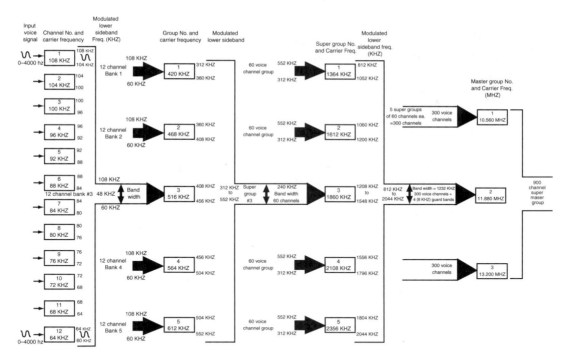

FIGURE 4-4 CCITT SSB-SC system layout.

the fifth Group will modulate a 612-kHz carrier. Only the lower sidebands from these modulators are used to create the Supergroup of 60 channels.

Earlier we stated that a lower sideband is the difference between two signals when they are mixed together. The frequencies of the signals in a Group are between 60 and 108 KHz. The carrier frequency for the first Group of a Supergroup is 420 KHz. When these two signals are mixed, the difference between 420 and 108 is 312 KHz, and the difference between 420 and 60 is 360 KHz. Thus, the lower sideband for Group 1 of a Supergroup is 312 to 360 KHz. The lower sidebands for all five groups of a Supergroup are 312–360 KHz, 360–408 KHz, 408–456 KHz, 456–504 KHz, and 504–552 kHz. Each of these sidebands is 48 KHz wide to accommodate the bandwidth of 12 channels (12 × 4 KHz = 48 KHz). The standard Group has a bandwidth of 48 KHz (60–108 kHz). Since a Supergroup is composed of five 12-channel groups for a total of 60 channels, it should be 240 KHz wide (60 channels × 4 KHz per channel = 240 KHz). The bandwidth of a Supergroup meets this criterion and is 240 KHz (552 KHz – 312 KHz = 240 KHz).

4.11 MASTER GROUP

A third modulation process can be added to the FDM process to create a Basic Mastergroup. The Basic Mastergroup contains five Supergroups. Since each supergroup is 60 channels, a Mastergroup contains 300 channels. It is formed by the same processes used above to form a Supergroup from five Groups. Frequencies 1364 KHz, 1612 KHz, 1860 KHz, 2108 KHz, and 2356 KHz, are the carrier frequencies used for each of the Supergroups within the Mastergroup. The band of frequencies in a Supergroup is from 312 to 552 KHz. When the first Supergroup is mixed with the 1364-KHz carrier signal, the lower sideband will be 812 to 1052 KHz (1364 − 552 = 812 KHz and 1364 − 312 = 1052 KHz). Since each Supergroup occupies a bandwidth of 240 kHz, the lower sideband of the first Supergroup is also 240 KHz wide (1052 − 812 = 240). An 8-KHz guardband is placed between each Supergroup. The lower sidebands for all five Supergroups in a Mastergroup are 812–1052 KHz, 1060–1300 KHz, 1308–1548 KHz, 1556–1796 KHz, and 1804–2044 KHz. The total bandwidth for a Mastergroup is 1232 KHz (2044 − 812 KHz = 1232 KHz) or (Supergroup bandwidth 240 KHz × 5 + (4)8 KHz guardbands) = 1232 KHz.

A fourth mixer can be added to an FDM system to create Supermastergroups. A supermastergroup can be made by combining three Mastergroups using carrier frequencies 10,560 KHz, 11,880 KHz, and 13,200 KHz. Each Mastergroup contains 300 channels; therefore a Supermastergroup contains 900 channels. Because there are a total of 15 Supergroups in a Supermastergroup, the system is often referred to as a *15-Supergroup system*.

4.12 L CARRIER

The Bell System used SSB-SC systems called *L carrier* in its long distance network prior to the widespread acceptance of digital carrier. However, L carrier contains only 10 supergroups instead of the 15 of CCITT Recommendation G.232. L carrier uses ten supergroup carrier frequencies to create a Mastergroup and does not create Supermastergroups. Ten supergroups of 60 channels each results in a total of 600 channels per L carrier system. The system is called an *L600 system*. This system uses frequencies slightly different than those used by the CCITT standard for creating Supergroups and Mastergroups (Figure 4-5). AT&T also uses a U600 system. This system has the same design as an L600 system but uses a different set of frequencies for the ten Supergroup carrier frequencies. The high bandwidth requirements (1.232 MHz) of the L600 and U600 systems require the use of coaxial cable or microwave radio as their transmission facility. AT&T continued to improve on FDM and developed an L5E system capable of multiplexing 13,200 voice circuits over one coaxial wire.

The L 600 system used one Mastergroup. The newer L5 system uses six Mastergroups. Each Mastergroup contains 600 voice circuits. Six Mastergroups provides 3600 channels or voice circuits. These Mastergroups were modulated to a higher order called Jumbo Groups. Each Jumbo Group contains 3600 circuits. The L5 system uses three Jumbo Groups for a total of 10,800 circuits. AT&T also has systems called *L5E*, which can multiplex 13,200 trunk circuits (voice circuits) over onto one pair (one transmit and one receive coaxial wire) of coaxial cable with amplifiers placed every mile. As mentioned earlier, coaxial cable is not the only medium used to carry FDM. Many FDM systems will use microwave technology to carry the L carrier systems, because microwave is cheaper. However, coaxial cable is quieter and does not require unsightly towers; it is used on many long distance routes.

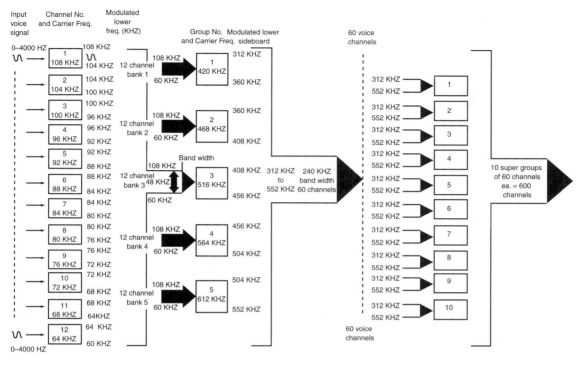

FIGURE 4-5 SSB-SC system.

4.13 FREQUENCY DIVISION MULTIPLEXING INPUT SIGNAL LEVELS

If the level of an input signal to an AM multiplier's channel unit is too high, it will cause crosstalk and noise. The input level to the channel is controlled by purposely introducing additional loss to the input signal. A resistive network is placed between the input signal and the channel input. This resistive network is called a *pad*. The resistive network is adjusted so that the input signal level to the channel unit is –16 dBm. The output of the signal at the channel unit of the receiving multiplexer is +7 dBm. The FDM system provides a +23-dB gain for each channel signal.

4.14 FREQUENCY MODULATION

AM signals are highly susceptible to noise caused by other electromagnetic signals. Another multiplexing technique-called *frequency modulation (FM)*—uses the same process used by FM radio broadcast stations. FM signals are less susceptible to noise than AM signals are. FM radio stations use radio transmitters that generate carrier frequencies between 88 MHz (88,000,000) and 108 MHz (108,000,000). In FM modulation, the input signal to a mixer causes the *frequency* of the *carrier signal* to change in step with the *amplitude of the input signal*. For example, at a particular amplitude, for a given input signal, the frequency of a 100-MHz carrier may change by 20 KHz in each direction (100,020,000 Hz and 99,980,000 Hz). *How often the carrier signal changes back and forth between these two frequencies is determined by the frequency of the input signal.* If the input signal has a frequency of 1000 Hz, the signal will change back and forth (between 100,020,000 and 99,980,000 Hz) 1,000 times in one second. The *bandwidth* in FM is determined by the *amplitude* of the input signal, and in this example the bandwidth is 40 KHz. The bandwidth of any signal is the difference between the highest and lowest frequency of the signal. The bandwidth of the signal above is (100,020,000 – 99,980,000) or 40 KHz. A detector in the receiver detects how often the carrier frequency changes. In this example, the frequency of the carrier signal is being changed 1000 times each second. This represents the frequency of the original signal on the input of the transmitter. This demodulated signal is then fed to the speaker of the radio for conversion to sound waves of the same frequency (Figure 4-6).

When using FM in telecommunications, the channel units are wired to cable pairs and inputs just like the AM channel units were. The channel units of the FDM simply contain FM technology instead of AM technology. The use of FM channel units was limited. Most FDM systems use SSB-SC AM technology. Remember that *FDM means multiplexing* and does not refer to the FM modulation technique. A modulation technique such as AM or FM is needed to prepare signals for multiplexing by converting voice input signals to different higher-frequency signals. FDM is the actual multiplexing technique to combine these different frequency signals.

4.15 TIME DIVISION MULTIPLEXING USING PULSE AMPLITUDE MODULATION SIGNALS

A third type of multiplexing is *time division multiplexing (TDM)*. Early TDM signals multiplexed *samples* of the human voice from different channel units over one common facility. These were analog voice samples, and the technique used to achieve these voice samples was called *pulse amplitude modulation (PAM)*. An engineer named Nyquist made some studies of voice sampling techniques and developed the Nyquist theorem. According to this theorem, if the sampling rate of a signal is twice the highest frequency of the signal, the signal can be reconstructed using the samples. Transmitting samples of a signal instead of the complete signal makes it possible to use a medium for transmitting many samples. A certain amount of time is set aside for the transmission of each sample. Each sample is transmitted over a common medium during its allotted time slot. The common medium is used to transmit all voice samples, but only one sample at a time. Each channel is multiplexed over the common facility by dividing the time available on the medium between all channels. This is TDM.

When a sample is taken of a voice signal, the voltage level of the signal at that point in time is stored in the channel unit. A 24-channel-unit TDM system would assign each channel to the TDM highway for 1/24th of a second. The transmitter on one end of the cable pair and the receiver on the other end of the cable pair were synchronized. Channel 1 of both ends was attached to the medium at the same time for 1/24th of a second. Then channel 2 on each end was attached to the cable pair, and so on up to channel 24. Then the system would start again on channel 1. When

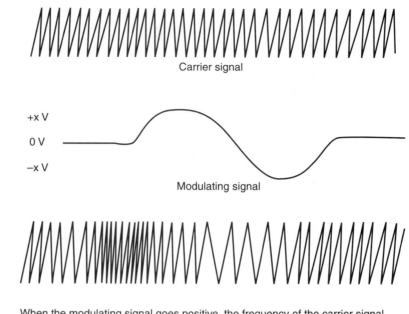

FIGURE 4-6 Frequency modulation.

When the modulating signal goes positive, the frequency of the carrier signal increases. When the modulating signal swings negative, the frequency of the carrier signal decreases.

channel 1 was attached to the cable pair, the transmitter on one end would read out the voltage of the stored sample. This voltage traveled over the transmit cable pair to the distant channel 1 receiver. The channel 1 receiver would take the received voltage samples and perform a dot-to-dot connection between the received voltage levels as it fed them through the hybrid circuit to the output of the channel unit. This reconstructed the original signal. Figure 4-7 illustrates samples taken from a signal. Note how the original signal is reproduced when a line is drawn connecting the PAM samples in the bottom drawing. Figure 4-8 provides an example of a three-channel PAM/ TDM system.

4.16 *TIME DIVISION MULTIPLEXING USING PULSE CODE MODULATION SIGNALS*

Early TDM systems used PAM, and the systems were analog TDM systems. Since these systems used amplitude modulation (PAM), they were susceptible to noise interference. A method was developed to convert the PAM signals into codes. Codes were established for 256 different amplitude levels. These codes, from 1 to 256, were stored in binary form (00000000 to 11111111). A 8-bit binary code allows for coding 256 distinct numbers (0 to 255) because the number of different combinations that 2^8 provides is 256. The conversion of an analog to a digital signal is done by first converting the signal to PAM and then converting the PAM sample into a digital code. The industry standard method for converting an analog into a digital signal is called *pulse code modulation (PCM)*. A TDM system using PCM signals transmits the digital code for the sample instead of the actual amplitude of the sample. The electronic device that uses PCM to convert an analog signal into a digital signal is called a *coder/decoder (codec)*.

All TDM systems now use PCM and not PAM. PCM/TDM is a type of multiplexing that requires the conversion of incoming analog voice signals into digital signals. These digitized voice signals are then multiplexed using TDM technology. PCM/TDM is digital transmission, and digital transmission has many advantages over analog FDM and PAM/TDM. Some advantages are:

1. Digital signals are less susceptible to noise
2. Digital circuitry lends itself readily to integrated circuit design, which makes digital circuits cheaper than analog circuits.
3. Digital-to-digital interface is easily achieved.

4.17 PULSE CODE MODULATION

Most of the calls handled by the PSTN are voice calls, which are analog in nature. It is necessary to convert the analog voice signal into a digital signal to achieve the benefits of TDM. The industry standard method for converting one analog voice signal into a digital signal is PCM. There are five basic steps to PCM: sampling, quantization, companding, encoding, and framing.

4.17.1 Sampling

Sampling refers to how often measurements are taken of the input analog signal. The sampling rate in telecommunications is represented by the Nyquist theorem, which

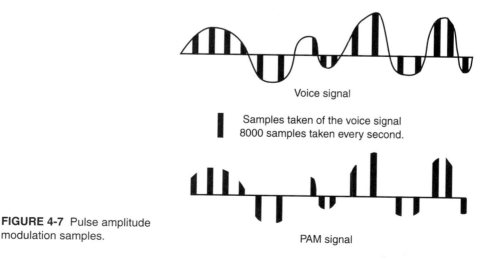

Voice signal

Samples taken of the voice signal
8000 samples taken every second.

FIGURE 4-7 Pulse amplitude modulation samples.

PAM signal

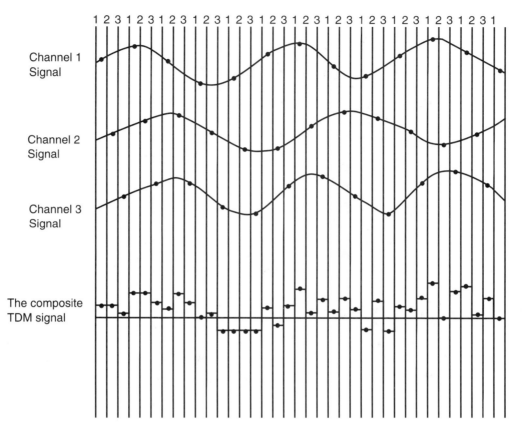

FIGURE 4-8 The creation of a time division multiplexing signal.

states that an analog signal should be sampled at a rate at least twice its highest frequency. In telecommunications, the network was designed to handle signals between 0 and 4000 Hz. Therefore, according to Nyquist, sampling at twice the highest frequency of 4000 Hz means a sampling rate of 8000 times per second. Every 1/8000 of a second (125 μs) a voltage measurement is taken on the input signal.

4.17.2 Quantization

In PCM, each voltage measurement is converted to an 8-bit code, and the 8-bit code is sent instead of sending the actual voltage. An 8-bit, two-state (1 or 0) code limits the conversion process to the recognition of 256 codes ($2^8 = 256$). The input signal can be any level. The input signal is not a discrete-level signal. There are billions of discrete levels between –1 V and +1 V, such as 0.0008362, 0.0794236, –0.998789, and so on. Since we only have 256 discrete codes available, the input signal must be adjusted so that it will be one of the 256 input levels that can be coded. This adjustment process is called *quantization*.

4.17.3 Companding

As noted, quantizing a signal will result in some distortion because we do not code the exact voltage of the input signal. This distortion is called *quantizing noise* and this distortion is greater for low-amplitude signals than for high-amplitude signals. To overcome quantizing noise, the signal is *compressed* to divide low-amplitude signals into more steps. When the signal is decoded at the receiving system, it is *expanded* by reversing the compression process. The combination of compression and expansion is called *companding*

4.17.4 Encoding

After quantization and compression of the input signal, it will be one of the 256 discrete signal levels that can be assigned an 8-bit code. The process of assigning an 8-bit code to represent the signal level is known as *encoding*.

4.17.5 Framing

The encoded 8-bit signal is time division multiplexed with 24 other 8-bit signals to generate 192 bits for the 24 signals. A single framing bit is added to these 192 bits to make a 193-bit frame. The framing bits follow an established pattern of 1s and 0s for 12 frames and then repeat in each succeeding 12 frames. This pattern is used by the receiving terminal to stay in sync with the received frames.

The integrated circuit chip that converts an analog input signal into a digital PCM signal is a coder/decoder or codec. The codec is a four-wire device. It accepts incoming analog voice signals and codes them into digital signals that are placed on the transmit leads to the TDM system. It also accepts digital signals over the receive leads from the TDM system and decodes them into analog outputs. A codec contains an analog-to-digital (A-to-D) converter, a digital-to-analog (D-to-A) converter, and a Universal Asynchronous Receiver Transmitter (UART). The UART is necessary to convert the parallel 8-bit output of the A/D converter to a serial 8-bit stream that can be placed on the TDM highway. The analog side of a codec is usually coming from a telephone line that is using only two wires for both the transmit and receive signals. A hybrid network is part of the codec circuitry. The hybrid network is a device that converts a two-wire into a four-wire circuit. Two wires are used as an input from the analog line. Two of the output wires are used as a transmit pair of the coder circuitry. The other two wires serve as the receive pair of the decoder circuitry. The Codec circuit card is part of the channel unit in a carrier system using TDM. If the TDM system is not interfacing to a two-wire analog circuit but is interfacing to a four-wire digital circuit, a different channel unit will be used. This special channel unit has the circuitry necessary to interface to the digital signal.

The basic building block for digital transmission standards begins with the DS0 signal level. DS0 is the 64-Kpbs signal generated by a codec converting an analog voice signal to a digital signal. DS0 is a single voice channel. To convert an analog signal into a digital signal, the analog signal is sampled 8000 times each second and each sample is converted into an 8-bit code. Therefore, when one analog voice signal is converted to a digital signal using PCM, the digital signal is 64,000 bps (8000 × 8). The basic digital multiplexing standard established in the United States was set by Bell Communications Research and is called the *Bell System Level 1 PCM Standard* or *Bell T1 Standard* (this is recognized by CCITT as CCITT Recommendation G.733). This standard is the standard for multiplexing 24 digital voice circuits over a pair of physical facilities. The CCITT Level 1 PCM standard for worldwide use is different from the Bell Standard because it involves the multiplexing of 32 DS0 Circuits (30 are voice channels and 2 are signaling/synchronization channels).

The Bell T1 Standard is a 24-channel PCM/TDM system using 8-level coding (256 codes). The T1 system can be equipped with channel units that interface to analog signals, or it can be equipped with channel units that will interface to a digital signal. When the T1 system uses channels that interface to analog signals, the channel unit contains a codec to convert the analog signal into a DS0 signal. The sampling rate for each channel is 8000 times per second, and each of the samples is converted into an 8-bit code. The output of the codec is 64,000 bits per second (bps) (8 bits per sample × 8000 samples per second = 64,000 bps). Each analog voice channel is converted to 64 kilobits per second (Kbps). This is referred to as *digital signal level zero (DS0)*. When multiplexing 24 channels, one sample (8 bits) is taken from each channel one after the other and sent over the transmission medium. After sending the 8 bits for the 24 channel, one framing bit is sent to start a new frame and maintain synchronization between the transmitting and receiving multiplexers (Figure 4-9). A frame consists of 193 bits (8 bits per channel × 24 channels + 1 framing bit = 193 bits). Since we are taking 8000 samples per second on each voice channel, it is necessary to transmit 8000 frames per second. The total bits transmitted when multiplexing 24 channels is 1,544,000 bps (1.544 Mbps). That is, 8000 frames per second × 193 bits per frame = 1.544 Mbps. This rate is also arrived at by the following method: 8000 samples per voice channel × 8 bits per sample × 24 channels + 8000 framing bits = 1.544 Mbps. Digital signal level 1 (DS1) = 1.544 Mbps. Finally, DS1 = 24 DS0 channels + 8000 framing bits.

It may help you understand multiplexing if you think of the time required to perform one operation. It would be possible to put two signals over a medium using TDM if we let each signal occupy the transmission medium for half a second. It is possible to allow ten signals to used a time-shared facility if each uses the facility for one-tenth of a second. The number of signals that can be multiplexed in a TDM system depends on the quality and design of the transmitter, receiver, and medium. If the receiver needs to see a signal for one-tenth of a second in order to properly recognize the signal, then the transmitter must place the signal on the medium for at least one-tenth of a second and only ten signals could be multiplexed. If a receiver can detect a signal change in 0.000000001 sec, then 1,000,000 signals can share the common facility, one at a time.

The number of events that can occur in a given time frame is the inverse of the time required for one event. The reverse is also true: the time required for one event is the inverse of the number of events in a given time frame. For example, if the time required for one event is one-tenth of a second, then we can do ten events in one second. Frequency (of events) = 1/time (of one event). Thus, $F = 1/1/10 = 10$. Time required for one event: $T = 1/F$, so that $T = 1/10$. In telecommunications, we usually state time in seconds and the number of events as the number that can occur in one second. Metric system prefixes derived from Greek, Latin, and French are used to designate quantities. The Greek prefixes used are:

K = kilo = 1000
M = mega = 1 million
G = giga = 1 billion
T = tera = 1 trillion.

Thus, 1.544 Mbps means the system is handling 1,544,000 bits per second. Latin and French prefixes are used for numbers less than 1:

d = deci = 1/10
m = milli = 1/1000
μ = micro = 1/1,000,000
n = nano = 1/1,000,000,000
p = pico = 1/1,000,000,000,000.

All numbers can be expressed as powers of 10. For example, 1000 (kilo) = 1×10^3 ($10 \times 10 \times 10$). Milli (1/1000) = $1/(10^3)$ and is also equal to 1×10^{-3}. The prefixes *kilo, mega,* and *giga* can be represented by 10 to a positive power: 10^{+3} = kilo; 10^{+6} = mega, and 10^{+9} = giga. The prefixes *milli, micro, nano,* and *pico* can be represented by 10 to a negative power: 10^{-3} = milli; 10^{-6} = micro; 10^{-9} = nano and 10^{-12} = pico. Note

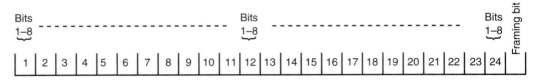

8 bits in each channel
24 channels × 8 bits per channel = 192 bits + 1 framing bit = 193 bits per frame

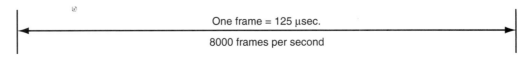

One frame = 125 μsec.
8000 frames per second

FIGURE 4-9 One frame of a DS1 signal.

that each of these prefixes can be stated as either a fraction or as 10 to a negative power—for example, milli = 1/1000 or = 1×10^{-3}. Since 1000 = 10^{+3}, 1/1000 can also be stated as $1/10^{+3}$. Therefore, milli can be stated as 10^{-3} or as $1/10^{+3}$. This is a basic mathematics theorem, which states that N^{-x} is equal to $1/N^{+x}$ and $1/N^{-x}$ is equal to N^{+x}. For instance, $1/10^{-3} = 1/0.001$ and $1/0.001 = 1000$ or 10^{+3}.

When calculating how long an event can be in order to do $\times$ amount in a given time frame, it is possible to change the sign of the power of 10 to its opposite sign to arrive at the calculation. If we wish to do 1.544 Mbps ($1.544 \times 10^{+6}$), then each bit must be no longer than $(1/1.544) \times 10^{-6}$. Further, 1/1.544 = 0.6476684 and $(1/1.544) \times 10^{-6} = 0.6476684 \times 10^{-6} = 0.0000006476684$. This is equal to 647.6684×10^{-9}. Since 10-9 power = $1/1,000,000,000$ = nano, the time frame is 647.6684 ns for each bit. If each bit sent on the TDM medium is only 647.6684 ns long, the system can transmit 1,544,000 bps. Thus, $F = 1/T = 1/(647.6684 \times 10^{-9})$ = 1.544 Mbps. Note that the same answer is achieved by using the reciprocal power method: Since $1/10^{-9} = 10^{+9}$, $F = 1/T = (1/647.6684) \times 10^{+9}$. Moreover, 1/647.6684 = 0.001544 and $0.001544 \times 10^{+9} = 1,544,000$.

Earlier, it was stated that the PCM sampling rate for a voice signal is 8000 times a second. The time for one sample = $1/f = 1/8000 = 0.000125 = 125$ microseconds (μsc). In PCM this sample consists of 8 bits. These 8 bits will be stored in a temporary store. 125 μsc later a new sample code will be stored by writing over the old stored code. The stored 8-bit code will be available for readout to the TDM for 125 μsc before it is lost. Because the TDM highway is running a 1.544 Mbps, it can read 193 bits in $193 \times 647.6684 \times 10^{-9}$ = 125 μsc. Recall from above that 193 bits are transmitted in each frame by sending 8 bits from channel 1, 8 bits from channel 2, and so on all the way up to 8 bits from channel 24 and then 1 framing bit for $(24 \times 8) + 1 = 193$ bits. Also, one frame equals a sample from every channel and these samples are being generated at a rate of 8000 per second. We must send 8000 frames per second to send every sample generated at the PCM level. Since each frame is 125 μsc long (1/8000 of a second), we can send 8000 frames per second.

Higher levels of multiplexing can build frames containing more bits because the receiver, transmitter, and medium used at higher levels of multiplexing can recognize smaller and smaller bits of time as a 1 or 0 digital state. All TDM systems must send 8000 frames per second, but the frame size contains more bits at higher and higher levels of TDM (it can contain more because each bit is smaller or takes up less time). The DS1 signal of a T1 carrier system is generated by reading 8 bits at a time from each channel unit. Higher levels of multiplexing read more than eight bits at a time from each input. Each higher level of multiplexing must have 8000 frames per second in order to stay in sync with the 8000-per-second requirement at the DS0 level. Each frame must contain one frame from each lower-level system, just as DS-1 contains one DS0 sample from each channel. For example, a DS-3 signal is composed of 28 DS1 signals and each frame of DS3 must contain one frame from each of the DS1 signals. Since each DS1 frame is 193 bits, each frame of DS3 contains 5404 bits plus additional overhead for control. The DS3 can receive 193 bits from DS1 system 1, then 193 bits from DS1 system 2, and so on. DS1 and DS3 both place one bit at a time on the TDM medium because it is a serial bit stream process. However, in DS1 the TDM bit stream will contain data from each channel as 8 adjacent bits on the TDM medium. A DS3 signal is composed of 28 T1 signals multiplexed together. The DS3 signal rate is 44.736 Mbps (28×1.544 Mbps + 1.504 Mbps additional overhead).

In the American T1 carrier system, the signals multiplexed are digital signals (PCM). The T1 receiver is capable of recognizing a bit as being a 1 or 0 in 0.000000647884th of a second (647.884 ns). To do higher levels of multiplexing, the time required for generating and recognizing a pulse must be smaller. Remember that frequency = 1/time. If we want a higher-frequency bit stream, each bit must take less time. In the European TDM system, the procedure is to multiplex 32 channels that are 64,000 bits each. This gives a total of 2,048,000 bits each second. The time for each pulse must be $1/(2.048 \times 10^{+6}) = 1/2.048 \times 10^{-6} = 0.4882813 \times 10^{-6} = 488.2813 \times 10^{-9} = 488.2813$ ns.

4.19 T1 CARRIER SYSTEMS

The multiplexing system of choice to connect two central offices is a North American DS1 system called *T1 carrier*. This 24-channel multiplex system uses the Bell Standard DS1 signal rate of 1.544 Mbps. These systems are generally connected to wire pairs in a cable that runs from one central office to the other. Because a DS1 signal level is a high-frequency signal, it cannot travel more than a mile and a half before the signal is lost. It is necessary to remove load coils from the pairs being used for T1 carrier to improve the high-frequency response of the cable pairs. It is also necessary to add regenerators to the T1 cable pairs every 6000 ft. Because cable pairs are loaded at the first 3000 ft from the central office and then every 6000 ft thereafter, as the load coils are removed from a cable pair to be used for a T1 carrier system, the pulse regenerators are put on the pairs at these points. The pulse regenerators are called *repeaters*. Repeaters are also added to the T1 line (or cable pair) at each central office. A T1 system is connected to a passive repeater at the central office. The central office repeater does not regenerate outgoing transmit pulse streams but merely passes the pulses from the T1 terminal to the transmit cable pair. The next repeater will be 3000 ft from the central office, then additional repeaters are placed at 6000-ft intervals. The repeater looks at the incoming signal (1s and 0s), and for each 1 received it sends out a fresh high-level output. Since the maximum distance between repeaters is 6000 ft, the T1 repeaters keep the signal at a high level.

Many T1 carrier systems are still in use, although the trend in larger cities has been to replace direct T1 carrier routes with an OC-48 fiber ring technology. An OC-48 fiber ring can carry 32,256 trunk circuits. The fiber optic cable links all central offices in a large city. An OC-48 multiplexer/demultiplexer is located in each exchange and is programmed to strip out the specific channels assigned to that exchange. T1 is used in places where fiber ring networks have not been installed. When a T1 carrier system is used, the on-hook and off-hook signals for each channel are sent over the system by a technique called *bit-robbed signaling*. Since two states are needed for the signal (either on-hook or off-hook), they can be represented by a binary bit (1 or 0). Because the signaling state does not change very often or fast, we do not need a signaling bit in every frame. The first signaling technique used for T1 substituted a signaling bit for each of the 24 channels in every sixth frame. This signaling bit for each channel replaced the least-significant voice bit in every sixth frame on each voice channel. This causes some distortion to the voice signal, but the distortion is not noticeable. Of course, this bit-robbing technique would destroy data if the signal on the channel is a data signal. For that reason, data on a PSTN T1 system is limited to 7 bits per channel in every frame, not just every sixth frame. Further, 7 bits per frame × 8000 frames = 56,000 bits per second. Thus a T1 system used in the PSTN can only accept a data rate of 56,000 bps on each channel. This is referred to as *56-kbps clear channel capability*. There are some proprietary T1 systems that use a means other than bit robbing to do signaling. These systems can use all 64,000 bps of the channel for data. A T1 system used as the vehicle for ISDN does not use bit-robbed signaling; thus, the signaling for all channels is carried by the 24th channel. The 24th channel does not serve as a voice channel. It serves as a common channel for signals and is called the D-channel. Thus, a T1 system used to carry ISDN provides a clear channel capability of 64,000 bps.

Figure 4-10 shows the layout for a Superframe. A Superframe is composed of 12 frames. Each of the frames is 193 bits (twenty-four 8-bit voice channels plus 1 framing bit). Notice that the framing bit pattern over 12 frames is 100011011100. These framing bits allow the receiver to synchronize off the incoming bit stream. The least-significant bit of every channel in the 6th frame is used for signaling and is called the *A bit*. The least-significant bit of each channel in the 12th frame is also used for signaling and is called the *B bit*. Another type of signaling developed for T1 signaling is called *Extended Superframe (ESF)* signaling. The ESF consists of 24 frames; thus, there are 24 framing bits. Only 6 of the 24 framing bits are used to provide synchronization for the receiver. Six bits are used to provide a CRC error-checking protocol. The remaining 12 bits are used to provide a management channel called the *Facilities Data Link (FDL)*. The ESF provides signaling bits in every 6th frame. The signaling bit in the 6th frame is called the *A bit*. The signaling bit in the 12th frame is called the *B bit*. The signaling bit in the 18th frame is called the *C bit*, and the signaling bit in the 24th frame is called the *D bit* (Figure 4-11).

When the transmitter of a T1 carrier system was first introduced, the maximum distance for sending the digital signal was 3000 ft. The sending of 1.544 Mbps looked like a 1.544-MHz signal to the cable pair. The design of the transmitter was changed to make a T1 signal look like a 0.772-Mhz signal by alternating the polarity of every 1 (Mark) transmitted. This technique is called *Alternate Mark Inversion (AMI)*. AMI is a bipolar signal. The signal changes back and forth between +3 and –3 V. A 0 is represented by no voltage at all. A 1 is represented by a +3-V signal. The next 1 transmitted is represented by a –3-V signal, and so on (Figure 4-12). AMI provides for an error-detection scheme called *bipolar violation (BPV)*. Since the T1 line signal level will always alternate between (1) either a +3-V signal, then 0, and then –3 V; or (2) +3 V, then –3 V, there should never be two consecutive +3-V signals or two consecutive –3-V signals. Figure 4-13 shows a bipolar violation occurrence. Compare the original signal of Figure 4-12 to the signal of Figure 4-13, which contains errors to the original signal.

The receiver is synchronized by each incoming 1. If a string of 0s are received, the transmitter could lose synchronization. To prevent a loss of synchronization due to many successive 0s, the B8ZS technique was developed. The encoding of a voice signal will never result in more than seven successive 0s. If data is transmitted at 56,000 bps using 7 bits per channel time slot, the 8 bit in each channel time slot can be made a 1. This also prevents more than seven successive 0s. *Binary 8 Zero Substitution (B8Zs)* was developed to keep a transmitter from sending more than eight consecutive 0s. When a string of eight consecutive 0s is encountered, the transmitter will induce a bipolar violation at the fourth and seventh 0s and AMI-compatible Marks at the fifth and eighth 0s. The receiver detects the B8ZS code and reinserts eight 0s for the B8ZS code. The receiver will receive enough 1s to stay in synchronization either from actual 1s in the bit stream or from the 1s forced into the bit stream by B8ZS.

4.20 EUROPEAN TDM 30 + 2 SYSTEM

The European TDM system multiplexes 32 DS0 channels together. This system uses channel 0 for synchronizing (framing) and signaling. Channels 1–15 and 17–31 are used for voice. Channel 16 is reserved for future use as a signaling channel. Because framing signals are part of channel 0, additional framing bits are not required. The DS0 signals are 64,000 bps. The total signal rate for the European 30 + 2 system is 2.048 Mbps (64,000 bps × 32 channels = 2,048,000).

4.21 HIGHER LEVELS OF TDM

The 24-channel T1 system developed in the early 1960s is the basic multiplex building block in North America. The next order of multiplexing is to combine two T1 systems into a 48-channel DS1C (T1C)system. Three T1 systems were combined into a 96-channel DS2 system in 1972. Twenty-eight T1 systems can be multiplexed into a 672-channel DS3 system, which is a 44.736-Mbps system. Six DS3 systems can be multiplexed into a 4032-channel DS4 system (274.176 Mbps). Wire can be used as the medium to carry DS1, DS1C, and DS2 systems. DS3 and DS4 must use coaxial cable, fiber optic cable, or digital microwave radio as a medium. Regenerators (repeaters) used on a fiber optic cable are placed 27 mi apart.

4.22 SONET STANDARDS

Fiber optics use *Synchronous Optical Network (SONET)* standards. The initial SONET standard is OC-1. This level is known as *Synchronous Transport Level 1 (STS-1)*. It has a synchronous frame structure at a speed of 51.840 Mbps. The synchronous frame

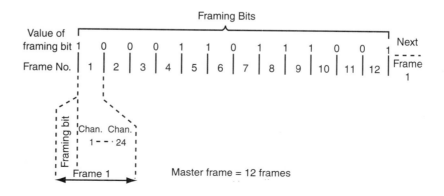

FIGURE 4-10 Superframe.

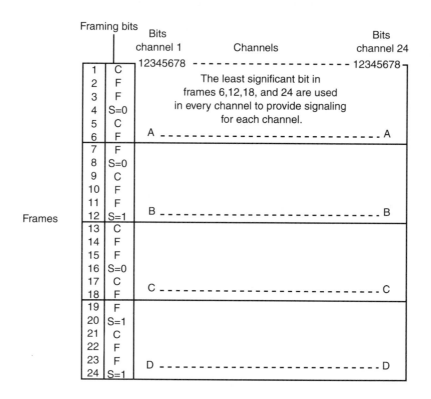

FIGURE 4-11 Extended superframe.

ESF uses 6 bits (S) for synchronization (001001). C bits are used for CRC error checking and F bits comprise the Facilities Data Link (FDL) channel.

structure makes it possible to extract individual DS1 signals without disassembling the entire frame. OC-1 is an envelope containing a DS3 signal (28 DS1 signals or 672 channels). With SONET standards any of these 28 T1 systems can be stripped out of the OC-1 signal. The North American SONET levels are:

OC-48 is 2488.32 million bits per second or 2.48332 billion bits per second. As noted, the prefix for billion is giga (G). Therefore, 2.48332 gigabits per second is the

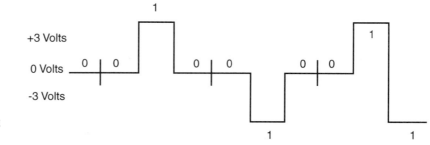

FIGURE 4-12 Alternate Mark Inversion.

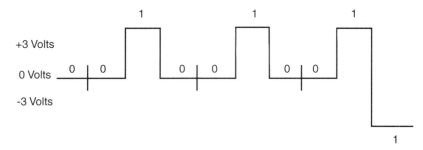

FIGURE 4-13 Bipolar violation.

signal rate of OC-48. OC-48 is 48 OC-1 systems and contains 48 times 672 channels = 32,256 channels. One fiber optic strand will carry all 32,256 channels. An additional benefit to using fiber optic technology is that it will not pick up noise from any outside source influence. The glass strands of fiber are carrying light pulses. It is impossible for glass fiber to carry electrical energy. Noise in wire cables is caused by electrical currents being induced into the wire from outside electromagnetic radiation. This cannot happen with fiber optic cables because fibers cannot carry electrical signals. In many of the BOC territories, OC-48 has replaced T1 carrier systems. Instead of having multiple T1 routes from class 5 exchanges into a class 4 exchange, OC-48 technology and fiber optic cables link offices together in a ring network. Fiber will connect all offices in the metropolitan area. Each office is equipped with an OC-48 multiplexer/demultiplexer. Each office will only strip out the channels assigned to that office. This is the beauty of SONET standards. The whole multiplexed stream does not have to be demultiplexed to pull out one T1 system or even one DS0 channel. The OC-48 can be instructed to pull out one DS1 bit stream, and this stream can be connected to a T1 system.

Today's digital central offices have fiber optic multiplex loops inside the switch running at OC-1 levels or higher. With an OC-1 level switching loop, it is possible to have an OC-48 system at each central exchange on the fiber ring strip out an OC-1 signal of 672 trunk circuits and connect them directly into the switch. This is because SONET technology also allows stripping out one DS0 signal (one trunk circuit). The signals on the OC-48 fiber ring can be stripped out as needed one at a time and then assigned an appropriate slot on the OC-1 loop of the switch. Software in the switch and SONET multiplexers makes it easy to assign DS0 channels for use between any exchanges connected to the fiber ring network.

Sprint has upgraded its fiber routes to handle as many as 516,096 trunk circuits (channels). This has been accomplished by using a technology called *wave division multiplexing (WDM)*. Instead of using strictly TDM technology and trying to extend this technology to multiplex many more signals in a TDM mode, wave division multiplexing technology combines FDM and TDM technology. The light spectrum of a fiber optic facility is broken down into multiple frequencies. Light waves are composed of many frequencies. A blue light has a different frequency than a red light, an orange light, a green light, and so on. Transmitters and receivers for fiber optic have been developed that transmit and receive at a specific frequency. Sixteen different frequencies are used. An individual OC-48 TDM connects to each of the 16 different transmitters in a *wave division multiplexer*. By splitting the light spectrum up into 16 different frequencies and attaching an OC-48 to each transmitter, the fiber optic facility can carry 516,096 channels (32,256 × 16). The next advance will be to use an OC-96 or OC-192 as the TDM device on a WDM. Each OC-192 can carry 129,024 channels. When OC-192s are placed on a 16-channel WDM, the fiber can carry 2,064,384 channels (16 × 129,024 = 2,064,384) (Figure 5-18).

4.23 SUMMARY

AT&T developed FDM systems for use in their long distance networks. The *L* designation for long distance was followed by a letter indicating the version of L carrier. L600 carrier was used extensively over coaxial tubes to provide 600 channels (or circuits) over two tubes (one in each direction for transmit and receive). Using AM techniques, voice circuits modulate the carrier frequencies of channel units. A channel bank consisted of twelve channel units. Each of these had a different carrier frequency to provide twelve distinct signals. Five of these twelve channel units modulated five different second-carrier frequencies to create a Supergroup containing 60 voice channels. Ten Supergroups could be combined by having each of them modulate ten different carrier frequencies to form a Mastergroup of 600 channels. Six Mastergroups could modulate six different frequencies to create one Jumbo Group of 3600 channels. Finally, by assigning three different frequency ranges to carrier frequencies at the Jumbo-Group level, it is possible to have 10,800 voice circuits carried by one coaxial tube. Of course, two tubes will be needed to provide for transmit and receive signals. L5E has expanded this FDM capacity to 13,200 circuits.

FDM is being replaced by TDM technology. TDM over fiber optic allows as many as 32,256 circuits to be multiplexed over the fiber. At the OC-1 level, 672 channels are available over 28 T1 systems. This envelope is enough to serve most class 5 central office trunk requirements. WDM combines FDM and TDM technologies to increase the number of circuits that can be multiplexed over a glass fiber. A big advantage of TDM over FDM is the TDM can be interfaced directly into a digital switching system.

REVIEW QUESTIONS

1. What is multiplexing?
2. What is space division multiplexing?
3. What is frequency division multiplexing?
4. What is time division multiplexing?
5. What is the purpose of a channel unit?
6. What is a four-wire system?
7. What is amplitude modulation?
8. What is frequency modulation?
9. What is the difference between FM and FDM?

10. How many voice channels are on one L-600 carrier system?
11. How many voice channels are on one L5E carrier system?
12. What type of modulation technique is used by L-600 and L5E systems?
13. What is the difference between PAM and PCM?
14. What is the industry standard method for converting an analog voice signal into a digital signal DS0?
15. What is the sampling rate of PCM?
16. How was the sampling rate of PCM determined?
17. What is quantization?
18. What device is used to do PCM?
19. What is the DS0 signal rate? What is the DS1 signal rate?
20. How long is each bit on the TDM path of a T1 carrier system?
21. How many bits are there in a DS1 frame?
22. How many frames per second does a DS1 system have?
23. When T1 is used on a cable pair, how many repeaters are needed?
24. What is the basic building block for SONET?
25. How many voice channels can be put on an OC-48 system?
26. What transmission medium is used for OC-48?
27. What are the major advantages of SONET standards?

GLOSSARY

Amplitude Modulation (AM) Used by AM broadcast stations. In amplitude modulation, the amplitude of the carrier frequency coming out of the mixer will vary according to the changing frequency of the input voice signal.

Bandwidth Refers to the width of a signal. the width of a signal is determined by subtracting the highest frequency of the signal from its lowest frequency.

Channel Unit Performs the function of interfacing the multiplexer to a specific type of input signal. The channel unit makes changes to the input signal so it can be combined (multiplexed) with other signals for transmission over the transmission medium.

Frequency Division Multiplexing (FDM) The process of converting each speech path to different frequency signals and then combining

the different frequencies so they may be sent over one transmission medium.

Mixer (Modulator) An electronic device that has two inputs. One input is a low-frequency signal that contains intelligence. The other signal is a pure sine wave at a high frequency. This high-frequency signal is called the carrier signal because it carries our intelligent signal after modulation. The mixer combines the two signals so that we end up with a high-frequency signal that has the intelligence of the low-frequency signal imposed on it.

Multiplexing In telecommunications, a process combining many individual signals (voice or data) so they can be sent over one transmission medium.

Pulse Amplitude Modulation (PAM) The process used to take samples of analog voice signals so they can be multiplexed using TDM. The samples appear as pulses in the TDM signal.

Pulse Code Modulation(PCM) The industry standard method used to convert an analog signal into a digital 64,000-bps signal. PCM takes a PAM signal and converts each sample (pulse) into an 8-bit code.

Sidebands The two additional signals generated by mixing two signals. The upper sideband represents the sum of the carrier and modulating frequencies. The lower sideband will have a frequency that is the difference between the carrier frequency and the modulating frequency. Both sidebands contain the intelligence of the modulating signal.

Space Division Multiplexing A process in which multiple communications are placed over many different wire pairs inside one cable. Each communication channel (voice or data) occupies its own space (occupies its own set of wires).

Time Division Multiplexing (TDM) The process of converting each speech path into samples, then combining the different samples by transmitting each sample at a different time, so they may be sent over one transmission medium.

Wave Division Multiplexer The hardware device attached to each end of a fiber pair that multiplexes 16 OC-48 systems onto the fiber pair.

Wave Division Multiplexing (WDM) Technology that splits the light spectrum into many different frequencies. A time division multiplexer is then assigned to each of the different light wave frequencies. WDM is used to multiplex multiple OC-48 (or OC-192) systems over one fiber pair.

5

The Medium

Key Terms

Branch Feeder Cables
Carrier Serving Area (CSA)
Crosstalk
Distribution Cable
Drop Wire
Exchange Boundary
Facility
Feeder Cable
Jumper Wire
Load Coils

Loading
Local Loop
Loop Extender
Loop Treatment
Main Distributing Frame (MDF)
Main Feeder Cable
Modified Long Route Design
     (MLRD)
Multiplexer
Outside Plant

Outside Plant Engineer
Plastic Insulated Cable (PIC)
Resistance Design
Revised Resistance Design (RRD)
Subscriber Line Carrier-96
     (SLC-96)
Voice Frequency Repeater
     (VFR)
Wave Division Multiplexing
     (WDM)

Every communication system requires a medium connecting the transmitter to the receiver. Using wire as the medium was the only choice available when the telephone was invented. Wire continues to be the medium of choice for connecting the telephone to the local switching exchange. The medium of choice for long distance transmission was coaxial cable and microwave radio until the 1980s. Microwave radio uses air as its medium and carries signals composed of electromagnetic radiation. Fiber optic cable has now become the medium of choice for the long distance networks. This type of cable is made of glass strands and is used to carry signals composed of light energy.

Neither air nor glass will conduct electricity. Wire is an excellent conductor of electrical energy. This is why wire is used to carry electricity to our homes and is also why we use it to connect our telephones to the central exchange. The telephone receives its electrical energy power from the central exchange, and the signals traveling from the telephone to the exchange are electrical signals. At the central exchange, we can convert these electrical signals into light signals or electromagnetic waves and then use other mediums such as fiber optic or microwave (Figure 5-1).

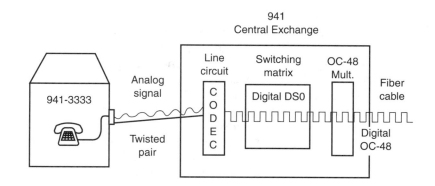

FIGURE 5-1 Analog phone line connected to a digital central exchange.

The wiring used to connect telephones in a home or business to the central exchange is called the *local loop*. This local loop will remain mostly twisted-pair copper wire for the next 10 to 20 years. Experiments are being conducted using fiber optic to serve as a local loop. Use of fiber to serve a telephone requires an expensive device to interface the telephone to the fiber, and the telephone must have a local power supply since it cannot receive power from the central exchange over the glass fiber. Cable television companies have begun to provide telephone service using their coaxial cables. Some LECs and IECs have established joint ventures with cable television companies to use their coaxial cable in some local service area locations.

5.1 THE LOCAL LOOP

The twisted-pair local loop is the weakest link and most costly portion of the telecommunications network. But describing the twisted-pair local loop as the weakest link does not mean it is easy to break. The term *weakest link* refers to the inability of the local loop to handle high-frequency signals. The twisted-pair local loop has a narrow-bandwidth. The other portions of the PSTN are composed of wide-bandwidth *facilities*. Voice and data travel across the PSTN on ribbons of fiber, which can handle many high-bandwidth signals, only to arrive at the local loop, which restricts communication to low-frequency analog signals. The heart of the PSTN contains fiber optic transmission mediums carrying high bit rate digital signals, but our connection to the PSTN is the low-bandwidth local loop. Think of the fiber optic transmission medium as a very large pipe and the local loop as a garden hose attached to the pipe. It is the size of the garden hose that determines how fast water can flow from a facet.

The twisted-pair local loop is like a small pipe or garden hose. It limits our communication to low-frequency audio signals. It is technically cost prohibitive to replace the existing wire cables with fiber optic cable and compatible phones in the local loop. To use fiber optic in the local loop, an interface device will be required at every phone to convert electrical energy into light waves. The device could be incorporated in the phone, but this would require customers to buy new phones. Twisted-pair copper wire has provided us with a low-cost medium to connect telephones to a central exchange, and the twisted-pair local loop has served as an excellent transmission medium for analog voice signals.

Increasing demand for the ability to place high-speed data signals over the local loop will eventually lead to the deployment of fiber optic facilities to replace all, or a portion, of the local loop. The first phase of fiber introduction into the local loop has been to use fiber as a main feeder cable to serve a segment of the exchange territory. *Multiple subscriber line carrier* systems (*SLC-96s*) are placed on a ***multiplexer*** in the central office. Each SLC-96 connects to 96 line circuits. Demultiplexers are placed at several locations along the fiber route and several SLC-96 field units are placed on the demultiplexer. The SLC-96 field units are powered by a local AC power source at the field terminal.

The SLC-96 field units convert the commercial AC power into DC voltage to power the telephone. Distribution wire cable pairs are connected from the SLC-96 units to telephones in the area. The field unit provides power over the wire pair to the telephone (Figure 5-2). Using fiber as a feeder cable only, allows customers to keep the telephones they currently have and does not require them to purchase an interface device. The telephone company may choose to take fiber directly to the house, but then the customer will be forced to buy new telephones or an interface device.

Originally both the local loop and toll loop were constructed using wire as a facility. At the turn of the century, plastic had not been invented. Since plastic insulated

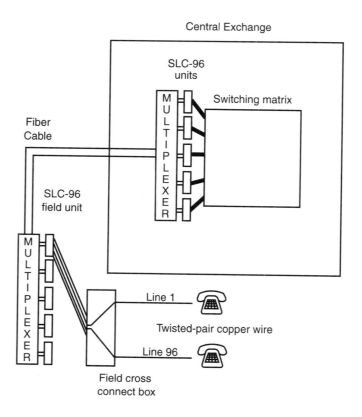

FIGURE 5-2 SLC = 96 units over a fiber.

wire was not available, it was necessary to string individual strands of wire on telephone poles and attach them to the poles with glass bulb insulators. This type of wiring is called *open wire*. Two wires were needed for each telephone line circuit. Iron wire was the wire of choice for the local loop because of its strength and low cost. Because iron is not as good a conductor of electricity as copper, a thicker wire was needed when using iron wire. The wire was a little thicker than the wire used on a heavy-duty clothes hanger. Copper wire was the wire of choice for trunk circuits between central exchanges. In the telephone industry, the engineering and construction of these facilities are the responsibility of **outside plant engineers,** located in the *outside plant engineering* department.

When more than one circuit was needed, a crossarm was added to the pole, making it look like a cross. A crossarm could support ten wires (five circuits). For every five circuits additional crossarms were added. As the number of telephones grew, the local plant in large cities became a mess (Figure 5-3). In small rural communities, open wiring systems were small. Usually, there were up to ten telephones on each line. Most residences in a small town were clustered together close to the central office. To serve the farms in the country required only a few lines leaving the central exchange in each direction.

Because only a few lines were needed to serve a small rural exchange with ten-party lines, open wire served many rural areas until the late 1960s. During the 1960s, most independent telephone companies aggressively pursued the elimination of party-line service. Party-line elimination was done in phases. First all party lines were reduced from ten- to four-party lines. A few years later, these lines were reduced to two-party lines and private lines. To convert a ten-party line to ten private lines required nine more pair of wires from the central office to the area converted. Party-line reduction and elimination required conversion of the outside plant from open-wire to aerial and buried cable.

5.2 INTRODUCTION OF PLASTIC INSULATED CABLE

With the development of insulated wires, it is possible to put many wires in one cable and they will be electrically isolated from each other by the insulation surrounding each wire. With the introduction of insulated wires, copper wire was chosen over iron. Copper conducts electricity better than iron. Choosing copper, instead of iron wire allows the use of a smaller-diameter wire. Smaller wire allows more wires to be placed inside one cable.

The first cables were made with copper wires coated with shellac. Each shellac-coated wire was then wrapped with paper to insulate the wires in a cable from each other. The wires were grouped together and surrounded with a lead covering to protect them from damage. Lead-covered cables served the local loop until the introduction of *plastic insulated cable (PIC)*. Each wire is now covered with a colored plastic material. The colors of the plastic can take on 50 different combinations of hues. This makes it possible to have a 25-pair cable with each wire in the cable identified by a unique color scheme (Figure 5-4).

FIGURE 5-3 New York City 1890—Open wire used for outside plant.

The Medium   Chap. 5

(a)

(b)

FIGURE 5-4 Plastic
insulated cable.

5.2   Introduction of Plastic Insulated Cable   123

5.3 PLASTIC INSULATED CABLE COLOR CODE

The colors used in PICs are blue, orange, green, brown, slate, white, red, black, yellow, and violet. The white, red, black, yellow, and violet colors are referred to as *mate* (or *primary*) colors. The blue, orange, green, brown, and slate colors are called *secondary* colors. The color of plastic around each wire will be composed of two colors: a mate color and the secondary color. The Tip wire of a pair of wires will have a wide band of mate color and a narrow band of secondary color. The Ring wire will have a wide band of secondary color and a narrow band of mate color.

The cable pair designated as pair 1 in a cable is composed of one wire that is predominately white with a narrow band of blue (this is the Tip lead of pair 1), and one wire that is predominately blue with a narrow band of white (this is the Ring lead of pair 1). The colors of pair 1 are referred to as white-blue and blue-white. White-orange and orange-white are the colors for pair 2. The mate color of white is used with the secondary colors of green, brown, and slate for pairs 3–5. Pairs 6–10 use red as the mate for blue, orange, green, brown, and slate. Black is the mate color for blue, orange, green, brown, and slate on pairs 11–15. Yellow is the mate for pairs 16–20, and violet is the mate for pairs 21–25 (Table 5-1).

Table 5-1 PIC COLOR CODE

Mate	Color	Pair Number in Cable	Tip Wire	Ring Wire
White	Blue	1	White-blue	Blue-white
Red	Orange	2	White-orange	Orange-white
Black	Green	3	White-green	Green-white
Yellow	Brown	4	White-brown	Brown-white
Violet	Slate	5	White-slate	Slate-white
		6	Red-blue	Blue-red
		7	Red-orange	Orange-red
		8	Red-green	Green-red
		9	Red-brown	Brown-red
		10	Red-slate	Slate-red
		11	Black-blue	Blue-black
		12	Black-orange	Orange-black
		13	Black-green	Green-black
		14	Black-brown	Brown-black
		15	Black-slate	Slate-black
		16	Yellow-blue	Blue-yellow
		17	Yellow-orange	Orange-yellow
		18	Yellow-green	Green-yellow
		19	Yellow-brown	Brown-yellow
		20	Yellow-slate	Slate-yellow
		21	Violet-blue	Blue-violet
		22	Violet-orange	Orange-violet
		23	Violet-green	Green-violet
		24	Violet-brown	Brown-violet
		25	Violet-slate	Slate-violet

If a cable contains more than 25 pairs, the first 25 are grouped together and wrapped with two plastic strings. One string is white and the other is blue. The second 25 pairs use the same color codes for the wires, but these 25 pairs are wrapped with a white-and-orange string. These plastic strings are called *binders*. Using the same scheme used in coloring the wires, it is possible to have 25 different binders from white-blue to violet-slate. Twenty-five binders, with 25 pairs each, makes it possible to have 25×25 or 625 pairs in a given cable. Larger cables are constructed by wrapping from 100 to 625 pairs in a white-blue binder, the next 100 to 625 pairs in a white-orange binder, and repeating the same process over and over.

Insulated cables have replaced iron wire and glass insulators on poles. This significantly improved not only the looks of the *outside plant* but also reduced maintenance. The next major improvement in outside plant was to remove the cable from elements of weather by burying the cable. Buried cable requires little maintenance. Most damage to buried cables comes from excavation activity when a road is being widened, a new water system placed, cable for television being placed, or new telephone cable being buried. Not surprisingly, the telephone company itself does a lot of damage when it buries a new cable. Cable cuts can be prevented by calling the telephone company a few days ahead and asking it to locate and mark the location of its cables. In many localities, all utilities have one number listed for cable locate call-ins. A contractor need only call one number and all utilities will locate and mark their buried facilities.

5.4 OUTSIDE PLANT RESISTANCE DESIGN

When designing outside plant, the engineer must provide a wire thick enough to provide proper power to the telephone. The older telephone transmitter needs 23 milliamps of current to work properly. With a central office battery supply of 48 V, SXS central office equipment was originally designed to work with a maximum loop resistance of 1000 Ω. Later, the SXS design was improved to handle loops up to 1200 Ω. The outside plant engineer would provide the thickness of wire needed to ensure that the resistance of the longest local loop did not exceed 1000/1200 Ω. Designing the local loop so that it does not exceed a stated resistance value for the longest loop is referred to as *resistance design*.

5.5 SXS DESIGN CRITERIA: 27 MILLIAMPS OF LOOP CURRENT

In the *Strowger automatic step-by-step* switching system, the relay that recognized when the called party answered (called a *Ring trip relay*) required 27 milliamps of current to operate. Each telephone had approximately 400 Ω of resistance, and the Ring trip relay had 350 Ω of resistance. The electrical circuit of the longest local loop consisted of 48-V DC applied from the negative terminal of the central office battery, through the Ring trip relay (350 Ω), over the Ring lead of the local loop (500 Ω), through the telephone (400 Ω), back over the Tip lead of the local loop (500 Ω), to the positive battery terminal. Total resistance in this series electrical circuit is 1750 Ω (350 + 500 + 400 + 500).

The flow of electricity in a circuit depends on the amount of voltage applied to the circuit and how much opposition or total resistance the circuit has to the flow of electricity. This is expressed as a formula called *Ohm's law*. Current is equal to total voltage applied to a circuit divided by the resistance of the circuit (current in amps = volts / ohms). In the local loop above, the current in the loop will be equal to: 48 V / 1750 Ω = 0.02748 amps. This is 27.48 milliamps and satisfies the requirements of the Ring trip relay in the central exchange. Of the 1750 Ω total resistance, 1000 was in the local loop.

5.6 TRANSMITTER DESIGN CRITERIA: 23 MILLIAMPS AND 20 MILLIAMPS OF LOOP CURRENT

Over time, improvements in the design of the central office and telephone set have allowed the use of longer loops. Central exchange Ring trip relay design in electromechanical switching systems was improved so the relay could work on 20 milliamps of loop current. This would allow the relay to work on loops as long as 1600 Ω, but the transmitter of the telephone required 23 milliamps of loop current and limited the loop to 1200 Ω. Engineers worked on the carbon granule transmitter and developed an improved transmitter that will work with 20 milliamps of current. This development took place about the same time that computerized *Stored Program Controll (SPC)* switching systems were introduced. A Stored Program Controll switching system is supplied power from a 52-V battery. The combination of higher voltage at the central office and the lower 20-milliamp transmitter current requirement allows the longest loop of a SPC exchange to be 1800 Ω under resistance design.

In today's environment the amount of current needed by central office equipment is no longer the requirement governing outside plant design. Electronic, SPC, central offices are designed to work with electrical currents of less than 20 milliamps. SPC switching equipment manufacturers will tout designs that will work on loops as long as 1700 and 1800 Ω when the telephone is equipped with a 20-milliamp transmitter.

The requirements of the telephone transmitter have become the governing factor in outside plant design. The Bell LECs engineer loops based on the newer 20-milliamp requirement. Most rural telephone companies engineer to the old transmitter requirement of 23 milliamps. When calculating resistance for a local loop in a SPC environment, a 52-V battery supply is used for calculations. Ohm's law can be used to find the maximum resistance possible at 20 milliamps, using a 52-V battery supply, as follows: Resistance = applied volts / amps of current needed. Therefore; 52 V / 0.020 amp = 2600 Ω. Allowing 400 Ω for the line relay at the central exchange and 400 Ω for the telephone will leave 1800 Ω (2600 − 800 = 1800) as the longest local loop we can use for a SPC exchange. Using a telephone with the newer 20-milliamp transmitter and a central office battery feed of 52 V allows the design of a longer loop (1800 Ω). This is the criterion used by manufacturers when they state that their switching system will support a local loop of 1800 Ω.

5.7 REVISED RESISTANCE DESIGN

The design concepts for outside plant have been revised, and outside plant engineers now use a concept called *revised resistance design (RRD)* to engineer the local loop. Under revised resistance design, the engineer uses the length of a local loop to determine how the local loop will be designed. Any loop up to 18,000 ft is engineered with nonloaded cable for a maximum loop resistance of 1300 Ω. Loops between 18,000 and 24,000 ft are engineered with loaded cable for a maximum loop resistance of 1500 Ω. Loaded cable is cable that has additional inductance added to the cable to improve its ability to handle voice signals on long loops. The technique used to load a cable is covered in Section 5.14.

5.8 MODIFIED LONG ROUTE DESIGN

Loops longer than 24,000 ft are engineered using *modified long route design (MLRD)*. MLRD involves placing *range extenders* (also called *loop extenders*) and voice frequency repeaters (VFR) on loaded cable pairs at the central office. The use of these devices is referred to as *loop treatment*. The loop treatment is wired between the line circuit of the switch and the cable pair of the local loop (Figure 5-5). The range extender boots the voltage supplied to the line from 52 V to either 78 or 104 V. The amount of voltage boost to use depends on the length of the cable pair served by that

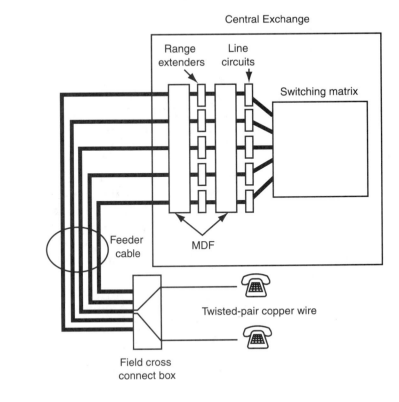

FIGURE 5-5 MLRD loop treatment.

particular loop extender. This voltage boost will ensure that the transmitter and DTMF pad receive adequate current. It also ensures enough current for the signaling requirements of the central office equipment. The VF gain circuit is used to amplify voice frequency (VF) signals to ensure the overall power loss of the circuit is between 4 and 8 dB. In rural areas, some telephones are so far from the office that both loop treatment and 19-gauge wire are used to serve the telephone.

MLRD can be utilized instead of RRD for loops between 18,000 and 24,000 ft. When these longer loops are designed according to RRD, 19- or 22-gauge cable will be required to meet the resistance requirement of 1500 Ω. MLRD can be employed on these loops to allow the use of smaller wire (such as 24 or 26 gauge) to serve the same area. Using the smaller wire with a range extender on the line is cheaper than using a thicker wire to serve the same telephone. The range extender is an electronic device that enables a switch to handle local loops exceeding the design requirements of the central office switching system. A range extender can effectively double the range of a SPC central office from 1800 to 3600 Ω.

5.9 THE SERVING AREA OF A CENTRAL OFFICE

The serving area of a central office depends on the geographic area it is situated in. The serving area of a central office in an urban area is smaller than the area served by a rural office. Exchanges (central offices) are often called *wire centers*. One design objective in metropolitan areas has been to limit the physical size of the wire center by limiting the number of wires (cable pairs) served by a center. All cable pairs terminate in a central office on the **main distributing frame (MDF)**. The more wire pairs coming into the exchange, the longer the MDF must be to accommodate all pairs.

The MDF is designed so that cables can be wired to terminal blocks mounted on the front and back sides of the MDF. Terminal blocks are mounted vertically on the back side of the MDF and horizontally on the front side of the MDF. All outside plant cable terminates on the vertical side of the MDF. All central office equipment (line circuits and trunk circuits) are wired to the horizontal side of the MDF. A short length of wire (called a *jumper wire*) is used to connect a cable pair to a line circuit. The use of a jumper wire allows the connection of any cable pair to any line circuit.

When an MDF serves more than 20,000 lines, many jumpers become very long. With many jumpers lying alongside each other, the possibility for *crosstalk* between adjacent jumper wires becomes a problem. Therefore, most central office exchanges are equipped to serve 10,000 lines, since this is the quantity of telephone numbers available to the *NNX* code assigned to an exchange. Large metropolitan areas, such as Wall Street, will have more than one *NNX* code assigned in the same switching center due to the sheer volume of telephones concentrated in a small geographic area. In this case, the switching center may serve two to five exchange codes and 20,000 to 50,000 lines.

Metropolitan areas have a high population density per square mile, and it is easy to have 10,000 people in an area of 9 to 25 sq mi. Central office exchanges in large cities are generally placed about 4 to 6 mi from each other. Thus, each exchange has a serving radius of 2 to 3 mi in a large city. This means all cables are nonloaded and designed for 1300-Ω loops. Exchanges in large cities are designed to serve a small geographic area to keep the quantity of lines served by each exchange to a manageable level and to keep the length of the longest line served under 3 mi. Placing central offices within 6 mi of each other meets this requirement. All loops are shorter than 18,000 ft and 26-gauge nonloaded cable can be used for all loops. Twenty-six-gauge wire is much cheaper than the coarser gauges (22, 24, and 19 gauge).

It is cheaper to use more exchanges and smaller wire in the loop than to use fewer exchanges and coarser cable for the long loops. Theoretically, it is possible to place one central exchange in the heart of a city and have it serve all telephones within a 30-mi radius using MLRD and 19-gauge cable for the longest loops. But the cost of cable would be tremendous. A 19-gauge wire contains almost five times as much copper and costs almost five times as much as a 26-gauge wire. The sheer number of wires coming into the exchange would be unmanageable. A much better design is to use many switching systems scattered around town and to establish small serving areas for each exchange (Figure 5-6).

Cities have a high density of population per square mile. When the city is subdivided into smaller exchanges, each exchange still serves thousands of telephones. The investment in the switching equipment of the central exchange can be spread over a large base of customers. Rural areas have a low population density per square mile. The size of a serving area for rural central office must be large enough to serve several hundred people in order to reduce the switching system cost per customer. This requirement means a rural exchange must serve a large area. We try to get as many lines as possible into each exchange. This reduces the capital expenditures needed for buildings and central office equipment (Figure 5-7). Designing the outside plant for a maximum loop resistance of 1500 Ω, using RRD, will allow for an exchange to serve telephones that are 17 mi away if 19-gauge wire is used. Nineteen-gauge wire has a resistance of about 42.5 Ω per mile (8.04 Ω per 1000 ft.) at 68° F. Since two wires are needed for the loop, the *total resistance for a two-wire loop using 19 gauge wire is about 85 Ω per mile.*

5.10 RESISTANCE DESIGN: DETERMINING THE GAUGE OF WIRE TO USE

The size of wire is referred to as the *gauge* of the wire. The larger the number, the smaller the diameter of the wire. A 22-gauge wire has a diameter of 0.025 in. The resistance of a wire varies inversely with the cross-sectional area of the wire. The cross-sectional area of a wire is cut in half when the gauge of the wire goes up by 3. A 22-gauge wire has half the cross-sectional area of a 19 gauge wire. Since a 19-gauge wire is twice as large as a 22-gauge wire, the 19-gauge wire has half the resistance of a 22-gauge wire. Twenty-two gauge copper wire has a resistance of 16 Ω per 1000 ft, and 19-gauge copper wire has a resistance of 8 Ω per 1000 ft. The resistance will vary slightly as the temperature increases or decreases. Twenty-two, twenty-four, and twenty-six gauge wire are the gauges used predominantly in the local loop.

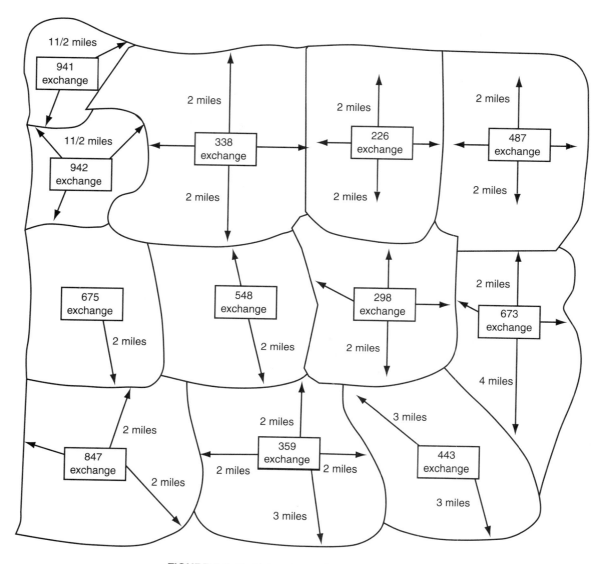

FIGURE 5-6 Typical exchange boundaries in a city.

Since two wires are needed in a local loop when using 22-gauge wire, the loop will have a resistance of $2 \times 16 = 32\ \Omega$ per 1000 ft or approximately $170\ \Omega$ per mile. Earlier, it was determined that the local loop of a SPC exchange cannot have a resistance greater than $1500\ \Omega$ using RRD. At $170\ \Omega$ per mile, using 22-gauge cable, we can go approximately 8.8 mi in all directions from the central office. The use of 22-gauge cable allows one central office to serve all telephones within 8.8 mi. Lines longer than 8.8 mi will receive loop treatment.

The total resistance for a two-wire loop using 19-gauge wire is about $85\ \Omega$ per mile. A loop using 22-gauge wire has a resistance of about $170\ \Omega$ per mile. Going up on the gauge of a wire by 3 $(22 - 19 = 3)$ doubles the resistance of the wire. The

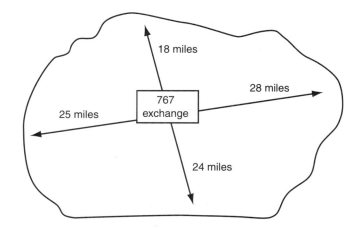

FIGURE 5-7 Exchange
boundaries in a rural area.

larger the gauge, the smaller the wire. A local loop using 24-gauge wire has a resistance of about 275 Ω per mile (26 Ω per wire per 1000 ft.), and *a loop using 26-gauge wire has a resistance of about 440 Ω per mile* at 68° F. A local loop constructed of 26-gauge wire is limited to serving a 3-mi radius. This is another reason that the serving area for a metropolitan office has a 3-mi radius. It allows the use of 26-gauge cable in the local loop. The smaller the wire (the higher the gauge number), the cheaper the cable. LECs can save a tremendous amount of money by using the smallest-diameter wire possible in the local loop.

A rural exchange can also use nonloaded 26-gauge cable pairs to serve customers within 3 mi of the exchange, but will require expensive 19-gauge loaded cable to serve a 17-mi radius. These long loops could be served with range extenders and VF gain devices (MLRD) to allow the use of 22-gauge cable. RRD could be used on loops that are 1500 to 2400 Ω and MLRD could be used on the loops between 2400 and 2800 Ω. The engineer selects the design that will result in the lowest cost to install. This will depend on the number of lines involved. Nineteen-gauge loaded cable can be used with range extenders and VF gain to serve a 34-mi loop. This effectively doubles the range of an 1800 SPC exchange to 3600 Ω. Outside plant engineers usually limit the maximum loop resistance of the longest loop to 2800 Ω. By using 19-gauge cable and loop extenders, a 2800-Ω loop can extend 32 mi from the central exchange. The use of 22-gauge cable (with loop treatment) will cut the serving distance in half (to 16 miles). The maximum distance that can be covered, using range extenders and VF gain, with loaded 24- or 26-gauge cable is 10.2 and 6.4 mi respectively. The outside plant engineer can select RRD or MLRD to serve each area of an exchange. The use of 22-gauge loaded cable will allow serving all lines within 8.8 mi of the exchange without the use of loop treatment (that is, range extenders and VF gain). Going beyond 8.8 mi requires the use of 19-gauge cable or design of the loop according to MLRD.

Another option available to the outside plant engineer is to use a subscriber carrier system for serving a particular area of the exchange. Subscriber carrier will require the use of two nonloaded cable pairs with repeaters placed every mile along the route. Subscriber carrier is available to serve from one line to hundreds of lines. Most subscriber carrier systems are designed to accommodate 96 lines and are called *subscriber line carrier–96 (SLC-96)*. A SLC-96 can be placed as far from an exchange as desired.

Subscriber carrier is often used to provide service to a rapidly growing part of the exchange serving area. If a new housing development springs up in an area that has only a few vacant cable pairs, subscriber carrier can be used to provide service until new cable can be added to the area or until the area can be served permanently with subscriber carrier. As mentioned early in this chapter, many SLC-96 systems can also be multiplexed over a fiber facility to serve a remote area of the exchange (Figure 5-2).

5.11 CARRIER SERVING AREA

So far we have discussed the serving area of a central office. The serving area for each central office is established based on the longest loop that the engineer wishes to have the exchange handle. The outermost customer establishes the *exchange boundary* for a central office. Another term used in outside plant engineering is the *carrier serving area (CSA)*. This is a distant area of the exchange that can support access to DS0 digital service and ISDN without special loop treatment. A CSA is served by a subscriber carrier system. A CSA cannot have any loops longer than 12,000 ft, from the subscriber carrier field terminal, using 24-gauge or coarser wire. If 26-gauge cable is used, the loop cannot be longer than 9000 ft. All loops must use nonloaded cable pairs.

Establishment of an area distant from the central exchange as a CSA is done through the use of subscriber carrier. The design rules for CSA apply to the cable pairs extending out from the subscriber carrier to telephones (Figure 5-8). The CSA requirements are also met by all lines within 2 mi of the central office wire center, but these lines are not part of a CSA because they are not served by subscriber carrier. These lines will support digital services and are administered the same way as lines in a CSA.

5.12 VOICE SIGNALS (AC) IN THE LOCAL LOOP

The resistance of wires limits how far a phone can be placed from a central exchange. We have stated earlier that the new carbon transmitter needs 20 milliamps of current to work properly. Using Ohm's law to find voltage, we can calculate the amount of voltage at the phone: volts = current × resistance. Thus, volts = 0.020 amp × 400 Ω (resistance of the phone) = 8 V. We apply 52 V DC to a line at the central office. As current flows through the line relay and local loop, we lose power and less voltage is available at the phone than is applied at the central exchange.

Resistance design of the outside plant ensures that the DC power loss over the local loop is within limits. But DC power loss is not the only factor to consider when using wire as a transmission medium. Voice signals are converted by the transmitter of the phone into variances in electric current. These variances look like AC signals. If

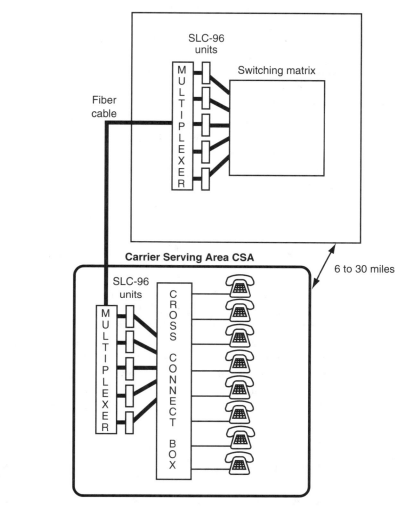

Central Exchange

SLC-96 units

Switching matrix

Fiber cable

MULTIPLEXER

6 to 30 miles

Carrier Serving Area CSA

SLC-96 units

MULTIPLEXER

CROSS CONNECT BOX

FIGURE 5-8 Carrier serving area.

a local loop has 30 milliamps of current flowing through it when no one is talking, the current will vary when someone talks through the transmitter. The current changes may be 29.95, 30.05, 29.96, 30.04, or almost any value between 27 and 33 milliamps. The varying DC signal looks like an AC signal that has a center point of 30 milliamps.

5.13 CAPACITIVE REACTANCE CONSIDERATIONS

Wire has properties other than resistance that must be considered when an AC signal is placed over the wire. A wire pair acts like a capacitor. A capacitor is made by separating two conductors with an insulating material. The two wires of our local loop (Tip and Ring) are very close together in a cable and are separated by the plastic insulation around each wire. The wires of the local loop have a capacitance of approximately 0.083 microfarad (µf) per mile. The longer the loop the greater the capacitance. When an AC signal is placed on a wire pair, the capacitance between the two wires will affect the signal. The capacitance between the two wires allows some of the voice signal to leak from one wire to the other. The amount of signal leakage that occurs is governed by the amount of opposition the capacitance between the two wires offers to the signal. Capacitive reactance measures the opposition that a capacitor has to an AC signal.

Capacitive reactance is inversely related to the frequency of an AC signal, and inversely related to the amount of capacitance between the wire pair. The higher the frequency of an AC signal, the lower the opposition of a capacitor to the signal. The higher the capacitance a device has, the less opposition it offers to an AC signal. The lower the frequency of a signal the higher the capacitive reactance. This is why a capacitor blocks a DC signal. The frequency of a DC signal is 0 and opposition to the signal is maximum. The longer the cable pair, the higher the capacitance will be. The longer a local loop is, the more signal loss we have due to leakage through the capacitance between the wires.

On local loops longer than 3 mi, capacitance of the cable pair becomes great enough to degrade a voice signal. Voice signals consists of frequencies between 300 and 3300 cps. On loops longer than 3 mi, capacitive reactance causes a significant power loss to signals above 1000 cps. The higher the frequency, the greater the loss. This effect causes the level of signal heard at the called party's receiver to go up and down as the frequency of the signal changes. This can be annoying. To prevent this from happening, more inductance is purposely put into the local loop to offset the effects of capacitive reactance. Inductive reactance and capacitive reactance tend to offset each other. The extra inductance is designed into a device called a load coil.

5.14 LOADING CABLE PAIRS TO IMPROVE VOICE TRANSMISSION

Cable pairs shorter than 3 mi do not require *loading*. For cable pairs longer than 18,000 ft, *load coils* are used. What type of load coil to use and how far apart to place them depends on the capacitance between the wires and the gauge of the wire. The type and thickness of insulation used will determine the capacitance between a pair of wires. This capacitance between the two wires (Tip and Ring) of a line circuit is called *mutual capacitance*. Cables manufactured for the local loop have 0.083 µf of capacitance per mile regardless of the gauge of wire in the cable. Some cables are designed for special high-frequency applications, such as for use with T1 carrier systems. These cables have a capacitance of 0.066 µf per mile.

Most loaded local loops will use 22- and 19-gauge wire. If a loop needs to be loaded, it is longer than 3 mi. Loops longer than 3 mi require the use of 22-gauge wire. Loops longer than 8.8 mi require the use of 19-gauge wire, or the use of loop extenders. Regardless of the gauge of wire used, the mutual capacitance is 0.083 µf. per mile. To offset this mutual capacitance, load coils having 88 millihenries (mh) of inductance are placed at 6000-ft intervals on the cable. The use of 88-mh load coils results in a cutoff frequency of approximately 3800 Hz. The cutoff frequency of a circuit is the frequency at which 70.7 % of the input voltage reaches the output of the circuit. A 70.7% reduction in voltage level is a change of 3 dB. Notice in Figure 5-10 that the signal has experienced a drop of 3 dB (from -3 to -6 dB) at approximately 3800 Hz.

The cutoff frequency of the local loop can be raised to 7400 Hz by using 44-mh coils and placing them at 3000-ft intervals. In other words, use half the inductance of an 88-mh coil and place twice as many coils on the loop. Therefore, 44-mh loading costs twice as much as 88-mh loading. Almost all loops are loading with 88-mh load coils. The use of 44-mh load coils is reserved for use on loops that must be specially conditioned to handle frequencies between 3800 and 7400 Hz. Since most local loop uses 88-mh load coils, loaded loops can not effectively pass signals above 3800 Hz.

Spacing load coils 3000 ft apart is called *B spacing*, spacing 4500 ft apart is *D spacing*, and 6000-ft spacing is *H spacing*. A load coil designated as 19H88 means it is an 88-mh coil designed for use on 19-gauge cable and 6000-ft spacing. When the outside plant engineer determines that cables serving a particular area are going to be more than 3 mi from the central exchange, the cable is loaded. The first load coil will be placed 3000 ft from the central office. The second load coil is placed 9000 ft from the central office, and a third load coil, if needed, is placed 15,000 ft from the central office. The first load coil is placed at half the recommended distance for the coil being used.

Load coils are enclosed in one case and connected to two-cable stubs that will protrude outside the case. When a cable needs to be loaded, the cable is cut. The load coil is spliced in between the end of the cable going to the exchange and the end of the cable that goes to telephones. To load a 200-pair cable, a load case containing 200 coils will be attached to a pole, for aerial cable, or placed in a pedestal, for buried cable, and then spliced into the cable. Note that the first load coil is placed 3000 ft from the central exchange. When a telephone call is established between two lines of the exchange, this results in a 6000-ft spacing. The first load coil of each line is 3000 ft from the exchange. When the exchange connects the two lines, these two load coils are 6000 ft apart. Thus, even though the first load coil is placed 3000 ft from the exchange, an H-88 load coil is used for the first load coil, because the spacing, between the two load coils of a connected call will be 6000 ft (Figure 5-9).

Loading a cable flattens out the power loss for all signals below 3000 cps (Figure 5-10). Signals between 3000 and 4000 cps will have increasing power losses, and signals above 4000 cps will have so much power loss that the signal will be hard to hear. To achieve maximum transmission, the telephone should not be closer than 3000 ft, or more than 9000 ft from the last load coil serving the telephone.

The single biggest disadvantage of using wire as a medium is the narrow bandwidth of wire due to mutual capacitance and loading. It can only carry signals having frequencies between 0 and 3800 cps efficiently. Data signals are far higher frequencies and cannot be carried over cable unless they are regenerated every mile. Putting regenerators on every local loop would be very costly. It is possible to convert voice into a digital signal at the phone. An ISDN phone does exactly that. ISDN phones do not work more than 3 mi from the exchange, unless the ISDN circuit is being provided over subscriber carrier or a specially conditioned cable pair.

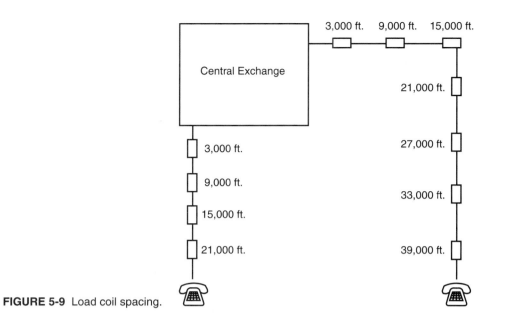

FIGURE 5-9 Load coil spacing.

Load coils improve the low-frequency response of a cable but add more loss to high-frequency signals. When the outside plant engineer knows a particular cable pair is going to carry high-frequency signals, an order is issued to take the load coil off of that pair, and the pair is spliced together without going through the load coil case. If the cable pair is going to be used for a digital signal, a signal regenerator is placed between the two pairs of wire at the point where the load coil is removed.

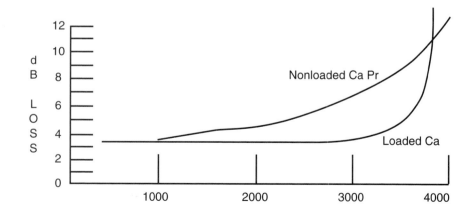

Frequency of Signal on Ca Pr

FIGURE 5-10 Signal loss comparison: loaded versus nonloaded cable.

5.15 DATA ON THE LOCAL LOOP

The local telecommunications network was designed to handle voice frequencies. Data can be sent over specially designed local loops and central office equipment. When a business needs to transmit high volumes of data, it will lease these special facilities and the local loop will be conditioned with special equipment such as equalizers and amplifiers. Low volumes of data can be sent over regular phone lines and facilities by using a device called a *modem*. The modem is placed between the digital output of a computer and the analog phone line. The modem converts digital signals received from the computer into analog signals between 300 and 3300 cps. These signals are handled by the telecommunications network with no problem (Figure 5-11).

When modems are used, they must be used in pairs with one modem at the transmitting end and another at the receiving end. A modem at the receiving end converts the received analog signals into the appropriate digital signal for the computer on the receiving end. Digital data signals can use wire for transmission but only for a short distance. If regular telephone wire (unshielded twisted pair) is used for data, the signals must be regenerated every 1 to 3 mi depending on the speed of data transmitted. *Integrated Services Digital Network (ISDN)* lines use digital subscriber line circuits, which will work up to 3 mi. The customer supplies an ISDN device and attaches it to the digital line. If digital data lines are leased from the LEC, the customer must provide a digital interface device instead of a modem. The digital interface device for data is called a *customer service unit (CSU)* or *data service unit (DSU)*.

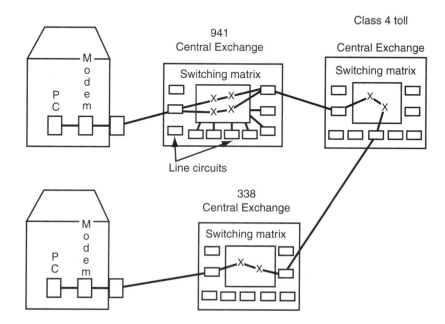

FIGURE 5-11 Modems allow the connection of a data circuit to the Public Switched Telephone Network.

5.16 POWER-LOSS DESIGN

The amount of power loss that any signal has over a medium is measured in decibels (dB). Decibels are used to compare two power levels that vary in a logarithmic fashion. A change in power level barely noticeable to the human ear is 1 dB. A decibel measures the relationship between two power levels; therefore, a decibel is not an absolute measurement. If one signal has twice the power of a second signal, the difference between the two signals is 3 dB. The difference between a 1-mW signal and a 2-mW signal is 3 dB.

Likewise the difference between a 1-MW signal and a 2-MW signal is 3 dB. In the first case, a change of one millionth of a watt is all that is necessary to double the power, but in the second case, a million watts is needed to double the power. In both cases the change in power is a change of 3 dB. The formula for calculating dB is: $dB = 10 \log P_2 / P_1$, where power = P. Anytime P_2 is twice the value of P_1, the value of $P_2 / P_1 = 2$. This represents a doubling of power. The log of 2 is 0.3010 and 10 times 0.3010 equals 3.01. Thus, according to the formula, 3 dB equals a doubling of power. Every 3-dB change is a double of power. A change from 1 to 2 mW equals 3 dB. A change from 2 to 4 mW equals 3 dB.

To provide an absolute type of power measurement in telecommunications, all signals are measured with respect to a signal level of 1 mW when a test set is used. The test set used to measure a signal compares the measured signal to a 1-mW reference level. Decibels measured with reference to 1-mW are called *dBm*. A signal of 0 dBm = 1 mW. A signal of 3 dBm equals 2 mW. A signal of 6 dBm equals 4 mW, and so on. When a signal is measured it will result in a measurement expressed as dBm. If the signal ratio is a calculation rather than a measurement, it is stated simply as dB.

For example, the change of a signal from 4 to 8 mW would be a change of 3 dB. If the signals were measured, the measurement at the 4-mW point would be 6 dBm, and the measurement at the 8-mW point would be 9 dBm, since both are measured with respect to a power level of 1 mW. A gain in power is stated as positive dB and a loss in power as negative dB. Note that going from 4 to 8 mW is +3 dB, and going from 8 to 4 mW is -3 dB. The formula also produces the same result: $dB = 10 \log P_2 / P_1 = 10 \log 4/8 = 10 \log 0.5$. The log of 0.5 = –0.3, and 10 times –0.3 = –3. If the power of one signal is half the power of another signal, the relationship between the two is –3 dB. Thus, doubling the power is 3 dB and half the power is –3 dB.

For the local loop, this power loss over the wire between the central exchange and the telephone should not exceed 8.5 dB with a signal that is 1000 cps. The loss at 500 and 2700 cps should be within 2.5 dB of the loss measured at 1000 cycles. In each central office a test number is assigned to a piece of equipment that will generate a test tone of 1004 cps for a few seconds, then transmit 500 cps for a few seconds, and then transmit 2700 cps for a few seconds. Each test signal is generated at 0 dBm (1 mW). After the 2700-cps signal has been sent, the circuit becomes quiet until the device is released.

A repair person can call into the test number with a special test set. The repair person will connect the test set to the Tip and Ring of a local loop at the customer's house. The test set can measure loop current from the central office to determine if it is at least 23 milliamps. The repair person can dial the test number to measure loss at the various frequencies. After the test number has sent all three tones and is quiet, the test set can be used to measure noise on the line. Noise should not be more than -60 dBm (30 dBrnC0). As noted earlier, dBm is decibels as referenced to a power level of 1 mW (0 dBm = 1 mW). Also, dBrnC0 is decibels as referenced to noise with C message weighting. Zero (0) dBrnC0 is equal to -90 dBm (90 dB less than 1 mW). This level (0 dBrnC0) is set as the starting (or reference) point for noise measurements.

We used the dB formula in Chapter 3 to find the power level at -90 dBm: dB = $10 \log P_2/P_1$. Rearranging the formula results in: antilog (dB/10) $\times P_1 = P_2$. Substituting known numbers results in: Antilog (-90/10) $\times$ 1 mW = P_2; antilog -9 $\times$ 0.001 = P_2; the antilog of –9 is 0.000000001 and the antilog of -9 $\times$ 0.001 W = 0.000000000001 W or 1 pW. Thus 1 pW equals -90 dBm or 0 dBrnC0. This is the starting point for noise measurements. Noise should not exceed 30 dBrnC0 on the local loop. This limit of 30 dBrnC0 is 1 nW of noise. It is necessary that noise be below this extremely low level because the power of the voice signals is extremely low (less than 1 mW).

Noise on the local loop can usually be traced to an unbalanced line, crossed lines, induction, or a ground fault. These problems are often caused by water entering a cable and damaging the insulation around the wires. Unbalanced lines can also be the result of a poor splice where two cables are joined. When party lines existed, they used grounded ringing, which caused unbalanced lines.

Induction noise results when the aluminum sheath surrounding all wires in a cable has been destroyed or improperly grounded. This can allow noise induction from power lines when a telephone cable is mounted on a power-line pole. Induction between wires within a cable would be a problem if the wires of a pair were not twisted around each other. When one wire is parallel and close to a wire carrying an electrical signal, it acts like the secondary winding of a transformer and will pick up the signal of the current-carrying wire. Twisting the wire pair prevents this source of crosstalk. The amount of twist per inch to use on cable pairs has been scientifically determined.

When current leaves the central exchange it flows out over the Ring wire and returns over the Tip wire. Twisting will first cause the Ring lead of one pair to lie next to an adjacent wire, then the Tip will lie next to the adjacent pair. Any induced signal in the adjacent wire from the Ring lead will be offset by the signal induced in an opposite direction by the Tip lead. Of necessity wires must lie close to each other in order to have many wires in one cable. Twisting the wire pairs and then twisting the pairs around each other eliminates induction and enables wire pairs to be placed close togather. Twisting also eliminates the inductive effect between the two wires of the same pair (Figure 5-12).

5.17 FEEDER CABLES, DISTRIBUTION CABLES, AND DROP WIRES

The cables that leave a wire center (central exchange) are called *main feeder cables*. These are large cables containing 1800 to 3600 pairs of wires. The *feeder cable* has a

FIGURE 5-12 Twisted-wire pair.

direct route to the area of an exchange it will serve. The feeder cable leaves the central exchange through an underground conduit. Conduit is a metal or plastic tube that comes in many diameters. The conduits used for feeder cables are usually plastic tubes with a diameter of 4 to 6 in. The conduit extends from the central exchange to a large manhole near the exchange. The conduit is placed under the streets in all directions from the central exchange. Feeder cables are pulled through these conduits to serve different areas. Manholes are placed at intervals along the route to allow for splicing cables together. Sometimes a long length of cable will be pulled through several manholes. The cable will be passed through the manhole from one conduit to another without splicing.

There is a limit on how long the cable pulled through a conduit can be. If you try to pull a very long cable through several manholes, the pulling strain on the cable can be great enough to damage the cable. The length of cable that can be pulled without damage will vary according to the size and construction of a cable. Few cables can be run as one continuous cable from the central exchange to their final destination. The cable run is broken up into smaller segments, and these segments are spliced together in the manholes. It will also be necessary to take some cable pairs out of a cable and run them in a different direction. A feeder cable may arrive in the manhole at the middle of an intersection of two streets as a 3600-pair cable. The outside plant engineer may specify splicing: (1) 900 of the 3600 pairs to a 900-pair cable in a westbound conduit; (2) 900 pairs to a 900-pair cable in an eastbound conduit; and (3) 1800 pairs to an 1800-pair cable in a conduit that continues in the same direction as the original feeder cable.

Manholes provide access to underground cable and serve as a junction point for many cables going in different directions. One or more of the cables coming into the manhole will be a feeder cable. The feeder cables start out as a large cable at the central exchange and are spliced into smaller cables called **branch feeders**. Each branch feeder serves a specific area of the exchange. Branch feeders will be spliced into smaller cables called **distribution cables** that are 25- to 400-pair cables. These distribution cables fan out in all directions from the branch feeder to serve homes and businesses (Figure 5-13).

In older downtown locations, the distribution cables are aerial cables usually placed on poles in the alley. Because an alley is between two streets, the distribution cable will serve customers on the west side of one street and customers on the east side of the other street. Access points are provided on aerial cable for attaching the wires from a home or business. The access points are called *ready access terminals,* and the wires going to a home or business are called a *drop wire* or *drop cable.* An aerial drop consists of two wires that are encased and separated from each other in a black rubber or plastic material.

Housing developments in a suburban location are usually served with buried distribution cable. The distribution cable is buried in the backyard along the property dividing line between houses. Access to the buried distribution cable is provided at

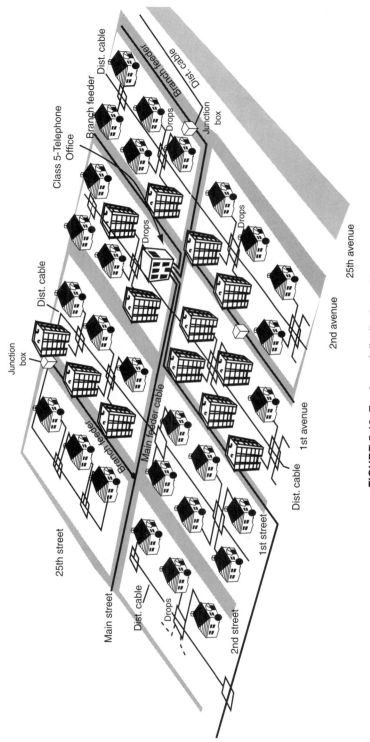

FIGURE 5-13 Feeder and distribution cables.

devices called *pedestals*. A pedestal is a steel can about 18 in. high and six in. square. Pedestals are usually placed at the intersecting property line for four houses. The drop cables from the four houses are buried directly in the ground on a line-of-sight path between the house and the pedestal. Buried drops usually have eight wires in the cable. The cable is impregnated with a petroleum jelly to keep out moisture. Larger pedestals are used to gain access for splicing buried branch feeders to buried distribution cables. These pedestals are steel cans 3 ft high and 9 in. square (Figure 5-14).

5.18 SUBSCRIBER CARRIER

Subscriber carrier is used in outside plant as a pair-gain device. In an urban area where a new subdivision or resort complex has been built, the engineer can choose between running new cable to the subdivision or using subscriber carrier. It is usually more economical to use subscriber carrier until the area builds up with more than 200 homes. Subscriber carrier is an electronic device operating like the carrier systems used to carry telephone calls between central offices. Carrier systems are devices that place many conversations over one cable pair. Placing many conversations over one wire is called *multiplexing*. The different ways that multiplexing is achieved are covered in Chapter 4. It is possible to place as many as 96 telephones on a subscriber carrier system. This system is often called *SLC-96* (pronounced "slick 96").

SLC stands for subscriber line carrier (or subscriber loop carrier). The number following SLC indicates how many lines (referred to as *channels*) it will handle. The subscriber line carrier uses two cable pairs. The engineer will use two existing cable pairs for the SLC-96 and thus saves (or gains) 94 pairs. Most SLCs use time division technology to put 96 telephones on two cable pairs. Time division technology is covered in greater detail in Chapter 4.

Smaller versions of SLC are available when needed. Single-channel and eight-channel systems, which will work over one cable pair, are used when it is necessary

FIGURE 5-14 Picture of pedestal.

to serve much less than 96 lines. These systems use amplitude modulation and frequency division multiplexing technologies, which are also covered in Chapter 4. These systems work like radio transmitters/receivers and require the use of a nonloaded cable pair. Single-channel systems must be used within 18, 000 ft of the exchange, but eight-channel systems can be used anywhere. The radio transmitters/receivers work like an AM radio, and the systems are called AML-1 and AML-8 to designate how many lines they will serve.

AML-1 can be used to add a customer when all cable pairs serving the area are already in use. An area close to the central exchange may have experienced some growth that used up all vacant (unused) cable pairs and no pairs are available to serve the new customer. It is also possible that all pairs may be in use when one of the existing pairs goes bad and cannot be fixed. Instead of running a new cable, the engineer can use AML-1. The engineer will select a cable pair that is currently in use by a neighbor of the affected customer. This cable pair will be used to serve both customers.

The drop wire serving the affected customer is attached to the same cable pair serving the neighbor. A radio transmitter/receiver is attached to each end of the circuit. A transmitter/receiver is attached to the drop wire of the new customer. The mate for that transmitter/receiver is installed in the central exchange and attached to the cable pair. The new customer will be served by the radio. The radio uses the neighbor's cable pair as its transmission medium. The old customer will be served by the physical cable pair. A filter circuit is placed on the drop wire of the old customer to prevent the radio signal from causing noise (Figure 5-15).

AML-8 will serve eight customers by using one cable pair for the system. The central office equipment for AML-8 has eight transmitters and receivers, each of which acts like a radio. Each radio operates on a different frequency. The AML-8 unit to be placed in the field is situated at the location where branch feeder cable meets distribution cable. One side of the AML-8 central office equipment is connected to eight line circuits, and the other side is connected to one feeder cable pair. The field unit is also connected to the same feeder pair; the other side of it is connected to eight different distribution pairs.

Each of the eight customers attached to AML-8 will have their own unique radio within the AML-8. This technology is similar to commercial broadcast radio technology. Many stations are transmitting at the same time over the same facility, but you can only hear the station that your receiver is tuned to. To provide two-way communication, each end of the AML-8 system has eight transmitters and eight receivers. This requires the use of sixteen different radio frequencies. The voice signals placed over AML-8 will not interfere with each other. The person assigned to channel 1 is tuned to the channel 1 frequency and can only pick up signals for channel 1 (Figure 5-16).

5.19 *MEDIUM FOR LONG DISTANCE NETWORKS*

When long distance networks were first established, the only medium available was wire. Toll trunks used a larger gauge of wire than the local loop. Most wires

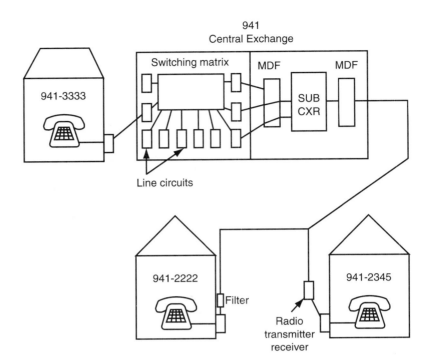

FIGURE 5-15 Single-channel subscriber carrier.

between cities were 16- or 19-gauge copper wire. These media were also equipped with *voice frequency repeaters (VFRs)* on each end. A VFR could amplify voice signals as much as 12 dB. The VFR was adjusted to compensate for all but 4 dB of a toll circuit's loss. Thus, all toll circuits would have a total loss of 4 dB regardless of the distance between cities. When radio carrier systems were introduced to multiplex 12 or 24 trunk circuits together, copper wire was also used as the medium for these multiplexing systems.

Later, AT&T developed a frequency division multiplexing carrier system that could multiplex 600 conversations or toll trunks onto one facility. This carrier system could not use regular copper wire but had to use coaxial cable as its medium. The cable used to connect a television set to the cable television company line is coaxial cable. AT&T continued to develop the capabilities of these systems and in 1978 had an L5E system that could place 13,200 trunk circuits on one coaxial cable. The coaxial cable used had 20 pairs plus 2 spare pairs. This cable could handle 10 L5E systems (132,000 two-way trunk circuits) between switching centers with repeaters or regenerators placed at intervals of 1 mi on the cable.

AT&T also developed carrier systems that used time division multiplexing technology. The basic system was a 24-channel system called T1. A 48-channel (T1-C) and a 96-channel (T2) system were also developed by AT&T. These systems could use twisted-pair copper wire as their medium. The wire pair had to be unloaded and regenerators placed at 1 mi intervals along the cable route. AT&T developed 672-channel TDM carrier system . This system operates at 44.736 Mbps and

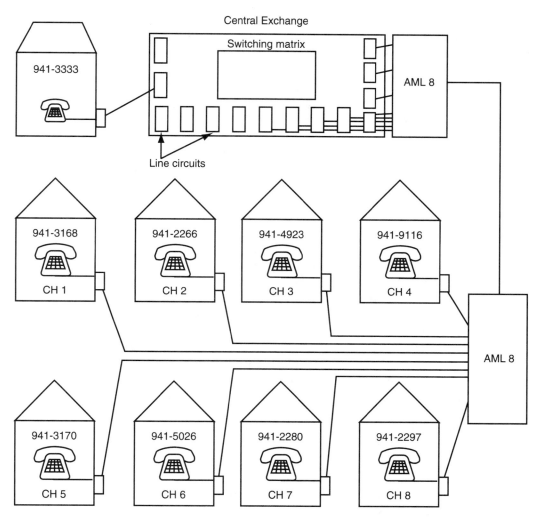

FIGURE 5-16 Eight-channel subscriber carrier.

consists of 28 T1 systems multiplexed together. The system is called a *T3 carrier system*, and the signal level (44.736 Mbps) is called a *DS-3 signal*. The high speed of this signal limits the medium choice to coaxial cable, microwave, and fiber.

Microwave radio systems were also developed by AT&T. These systems could multiplex up to 28,244 trunk circuits. Microwave signals are referred to as *line-of-sight transmission*. The signals are very high frequency signals (4 to 11 gigahertz) that are focused by the transmitting antenna into a narrow beam and transmitted in a straight line. The curvature of the earth and the height of an antenna limit how far apart transmitters and receivers can be. The microwave towers used in the PSTN are placed 20 to 30 mi apart depending on the terrain. The signal received at one of the relay towers is demodulated and fed into a transmitter, which transmits to the next tower down the line.

Microwave signals can be impacted by rain, but different frequencies are affected by different rain patterns. To prevent signal fade due to rain or other atmospheric conditions, two different transmitters (thus two different frequencies) are connected to each dish on the microwave tower. Equipment at the receiver monitors and compares the two received signals. The signal of highest quality is demodulated and used as the input to the transmitters, which feed the signals to the next tower in line. Microwave and coaxial cable have given way to fiber optic cable as the medium of choice in the toll network. AT&T still has L5E coaxial systems and microwave systems in use but has converted much of its long distance network to fiber. Sprint's long distance network uses fiber exclusively as the transmission medium.

5.20 FIBER OPTIC CABLE

A fiber optic strand is made by surrounding a thin fiber of glass with another layer of glass called *cladding*. The cladding acts like a mirror and reflects light down the fiber. The glass core and cladding are surrounded by a jacket to prevent stray light from entering the fiber. A fiber optic communication system consists of a light pulse generator, the glass fiber, and a receiver that can detect changes in the light pulses.

Fiber optic systems are noise free. Noise is caused by electromagnetic induction on top of electrical signals. Electrical signals are susceptible to noise but fiber does not use electrical signals; it uses light signals, which cannot be impacted by electromagnetic induction. The fiber has been improved over time, and today most high-capacity systems use a fiber known as *single-mode fiber*. This fiber does not provide multiple paths for reflected light, and the light travels along the axis of the fiber. Single-mode fiber uses lasers as the light source.

AT&T introduced fiber optic technology in 1979 using graded-index fiber. This system multiplexed 672 trunk circuits onto one fiber using a signal rate of 44.736 Mbps (digital signal level 3 or DS3). In 1984 AT&T introduced a single-mode fiber system that multiplexed 2688 trunk circuits on one fiber. In 1986, a single-mode fiber was handling 6048 trunk circuits with repeaters spaced at 20-mile intervals. Using a standard OC-48 time division multiplexer, the standard single mode fiber today is carrying 32,256 trunk circuits using a signal rate of 2.488320 gigabits per second. Repeaters are spaced at 27-mi intervals. Repeaters on fiber optic cables placed in the ocean are spaced about 75 mi apart. Repeaters can be spaced farther apart at lower speeds.

Sprint and AT&T have upgraded their fiber routes to handle as many as 516,096 trunk circuits (channels). This has been accomplished by using a technology called *wave division multiplexing*. Instead of using strictly time division multiplexing (TDM) technology and trying to extend TDM technology to multiplex many more signals in a TDM mode, wave division multiplexing (WDM) technology combines frequency division multiplexing (FDM) technology and TDM technology. The light spectrum of a fiber optic facility is broken down into multiple frequencies. Light waves are composed of many frequencies. Blue light has a different frequency from red light, orange light, green light, and so on.

Transmitters and receivers for fiber optic have been developed that transmit and receive at a specific frequency.

At the present time, WDM uses 16 different frequencies. An individual OC-48 TDM connects to each of the 16 different transmitters in a wave division multiplexer. By splitting the light spectrum into 16 different frequencies and attaching an OC-48 to each transmitter, the fiber optic facility can carry 516,096 channels ($32,256 \times 16$). The next advance will be to use an OC-96 or OC-192 as the TDM device on a WDM. Each OC-192 can carry 129,024 channels. By placing OC-192s on a 16-channel WDM, the fiber can carry 2,064,384 channels ($16 \times 129,024 = 2,064,384$) (Figure 5-17).

A fiber optic strand carries signals in one direction only. For transmission in both directions at the same time, two fibers are needed for each fiber optic circuit. The basic fiber cable for long distance will have 72 fiber pairs and one steel wire. The steel wire is used to pull the cable between manholes. Six of the 72 pairs are reserved as spares. With 66 available pairs, this cable can handle 2,128,896 trunk circuits ($66 \times 32,256$) when OC-48 multiplexers are used. The same cable can carry 34,062,336 channels ($16 \times 66 \times 32,256$) when the facility uses OC-48 TDM and WDM technology. By using OC-192 TDM and WDM, the same 72-pair cable can carry 136,249,340 calls.

WDM

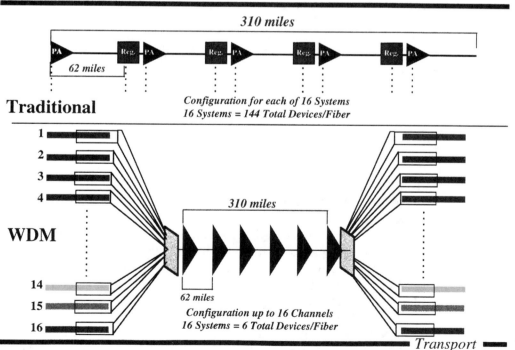

FIGURE 5-17 Wave division multiplexing. (Courtesy of Sprint.)

It is estimated that a single-mode fiber can handle 3 million trunk circuits when the transmitting lasers and receiving devices are developed that can generate and detect a signal of 250 gigabits per second (250 billion bits per second). This is ten times the signal rate of an OC-48 multiplexer. This signal rate has the capacity to handle more than 3 million voice circuits or transmit 25,000 textbooks in one second. The OC-48 capacity of 32,256 voice circuits over one fiber is equivalent to transmitting about 250 textbooks per second. By combining a 250-gigabit TDM multiplexer and WDM technology, a fiber could carry 50 million circuits.

As improvements are made in the laser transmitters and receivers to achieve greater capacity, improvements in their design has also lead to wider spacing for repeaters. One of the major improvements made in fiber optic transmission systems was the simple adoption of standards. Prior to standardization, different vendors' systems could not be connected directly to each other. The adoption of Synchronous Optical Network (SONET) standards was a major boost to the deployment of fiber transmission facilities and networks. The basic SONET signal rate in the United States is OC-1 (672 trunk circuits).

A T3 carrier system uses a DS3 signal rate of 44.7 Mbps to multiplex 672 trunk circuits. An OC-1 SONET system uses a STS-1 (synchronous transport signal–1) rate of 51.8 Mbps to multiplex 672 trunk circuits. The difference in signal rates is due to the additional overhead in the STS-1 signal. The SONET level of OC-48 (48 OC-1 systems or 32,256 trunk circuits) (48×672) was the standard for telecommunications but is rapidly being replaced by OC-192s on a wave division multiplexer. SONET standards allow the interconnection of different vendors' equipment and allow stripping out a DS0, DS1, or DS3 signal without having to demultiplex the whole bit stream. When WDM is used, it is necessary to use a WDM receiver that will demodulate one complete frequency in order to have access to the OC-48 or OC-192 TDM. The required channels can then be pulled out of the OC-48 or OC-192 TDM stream.

5.21 SUMMARY

The medium used to convey voice signals from the telephone to the central exchange is mostly twisted-pair copper wire. This facility is called the local loop. The local loop was designed to effectively carry signals in the 0 to 4000 Hz range. Most of the intelligence contained in an analog voice signal falls in this range of frequencies. When the loop is longer than 18,000 ft, the cable pairs must be loaded to compensate for the signal loss caused by the mutual capacitive reactance of the cable pairs. Loops shorter than 18,000 ft. do not have to be loaded. Load coils adversely affect digital signals. When digital signals must be carried by the local loop, the length of the loop is limited to 18,000 ft, unless the cable pairs are unloaded and digital signal regenerators are placed on the cable pair.

Most loops in a city are less than 18,000 ft. These loops do not have load coils and can carry low-speed digital signals without line treatment. High-speed digital signals, such as the 1.544 Mbps of a T1 carrier, will require signal regenerators every mile along the route. Wire cable has been replaced by fiber optic cable for most T1 systems and other high-speed data signals. The medium of choice in the long distance area is fiber optic cable. Most LECs and IECs have replaced T1 multiplexing systems with SONET OC-48 multiplexers and fiber optic cable.

Because of the tremendous investment LECs have in the twisted-pair local loop, they will seek to employ technologies that extend its lifespan. Exchange boundaries for exchanges in a city have a radius of less than 3 mi from the central exchange. Thus, these lines can be served by twisted-pair copper wire and still carry digital signals. Most LECs will phase fiber into the local loop by first deploying fiber as feeder and distribution cables. By multiplexing many SLC-96 systems over one fiber, the LECs can establish a CSA and provide DS0 capabilities to customers located many miles from the central exchange. These capabilities are critical to the deployment of ISDN and other digital services in the remote areas of a rural exchange.

The rural exchange must serve a much larger area than an exchange in a city and requires the application of RRD and MLRD design concepts to determine the size of wire cable to use in serving the area. Rural areas cannot have ISDN and other digital services without unloading these long loops and placing signal regenerators along the cable route. Establishing a CSA through the use of subscriber carrier is a more cost-effective way to provide digital capabilities to remote areas. Because a large investment would be needed to convert all long loops from wire cable to a CSA, it will be some time before all rural areas are converted to fiber feeder cables and a CSA. In the past, many rural customers did not desire digital services, though today more of them are demanding these services. As the demand increases, the LECs will selectively deploy fiber and a CSA concept to the areas with a high demand for digital services.

REVIEW QUESTIONS

1. What is the medium of choice for the local loop? Why?
2. What is the medium of choice for the toll network? Why?
3. What device governs design of the local loop? Why?
4. Outside plant engineers use what principle in design of the local loop?
5. What are the two wires connecting to our telephone called?
6. What does *PIC* stand for?
7. Where does the power for the telephone come from?
8. How much voltage does the central exchange apply to the local loop?
9. How far can a telephone be located from the central exchange before loop treatment is needed?
10. What does *SLC-96* stand for?

11. When would subscriber carrier be used?
12. What is mutual capacitance?
13. How does the outside plant engineer offset the effects of capacitive reactance in the local loop?
14. What is D spacing?
15. At what points is a local loop cable loaded?
16. How many simultaneous conversations can OC-48 support?
17. What does *SONET* stand for?
18. Why are wire pairs in a cable twisted?
19. What is the primary source of noise and crosstalk in a local loop cable?
20. Why is fiber optic cable not susceptible to noise from electrical interference?

GLOSSARY

Branch Feeder Cables Cables that connect distribution cables to main feeder cables.

Carrier Serving Area (CSA) The establishment of DS0 capabilities to a remote area of the exchange by using subscriber carrier to establish a remote wire center.

Crosstalk A condition where a circuit is picking up signals being carried by another circuit. An undesirable condition. For example, a customer hears other conversations or noise on their private line. This is usually caused by a breakdown in the insulation between wires of two different circuits.

Distribution Cable A cable that is fed by a branch feeder cable from the central exchange and that connects to telephones via drop wires. The distribution cable is usually a small cable of 25 to 400 pairs.

Drop Wire A pair of wires connecting the telephone to the distribution cable. One end of the drop wire connects to the protector on the side of a house or business, and the other end connects to a terminal device on the distribution cable.

Exchange Boundary Pertains to the limits of the serving area. The outermost customers being served by a particular exchange establish the boundary of the exchange.

Facility A term typically used to refer to the transmission medium.

Feeder Cable Includes various types. Cables leaving the central exchange are called main feeder cables. The main feeder cables connect to branch feeder cables. The branch feeder cables are used to connect the main feeder cables to distribution cables.

Jumper Wire A short length of twisted-pair wire used to connect wire pairs of two different cables. A jumper wire is a very small gauge of wire (usually 24 or 26 gauge). Except for its small size and much longer length, a jumper wire is similar to the jumper cable used to start a car with a dead battery.

Load Coils The devices that introduce additional inductance into a circuit. Load coils are added to cable pairs over 18,000 ft to improve the ability of the pair to carry voice signals.

Loading Purposely adding inductance to a cable pair over 18,000 ft long to offset the mutual capacitance of the cable pair.

Local Loop A term used to describe the facilities which connect a telephone to the central exchange. These facilities usually consist of a twisted pair of wires.

Loop Extender A hardware device. Can be connected between the line circuit of the switch, and the cable pair serving a remote area of the exchange, in order to extend the range of a central exchange. A loop extender adds a voltage boost to the circuit. By doubling the voltage applied to a line, the distance served can be doubled.

Loop Treatment A term indicating that a local-loop cable pair has been specially conditioned. The most prevalent use of this term is to describe the addition of extra equipment to a local-loop cable pair. To extend the range of a central exchange, a device called a loop extender can be connected between the line circuit of the switch and the cable pair serving a remote area of the exchange. A loop extender adds a voltage boost to the circuit. By doubling the voltage applied to a line, the distance served can be doubled. Another device called a voice frequency repeater (VFR) is also added to long local loops. The VFR will amplify voice signals in both directions to compensate for the extra decibel loss of long loops.

Main Distributing Frame (MDF) In a central exchange, the place where all cables that connect to the switch are terminated. Outside plant cables that connect to telephones and to other central offices are also terminated at the MDF. Short lengths of a twisted-wire pair (called a jumper wire) are used to connect wires from the switch to wires in the outside plant cable. In a PBX environment, the MDF is where all cables from the switch, IDFs, and cables from the central exchange are terminated and jumpered together.

Main Feeder Cable Cable that leaves the central exchange and connects to branch feeder cables.

Modified Long Route Design (MLRD) Outside plant design that involves adding loop treatment to cable pairs serving remote areas of an exchange. Local loops longer than 24,000 ft are designed using MLRD.

Multiplexer A device that can combine many different signals or calls (data or voice channels) so they can be placed over one facility. A multiplexer also contains a demultiplexer so that it can demultiplex a received multichannel signal into separate voice channels.

Outside Plant Cables, telephone poles, pedestals, and anything that is part of the telecommunications infrastructure and not inside a building.

Outside Plant Engineer The person charged with properly designing the facilities that connect telephones to central exchanges and the facilities that connect central exchanges together (outside plant designer).

Plastic Insulated Cable (PIC) Cable that contains wires electrically isolated from each other by plastic insulation around each wire.

Resistance Design Designing a local loop, which uses twisted-pair copper wire, so that the resistance of the wire serving a telephone does not exceed the resistance design limitations of the central exchange line circuit. Each switching system manufacturer will state how much resistance its switch is designed to support.

Revised Resistance Design (RRD) A design approach in which the outside plant engineer uses the length of the local loop, as well as the design limitations of the central exchange, to determine the local-loop design criteria. Loops up to 18,000 ft are designed to use nonloaded cable pairs with a maximum loop resistance of 1300 Ω. Loops between 18,000 ft and 24,000 feet are designed to use loaded cable pairs with a maximum loop resistance of 1500 Ω. Loops longer than 24,000 ft are designed according to Modified Long Route Design (MLRD).

Subscriber Line Carrier–96 (SLC-96) SLC-96 uses TDM to multiplex 96 lines over two cable pairs. Multiple SLC-96 systems are often multiplexed onto a fiber facility.

Voice Frequency Repeater (VFR) A device that will amplify voice signals in both directions to compensate for the extra decibel loss of long loops.

Wave Division Multiplexing (WDM) Technology that splits the light spectrum into many different frequencies. A time division multiplexer is then assigned to each of the different light wave frequencies. WDM is used to multiplex multiple OC-48 (or OC-192) systems over one fiber pair.

6

The Telephone Set

Key Terms

Alerters
Analog Signal
Bridged Ringer
Decimonic Ringing
Digital Signal
Dual-Tone Multifrequency Dial
   (DTMF)
Feature Phone
Grounded Ringer
Harmonic Ringing

Hookswitch
Hybrid Keysystem
Hybrid Network
Keysystem
Magneto
Private Branch Exchange (PBX)
Proprietary Telephone
Receiver
Ring Detector
Ringer

Sidetone
Station Equipment
Straight-Line Ringer
Superinposed Ringer
Transmitter
Tuned Ringer
2500 Telephone Set
Varistor

A dispute over the invention of the telephone set marks the beginning of telecommunications. Several inventors were working on devices that could convert audio waves into electrical signals. Two of these inventors were Elisha Gray and Alexander Graham Bell. Both individuals filed for a patent on February 14, 1876. The Patent Application and Disclosure filed by both men described a variable resistance transmitting device as part of the telephone. Gray's description was included in the body of the formal disclosure document. Bell's formal document did not originally contain the disclosure of a variable *transmitter*. The remarks about a variable transmitter were handwritten in the margin of the document. This was a very critical component of the invention. Did Bell learn of Gray's device and hurriedly add these remarks in the margin? Evidently some of the Supreme Court justices ruling on the patent believed he did. The Supreme Court was split in its decision upholding Bell as the inventor of the telephone, but the majority did rule that he is to be recognized as the inventor of the telephone. When someone invents something, the person is always faced with other people or companies trying to steal the technology. It may well be that people were trying to steal the invention from Bell. On March 10,1876, Bell and his assistant, Thomas Watson, successfully demonstrated a working model of the variable resistance transmitter.

Bell's original concept for the telephone did not use this type of transmitter but used a small electromagnetic receiver as both the transmitting and receiving device (a combined transmitter/receiver is referred to as a *transceiver*). You can test this device yourself. Instead of talking into the transmitter on your next telephone call, talk into the *receiver*. Cover the transmitter of the telephone with your hand and speak into the receiver. You can make yourself heard, but the other person will tell you the volume was very low. Using the receiver device as a transmitting device produced very small levels of current and could not produce a loud signal in the far-end receiver. By using a variable resistance transmitter such as proposed by Gray (and Bell's marginal notes), a much higher signal level was attained. The variable resistance transmitter consisted of an acid and water solution between two wires attached to the transmitter. When sound hit the diaphragm, it moved the wire attached to it. This movable wire would move up and down in the acid solution. The more immersed it became, the less resistance there was between the two wires.

Unlike the transceiver, which can generate its own signal and does not need external power, the variable resistance transmitter cannot generate a signal. It must have power supplied by a battery. The battery supplies power to the transmitter. The voice signal hitting the diaphragm will cause the resistance of the transmitter to vary. The varying resistance varies in unison with the speaker's voice. The varying resistance will cause current flow from the battery to vary. The varying electrical signal is similar to the varying sound signal. For this reason, it is called an *analog signal*. The electrical signal is analogous (similar) to the sound signal. This varying signal is coupled by wires from the transmitter to a receiver. Varying current in the receiver will cause a varying sound output that is in unison with variances in the electrical current. Because the electrical signal is analogous (or similar) to the sound wave, we call it an analog signal. All electrical signals with varying amplitudes are called analog signals in telecommunications.

6.1 THE CARBON GRANULE TRANSMITTER

The liquid acid variable resistance transmitter was not practical for use by the general public. Bell continued to investigate different devices for use as a transmitter. In 1877 Thomas Edison invented a variable resistance transmitter that used carbon powder. In 1878, Henry Hummings, a British inventor, developed a transmitter using carbon granules. In 1886, Edison improved Humming's design. When air waves from your voice hit the transmitter, they cause a thin metal diaphragm to move back and forth at the frequency of the voice signal. This movement puts more or less pressure on the carbon granules. Higher pressure packs the carbon granules tighter and closer together and lowers the resistance of the transmitter. Less air pressure leads to looser granules and a higher resistance. The carbon granule transmitter is the transmitter used in most telephones. The transmitter and receiver for early telephones were wired in series between the two wires of the local loop. With this arrangement, the signal of the transmitter was reduced by loss in the receiver before the signal reached the local loop. The design of the telephone was changed to improve the amount of transmitter signal reaching the local loop. A transformer with three windings was added to the telephone. One winding connected to the two wires of the local loop, one winding connected to the transmitter, and one winding connected to the receiver. The design of the transformer couples voice signals between windings in such a manner that maximum signal transfer occurs between the transmitter winding and the winding attached to the local loop. The transmitter winding was also physically attached to the loop winding in order to pass the DC voltage from the loop to the transmitter.

6.2 SIDETONE

The transformer in a telephone couples a small portion of the transmitter signal to the receiver. This signal coupled to the receiver is called *sidetone*. Sidetone is necessary so that people talking into the transmitter hear what they are saying. Try talking and covering one ear with your hand. You will hear yourself only in the uncovered ear. Since you cannot hear with the covered ear you sound strange, as if you are in a well. If no sidetone were coupled to the receiver, when the receiver covered that ear, you would sound strange to yourself when talking. The receiver is coupled to the local loop winding so that maximum signal transfer occurs on signals coming from the central exchange to the phone. A device called a *balancing network* is also included between transformer windings. This device matches the impedance of the phone to the impedance of the cable pair to improve signal energy transfer. The combination of the transformer and balancing device is known as a *hybrid network.* The hybrid network converts the two-wire line circuit of the cable pair to the four-wire circuit of the transmitter and receiver.

6.3 INVENTION OF THE RINGER

While work was proceeding on improvements to the transmitter of the telephone, Bell was also working on some way to alert people that someone wanted to talk to them over the phone. His assistant Thomas Watson invented the device called a *ringer*. Each phone was equipped with a hand-cranked electric *magneto* and a ringer. When a call was placed, the originating party would generate a high voltage AC signal by using the hand-crank magneto. This signal caused the ringer of all phones connected to the line to ring. Watson patented his ringer invention in 1878. Improvements have been made in his original ringer but the ringer used in many of today's telephones are basically the same device Watson invented. If the ringer in your telephone is the old-style electromechanical ringer, it is Watson's ringer.

6.4 RING DETECTORS

The current trend in telephone design is not to use electromechanical ringers. The latest trend uses a solid-state transistorized device, called a *ring detector*, to detect ringing signals. When a ringing signal is detected, the ring detector sends a signal to activate a sound from a small speaker. The sound from the speaker is usually a loud shrilling tweeter. Some devices being marketed today have microprocessors and ROM chips to include different messages that can be sent to the speaker. When the ring detector senses an incoming ringing voltage from the central exchange, it sends a signal to the microprocessor chip. The chip will select one of the messages stored in ROM and send it to the speaker. These messages can be almost anything you desire, from jingles to poems (for example, "pick up the phone, pick up the phone, you know you are home so pick up the phone"). The electromechanical ringer is being replaced by speaker type audible alerters.

6.5 STATION EQUIPMENT

The telephone set belongs to a broader classification called station equipment. *Station equipment* is basically anything that connects to a local loop (telephone, answering machine, teletype, fax machine, modem, and personal computer). Telephones come in a variety of designs, colors, and styles. Many telephones are designed to work with the local telephone company's central office (some phones are designed to only work on private switching systems). Most telephones work on one cable pair or line from the central exchange and are referred to as single-line phones. The telephone receives its power from the central office 52-V DC battery. If the single-line phone has the older rotary dial, it is referred to as a *500 telephone set*. If the phone uses a touchtone keypad (dual tone multifrequency or DTMF) for dialing, it is called a *2500 telephone set*. The current telephone set design has been arrived at after years of improvements. The primary functions of the telephone are to: (1) signal the local central exchange that a call is being originated or has been answered, (2) signal the customer of an incoming call using a ringer or *alerter*, (3) provide a signal to the telephone exchange that will indicate who the caller wishes to contact, (4) convert voice signals into electrical signals, (5) convert electrical signals into sound signals, (6) provide a means to match the impedance of the telephone to the local loop, and (7) provide just the right level of sidetone to the receiver regardless of loop length.

6.6 TELEPHONE STANDARDS: LSSGR

There are no official standards for the telephone. However, the phone must be able to work with the equipment in the central exchange. The central exchange and line interface requirements can be found in the *Local Switching Systems Generic Requirements (LSSGR)* by AT&T and Bell Communications Research. These recommendations serve as a de facto standard for the telecommunications industry. The line interface requirements will specify how long the local loop can be, what type of signaling can be used, and what type of ringer is required. The length of the local loop determines how much current will exist in the local loop, when the phone is off hook. Telephones containing a carbon granule transmitter require a minimum of

23 milliamps of current for the transmitter to work properly. The phones containing new transmitters require 20 milliamps. The DTMF pad is also powered over the local loop. It requires a minimum of 20 milliamps to work properly.

6.7 FEDERAL COMMUNICATIONS COMMISSION REGISTRATION

When a telephone is designed, it must be registered with the Federal Communications Commission (FCC). This requirement was brought about as part of telecommunications deregulation in the 1980s. Prior to deregulation, the local telephone company included one telephone as part of a subscriber's basic service and would rent additional phones for $1.75 a month. The telephone company owned all phones. Maintenance of phones was included as part of the basic subscription fee. With deregulation, a person no longer has to rent phones from the local telephone company. You can buy your phone from anyone and you are responsible for all maintenance. When telephones were provided by the local phone company, it made sure the quality of phones was high, and manufacturers were held to high standards of quality. To ensure manufacturers would continue to provide high-quality telephones, regulation was turned over to the FCC.

6.8 RINGING SYSTEMS

Different types of ringing signals are used in central exchanges. Some central exchanges use ringing frequencies of 20, 30, 40, 50, and 60 cps (referred to as *decimonic ringing*). Some exchanges use 16 2/3, 25, 33 2/3, 50, and 66 2/3 (usually referred to as *harmonic ringing*). These multiple ringing frequency schemes were used primarily by independent telephone companies serving rural areas. In rural areas, using these ringing schemes, it was possible to assign different ringing frequencies to each party on a party line. When a call arrived for a particular party on the line, the appropriate ringing frequency was sent over the line and only one phone would ring. Bell used mostly 20-cycle ringing exclusively because most lines in a Bell office were either one- or two-party lines. For party lines, Bell used a technique called *superimposed ringing*. The ringing signal would be imposed on top of either a negative 130-volt or positive 130-V DC battery supply. The telephone had a diode wired in series with the ringer. If the proper 130-V polarity was present for that phone, only that phone would ring. This type of ringer was called a *superimposed ringer*.

The ringer in the telephone must be compatible with the ringing signals used by the central exchange. Each telephone ringer was assigned a code by the FCC. Before buying a telephone, it was necessary to check with the local telephone company to find out what code ringing signal was assigned to your telephone number. Different ringing frequencies were originally designed for party-line service. Each party on the line had a distinct *tuned ringer* that would only ring on one frequency of ringing signal. If each phone on a party line has a different ringer, it is possible to send out different ringing signals for each phone on the party line. Only the called phone would ring on a party line using multiple ringing signals.

The elimination of party-line telephone service eliminated the need for multiple ringing frequencies. This also eliminated the need for different types of ringers in the phone. Telephones manufactured today are made with a ringer that will ring on any frequency of signal. This ringer is called a *straight-line ringer*. This eliminated the need to check ringer codes when buying a telephone. Most new telephones have eliminated the mechanical ringer and use a ring detect circuit instead. A ring detect circuit is an integrated solid-state electronic device designed to detect the presence of a 90-V AC signal at any frequency. This circuit is often called an *alerter*. When the circuit detects a ringing signal it causes a beeper, buzzer, or tone to sound. The central exchange must retain multiple ringing signals in rural areas because of the need to remain compatible with all the old telephones still in service. As time passes most exchanges will phase out the old ringing signals. The exchange will use mostly 20-cycle ringing signals in the future.

6.9 THE DIAL AND DUAL TONE MULTIFREQUENCY PAD

When automated switching systems were invented, the dial was added to the telephone. The dial would interrupt the current flowing in the line and the switching systems counted the interruptions of current to determine dialed digits. Telephones being manufactured today are made with a dial that use tones for dialing signals. This dial is called a **DTMF dial** or **dual-tone multifrequency dial** and was introduced in 1964 by AT&T. The dialpad is a keypad containing 12 or 16 buttons. The buttons are arranged into four rows and three/four columns. A separate tone is assigned to each row and column. When any button is depressed it sends out a unique set of tones to the central exchange.

	1209 Hz	1336 Hz	1477 Hz	1633 Hz
697 Hz	1	2	3	A
770 Hz	4	5	6	B
852 Hz	7	8	9	C
941 Hz	*	0	#	D

Depressing button 1 will cause a 697-cycle (Hz) tone and a 1209-cycle tone to be sent to the central exchange. The central exchange decodes the different combination of tones received into the equivalent dialed digit. Because of the need to be compatible with all the old telephones with rotary dials, central exchanges will still work with rotary dials. There are some pushbutton telephones which do not send DTMF tones. They convert the button pushed to rotary dial pulses. Some phones allow you to select whether tones or rotary dial pulses will be sent by the keypad. Rotary-type dial pulses will not work with voice messaging systems that ask you to press a button on your phone. These messaging systems use DTMF tones to inform the system of what action it should take. Rotary dial signaling pulses cannot pass through a central exchange, but DTMF tones will pass through the exchange to the called number. If you call your answering machine from another phone, it will respond to DTMF tones. The machine has a DTMF receiver that will decode the tones sent and follow the commands you give it using DTMF tones.

The major components of a telephone are: (1) the transmitter, (2) the receiver, (3) the hybrid network, (4) the dial or DTMF pad, (5) the ringer, and (6) the *hookswitch* (or *switch-hook*). A block diagram of the phone is shown in Figure 6-1. A circuit diagram for a rotary dial telephone (500 set) is included in Figures 6-2 and 6-3. A circuit diagram for a touch tone (2500) set is displayed in Figure 6-4.

6.10 TELEPHONE CIRCUIT DESCRIPTION

The hookswitch contacts S1 and S2 are in series with the Tip and Ring of the connecting line from the central exchange. Contacts on S1 and S2 are wired so that when the phone is hung up, the ringer is the only device attached to the line. The ringer can be bridged across the line by connecting the yellow wire to the green wire. When a ringer is connected between the Tip and Ring wires, it is referred to as a *bridged ringer*. The ringer will then be connected between the Tip and Ring wires. When a telephone was used on an eight-party line, one side of the ringer was connected to ground. On an eight-party line, the yellow lead from the ringer was connected to a ground wire. When a ringer was connected between one of the wires of the local loop and ground, it was referred to as a *grounded ringer*. Later designs of telephones include a hookswitch contact in series with the ringer. When the handset of the telephone is picked up (the phone goes off-hook), these hookswitch contacts open the ringer from the line and attach the transmitter, receiver, dial, and hybrid network to the line. With this arrangement the ringer will not be on the line during transmission and reception of voice signals, and will not reduce those signal levels.

A varistor (V22) is placed across the Tip and Ring leads to limit the amount of voltage applied to the transmitter. A shorter local loop will have less voltage drop in the loop and supply a higher voltage to the telephone than a longer loop with its higher loop loss. The *varistor* serves to decrease the voltage supplied on a short local loop so that it will not be significantly higher than the voltage a phone on a long loop receives. A telephone attached to the longest local loop will receive about

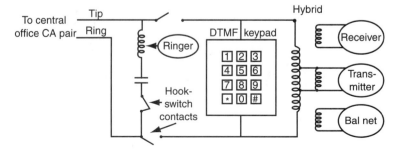

FIGURE 6-1 Diagram of a telephone set.

Hookswitch Contacts shown when phone is on hook (hung up)

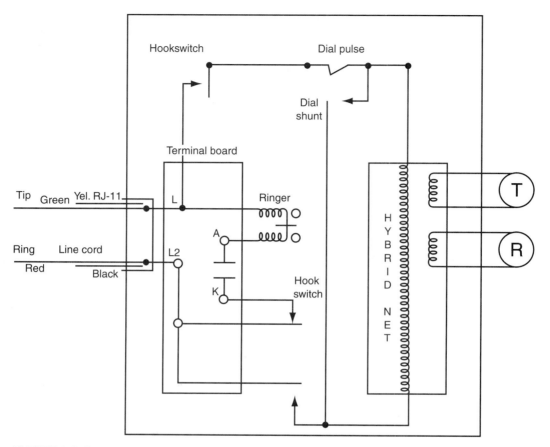

FIGURE 6-2 Rotary dial telephone schemantic. Dial pulse springs open as many time as the value of the digits dialed, then remains closed. Dial shunt springs closed when dial is pulled to a finger stop and remains closed until the dial returns to normal. This prevents pulses from being heard in the receiver.

9 V DC after loop loss. If a varistor were not used in the telephone, on a short loop it would receive about 20 V. This high voltage would cause signal currents of more than 60 milliamps in the local loop and would cause crosstalk in the cable. The varistor will limit the voltage across the phone to approximately 14 V and will limit current through the transmitter to 60 milliamps of current on the shortest loops. Thus, current levels for all loops will be between the minimum requirement of 20 milliamps to a maximum of 60 milliamps. These values are based on newer telephones with an internal resistance of 250 Ω. The current flow will be below the maximum recommended value of 60 milliamps. A second varistor (VR3) is placed across the receiver to limit the voltage of the signal across the receiver so that the sound generated by the receiver is not loud enough to damage an eardrum.

The rotary dial contains two separate set of contacts. One set of contacts (D1) are normally closed and will open when a number is dialed. These contacts open and close once for each digit dialed. If a 5 is dialed, they open and close five times and remain closed at the end of the dialed number. Another set of contacts (D2) are

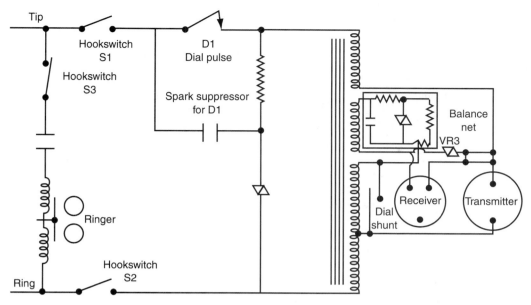

FIGURE 6-3 500 set (rotary dial) schematic.

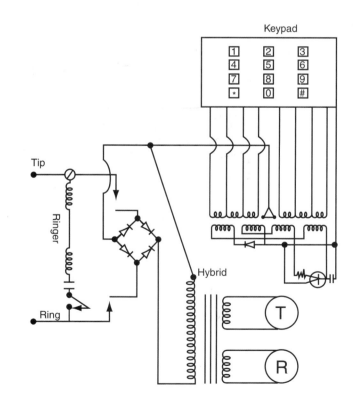

FIGURE 6-4 Touchtone set
(2500) schematic.

Touch tone telephone

called *shunt contacts* and are normally open. When the rotary dial is pulled to the finger stop, they operate to place a short (or shunt) across the transmitter and receiver. The shunt contacts keep a short across transmitter and receiver until the dial returns to normal. The dial pulses generated by D1 will not pass through the receiver but will bypass it. The resistor and capacitor (R & C) across the dial contacts (D1) act as a spark suppresser to prevent damage to these contacts. To dial a number, we put our finger in the hole of the number desired and pull the dial around until our finger hits the finger stop. When we pull our finger out of the dial, a spring returns it back to normal. The spring is adjusted to return the dial to normal in one second when a 0 is dialed. The dial contacts (D1) will be opened from one to ten times (depending on the number dialed) as the dial returns to normal. The dial will open the contacts, allow them to close briefly, open them again, allow them to close briefly, open them again, and so on. Each open and closure is called a *pulse*.

Dialing a zero results in ten pulses in one second, and we say the speed of the dial is ten pulses per second (pps). At ten pps, each pulse is one tenth of a second (100 milliseconds). Each time the contacts open they will be opened for 60 milliseconds and closed for 40 milliseconds. The open occurs when the contacts break open and the open is called a 60% break. The open time is 60% of the pulse time. The dial contacts are adjusted to provide a 60-millisecond (ms) open interval for each pulse. With a 100-ms pulse, the contacts will close briefly (40 milliseconds) between each open. Thus, the dial is adjusted to provide a 60/40 pulse. There is a small distance between the finger stop and the one hole of a dial. This design ensures that a small amount time is provided between digits. During this time the contacts remained closed. This time frame is designed to be more than 500 ms. When the equipment in the central exchange is receiving dial pulses, it recognizes each open interval (60 ms) as a pulse belonging to one digit and will count these pulses to determine the digit dialed. The switching equipment uses the length of the closed interval to determine when the end of a digit occurs. With a closure of 40 ms, the switch continues counting each open condition. When the closure is 500 ms, the switch recognizes this as a signal that the previously counted pulses should be stored as a digit and that the next open will be the start of a new digit. This 500-ms time frame between digits is referred to as the interdigital time frame (Figure 6-5).

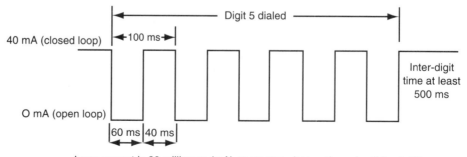

Loop current is 20 milliamps (mA) or greater when a phone is off-hook. When a number is dialed, the dial opens the loop. Figure 4-2 shows how the current changes when the number 5 is dialed.

FIGURE 6-5 Dial pulse string.

When additional features have been added to the basic single-line telephone such as last number redial and speed dialing lists, the phone is called a *feature phone*. The feature phone has a microprocessor and memory contained in the telephone but is still powered by the 52-V DC voltage from the central exchange. This electronic phone can be designed to include special features such as an answering machine. The answering machine uses an electronic solid-state ring detector. The ring detector signals the microprocessor on each ring received. The microprocessor will count the number of rings received and can be programmed to answer on two rings, four rings, or not answer at all. The answering machine also includes a DTMF receiver chip. The owner can call into the answering machine from a remote location and use the DTMF dial, at the remote location, to provide directions to the answering machine. The directions are usually provided by sending two digits as a code. The DTMF receiver in the answering machine receives the tones sent by the DTMF dial and is programmed to provide a specific function depending on the two-digit code received.

When the answering machine answers a call, it is programmed to look at the first code as a password. If it receives the correct password, the answering machine will honor any additional functions requested. Some functions are: turn on the machine, turn off the machine, play back all messages, play back only new messages, record a new announcement, fast forward, fast reverse, rewind, change the password, and so on. The use of DTMF tones to signal the answering machine allows a person to perform any administrative function on the machine from any remote location. The machine has programming that will automatically answer an incoming call, wait for and recognize a password code, return a recorded announcement, respond to DTMF signals and interpret the received code, and perform the function assigned to that code.

Software is available for your personal computer to allow it to function as an answering machine. The PC is equipped with a special modem and a multimedia sound card. This modem can recognize the start of an incoming call from the initial signal received when it answers a call. The initial signal will indicate whether the incoming call is a voice call, an e-mail, or a fax. The modem will answer the call accordingly. When it detects that the incoming call is a voice call, it sends an answering message to the caller. This message is recorded using the multimedia card of the PC and its attached transmitter (microphone). The message is converted by the card into digital format and stored in a file on the hard drive. The system answers a voice call by calling up this file and playing it back through the multimedia card and modem to the phone line. The message usually states that if you wish to leave a message for person A press 1 on your phone, press 2 to leave a message for B, and so forth. The modem and software detect which digit was received and will store the message received, in a file under that person's mailbox (directory). The multimedia card converts the received message into digital file for storage. Many businesses are beginning to use a PC with multimedia, a special modem, and appropriate software to answer calls after hours or to serve as a voice mail system.

6.12 PBX PROPRIETARY TELEPHONES

PBX switching systems can be equipped with line circuits to interface to a regular 500/2500 telephone set, and they can be equipped with special line circuits to interface to a proprietary telephone set. The *proprietary telephone* set is a specially designed electronic feature phone. These phones are proprietary telephones because they contain proprietary software developed by the switching manufacture. The CPU in the telephone set is programmed to talk to the CPU of the switching system. Because they contain special programming to talk with a particular switching system, the proprietary telephone for one manufacturer will not work on another manufacturer's switch. These telephones work with a special line interface circuit and provide more features than a 2500 set. They are designed to provide access to several lines and features. They are often called *multiline telephones* because they have access to more than one line. These telephones require four station wires between the line circuit and the telephone. Two wires are used for the Tip and Ring leads to carry voice as well as DTMF signals. The other two wires serve as signal leads carrying *digital signals*. One signal lead carries signals from the PBX line circuit to the telephone and the other wire carries signals from the telephone set to the PBX line circuit. These signaling leads provide a duplex signaling path, allowing the CPU of the PBX, and the CPU in the telephone to talk to each other. Using this out-of-band signaling path, digital messages can be sent between CPUs to provide instructions from one CPU to the other. If a button on the telephone set is pushed, the CPU in the telephone sends a message to the CPU in the PBX to inform it of the button pushed. The CPU in the PBX looks in the line translation's database for the line circuit associated with this phone to determine what line or feature has been assigned to that button. It will then activate that line, function, or feature.

A PBX has many features available. An electronic, multibutton, proprietary telephone allows those features to be accessed by merely pushing the button assigned to a feature. The electronic telephone enhances the operation of a PBX because it has features that complement those in the PBX. The programs running the CPU in the switch and the CPU in the telephone are proprietary. This is the major reason why the electronic telephone of one vendor will not work on another vendor's switch. When the user of an electronic telephone wishes to access a station line, the person merely pushes a button associated with a line. The CPU of the PBX receives a message from the telephone over the signaling lead. The CPU is informed by the line table that this button is assigned as a particular station number. The CPU issues instructions to the switching matrix of the PBX. Acting on these instructions, the switching matrix connects dial tone to the line circuit's voice path and attaches a DTMF register to received dialed digits from the phone. The tones of the DTMF pad on the phone are transmitted over the voice path to the register in the switch. The switch processes the call based on the digits received.

6.13 HANDS-FREE PHONES VERSUS SPEAKERPHONES

Most electronic telephones have a feature called *hands-free.* These telephones have a small speaker that serves as a receiver if the phone is on-hook. After pushing a button to access a line, dial tone is heard in the speaker. If the phone is taken off-hook, the receiver of the handset will replace the speaker. Thus you can listen to the progress of a call being established through either the handset or the hands-free speaker. The speaker will allow you to hear the called phone ringing. You will also hear the voice of the called party when he or she answers but that person cannot hear you through the speaker. You must speak into the transmitter of the handset. A hands-free telephone allows you to place a call without picking up the handset, but once the call is connected, you must talk into the transmitter of the handset.

Some people erroneously think that the speaker on a hands-free telephone will also transmit your voice, but it will not. It is hands-free in a receive mode only. The telephone that will transmit and receive in a hands-free mode is called a *speakerphone*. The speakerphone has a speaker that acts as a receiver and also has a small transmitting device usually placed on the front edge of the telephone. You can recognize the location of this transmitter. It will be covered by the case of the telephone. The case has two to three small slits in the plastic hosing to allow voice signals to reach the transmitter. The hands-free telephone allows you to listen without lifting the handset. The speakerphone allows you to talk and listen without lifting the handset. Some speakerphones have electronic circuitry to detect when the user is talking. If the user is not talking, the transmitter output is very low. When the transmitter detects a voice, it reduces the output of the speaker. The operation of this electronic circuit prevents feedback to the speaker and eliminates squealing from the speaker.

6.14 INTEGRATED TELEPHONE AND DATA TERMINAL

Some electronic feature telephones include a *visual display unit (VDU)* and a keyboard. The phone is usually equipped with two telephone lines and numbers. The VDU provides a visual display of speed calling lists and system options. The user can use arrow keys on the keyboard to select an option and then press the enter key to activate that option. The keyboard and VDU enable the phone to serve as a dumb terminal. The phone has an EIA-232 connection for connection to a modem. The user can use one of the two telephone numbers for placing voice calls. You would use the modem (which has a connection to its own telephone number) for placing calls to a computer. The dumb terminal and modem allow using this device for e-mail.

6.15 INTEGRATED SERVICES DIGITAL NETWORK (ISDN) TELEPHONES

The telephones discussed above are designed to convert voice signals into analog electrical signals. ISDN telephones convert voice signals into 64,000-bps digital signals. The ISDN phone contains a codec to perform this conversion. The ISDN phone connects to the central exchange using a *digital subscriber line (DSL)*. The DSL operates at 192,00 bps and provides the user access to 144,000 bps. The ISDN telephone separates the 144,000 bps into three logical channels. Two of these channels are called *Bearer (B)* channels and one is called the *Delta (D)* channel. The ISDN telephone uses one of the B-channels for the telephone's voice circuitry. The other B-channel can be used to service another device or can be used to serve a second telephone number. The D-channel carries signaling between the telephone and the central exchange. It can also carry packet data. ISDN telephones also have an EIA-232 port. This port allows connecting the serial port of a PC to the ISDN phone. The PC can use the second B-channel to send data over the switched network. This allows for the transmission of data without the use of a modem if the receiving end is also using ISDN lines. ISDN telephones are discussed in more detail in Chapter 11.

6.16 KEYSYSTEMS

When more than one telephone number is needed by a business, either a multiline phone or keysystem can be used. Multiline telephones can handle three lines. These telephones have four buttons (one button for each line and a hold button). If three or more lines are needed by a business, a keysystem is preferred over a multiline phone. *Keysystems* use special electronic telephones that can access several telephone lines and are also referred to as *multiline phones*. Many keysystems have names indicating the number of central exchange lines and telephones the system can handle. A 308 keysystem can have three local telephone numbers and eight multiline phones attached to the system. A 612 keysystem can handle six exchange lines and twelve multiline telephones. Because of the extensive use of electronic components and microprocessors, today's keysystems are called *Electronic Key Telephone Systems (EKTS)*. The part of the keysystem that controls the system is called the *keysystem unit* or *KSU*. Telephone lines from the central office are connected to the KSU. All keysystem multiline telephone sets are also connected to the KSU. The KSU contains a space division switching matrix under control of a microprocessor. The size of the switching matrix determines how many exchange lines and telephones the system can handle. The switching matrix in the KSU connects telephone lines from the central office to the multiline telephone which answers the call.

Telephones for a keysystem are special purpose phones and each keysystem manufacturer makes phones that will only work with their keysystems. A single-line telephone (2500 set) will not work on a keysystem because the keysystem uses proprietary signaling protocols. This is the same reason multiline phones from different manufacturers will not work on keysystems manufactured by another company. Each keysystem manufacturer uses a different signaling protocol. Keysystems have just begun appearing from a few manufacturers that do allow the use of a regualr 2500 set instead of a proprietary telephone.

A 2500-type telephone has only one pair of wires connecting it to the central office. Everything is done over this pair of wires. This one pair of wires carries the off-hook signal, dial signals, ringing signals, on-hook signals and the voices of a conversation. A keysystem uses four wires to connect a telephone to the KSU. Two wires are used to carry voice and DTMF dial signals. Two additional wires are used to carry signals between the telephone and a microprocessor located in the KSU. The phone will have one button for each exchange line coming to the keysystem. For a 308 system, the multiline phone will have three buttons for outside lines. It will also have a hold button and an intercom button. When the multiline phone is taken off-hook nothing will happen unless a line button on the phone is depressed. When the phone is off-hook and a line button depressed, a microprocessor in the phone will send a signal over the signal pair of wires to the microprocessor in the KSU. This signal will cause a switching matrix in the KSU to connect the telephone to the desired exchange line. On outgoing calls, a dial tone will be received from the exchange, and the station can send DTMF tones over the voice pair of wires using the DTMF pad on the multiline phone.

6.17 KEYSYSTEM AUTOMATED ATTENDANT

In any keysystem, one of the stations will be designated as the attendant position or station. The system is configured to ring this station on all incoming calls. The attendant can answer incoming calls, put them on hold, and then use the intercom to signal a station to pick up the appropriate incoming line. Many keysystems are capable of automatically answering incoming calls and requesting callers to identify which department they want. An auto parts store using this system will prompt callers to press 1 for service, 2 for parts, and 3 for tires, and instruct callers to stay on the line if they are not calling from a touchtone phone. The microprocessor will detect the DTMF tones received and ring the appropriate station. If no tones are received within a few seconds, the call will ring in on the attendant's phone. Each station is assigned an intercom number. To call another station, the user will press the intercom button and then press the digit on the DTMF pad for the station desired. A DTMF receiver in the intercom circuitry will detect the station number dialed, and ask the microprocessor of the KSU to connect the two stations. The CPU of the KSU operates the switching matrix in the KSU and rings the called station.

6.18 KEYSYSTEM CABLING

Keysystems come in a variety of sizes and with a wide assortment of features. Early electromechanical keysystems such as Western Electric's 1A2 Keysystems required a 25-pair cable between each telephone station and the KSU. The *Electronic Key Telephone Systems (EKTS)* manufactured today only need two pairs of wires between the station and KSU. The reduction in cabling has been a boon to keysystem use. Keysystems are found in almost all small businesses that require more than two exchange lines. Auto parts stores, small department stores, home improvement stores, as well as the offices of doctors, lawyers, and dentists are prime candidates for a keysystem. Many businesses only need a small EKTS, but it is possible to purchase an EKTS that will handle as many as 24 central office trunks with 100 stations. However, it is usually advisable to use the more efficient (but much more costly) PBX when a business needs this many central office lines and/or stations.

6.19 LARGE KEYSYSTEMS (HYBRIDS)

Today's EKTS offers many PBX-type features. Large keysystems are usually called *hybrid keysystems*. They are considered a hybrid of a keysystem and a PBX. With a large keysystem, it is impossible to put keys for all central office lines on the phones, so central office lines are accessed by dialing a 9 (the type of access used by a PBX). A regular small keysystem offers access to central office lines by pushing a button on the keysystem phone associated with the line desired. The CPU of the EKTS is controlled by software which is burnt into a read-only memory (ROM) chip. A small portion of the CPU is controlled by an electrically erasable programmable read only (EEPROM) chip. This chip contains features that can be programmed by the customer from the attendant's telephone. The customer can program which central office lines can be accessed by each phone and which phones will ring on incoming calls. Each line on each phone can be programmed individually to ring or not ring. All programming is changed in the EEPROM chip controlling the CPU of the KSU. Each station also has a microprocessor (CPU) for controlling signaling to the KSU. This microprocessor is also controlled by ROM and EEPROM. The EEPROM is used to store speed-calling entries that can be programmed by the station user. A station user also has access to system speed calling lists stored in the KSU's EEPROM by the attendant station.

Early keysystems required all calls to be answered by an attendant. After answering a call and determining who the call was for, the attendant would put the caller on hold an ring the intercom station of the person needed. Once the station user answered the intercom, the attendant would advise them which incoming line they should answer. Today's keysystems have automated this attendant function and allow callers to direct their own calls. The KSU answers the call with a recorded announcement instructing the caller to depress a DTMF key on the phone for the department required (for example, "Thank you for calling Cole Auto, please press 1 for tires, 2 for parts, 3 for service, and so on").

6.20 KEYSYSTEM FEATURES

The features available in keysystems depends on how old the technology of the particular system is. Some features are provided by the KSU and others are provided by the station instrument itself. Today's current technology offers:

1. The ability to pick up or access any central office line.
2. The ability to put any line on hold.
3. Automatic recall of a line that has been on hold for a while. The time a line can be on hold before automatic recall is activated is programmable by the attendant.
4. Last number redial (this is a feature provided by the telephone).
5. Speed calling (some by the telephone and some by the KSU).
6. Hands-free answer (by the telephone).

7. Intercom (some systems provide several conference lines).
8. Music on hold.
9. Power fail transfer (the KSU transfers line one to a 2500 phone when a power failure occurs).
10. Privacy. Unlike the old 1A-2 keysystem, today's keysystem will not allow multiple phones to pick up on the same line unless the first station to use the line authorizes it via the establishment of a conference call.
11. Programmable features and restrictions such as the ability to restrict which stations can use outside lines and which stations can place toll calls.
12. Some systems provide station instruments that have displays for time, date, number dialed on current call, amount of time on current call, station number of caller, and speed call list. Many of these features are provided by the phone itself, but some—such as calling station—require input from the KSU.
13. Distinctive ringing provides a different ringing signal for incoming lines and intercom calls.
14. All keysystems provide supervisory signals to lights on each phone for each line and intercom button. The lights are usually off when the line is not in use but are lit steady when in use. If a line is on hold the light will blink. The lights are usually inside the button used to access a line or intercom.
15. Paging is offered by most keysystems. A person can initiate a page from any station to any individual station or have the page broadcast to all stations.

All current keysystem telephones have a hands-free speaker. Today's telephones no longer have a ringer and the ringing signal is converted to an audible twitting sound emanated from a small speaker. The EKTS takes this speaker technology one step further. The speaker allows a caller to place a hands-free call and hear the call progress in the speaker. The speaker will not pick up the station user's voice. Once the called party answers a call from the keysystem user, it is necessary for the handset to be used by the keysystem caller. It is possible to buy a true hands-free speaker phone that will pick up the station user's voice and allow conversation both ways over the speaker. These "speakerphones" cost more than a "hands-free phone" and are usually only provided to senior management stations.

6.21 WHEN SHOULD YOU USE A KEYSYSTEM?

A keysystem provides a business with access to multiple central office trunk lines and many features typical of a *Private Branch Exchange (PBX)* at a small portion of what a PBX will cost. A keysystem is the best solution to the telecommunication needs of a small- to medium-sized business. Keysystems are not designed to handle a large volume of calls at the same time. A PBX or hybrid keysystem is required for large calling volumes. When a business has more than ten incoming trunks from the central office, a hybrid is preferred over a regular keysystem because a regular keysystem requires one button on the phone for each trunk. When a business has more than 24 trunks or more than 100 stations, a PBX is preferred. Regular keysystems are the perfect station equipment for small retail stores, restaurants, doctor's offices, and dentist's offices.

6.22 SUMMARY

This chapter has focused on the various analog telephone sets. These devices are used to convert voice signals into analog electrical signals. Most of these telephones are connected over a local loop to a digital central office. The central office interface to a local loop cable pair is called a line circuit. In a digital central office, the line circuit includes a codec to convert the analog signal into a digital signal. Telephones are being manufactured that include the codec in the telephone. Most of these are manufactured for use on PBX switching systems. The telephone, manufactured with a Codec, for use on the local loop is called an Integrated Services Digital Network (ISDN) telephone. The ISDN telephone converts a voice signal into a digital signal for transmission over the local loop but the local loop must be a DSL. This limits the range of an ISDN line to three miles unless repeaters are added to the DSL. ISDN technology is discussed in greater detail in Chapter 11.

REVIEW QUESTIONS

1. Who is credited with inventing the telephone? When?
2. What was the key element in Elisha Gray's design and Bell's design.
3. What was Thomas Edison's contribution to the telephone?
4. How much current does a carbon granule transmitter need?
5. What type of voltage is required for power to the standard single-line telephone set (2500 set)?
6. Which standards agency sets the standards for telephones?
7. What larger classification of telephone equipment does the telephone belong to?
8. Why must station equipment be registered with the FCC?
9. What is a multiline phone?
10. What advantages does a keysystem offer over a simple multiline phone?
11. What is an easy way to tell if a keysystem is a hybrid system?
12. What is the difference between a hands-free phone and a speakerphone?
13. What is the difference in cabling for the old 1A2 and today's EKTS?
14. What type of switching matrix does an EKTS use?
15. How are system features programmed?
16. How are station features programmed?

17. What is the purpose of a hybrid circuit in the telephone? *conv. 2 w/ 4w*
18. What is sidetone and what purpose does it serve? *signal coupl. to Receiver/hear w/y say*
19. How many tones are sent to the central office when one DTMF button is depressed? *2*
20. Why is a varistor used in the telephone? *decrease voltage on short loops*

GLOSSARY

Alerters A ring detector circuit that results in a chirping sound from a speaker.

Analog Signal An electrical signal that is analogous (similar) to a voice signal. The signal continuously varies in amplitude and frequency. An analog signal has an infinite number of values for voltage, current, and frequency.

Bridged Ringer A term indicating that the ringer is wired between the Tip and Ring leads. Bridged ringers were used on private lines as well as on two- and four-party lines.

Decimonic Ringing A ringing system used by Independent telephone companies. The system could send out five different ringing signals (10-, 20-, 30-, 40-, and 50-cycle signals).

Digital Signal An electrical signal that has two states. The two states may be represented by voltage or current. For example, the presence of voltage could represent a digital logic of 1, while the absence of voltage could represent a digital logic of 0.

Dual-Tone Multifrequency (DTMF) Dial The dial on a 2500 set that sends out a combination of two tones when a digit on the keypad is depressed.

Feature Phone A telephone containing electronics that allow it to provide features. When additional features have been added to the basic single-line telephone such as last number redial and speed dialing lists, the phone is called a feature phone.

Grounded Ringer A term indicating that the ringer is wired between the Tip lead and ground or between the Ring lead and ground. Grounded Ringing was used on eight- and ten-party lines.

Harmonic Ringing A ringing system used by Independent telephone companies. The system could send out five different ringing signals (162/3-, 25-, 332/3-, 50-, and 662/3-cycle signals).

Hookswitch Often called a switch-hook. This is the device in the telephone that closes an electrical path between the central office and the telephone, by closing contacts together, when the receiver is lifted out of its cradle.

Hybrid Keysystem A keysystem designed to handle more than 24 central office lines and 40 telephone sets. Central office lines are accessed by dialing the digit 9 for access.

Hybrid Network A network that consists of a transformer and an impedance matching circuit. The transformer performs a two- to four-wire conversion and vice versa. In a telephone, the hybrid network connects the two-wire local loop to the four-wire transmitter/receiver.

Keysystem A telephone system used by small businesses to allow several central office lines to terminate at each telephone attached to the system.

Magneto A device that generates an AC voltage by turning a coil of wire inside a magnetic field. The old hand-crank telephone had a magneto attached to the hand-crank.

Private Branch Exchange (PBX) The name given to manual switchboards in a private business location. A PBX may also be small SPC switching system used by large businesses.

Proprietary Telephone A telephone designed by its manufacturer to only work with certain keysystems or PABXs made by that same manufacturer.

Receiver The device responsible for decoding or converting received information into an intelligible message.

Ring Detector An electronic solid-state transistorized device designed to detect the presence of a ringing signal (a 90+-V AC signal).

Ringer An electromechanical device (relay) that vibrated when a 90+V AC signal was received by the telephone. The vibrating device would strike metal gongs to create a ringing sound.

Sidetone In a telephone set, some of the transmitted signal is purposely coupled by the hybrid network to the receiver so you can hear yourself in the ear covered by the telephone receiver. This signal is called sidetone.

Station Equipment The largest segment of station equipment is the telephone but the term station equipment has been broadened to include anything a customer attaches to a telephone line. The most common piece of station equipment is the telephone. A modem, CSU/DSU, personal computer, and keysystem are all referred to as station equipment.

Straight-Line Ringer A ringer that will operate on any ringing signal. This type of ringer is used on single-party private lines.

Superimposed Ringer The type of ringer used by Bell on four-party lines. The ringer included a diode. Ringing signals were superimposed on top of a DC voltage. The diode only allowed the signal with the correct polarity of DC for that station to pass to the ringer.

Transmitter The device responsible for sending information or a message in a form that the receiver and medium can handle. The device in a telephone that converts the air pressure of a voice signal into an electrical signal that represents the voice.

Tuned Ringer A ringer that will operate on only one ringing signal. These ringers were used on party lines. They were tuned to only ring on the signal assigned to that party.

2500 Telephone Set The standard single-line telephone, which contains a DTMF dialpad.

Varistor A specially designed resistor. As the voltage across the resistor increases, the resistance increases.

7

Automated Local Exchange Switching

Key Terms

Blocking
Circuit Switching
Class 5 Exchange
Concentration
Exchange Code

Extended Area Service (EAS)
Local Exchange
Numbering Plan Area (NPA)
Public Switched Telephone
    Network (PSTN)

Register (Dial Register)
Space Division Switching
Tandem Exchange
Time Division Switching
Translator

As the number of telephones in use grew, it became necessary to explore ways to automate the switching process. If we did not have automated switching today, it would require that all of our population work as switchboard operators or that we reduce by 90% the number of people allowed to have or use phones.

Our economic growth is tied directly to growth and development in the areas of transportation: transportation of people; transportation of goods; and transportation of knowledge. The transportation of knowledge is critical to our decision-making capabilities and intellectual growth. The amount of information available to us, and the speed with which information can be accessed, are directly correlated with progress in the telecommunications infrastructure of our country. The ability to learn has always been directly related to the access a person has to reference materials and instructors. The growth of telecommunications has expanded the base of intellectual resources beyond what is available from the local library and local intellectuals. The citizens of other countries are as smart as the citizens of the United States, but in many cases the synergy of their intellectual resources has been less than optimal due to a poor or nonexistent telecommunications infrastructure. On the other hand, some countries have been able to achieve a high level of education for their citizens despite the lack of a telecommunications infrastructure. These countries have been favored by certain factors, including in some cases small size—which may bring intellectuals close together and facilitate the exchange of information.

Instant access to news of world events is possible due to the advances in telecommunications. For example, news coverage of the Gulf War provided us with an immediate presence on the war scene. This event dramatically demonstrated the advances that had been made in satellite and radio communications. Who can forget the live pictures of air raids and anti-aircraft fire that were captured in Baghdad and beamed immediately into our living rooms? Radio-controlled and laser-guided missiles, along with the use of "spy satellites" to capture the actions and movements of enemy troops, played a crucial role in the swiftness with which this action was brought to a successful conclusion. Intelligence gathering has been made much easier with the advances in telecommunications technology.

The benefits we enjoy from this technology are not limited to military use. It is true that most successful military ventures belong to those who have the "intelligence edge." It is equally true, in the business world, that those with the best, most accurate, latest, and fastest information sources win most of the battles. Providing your company with the best telecommunications technology available not only gives it a competitive edge; this strategy is necessary to survive.

Business associates, friends, and family can now easily talk with each other over extended distances on a daily basis due to advances in our national telecommunications infrastructure. Telecommunications progress allows immediate access to numerous sources of information and enables individuals to make more informed and quicker decisions. The explosion of the Internet is the latest development to make more resources available for learning and research.

Our telecommunications infrastructure is composed of many different networks, each designed for a specific purpose. Some networks are used to transport television and radio signals around the world. Others, such as the Public Data Network, are used only for data. In this chapter, we will explore the Public Switched Telephone Network (PSTN) in more detail. The chapter will focus on the nodes of the PSTN. These nodes are switching systems (central offices—that is, central exchanges). The PSTN is designed to handle dial-up voice and/or data calls. Customers can also lease circuits in the PSTN from the LECs, IECs, and other common carriers, to establish dedicated nonswitched circuits or a private network.

7.1 AUTOMATED CENTRAL EXCHANGES

Innovations like automated switching systems, fiber optics, and multiplexing techniques for placing numerous conversations over one medium represent the greatest advances in telecommunications. We can thank AT&T Bell Labs for most of these developments in telecommunications technology. Bell Labs invented the transistor. The integrated circuit chips used in computers, and the signal logic processors used in a modem, would not exist if Bell Labs had not invented the transistor. Bell Labs not only invented technology but also improved on the technology invented by others. Bell Labs developed and enhanced the fiber optic technology used by IECs and LECs. It also invented and improved the automated electromechanical crossbar and the computerized, Stored Program Control, switching systems.

The automated switching systems allow us to reach anyone in the world simply by dialing the person's telephone number. This automated system can interpret dialed digits as a specific address location on the *Public Switched Telephone Network (PSTN)* and connect one caller to anyone else. Automated switching systems were placed in the central exchanges to automate the functions previously handled by operators. Throughout this book, references are made to these switching systems as either central offices, exchanges, or central exchanges. All three terms mean the same thing and are interchangeable.

7.2 THE LOCAL EXCHANGE

All telephones connect to switches (central exchanges) usually located within a few miles of the telephone. We call the switch that a telephone connects to the *local exchange* or local office. A local switching exchange is also referred to as a *class 5 exchange*, or the *end office*, because it is the lowest and last switching system in the PSTN five-level hierarchy. The class 5 exchange is the node on the PSTN used by a telephone to access the PSTN. All telephones connect to the end offices using line circuits in the end offices. The end office connects to other class 5 and class 4 offices using trunk circuits. The switching system, for an end office, must be capable of connecting any of its line circuits to any other line circuit for a local intraoffice call. An *intra*office call is a call completed within one class 5 exchange. Calls that cannot be completed within the originating exchange but require connections via another central office are called *inter*office calls. *Intra* means "within" and *inter* means "between."

Figure 7-1 illustrates an intraoffice call between two telephones serviced by the same local exchange. The exchange prefix for both telephone numbers is 941. Telephone number 941-3333 has called 941-2222. The switch translates the dialed number into switching instructions and connects the line circuit serving 941-3333 to the line circuit serving 941-2222.

7.3 TOLL CALLS OVER THE PUBLIC SWITCHED TELEPHONE NETWORK

All *inter*state toll calls must involve a toll office and an IEC to complete the call. With the passage of the Telecommunications Act of 1996, LECs can also perform IEC functions and can complete interstate calls. When a customer places a long distance telephone call, the local exchange will connect the caller to the class 4 toll office serving the originating class 5 exchange. If the toll call covers less than 30 mi, the class 4 toll exchange will also be serving the terminating class 5 office. Most of these short distance calls are within the same LATA. In this case the caller is connected from the class 5 office to the toll office, and the toll office will connect the call to the class 5 office serving the called number (Figure 7-2).

The use of a five-level hierarchy for the toll network allows the PSTN to use far fewer trunk circuits in the network than if a four- or three-level network was used. There are almost 20,000 class 5 central exchanges in the United States. If some type of hierarchy did not exist, each class 5 exchange would need multiple trunk circuits to all other class 5 exchanges—a costly and antiquated network design. Just

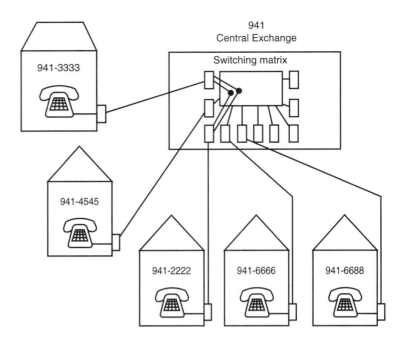

FIGURE 7-1 Intraoffice call.

as the first central exchange was put in place to eliminate having to wire each phone to every other phone, the hierarchy design was implemented to eliminate the need for direct connections between exchanges.

The design of any network involves a trade-off between the number of transmission facilities (circuits) and the number of switching stages used. It would be physically impossible or at least extremely costly to have a direct connection between each and every local switching exchange in the United States. A much better arrangement is to connect several class 5 offices to a class 4 office. The five-level hierarchy design used by AT&T was put in place to handle toll calls. All calls between exchanges were toll calls. The class 4 office was called a *toll office*. When someone dialed a 1 or 0 for a long distance call, the person was connected from the class 5 office to the class 4 toll office via an interoffice toll trunk.

There are about 950 class 4 toll offices in the United States. Most class 4 toll offices are owned by one of the BOCs. If the call is an inter-LATA call (from one LATA to another), it is handed to the originating caller's Preferred IEC. With divestiture in 1984, the separation of AT&T and the BOCs left the ownership of the class 3 Primary Centers with AT&T. Therefore, inter-LATA calls handled by AT&T are passed from the LEC's class 4 toll center to AT&Ts class 3 Primary Center. There are about 150 AT&T Primary Centers located in the United States. If AT&T's Primary Center cannot complete the call, it will pass the call on to a class 2 Sectional Center. AT&T has about 50 class two Sectional Centers. If the class 2 Sectional Center cannot complete the call, it passes the call to a class 1 Regional Center. AT&T has about seven class 1 Regional Centers in its network.

When Sprint built its toll network, it was able to take advantage of the latest in switching and transmission technologies to create a flatter network. Sprint has about 48 Toll Centers, which are combined class 3, class 2, and class 1 exchanges.

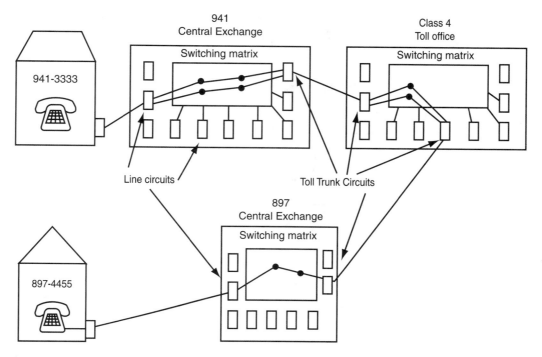

FIGURE 7-2 Interoffice call.

These 48 Toll Centers are computer-controlled switching systems and are interconnected with fiber optic cables. When a toll call is handed from the LEC to Sprint, it is passed from the LEC's class 4 toll office to one of Sprint's 48 Toll Centers.

If the called telephone number is more than 1000 mi away from the originating exchange, the caller must use several layers of the PSTN network. Figure 7-3 is a diagram showing the completion of a toll call from Wausau, Wisconsin, to Raeford, North Carolina, using AT&T's five-level network.

7.4 EXTENDED AREA SERVICE IN THE PUBLIC SWITCHED TELEPHONE NETWORK

Over time, customers demanded that it be possible to place calls between some exchanges without having to pay a toll charge. Sometimes a town or city has several central exchanges located around the city. Customers demanded the ability to place calls between these exchanges without a toll charge. This free interoffice calling between local exchanges is called *Extended Area Service (EAS)*. EAS service is not really free. The cost for this service is included in the local service rate charge. As additional EAS points are added to a particular local exchange, the local service rate for the exchange is increased. The more telephones that can be called without a toll charge from a local exchange, the higher the local rate is. This is why the rates for telephone service in a city are higher than the rates in a small town.

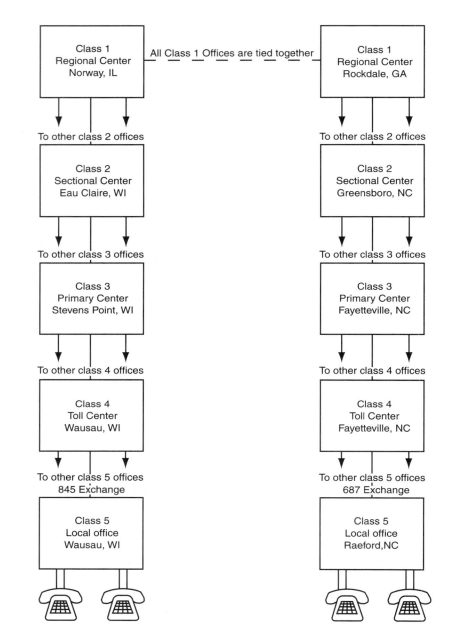

FIGURE 7-3 The five-level
PSTN network.

In a small town there is only one central exchange. This exchange serves the town and the adjacent rural area. If another town is close by, EAS calling may be allowed to it. When two towns are close to each other and have a high degree of community interest between them, they will petition the PUC for EAS. Most small towns do not have EAS. When they do have EAS, only a few exchanges will be involved. When the number of exchanges involved in an EAS plan is small, these class 5 central exchanges are connected directly to one another via EAS trunk circuits (Figure 7-4).

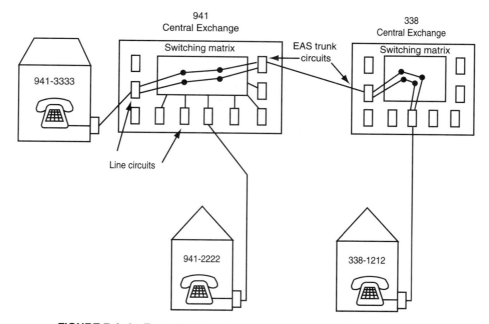

FIGURE 7-4 An Extended Area Service call direct to another exchange.

In a larger city, there are many local exchanges within the city that must be able to connect calls between them without a toll charge. A person originating a call from an exchange in the southern part of New York City must be able to call a telephone in a northern exchange (or any of the 50-plus exchanges serving New York City) without a toll charge. When many class 5 switching systems exist in an EAS network, it is cheaper to use a class 4 switching exchange and connect the EAS calls between class 5 exchanges via the class 4 exchange than to use direct circuits between each office. The use of a class 4 switch to connect calls between class 5 exchanges significantly reduces the number of trunk circuits that would be needed to connect many class 5 exchanges directly to each other. This reduces the cost of the EAS network.

The class 4 switch used for EAS calls is called the *class 4 EAS Tandem exchange*. Each of the local exchanges is connected to the **Extended Area Service (EAS)** Tandem using EAS trunk circuits. Local calls between exchanges will be connected from the originating exchange to the EAS Tandem and the EAS Tandem will connect the call to the appropriate class 5 exchange containing the called number (Figure 7-5).

7.5 CIRCUIT SWITCHING VIA TANDEM EXCHANGES IN THE PUBLIC SWITCHED TELEPHONE NETWORK

The switching systems in the PSTN connect (or switch) input circuits to output circuits. These switches have the ability to connect any circuit attached to the switch to any other circuit attached to the switch. This type of switching is called *circuit switching. The PSTN is a circuit-switched network.* As stated earlier, the central exchange

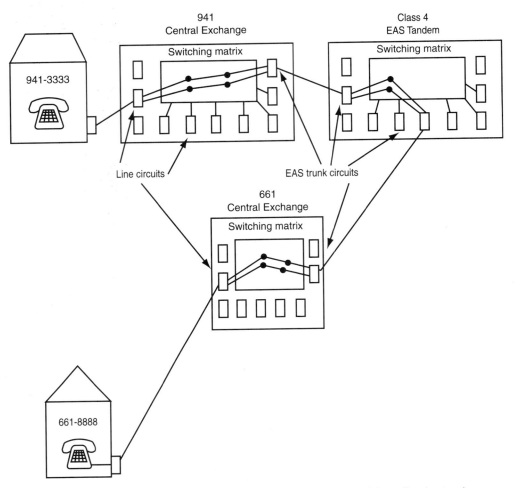

941
Central Exchange

Class 4
EAS Tandem

941-3333

Switching matrix

Switching matrix

Line circuits

EAS trunk circuits

661
Central Exchange

Switching matrix

661-8888

FIGURE 7-5 Extended Area Service connections via an Extended Area Service tandem.

was developed to allow any telephone to be connected to any other telephone linked to the exchange. When a local call is made within a switch, it is called an *intraoffice call*. The class 5 switching system connects one line circuit to another line circuit to complete an intraoffice local call. Because the call is completed within the switch, it is also called an *intraswitch call*. When a called number is not found within the originating switch, the call must be sent to another switch for completion of the call. If the call is a toll call, it will be sent to a class 4 toll switch for handling. The class 5 switch will connect the line circuit of the caller to a trunk circuit going to the class 4 toll office.

The class 4 to class one toll exchanges do not have subscriber line circuits attached to them. These exchanges only have trunk circuits that connect to other central exchanges. All class 1, 2, 3, and 4 exchanges are called *Tandem exchanges*. Here *tandem* means "in conjunction with" or "associated with." Each class 5 office connects to a certain class 4 office. Each class 4 office connects to a specific class 3 office. Several class 5 offices connect to each class 4 office. Several class 4 offices connect to each class 3 office.

Automated Local Exchange Switching   **Chap. 7**

Likewise for class 3, 2, and 1 offices in a five-level hierarchy network. Each of these is associated with specific offices below and above its specific level. The PSTN has been flattened with the introduction of fiber optics and computerized switching. Figure 7-6 illustrates the connection of our theoretical long distance call from Wausau, Wisconsin, to Raeford, North Carolina, using the flatter Sprint Network.

7.6 ONE-STAGE VERSUS MULTIPLE-STAGE SWITCHING NETWORKS

Just as multiple levels of switching in the multilevel toll network have reduced the number of interoffice routes required in the PSTN, multiple switching stages in the local exchange reduce the number of circuits required between stages. A one-stage switching network in the local exchange would require many more circuits than a four-stage switch. A one-stage switch requires that every circuit must exist as an input and an output on the switching stage. This makes the switching stage very large. Some small switching systems designed for private use as a *Private Automated Branch Exchange (PABX)* use a single-stage network when the number of lines handled by the PABX is small (Figure 7-7).

Figure 7-7 is a single stage network that shows circuit 30 connected to circuit 70. All circuits are wired to both the inlet and outlet ports. Any circuit can then be connected to any other circuit by activating the crosspoint at the intersection of the two circuits in the network. This figure only shows every tenth circuit for purposes of clarity.

A switch used as a central office, or a large PABX, will have thousands of line circuits and will use a multiple switching stage network. By using many stages, the size of the stages can be reduced. A portion of the total circuits served by the central exchange will be connected to each stage. This makes the stages smaller. The stages are connected via interstage circuits to allow for calling between circuits that are on different stages. The number of interstage circuits to use depends on the calling volume the switch will handle (Figure 7-8).

Figure 7-8 illustrates a four-stage network with 100 circuits attached to a stage 1 input and 100 circuits attached to the output of a stage 4. There are numerous stage 1, 2, 3, and 4s located within the switching system. Notice that there are only 40 outlets from stage 1 to stage 2. There are actually more than 40 inlets and outlets on switching stages 2 and 3, but they connect to other switching matrices. Only the path for connecting our call is shown (circuit 50 to circuit 400). Each stage is larger than drawn. For example, stage 1 will have outlets connected as inlets to other stage 2 matrices in order to access all possible outlets connected to the switching system. The controller for the switching system (marker in a XBAR or CPU in a SPC switching system) will activate the crosspoints needed in each stage to connect the caller to the desired output.

7.7 BLOCKING VERSUS NONBLOCKING SWITCHING NETWORKS

A single-stage network provides only one possible path for each connection. Each inlet connects to a specific outlet via one specific crosspoint. If that crosspoint or path

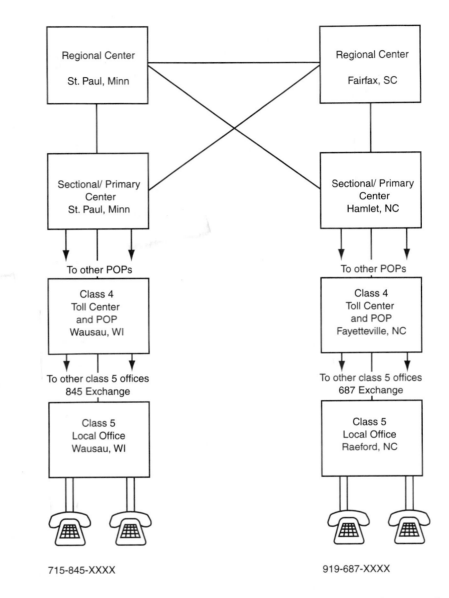

FIGURE 7-6 The Flatter
Four-Level Network.

715-845-XXXX   919-687-XXXX

is defective, the path must be disabled and the input/output for that path cannot be assigned. If a call cannot be completed due to the lack of a path through the network, the call is said to be **blocked** by the network. Single-stage networks are nonblocking networks because a path through the network exists for every connection to the network. That path will never be used for another call and will always be available for the particular inlet/outlet connection.

A multiple-stage network provides many paths over which a call can be completed. The outlets on each stage connect to inputs of other switching stages. The switching stage that telephones connect to is called a *line link frame*. It has 100 inlet connections and 40 outlet connections. Since everyone does not use

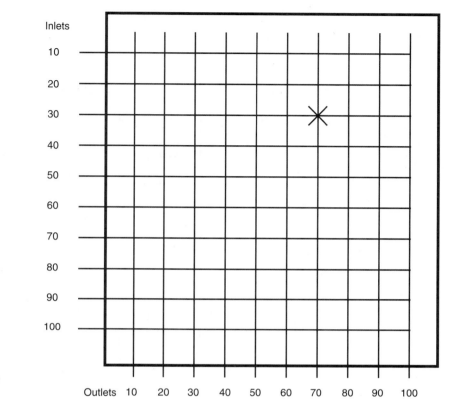

FIGURE 7-7 Single-stage network.

their phone at the same time, there is no need to provide 100 outlets. The use of fewer outlets than inlets is called *concentration.* In our example (Figure 7-8), the line link frame concentrates 100 inlets into 40 outlets, providing a concentration ration fo 2.5 to 1. SXS switching systems also used concentration. A linefinder served 200 lines and was equipped with from 10 to 20 linefinders. This provided a concentration ration of between 20 to 1 and 10 to 1.

The use of concentration reduces the cost of the switch. The amount of concentration to use is determined from past studies on the amount of telephone usage by customers. From this study a determination is made on how many servers are required to serve that amount of usage with a minimum number of blocked calls. Calls that do not get connected to their desired destination, but receive a fast busy signal are referred to as *blocked calls* because their completion was blocked by the switching system. The basic function of any switch is to connect inputs to outputs by providing circuit paths between the two. If the switching network contains fewer paths than the inputs and outputs attached to it, it is a *blocking network* (Figure 7-8).

Figure 7-8 is a blocking network. It is impossible to connect all input circuits of any stage 1 to all output circuits of any stage 4 due to the concentration at stage 2. The drawing in Figure 7-8 has been simplified a great deal for purposes of clarity. In any *space division switching* matrix (such as the one in Figure 7-8), concentration is used because we know that the customers on any given input stage will spread their calls out over all outgoing circuits. Switching paths are designed to accomplish this spread

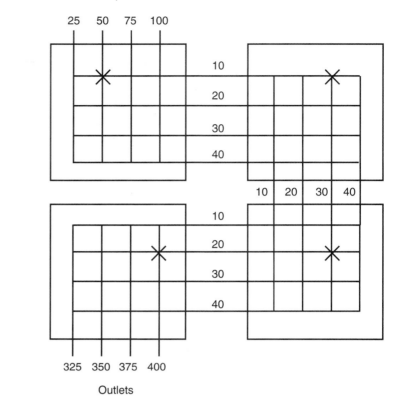

FIGURE 7-8 Four-stage network.

calling pattern to allow any inlet to reach any outlet. Since we know the calls will be spread out, it does not make economic sense to provide as many paths between inputs and outputs as there are input and output circuits. Of course, when everyone does try to place a call to one output (such as the telephone number of a radio station giving away $1 million to the 15th caller), the system blocks calls because insufficient paths are available to reach the output.

Some networks are called virtually nonblocking networks because they contain enough paths that blocking rarely occurs. Even if a network is nonblocking, blockage can still occur in the switch because a switch contains more than just the switching network. An example of such blockage would be when making a long distance call. The network of the switch may not block the call, but there will not be as many long distance trunks as there are phones attached to the network, and when all available toll trunks are busy the call is blocked.

The word *network* is overused in telecommunications and has several meanings. The network of a switching system is also referred to as the *switching fabric*. This network (as in switching network) is a device to provide a circuit path between any of its line or trunk circuits. Public Switched Telephone Network or PSTN refers to the interconnection of switches on the network (meaning on the PSTN). The telephone contains a device called the *hybrid network*. This network converts the two-wire circuit of the local loop into a four-wire circuit (two wires for the transmitter and two for the receiver). In short, the meaning given to the word *network* depends on the context in which it is used.

7.8 ESTABLISHMENT OF THE NORTH AMERICA NUMBERING PLAN

Establishment of seven-digit dialing was part of the North American Numbering Plan. North America was divided into many **numbering plan areas (NPAs)**. Each NPA was assigned a three-digit number called the *area code*. Within each NPA, each central office exchange is assigned a three-digit number called the **exchange code**. This three-digit number represents the first three digits of the seven-digit telephone number. In numbering plan assignments, if a number could be any value from 1 to 0, it was designated by the variable X. If a number could be any value from 2 to 9, it was designated by the variable N. If a number could be only a 1 or 0, it was designated by the variable 1/0. Exchange codes were of the form NNX. Area codes were of the form $N1/0N$.

The area code had to be of the form $N1/0N$ to assist electromechanical switching systems (such as SXS and XBAR) in the switching of calls. The use of a 1 or 0 as the second digit of an area code allowed electromechanical switches to distinguish between an area code and an exchange code. Area codes could be identified because the first and third digits could be any number except a 1 or 0, while the second digit must be a 1 or 0. Central office exchange codes were designated by numbers of the form NNX. A central office exchange code could not have a 1 or 0 as either a first or second digit. Additionally, each NPA or area code could *not* have more than one central office exchange code with the same code. Every exchange within an area code had to have its own unique code. Central office numbers (exchange codes) could be reused in many different area codes, but within any particular area code or NPA, each central office exchange must have a unique NNX identifier.

When the North American Numbering Plan was implemented, existing exchange names were converted to exchange codes. The Willow exchange, for example, was converted to 941. Look at the telephone dial and you will see that the letter W is also digit 9. The letter I is digit 4. Exchange names were converted to exchange codes according to which number on the dial contained the letter. Only the first two letters of the exchange code were converted using this technique. Have you ever heard of Glenn Miller's song "Pennsylvania Six-Five thousand" (Pennsylvania 65000)? This was the telephone number for the Pennsylvania Hotel (which later became the Statler Hotel) in New York City. Miller played in the hotel's Cafe Rouge, and people dialed the hotel's number for reservations. This famous number was 65000, located in the Pennsylvania exchange of New York City.

With the North American Numbering Plan, Pennsylvania 65000 became 736-5000. This is not nearly as glamorous sounding as using exchange names. Note that $P = 7$ and $E = 3$ on the telephone dial. This method of assigning exchange names allowed people to dial a number using the first two letters of the exchange name. Exchange names were also used in the directory (for example, PE 6-5000). After people had accepted seven-digit dialing, the abbreviated names were replaced by the corresponding digits. Local central offices are referred to by their exchange names and/or exchange code. The central office serving telephone numbers in a certain part of South Kansas City is called the Willow Exchange and is also called the 941 exchange.

Using area codes of the form $N1/0N$ made it possible to have 162 ($9 \times 2 \times 9 = 162$) area codes or NPAs. North America ran out of NPA area codes in 1995 and the requirement that the second digit of an area code be a 1 or 0 was dropped. Fortunately, by this time all electromechanical *toll* offices had been replaced by Stored Program Control (SPC) offices, and dialed numbers were no longer sent over the voice switching path. Switching connections between offices are now established over SS7 networks. These SS7 signaling networks will be discussed in greater detail in Chapter 10. The new area code numbering system was put in service in 1995. Alabama had run out of central office codes. The area code for Alabama was 205. A new area code of 334 was added on January 15,1995, for Montgomery and the southern half of Alabama. This change allowed the northern half of Alabama (which includes Birmingham) to reuse the exchange codes that had been reassigned to the 334 area code. The new 334 area code will be able to grow by using the exchange codes that had been left behind in the old 205 area (Figure 7-9).

7.9 STEP-BY-STEP OR STROWGER AUTOMATED SWITCHING

In 1892, Almon B. Strowger invented an automated switching device. His switching system is called the *step-by-step (SXS)* or *Strowger switch*. However, his device was not accepted for use by the telephone companies until around 1920. Between 1878 and 1920, the design of telephones changed to eliminate the local battery supply at the phone and the hand-crank generator. The central exchange design was changed to include a 50-V DC battery supply (in place of the earlier 24-V battery), a line relay for each line, and a 90-V AC ringing supply. The battery used in the Strowger central office consisted of 23 active cells (50 V). The battery used in today's computer-controlled switches consist of 24 active cells (52 V). The batteries in a central office are continuously being charged by a battery charger. The battery charger converts the AC power supplied by the electric company to an output of 50 (or 52) V DC, which is connected to the central office 50-(or 52-) V battery. Each telephone is powered from this 50- or 52-V DC supply via the wire pair from the central exchange to the telephone.

Each wire pair leaving a central office is referred to as a *line* if it connects to a telephone. The 50- (or 52-) V supply passes through a relay in the central exchange before being placed on the wire pair. This relay is called a *line relay*. There is one line relay for every line coming into the central exchange. In a manual switchboard system, these relays were located outside the operator switchboard and were wired between incoming phone lines and the switchboard. Strowger wired the line relays to a mechanical switching system instead of the operator switchboard. The line relays were connected to a device called a *linefinder*. The purpose of a linefinder was to automatically find a line that was requesting service.

A telephone requested service by taking the receiver of the telephone off hook. Once the linefinder found the line requesting service, it connected this line to the input of a first selector. The first selector would provide dial tone and collect the first dialed digit. On receiving of the first dialed digit, the first selector connected the originating line to another selector, which would collect the next dialed digit. As each digit was dialed, the selector that a line was attached to would take one step for each value of the digit. If a 5 was dialed, the selector switch would take five steps. The Strowger SXS system was called a *direct-controlled switching system* because each switching stage (that is, selector) was under direct control of the originating telephone's dial.

Since all new switching systems employ SPC and a digital switching matrix, SXS systems have become history in all but a few rural areas. Most college and university enginerring curriculums limit the discussion to the latest computer-controlled switching systems. This book supports that approach by focusing on the latest switching designs. Chapter 8 expands on the technology behind the SPC switching systems in use today. The present chapter provides some detail on crossbar (XBAR) switching, because XBAR is the system on which computer-controlled switching is designed. For reference purposes, Figure 7-10 presents a diagram of a local intraoffice call through the various switching stages of a Strowger SXS switching system.

7.10 STEP-BY-STEP SPACE DIVISION SWITCHING

The automated switching system invented by Strowger is referred to as step-by-step (SXS) or Strowger switching in honor of its inventor. In a step-by-step switching system, each switching stage connects a pair of input wires to a pair of wires going to the next switching stage and finally through a connector switch to the pair of wires going to the line being called. The wires connected for a particular call remain connected for that call, and only one conversation takes place over these wires. Each call established through a step-by-step office uses separate wiring paths through the office for each call. The use of separate and individual wiring paths for each conversation path is referred to as *space division switching* technology.

A pair of wires connecting two devices to each other is referred to as a *circuit*. If the two devices (such as two phones) are connected directly, the circuit is known as a dedicated point-to-point circuit. When circuits are connected to a switch so that any one of the circuits can be connected by the switch to any other circuit, the switch is known as a *circuit switch*. Circuit switching is the type of switching used in the PSTN.

7.11 DISADVANTAGES OF DIRECT CONTROL STEP-BY-STEP SWITCHES

SXS switches have several drawbacks. They are incapable of offering many of the options customers want today, such as three-way calling, call waiting, caller ID, and ability to select a long distance carrier of choice. Another problem with SXS switching is the need to have dial pulses. Stepping switches need electrical current flow to operate. Turning the flow of current off and on in a controlled manner using a rotary dial controls the stepping switch (Figure 7-11). But telephones that we call touchtone telephones do not turn the flow of electrical current off and on to transmit a digit. These telephones transmit different tone combinations to represent each digit. When touchtone telephones were introduced, they could not be used on a SXS switching system. A device was invented to store the tones generated by a touchtone phone, and this

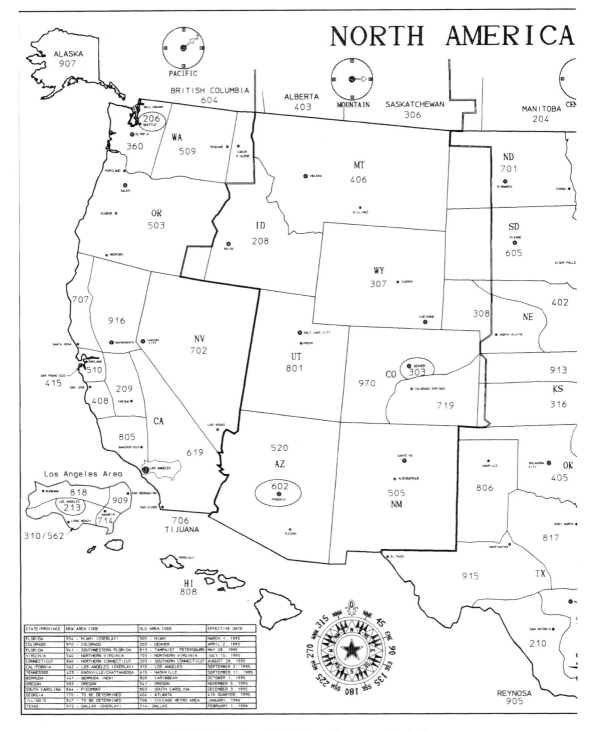

FIGURE 7-9 Area Code Map. (Courtesy of Sprint.)

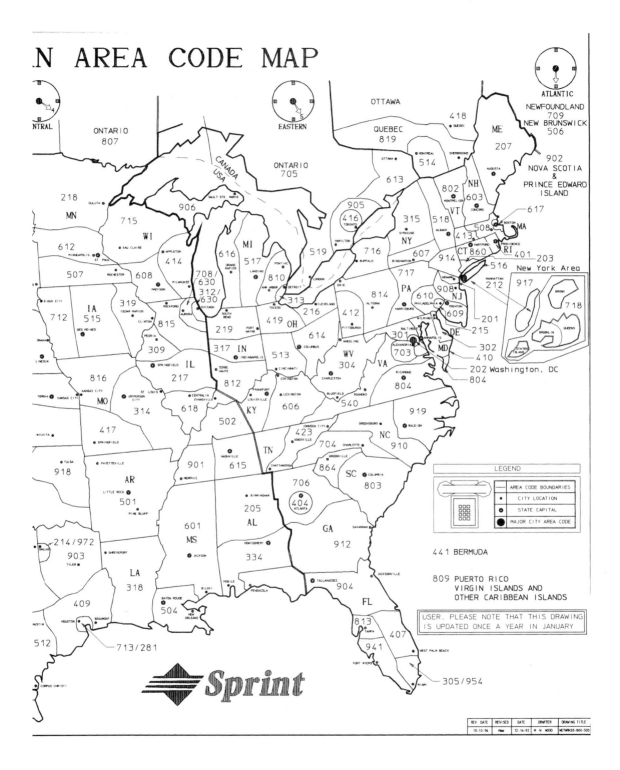

device converted the received tones into rotary dial pulses. These devices were placed between the linefinders and the first selectors. Once these devices had been added to a linefinder group, the lines in that group could have touchtone phones.

Because the telephone companies incurred additional expense in adding touchtone to dial pulse converters between all linefinders and first selectors, they charged an extra fee ranging from $0.75 to $1.25 each month for a touchtone phone. SXS switching has been replaced in all metropolitan areas with computer-controlled switching systems but is still found in small communities. Telephone companies in large cities no longer incur extra expense to provide touchtone service, and the additional charge for touchtone service should no longer exist.

Figure 7-11 illustrates dial pulses. These pulses can be depicted as either pulses of current or pulses of voltage. The top diagram shows that when a telephone is off-hook, a certain amount of current flows in the circuit to the central office. The amount of current depends on the length of the local loop. In our example, 40 milliamps of current is flowing when the telephone is off-hook. To dial the number 3 using a rotary dial, the caller inserts a finger in hole number 3 of the dial and pulls the dial around until the finger hits a finger stop. The finger is removed from the dial and the dial spins automatically back to its normal at-rest position. As the dial returns to normal, the loop is opened three times. Each time the loop is opened current flow will cease (0 milliamp). After opening the loop three times the dial will be back at normal and the loop remains closed with a

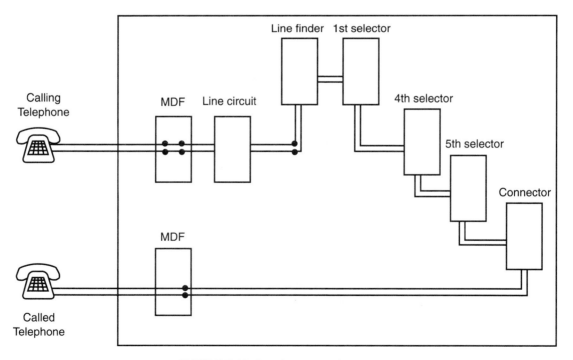

FIGURE 7-10 Step-by-step exchange.

steady current flow of 40 milliamps in the local loop and central office equipment. The equipment will recognize each absence of current flow as one pulse of a digit. The equipment counts the number of open-circuit conditions encountered and stores the count in a register as the dialed digit.

Figure 7-11 also shows the dial pulses as voltage conditions. Each time the dial opens the circuit, a voltmeter would read the 48 V of the central office battery placed on the loop. When the dial contacts close the loop, current will flow and voltage drops occur across each component in the circuit (that is, the central office equipment, the local loop, and the telephone). The voltage drop read across the telephone will be between 8 and 13 V (depending on the length of the local loop). In Figure 7-11, we are reading the voltage across the telephone, and in our example the voltage changes from 10 V (closed circuit) to 48 V (open circuit) for each open-circuit condition (that is, each pulse).

7.12 CROSSBAR SWITCHING

In 1938, the Bell telephone companies begin the deployment of a new switching technology called the *crossbar switch (XBAR)*. This switch was improved over the next ten years, and in 1948 the no. 5 crossbar switch was introduced. This switch consisted basically of line link frames, trunk link frames, and common control equipment. Common control equipment is a group of equipment used to establish call setup. After the call is set up, the equipment is released and is available to set up other calls. Instead of using linefinders, the crossbar used linescanners to determine when a line connected to a line link frame has gone off-hook. The scanner would send a signal to a device called the *marker*. The marker would assign another common control component called a *register* to the line. A set of crosspoints was made in the line link frame connecting the line to the register. This set of crosspoints was operated by the marker. The register is a storage device. In the crossbar switch, the register consisted of electromechanical devices

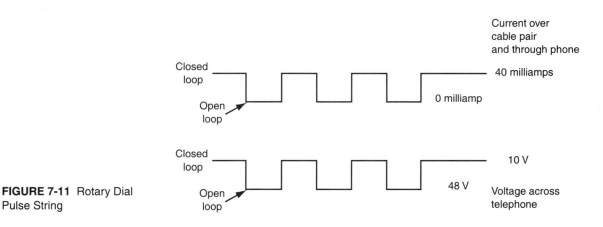

FIGURE 7-11 Rotary Dial
Pulse String

called *relays* to store dialed digits. The register received all dialed digits from the telephone and stored them for later use (Figures 7-12 and 7-13).

On completion of dialing, the register would request that the marker assign another common control component (translator) to the register. The **translator** looked at the dialed digits stored in the register and translated the dialed number into a series of switching instructions for the marker. The marker used the instructions from the translator to operate the necessary crosspoints in line link and trunk link frames to connect the calling line to the called line. Because each conversation path occupied its own wire path through the crossbar switching system, it is referred to as a space division switching technology.

The register, marker, and translator were all part of the common equipment used by all calls. Once they determined which crosspoints were to be activated in the switching matrix, they issued instructions to the switching matrix to operate the needed crosspoints. The crosspoints stayed operated for the duration of the call, but the common equipment was free to set up additional calls. The marker controlled all operations in the switch. Using pooled common equipment for switching control significantly reduced the size of electromechanical switching devices. This meant less maintenance was required in a XBAR office compared to a SXS office, and XBAR equipment cost less than SXS.

7.13 DUAL-TONE MULTIFREQUENCY DIAL REGISTERS

When touchtone telephones were introduced, the registers of the XBAR switch were replaced with registers that could recognize both dial pulses and touchtones. Included in the new registers were tuned circuits. There was a tuned circuit for each of the touchtones a telephone could send. When a button was pushed on the telephone, two tones were transmitted from the telephone to the register. The tuned circuits would determine what two tones were received and convert the tones into a dialed digit for storage in the register. If the telephone was a rotary dial phone, a stepping relay would count the dial pulses of each digit dialed for storage in the register. Thus it did not matter what type of telephone was on a line, because the register was capable of storing digits received from either type of phone.

7.14 DUAL-TONE MULTIFREQUENCY DIAL OR TOUCHTONE DIALPAD

Telephones manufactured today are made with a dial that use tones for dialing signals. This dial is called a *dual-tone multifrequency dial* or *DTMF dial*. The dial is a keypad containing 12 or 16 buttons. The buttons are arranged into four rows and three or four columns. A separate tone is assigned to each row and column. When any button is depressed, it sends out a unique set of tones to the central exchange.

	1209 Hz	1336 Hz	1477 Hz	1633 Hz
697 Hz	1	2	3	A
770 Hz	4	5	6	B
852 Hz	7	8	9	C
941 Hz	*	0	#	D

Depressing button 1 will cause a 697-cycle (Hertz) tone and a 1209-cycle tone to be sent to the central exchange. The receiver in the register at the central exchange decodes the different combination of tones received into the equivalent dialed digit. The dial register stores the dialed digit in a storage device. The storage device for older electromechanical switches, such as SXS or XBAR, was relays. The storage device for SPC switching systems is RAM. The DTMF pad used on a telephone has 12 buttons: *, #, and 1 through 0. It would have been possible to use a single tone for each button pushed rather than the dual-tone design. This would require the ability to generate and detect 12 tones (16 if

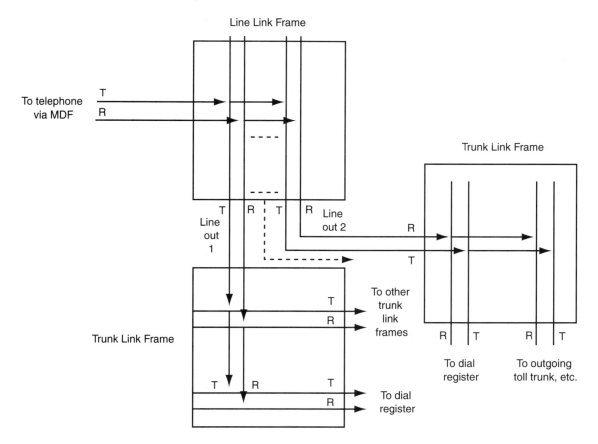

FIGURE 7-12 Attaching a register in XBAR

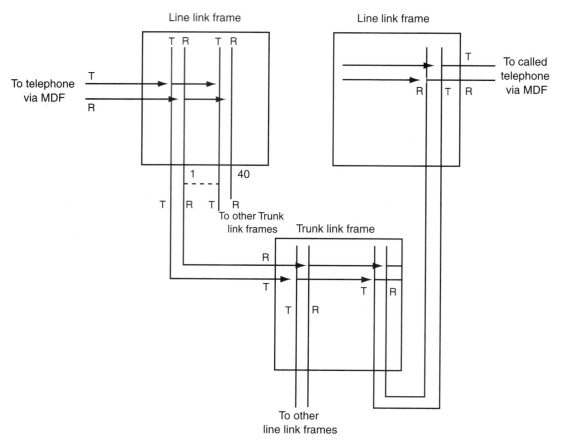

FIGURE 7-13 Local call in XBAR

you include *A, B, C, D* buttons). The dual-tone design requires fewer tone detectors in the dial register of the central office than a single-tone design would. The use of fewer tones also reduces the sensitivity requirement of the receiver. A dual-tone receiver is better than a single-tone design in its ability to recognize tones distorted by abnormal conditions on a local loop that causes noise on the cable pair.

All push-button telephones are not DTMF telephones. There are some push-button telephones that do not send DTMF tones. They convert the button pushed to rotary dial pulses. You can recognize this type of push-button telephone because you can hear the clicks of the rotary dial in the receiver of the phone. You can also tell when a push-button phone is not a DTMF phone since this type of telephone will not work with voice messaging systems. It will work with the central office switching system because the dial register in the central office is designed to work with either a rotary dial or a DTMF dial. Voice messaging systems must receive DTMF tones to inform the system of what action it should take.

Rotary dial signaling pulses cannot pass through a central exchange. DTMF tones will pass through the exchange to the messaging system attached to the called number. Because a dial register in a central office can recognize both rotary dial pulses and DTMF tones, both a rotary dial telephone (called a *500 set*) and a touchtone phone (called a *2500 set*) will work with any central office exchange. DTMF tones must be present for at least 40 ms in order for the register at the central office to recognize the tones. A pause of 60 ms is required between digits. The total time for transmission of one digit is 100 ms; therefore it is possible to transmit ten digits in 1s. DTMF is faster than rotary dial, which takes an average of 11/2s per digit or 15 s for 10 digits.

7.15 STORAGE OF DIALED DIGITS IN A REGISTER

In a common control switching system, the dialed digits were received by a device called the *dial register* or simply the **register**. The register recognized a rotary dialed digit by counting the number of open-circuit conditions (pulses) sent by the dial (Figure 7-11). The pulses were counted by a stepping relay. It stepped once for each open circuit (pulse). The register also contained tuned circuits so it could detect the different tones sent by a touchtone keypad. In the dial register of a XBAR switching system, the output of the tuned circuits for a DTMF call (or the output of the stepping relay for a rotary dial) was wired through an additional stepping switch, called the *selector switch*, to 40 storage relays (Figure 7-14).

The selector switch would step once at the end of each dialed digit and select the storage location for the dialed digit. It could take ten steps to ten different levels. Each level of the selector contained four storage relays. Initially the selector switch would be setting on level 1 and would connect four signal leads from the stepping relay/tuned circuits to four relays labeled W_1, X_1, Y_1, and Z_1 (storage location 1). The stepping relay and tuned circuits would cause an output to occur on the four signal leads (W, X, Y, Z) that would be different for each dialed or touchtone digit. This caused a different combination of relays to be operated for each digit sent by the telephone.

If the register recognized the digit as a 1, it operated the W and X relay. If it recognized the digit as a 2, the W and Y relays were operated. If the dialed digit or touchtone digit was a 3, the W and Z relays were operated. For a digit 4, the X and Y relays were operated, and so on.

As stated above, the additional selector switch initially connected the output of the tone detector circuits and stepping relay over signaling leads W, X, Y, and Z to relays W_1, X_1, Y_1, and Z_1. After the first digit was detected and stored, the selector switch stepped to level 2 for storage of the second dialed digit. The W, X, Y, and Z signaling leads were then connected to relays W_2, X_2, Y_2, and Z_2 for storage of the second digit sent by the telephone. After each digit was received, the selecting switch stepped to the next set of four storage relays. The selecting

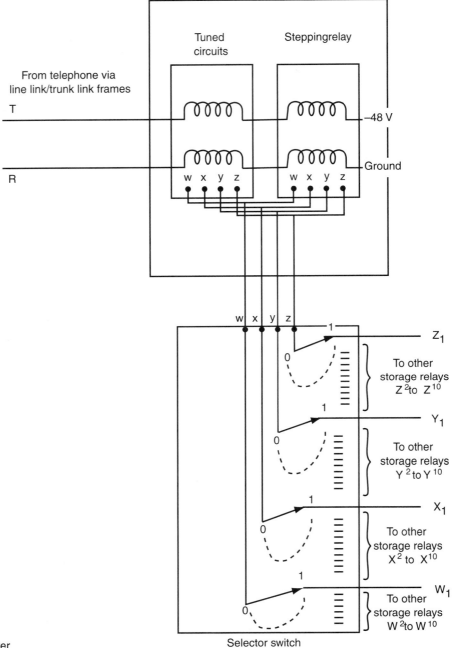

FIGURE 7-14 Dial register

switch could take ten steps. This allowed it to direct the storage of dialed digits to ten different storage locations for storage of up to ten dialed digits. The tenth storage location used relays W_{10}, X_{10}, Y_{10}, and Z_{10}.

Dialed Digit	W (*) Relay	X (*) Relay	Y (*) Relay	Z (*) Relay
1	Operated	Operated		
2	Operated		Operated	
3	Operated			Operated
4		Operated	Operated	
5		Operated		Operated
6			Operated	Operated
7	Operated			
8		Operated		
9			Operated	
0				Operated

7.16 XBAR COMMON CONTROL: THE MARKER AND TRANSLATOR

After receiving one digit, the register asked the marker to connect the translator to it. The marker was the brains of a XBAR switching system. The translator looked at the first dialed digit to see if it was a 1 or 0. If the first digit was a 1, the translator would tell the marker to connect the calling line to the *Direct Dial* Automated Long Distance (DDD) switching system. The DDD machine then received all additional dialed digits and determined how the call should be handled. If the first digit received by the register was a 0, the translator told the marker to connect the calling line to an idle trunk terminating on the operator's position. If the first digit received by the register was neither a 1 nor a 0, the translator informed the register to collect two additional digits and return after the third digit was received.

After receiving three digits, the register would ask the marker to connect it to the translator. The translator would look at the three digits to see if they matched the NNX exchange code assigned to this exchange. If they matched, the translator would direct the register to collect four more digits and come back after they had been collected. When the register came back and presented all seven dialed digits to the translator, the translator converted the dialed number into the equipment location of the terminating line circuit and instructed the marker to connect the originating line circuit to the called (or terminating) line circuit.

If the first three digits of the dialed number did not match the central office exchange code of the originating office, the translator instructed the marker to connect the calling line to an idle trunk circuit going to the desired central office. When the first three digits of a dialed telephone number did not match the exchange code of the originating office, the call had to be destined for a nearby central office. If the call was destined for an office many miles away, the caller would have dialed a 1 as the first digit and would have been connected to a toll exchange. The fact that the call remained attached to the register in the local class 5 exchange means the call was a local call but not a call within the same exchange. These types of calls are termed EAS calls. As mentioned earlier, EAS calls are not charged toll charges. The local service rate is raised by tariff filings as additional EAS points are added to a local exchange. The LEC calls the EAS points free calling areas, but they are not free. Calls to EAS points are paid for by way of higher local service rates. Local exchange rates are based on how many telephones can be called within an exchange. As the number of accessible phones rises, so do the rates.

7.17 COMPUTER-CONTROLLED SWITCHING

The first computer-controlled switches, introduced in 1960, retained the XBAR switching matrixes and simply replaced the electromechanical common control units of the XBAR switch with electronic circuitry. The marker was replaced by a computer. The computer controlled the XBAR switching components. The previous section on XBAR switching was included in this book because computer-controlled switching was initially based on the XBAR switching system. The early computer-controlled switching systems were space division switching systems that employed a XBAR switching matrix or a reed relay switching matrix. Computer-controlled switching is also referred to as *Stored Program Control* because switching is controlled by a software program. The computers used to control switching in 1965 were solid-state design using transistors and printed-circuit technology. The computer was composed of numerous circuit cards placed in approximately six cabinets. Each cabinet was approximately 7 ft high and 2 ft wide. The memory for the computer was composed of ring core memory. The operating program resided in one ring core field, and a second ring core field contained the database necessary to translate dialed telephone numbers into an equipment location (Figure 7-15).

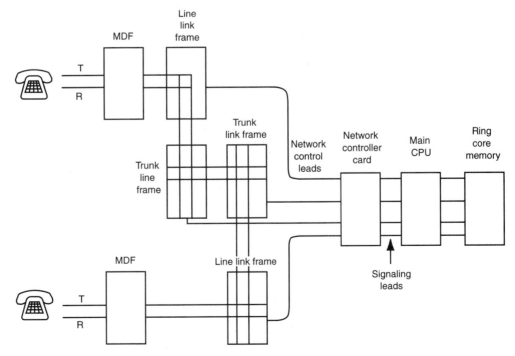

FIGURE 7-15 Early SPC with analog XBAR matrix.

7.18 SWITCHING CONTROL BY A MICROPROCESSOR

Since 1965, advances in electronic circuit design have resulted in the microprocessor on a chip used to control today's switching. The switching programs are relatively simple and can be controlled by an 8086 microprocessor. The ring core memory has been replaced by random access memory (RAM), tape drive storage, and disk storage. Additionally the XBAR switching matrix has been replaced by *time division switching* technology. The marriage of computer control and time division switching technology is used on all automated switching systems manufactured today. This chapter has provided only an overview of past switching systems because they are rapidly becoming extinct. All switching systems manufactured today are computer-controlled switching; they will be discussed in greater detail in Chapter 8.

7.19 STORED PROGRAM CONTROL TRANSLATORS

The translators in SPC toll tandem offices look up numbers in a database and are much smarter than their electromechanical predecessors. The setup of switching paths is no longer done over the voice network but is done over a special signaling network called *Signaling System 7 (SS7)*. SS7 can establish a complete switching path in the voice network between several central office exchanges in milliseconds. The development of SS7 has meant that the originating toll office can wait until all digits have been dialed before calling in the translator. When a toll call is being placed, the caller dials a 1 as the first digit. The end office (class 5 local exchange) register stores the first digit as a 1 and the translator of the local exchange instructs the switching network to connect the calling line to a toll trunk going to the class 4 toll office. All ten digits of the called number are received by the register in the toll office switching system. The translator knows the first three digits are an area code. The middle digit no longer has to be a 1 or 0 for the translator to recognize it as an area code. That requirement existed when we were allowed to place long distance calls within an area code without having to dial the area code of our home NPA.

Now that the area code must be dialed on all long distance calls, even those within the same NPA, all long distance calls are ten-digit calls. The translator in a toll switch is programmed to recognize the first three digits stored in a register as an area code regardless of what the digits are. This has eliminated the requirement for an area code to be of the form *N1/0X*. This means an area code can be of the form *NXX*. This provides 800 possible area codes instead of the previous limitation of 160. The PSTN was running out of area codes when the new codes were added in 1995. Once the translator in a toll switch has determined where the call should be sent, the SS7 network is notified to establish the switching path necessary to connect the call.

7.20 IN-BAND SIGNALING

The ability to establish call setups using the SS7 network has been a tremendous benefit to the LECs and IECs. Prior to the establishment of the SS7 network, all call setup was done over the voice network. Call setup and signaling used the same path that our conversation for the call would use. This type of signaling was called *in-band signaling*. Signaling was done within the voice band. The signaling used over the toll system was a dual-tone multifrequency system called *MF signaling*. MF signaling was developed prior to the introduction of DTMF for telephones. The two systems cannot use the same tones for signaling; DTMF tones for touchtone telephones are significantly different from MF tones. MF signaling used combinations of tones per the following chart:

Digit Sent	Frequencies Sent
1	700 Hz + 900 Hz
2	700 Hz + 1100 Hz
3	900 Hz + 1100 Hz
4	700 Hz + 1300 Hz
5	900 Hz + 1300 Hz
6	1100 Hz + 1300 Hz
7	700 Hz + 1500 Hz
8	900 Hz + 1500 Hz
9	1100 Hz + 1500 Hz
0	1300 Hz + 1500 Hz
KP	1100 Hz + 1700 Hz
ST	1500 Hz + 1700 Hz
IDLE	2600 Hz single tone

*If a trunk circuit was not in use, the trunk would place a 2600-cycle tone on the facility. A failure of the facility would cause a loss of tone and all circuits would go busy to prevent their use on calls.

When dialing signals were transmitted, the dialed digits were preceded by *KP* and followed by *ST*. For example KP3125551212ST would be sent over the toll network to reach telephone number 555-1212 in the 312 area code.

This in-band signaling system was hit with heavy fraudulent use in the 1970s. Some people learned the signaling scheme for MF signaling and made equipment that could duplicate these tones for each digit. These devices were called *blue boxes*. A person using them would make a direct distance dialed (DDD) call to an 800 number. The originating toll office would register that an 800 number was being called and would not bill the call when it was done. After calling the 800 number, a speaker on the blue box device was held close to the transmitter of the phone and a key was depressed to transmit a 2600-cycle tone to the speaker. The 2600-cycle tone was heard by equipment in the toll center, and the tone forced the toll circuit to go idle and drop the connection to the 800 number. The circuit between the originating phone and toll office was not disconnected, nor was the trunk between the class 5 office and the class 4 toll center. These facilities were using loop current signaling, not MF signaling, and the circuit to the toll center would stay up until the phone was hung up. The trunk between the class 4 toll center and the class 3 toll center used MF signaling and was forced idle by the 2600 tone. All connections from the class 4 toll center on out were disconnected.

After forcing the toll trunk idle, the caller would still be connected to the toll office and a toll trunk going into the class 3 toll office. The caller would now press the *KP* key. The tones representing *KP* were sent to the speaker of the blue box and were picked up by the transmitter of the phone. The *KP* tones went out over the voice path to the class 3 toll office. When the class 3 toll office saw *KP*, it assumed this was the start of a new call and attached a dial register. The caller now keyed in the desired area code and telephone number, followed by *ST*. The dial register accepted the dialed digits and the call was completed to that number. At the end of the call nothing was billed because the billing register in the DDD machine of the originating office still had the 800 number stored as the called number.

7.21 OUT-OF-BAND SIGNALING, CCIS AND SS7

This fraud involving in-band signaling is what led AT&T to develop out-of-band signaling called *Common Channel Inter-office Signaling (CCIS)*. CCIS separates signaling from the associated voice call and carries the signaling for a call over a different facility than used by the voice call. Signaling signals for many different voice calls are combined and sent over a common channel. CCIS and its offspring SS7 are covered in more detail in Chapter 10.

7.22 SUMMARY

The greatest advances in telecommunications over the past 25 years have occurred in the areas of switching systems, fiber optics, and multiplexing technologies. These technologies have existed for more than a half century, but significant changes and improvements have unfolded in the last 25 years due mainly to research work at AT&T Bell Labs. The development of transistors and very large scale integrated circuits has made it possible to implement technologies for switching and multiplexing that were not feasible with vacuum tubes and mechanical switching. The SXS switch was a critical and necessary step toward the automation of switching but could not possibly serve the demands made on today's switching systems.

Thanks to continuing research and development by the telecommunications industry, we have switching systems and transmission facilities that allow us to communicate with one another instantly. These developments led the way in the trend toward information explosion. In fact, many things we take for granted today—such as personal computers, real-time stock quotes, up-to-date news, the Internet, the World Wide Web, and so on—would be impossible without the technological developments and inventions given to our society by AT&T Bell Labs and the rest of the telecommunications industry.

REVIEW QUESTIONS

1. What have been the greatest developments in telecommunications over the past 25 years?
2. Who invented the step-by-step switching system? When was it invented?
3. What advantage does a common control system offer over SXS?
4. What is the PSTN?
5. Why was it necessary to modify the North American Numbering Plan to include new area codes?
6. When was the crossbar switching system introduced?
7. Can a SXS receive DTMF digits?
8. What is the name given to the device that receives dialed digits from a rotary phone or tones from a DTMF keypad?
9. What is the name given to switching systems using microprocessors for the control of swiching functions?
10. When was the North American Numbering Plan changed to include area codes of the form *NXX*?
11. Define the difference between direct control and common control.
12. What are the primary common control components in a XBAR switch?
13. Which component controls all functions in the XBAR?
14. What is the primary purpose of a translator?
15. How many tones are used by a DTMF pad to transmit one digit?

GLOSSARY

Blocking A term used to describe a situation where a call cannot be completed because all circuit paths are in use.

Circuit Switching The process of connecting one circuit to another.

Class 5 Exchange Also called the end office. The lowest-level switch in the PSTN hierarchy;the local exchange.

Concentration The use of fewer outlets than inlets.

Exchange Code The first three digits of a seven digit telephone number.

Extended Area Service (EAS) Tandem A class 4 switching system used in a metropolitan area to connect class 5 switching systems together on EAS calls. Extends the rate base of a local exchange so that calls to neighboring exchanges are not toll calls but are local calls.

Local Exchange The switching system that telephones are connected to.

Numbering Plan Area (NPA) An area represented by an area code.

Public Switched Telephone Network (PSTN) A network of switching systems connected together to allow anyone to call any telephone located in the United States. A public network is accessible to any members of the general public. The PSTN is composed of many nodes (switching systems) and many transmission links connecting these nodes. Stations are the end points of the network and connect to the PSTN via their local exchange.

Register (Dial Register) The device in a common control switching system that receives dialed digits.

Space Division Switching A form of switching (SXS and XBAR) where each conversation occupies its own distinct and separate wire path through the switching system.

Tandem Exchange Class 1, 2, 3, and 4 exchanges that are part of the PSTN.

Time Division Switching A form of switching (Stored Program Control or SPC) where each conversation occupies its own distinct and separate time slot on a common wire path through the switching system.

Translator In a Stored Program Controll switching system, a database that converts dialed digits into switching instructions.

8

Stored Program Control Digital Multiplex Switching Systems

Key Terms

Call Store
Computerized Branch
  Exchange (CBX)
Controller
Custom Local Area Signaling
  Services (CLASS)
Data Store
Digital Multiplexed Switching
  System
Digital Trunk Controller (DTC)
Distributed Processing
Electrically Programmable
  Read-Only Memory Chip
  (EPROM)

Generic Program
Line Card
Line Circuit
Line Controller
Line Drawer
Line Translations Database
Main Processor
Overlay Program
Parameters Software
Program Store
Pulse Code Modulated/Time
  Division Multiplex
  (PCM/TDM)

Random Access Memory
  (RAM)
Register (Dial Register)
Standby Processor
Stored Program Control (SPC)
Time Division Multiplex (TDM)
  Loop
Time Slot Interchange (TSI)
Tranlations Software
Trunk Controller

The automated switching system was invented in 1892 by Almon B. Strowger. Bell Labs invented the panel switch around 1900 and the XBAR switch in 1938. All these switching systems are called *electromechanical switches* because they use relays to control switching. The developments made in computer technology during the 1950s (from vacuum tubes to transistors) led to the use of computers to control switching. The computer-controlled switches of the early 1960s retained the XBAR switching matrices. In the late 1960s, transistors were used for the switching matrix, and the switches were called *electronic switching systems.* The electronic switching matrices evolved from space division matrices to time division matrices. The first time division matrix used PAM and was an electronic time division switching system. Advances in TDM technology led to the replacement of PAM by PCM and the establishment of an electronic time division switch commonly called a *digital switching system.*

8.1 SINGLE PROCESSOR VERSUS MULTIPLE PROCESSORS

The switching network in a digital switching system is controlled by a computer (CPU). The CPU is controlled by a stored program. Digital switching systems are called *Stored Program Control (SPC)* switching systems. Many SPC switching systems use more than one computer to control multiple processes at one time. This multiprocessor architecture is called *distributed processing*. The switch has a main processor that controls everything, but it will have additional processors, called *controllers*, dedicated to performing certain tasks. The controllers are distributed around the switch. There are network controllers to control the switching stages of the network, *line controllers* to control the line interface circuits, and *trunk controllers* to control trunk interface circuits. A network will have more than one network controller as the network gets larger. There are many line controllers in a switch (approximately1 line controller for every 640 lines). The main processor controls the line, trunk, and network controllers. Communications between the controllers and the main processor are over messaging links that connect controllers and the main processor.

The main processor is a general purpose machine (or processor). It controls everything that goes on in the switch. The main processor is controlled by a program residing in a *random access memory (RAM)* card adjacent to the processor (CPU chip). The line, trunk, and network controllers are special purpose machines (or processors), since they are designed to perform repetitive special purpose functions. These controllers are under the control of software stored in *electrically programmable read-only memory (EPROM)* on the controller circuitboard but also receive directions from the main CPU over the messaging link. EPROM is used because the controller programs are small enough to fit on an *EPROM chip*. As improvements are made in the controller program, the program is burned into a new EPROM at the switching manufacturer's factory. These new EPROMs are sent to all switching sites to replace the old controller PROMs. Small switching systems, such as the *Computerized Branch Exchange (CBX)* used in a small business, do not use distributed processing because one processor is capable of performing all control functions. A CBX is generally equipped with 100 to 2000 lines, and many fewer calls are handled at one time than in a central office equipped with 10,000 to 50,000 lines. This lowers the processing time needed in a CBX and leads to a single-processor architecture.

If central office switches used one processor to control all functions, they would not be able to handle all functions at one time. Any processor can only handle so many functions in a given time frame, depending on the speed of the processor and the clock speed used. Every process requires a specific amount of processing time. The amount of time is limited in every computer-based system. To provide many additional features (all of which require processing time), it is best to take load off the main processor by distributing routine, mundane, chores to other processors located around the system. This frees up the main processor and allows the system to run much faster in processing telephone calls. In addition to using distributed processor architecture, central exchanges also use duplicate processor architecture. The main processor has a duplicate (*standby*) *processor* that runs in parallel with the main processor. If the main processor fails, the standby processor will take over and become the main processor.

The controllers (distributed processors) are also duplicated and if a controller fails, its mate will take over. Controllers are usually assigned to handle a certain group of lines, trunks, or network shelves. If the controller for one group fails, the controller handling the adjacent group of equipment will control both groups. In addition to duplicating the main CPU and controllers of the switch, the switching networks are duplicated in central exchanges. A call will be established over two separate paths and one path selected to handle the call. If that path should fail, the other path will become the active path. The architecture discussed above provides a fail-safe switching system.

8.2 TIME DIVISION MULTIPLEX SWITCHING NETWORKS

Digital switching systems are computer-controlled, time division multiplex (TDM) switching systems. Time division switching provides a common wire path that can be used by many simultaneous calls. Although many calls can use the common transmission medium, each call occupies the medium in a shared time arrangement. If the time division bus is a 30–voice channel TDM, each voice channel can use the bus for 1/30 s. The TDM system can use PAM or PCM signals. Early systems used PAM, but all systems sold today use PCM technology.

Pulse Code Modulated/Time Division Multiplex (PCM/TDM) technology was first used in carrier systems to multiplex many calls over two pair of wires between towns. When this technology was moved to the *switching system*, it was necessary to move the hybrid and codec circuitry from the carrier system interface (the channel unit) to the line circuit interface. SXS and XBAR switching systems had line circuits consisting of two relays (a line relay and a cutoff relay). In most SPC switching systems, the line relay has been replaced by electronic circuitry that detects the presence or absence of loop current. The line circuits of a SPC switch have the following components: line relay or current detector circuit, cutoff relay, hybrid network, codec, test access relay, and ringing relay. All relays are miniature relays.

All relays and other components are mounted on a small printed circuit card. The circuit card plugs into card slots in a line shelf. The card slots connect to a printed circuit backplane. The backplane circuit tracks will connect to the TDM highway and are also connected to cables wired to the horizontal side of the *main distributing frame (MDF)*. The other side of an MDF has cables connected to it that serve the local loop. Jumper wires can be run to connect any line circuit to any cable pair, at the MDF. A jumper wire is a short piece of wire much like a jumper cable used to start a stalled car, except a jumper wire is a much smaller gauge of wire and will be longer than 10 ft. The jumper wire is run between wire pairs in two different cables to connect the pairs. The cables are terminated to wiring pins on devices called *quick connect blocks*, and jumpers will run between the appropriate wiring pins on each block to tie the pairs together (Figure 8-1).

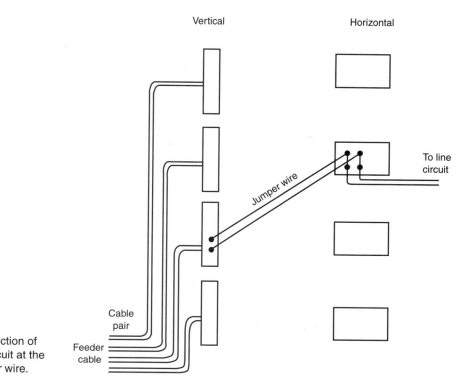

FIGURE 8-1 Connection of
cable pair to line circuit at the
MDF, using a jumper wire.

8.3 PCM AT THE LINE INTERFACE CIRCUIT

The hybrid network and codec are part of the line circuit in a SPC switching system. All analog voice signals received from the telephone are converted into digital signals at the line circuit of an SPC switching center. The codec on the line card circuit will connect to the telephone via the two-wire local loop, and to a TDM transmit bus and a TDM receive bus in the line card shelf. The codec will convert the analog voice signals received from the telephone into 8-bit digital codes. Under control of the SPC, a clock pulse is applied to a gate switch on the line card to gate the 8-bit PCM code onto the 30-channel TDM highway. The gate will be closed at the appropriate time slots to gate all 8 bits, one behind the other, onto the TDM transmit bus.

Another gate will gate the TDM receive bus to the decoder portion of this codec at the appropriate time slots to receive PCM signals routed to this line circuit by the SPC processor. The gate is closed for eight successive clock cycles to receive 8 serial bits. The decoder circuit in the codec converts the 8 bits received into an analog signal level. The analog signal is connected by the hybrid circuit to the cable pair for transmission to the telephone. The line circuit serves as an interface between the two-wire analog telephone line and the four-wire TDM system (Figure 8-2).

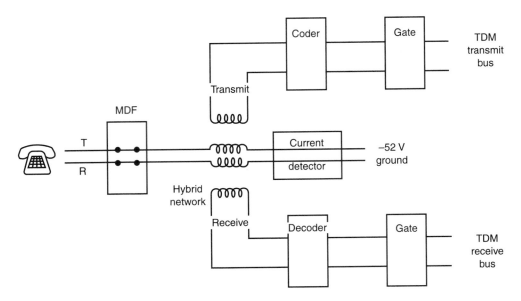

FIGURE 8-2 Hybrid and codec portion of SPC line circuit.

8.4 LINE CIRCUIT FUNCTIONS

In a digital switching system, the *line circuit* has eight functions. It is intended to provide:

1. A 52-V DC voltage to the cable pair as power for the transmitter of the handset and the DTMF dial of the telephone.
2. A means of connecting a 90- to 150-V AC ringing signal to the cable pair when the telephone needs to be rung.
3. The circuitry that can detect when a telephone attached to the cable pair has gone off-hook or hung up.
4. A means of testing the cable pair and/or line circuit by local or remote test equipment.
5. Circuits to protect the line circuit from damage due to lightning or power-line contacts with the cable pair.
6. A hybrid network to interface between the two-wire analog circuit of the cable pair and the four-wire codec.
7. A codec to convert the analog signal from a telephone into a PCM signal and place that signal on the transmit bus of the TDM highway. The codec also receives PCM signals from the receive bus of the TDM, converts them into analog signals, and sends them via the hybrid to the telephone.
8. The gate circuits that gate the digital transmit output of the codec to the transmit TDM bus and provide gate circuits to gate the TDM receive bus to the codec decoder circuit.

Line circuits reside on a printed circuitboard called a *line card*. In a PABX system there may be eight line circuits on one line card. In some central office switching systems there are also eight line circuits per line card. In a DMS-100® central office switch there is only one line circuit on a line card. In a DMS-100® central office, 32 line cards are plugged into a device called the *line drawer*. There are 20 line drawers in a line shelf. Two line shelves are mounted in a line bay. A line bay is about 6 ft high, 18 in. deep, and 3 ft wide. Each line shelf in a central office contains 640 lines. A line bay contains 1240 lines (Figure 8-3).

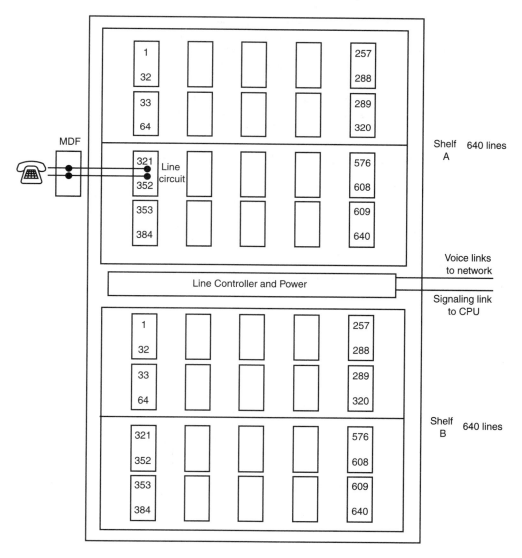

Line Bay 1

FIGURE 8-3 Typical line bay.

8.5 LINE CIRCUIT OPERATION

When a telephone goes off-hook (the handset of the phone is lifted) an electrical circuit is completed that allows electrical current to flow from the line card, out over the Ring lead of the cable pair, through the phone and back over the Tip lead wire to the line card. This current flow is detected by electronic circuitry, or a relay, on the line card. The detector will operate and send a signal to the line controller. The line controller sends a message to the stored program control processor (SPCP). The SPCP is commonly called the *main processor*. The message sent by the line controller consists of an interrupt signal along with the equipment address of the line circuit generating the request for service. The main processor will determine which line circuit has gone off-hook and will issue instructions to the switching system's network controller, line controller, and trunk controller to connect the line to a *dial register*. The *registers* are mounted in a trunk shelf and are controlled by the trunk controller in their shelf (Figure 8-4).

The SPCP keeps track of which equipment is in use so it knows which registers are idle. It tells the network and trunk controller the location of the idle register to use. The SPCP will also tell the line controller and network controller which multiplex loop to use and which time slot on the loop to assign to the line circuit. The SPC tells the network controller and trunk controller which multiplex loop and

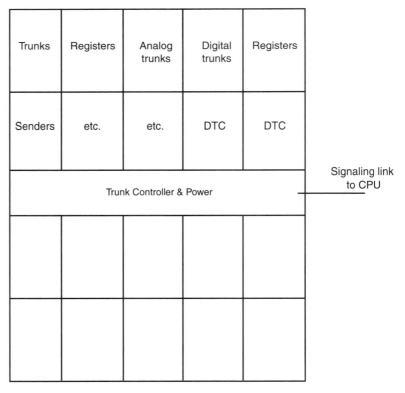

FIGURE 8-4 Typical
trunk frame.

Trunk Frame

time slot to assign to the register. The line controller will operate the appropriate gates on the appropriate line card at the time slot assigned to them by the main SPCP. The trunk controller will operate a gate on the appropriate trunk card to attach the register to the assigned loop at the appropriate time slot. The network controller will connect the time slot of the line circuit to the time slot of the register.

A trunk shelf contains the printed circuit cards for trunks, dial registers, tone generation, and so on, plus a controller card. There are two trunk shelves in a trunk bay. The trunk bay is about 6 ft high, 18 in. deep, and 3 ft wide (the same size as a line bay). Trunk shelves may contain analog trunk circuits or digital trunk circuits. Most trunk circuits will be digital because it is easy to connect digital trunk circuits directly into the multiplex bit stream of a multiplex loop in the switch. The digital trunk circuit interface is a printed circuit card designed to interface the DS1 bit stream of a T1 line directly to the PCM/TDM multiplex loop of the switch. This card simply maps the 24 DS0 channels in a T1 bit stream to 24 DS0 channels on the multiplex loop. These cards are called **Digital Trunk Controller (DTC)** cards. DTC cards are also available to interface the 672 DS0 channels of T3 or OC-1 to the multiplex loops of the switching system. No concentration is used on the connection of digital trunk circuits to the switching system. Every DS0 is permanently mapped to a specific time slot on a multiplex loop of the switch (Figure 8-5).

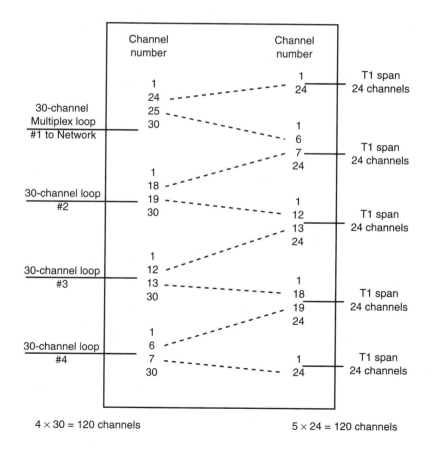

FIGURE 8-5 Digital trunk controller card.

8.6 DESCRIPTION OF TIME DIVISION MULTIPLEX SWITCHING

The network controller, line controller, and trunk controller will assign time slots by activating gates on the line card and the dial register at the appropriate times. The network controller and trunk controller are also told to connect a dial-tone signal circuit pack to the line circuit and are charged with the task of connecting the two time slots. Suppose the line card was assigned to TDM slot 1 and the dial register was assigned to TDM slot 17. The gate on the line card would close the codec output to the TDM transmit bus at time slot 1. This time slot is 40 ns long in order to pass all 8 bits, one after the other, to the transmit bus. The dial register will be connected to the TDM receive bus at time slot 17 for 40 ns.

8.7 TIME SLOT INTERCHANGE

The network has a device called a *time slot interchange (TSI)*, which is controlled by the network controller. The network controller will activate a gate on one side of the TSI to connect the TSI to the TDM transmit bus at the appropriate (assigned) time slot. Later, the network controller will activate a gate on the other side of the TSI to connect the TSI to the TDM receive bus at the assigned time slot. To accomplish connecting the transmit slot 1 to the receive slot 17, the TSI will be connected to the TDM transmit bus during time slot 1. The 8 bits coming over the transmit bus are stored in a memory on the TSI. When time slot 17 on the receive bus arrives, the TSI will read the 8 bits out of memory and place them on the receive bus. In the example above, the dial register receives 8 bits from the TSI over the receive bus (Figure 8-6).

A TSI will be needed for each connection made in a TDM switching system. The TSI on the network card is gated to the TDM's transmit and receive buses, just as the codecs on the line cards are gated to the TDM bus. Thus, a TSI can serve any time slot, but a TSI must be dedicated to one call at a time and remained connected to the time slots assigned for that call, for the duration of the call. This requirement means the TSI is a space division apparatus. In a TDM switching system, a slot on a multiplexed loop is referred to as a *time division switch (TDS)*. The TSI is a *space division switch (SDS)*. The TSI is inserted between the transmit bus of a multiplex loop and the receive bus of a multiplex loop. This results in an architecture called TDS-SDS-TDS.

In the above example, when the register was connected to the line, a dial-tone card was also connected to the line. The network controller and a trunk controller will connect a tone card (in the trunk shelf) to time slot 17 on the TDM transmit bus and will connect the TDM receive bus to the line card during time slot 1. The tone card has dial tone stored as a digital code. The digital code for dial tone will be stored in many 8-bit storage locations in a ROM chip. In the example above, to supply dial tone to the originating telephone, this digital code will be read from the ROM chip and sent, 8 bits at a time, over the transmit bus during

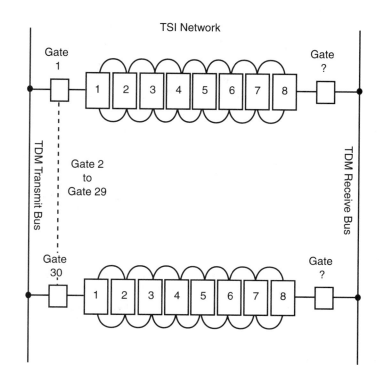

FIGURE 8-6 TSI network.

time slot 17 to the TSI assigned to time slot 17. At time slot 1 on the receive bus, these bits will be sent to the line card. The decoder in the line card will convert the bits received to an analog signal and dial tone will be heard at the telephone. When the DTMF digits are keyed in at the telephone, the codec will convert these tones to a digital signal. When the dial register receives the digital codes, it stores the codes as dialed digits.

8.8 TIME DIVISION MULTIPLEX LOOP

If the call is a local call, at the end of dialing the dial register will present the dialed number to the SPCP for translation of the dialed number into an equipment location. The SPCP looks in the *line translations database* to determine which line circuit is assigned to the number dialed. The SPCP will send the equipment location identifier (line circuit location) of the called number to the network controller, and to the line controller of the identified line. The SPCP will instruct the line controller of the originating line; the line controller of the terminating line, and the network controller for the assigned network to connect the calling line to the called line, using designated time slots on the assigned TDM loop and TSIs in the assigned network matrix. The switching network will connect the designated time slots to the assigned TSIs as discussed above.

One TDM facility (called a *time division multiplex* or *TDM* loop) contains many voice channels. When digital switching was first introduced, the loops contained 30 voice channel slots on a transmit bus and 30 voice channel slots on a receive bus. Remember that for full-duplex operation we need both a transmit path and a receive path. When talking about a TDM loop or time slots, it is understood that both the transmit and receive paths are included. Each line will require a voice channel slot on both the transmit and receive buses. Therefore, a total of two time slots on the transmit as well as two time slots on the receive buses are required. If a call originates and terminates over the same TDM loop, the 30-channel TDM loop can handle 15 calls at one time.

A TDM loop contains both a transmit and a receive bus. To handle more than 15 calls at a time, additional TDM paths or loops are added. A line shelf in a central office contains 640 lines. We know that all these lines are not in use at the same time. There are usually less than 90 lines in use at the same time, and they will probably be calling lines on other loops. To handle 90 calls at the same time (when the call *does not* originate and terminate on the same loop), 90 time slots are needed. This requirement is met by connecting the line shelf to the network using three 30-channel TDM loops.

When a switching system contains multiple multiplex loops, the TSI on each loop must be capable of connecting the output of the TSI to the receive bus of any loop in the system. To achieve this capability, multiple-stage networks are used. This architecture is referred to as *TDS-SDS-SDS-TDS*. This is a four-stage network. All digital switching networks contain a combination of TDSs and SDSs. The more lines and trunks a switch has connected to its input ports, the more TDM loops it needs and the more switching stages it has (Figure 8-7).

As multiplexing techniques have advanced, so has the TDM loop. The TDM loops have grown from 30- to 512-channel loops by using higher bit rates on a fiber loop. Regardless of the techniques used, the switching system will still need at least one time slot interchange units for each channel on the loop. If the loop has 256 channels, 256 TSIs are needed. If the network is a four-stage network, 512 TSIs will be needed. When multiple-stage networks are used the buses between stages are set up for parallel instead of serial transfer data. The TDM inputs to a TSI are serial, but the connection from one TSI to another is over a parallel bus to achieve faster speed within the network.

8.9 ELECTRONIC SWITCHING GATE

Switching in an electronic switch consists of turning gates on or off at the appropriate time to connect an input or output device to the TDM loop in the assigned time slot. These gates are composed of semiconductor devices (transistor devices) that work like a relay but are much faster and have no moving parts. Switching is controlled by clock pulses that will cause the transistor device to turn on or off. When it is turned on, its input is connected to its output. This is

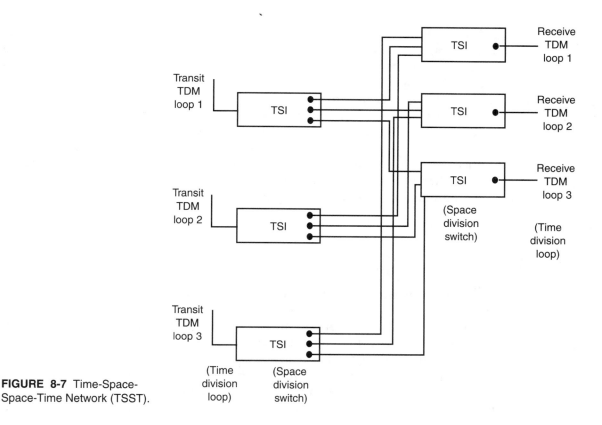

FIGURE 8-7 Time-Space-
Space-Time Network (TSST).

what happens when you turn on a switch, such as a light switch. Unlike a light switch, the transistor switch has no moving parts; it will not wear out and does not need maintenance. The type of transistor used as a switch is usually a *Silicon Controlled Rectifier (SCR)*.

The switching stage of a digital switching system is referred to as the *switching network* and is often simply called the *network*. As we saw in Chapter 1, the word *network* can have many different meanings depending on the context it is used in. A telephone *network* refers to the hybrid network used to connect the local-loop cable pair to the transmitter and receiver of the phone. The PSTN *network* refers to the interconnection of all public switching systems. When central office personnel use the word *network*, they mean the switching network in the central office. Central office personnel also use the term *switching fabric* to describe the switching network of the SPC switch.

8.10 STORED PROGRAM CONTROL

Using a computer to control the switching system is common control, under the direction of the program controlling the computer, and—as noted earlier—is called *Stored Program Control* or *(SPC)*. Every program requires a storage device, and the program must be in main memory for execution by the processor. The processor used in computer-controlled systems is usually a 16-bit processor similar to the Intel 8086. In small switching systems such as a CBX, small cartridge tapes are used as a permanent storage device for the programs and database. Central offices use large tape drives or large disk drives to store programs and office data. If a power failure occurs in a CBX that does not use battery backup, the system will boot up from ROM on the CPU chip, when power is restored. All central offices use batteries to supply power to the switch. They never lose power to the CPU. After power-on boot-up, the CPU will load the call processing program, features, and database for the office from a tape or disk into memory.

8.11 MEMORY

The early computer-controlled switching systems used two ring-core memories (one for the program and one for the database). This type of memory did not lose its contents on a power failure because the ring cores were hard wired. The development of large RAM memory allowed the conversion from ring core to RAM and significantly reduced the physical size of the memory. Two ring core fields required equipment 8 ft wide and 6 ft high. RAM can be placed on one or two printed circuit cards that are 8 in. long by 4 in. wide. Today, each memory card contains about 500 K of RAM. Older CBX systems have 192 K of RAM on a card. RAM in a central office does not lose its memory during an AC power outage because the RAM is powered from the central office batteries. Many CBX systems do not have a battery backup system and RAM loses its contents when a power outage occurs. When power is restored, the CBX will reload the generic program and the office database back into RAM from either a tape drive or disk drive.

8.12 PROGRAM STORE: CALL STORE AND DATA STORE

SPC switching systems divide the RAM memory into three logical units: program store, data store, and call store. *Program store* is used to store the generic program for the switch. The *generic program* contains the call processing program and all the features of the switch. A section of memory called *data store* is used to store the database for the switch. The database contains the data for each telephone number, in a line attribute table, and the data for trunks, in a trunk attribute table. The SPC uses the database to translate dialed telephone numbers into line or trunk equipment locations. The network switch connects calls based on equipment locations. If the call is a local call, it is connected within the switch; the network connects one line circuit to another line circuit. If the call must leave the switch, the network connects a line circuit to a trunk circuit. A section of memory designated *call store* is used as a temporary scratch pad to set up calls, get calls switched to their destination, and keep track of calls in progress. Call store is used to store the duration of a call, for billing purposes on a long distance call, and is also used as a temporary store, when making updates to the database. Even though the database is stored in data store, changes to it are made in call store and then loaded into data store (Figure 8-8).

The generic program is common to all switches of the same type. The generic number changes as additional features are added. Thus generic 30 can be used on all switches but will not have all the features of generic 31. The initial

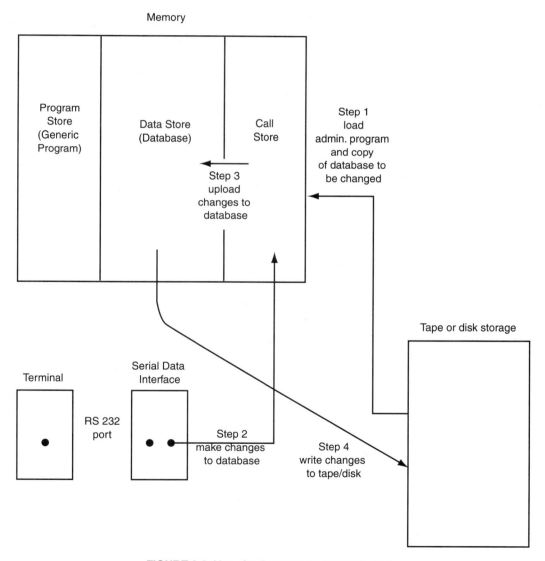

FIGURE 8-8 Use of call store to update data store.

call processing software contained basic features and was labeled generic 01. As additional features are added, the generic designation is changed. Each new generic contains additional features. Generic 30 has features not provided by generic 29. Sometimes an error is uncovered in a call processing program. The problem is fixed temporarily by modifying the software. The modification software is called a *patch*. The patches are implemented as a temporary fix. The software of the next generic released will include a permanent fix to replace the patch. When a telephone company encounters a problem with the software, a patch will be made and the customer is informed which generic will contain a permanent fix—for example, " that problem will be fixed in generic 33."

Since call processing is much the same from one generic to the next, the major reason for new generics is to provide new features such as call waiting, caller ID, equal access, and so on. The generic program contains all the features of the switch and the call processing software.

The generic program can be used in any central exchange. What makes each exchange different is its database. Just as any IBM PC can use Microsoft Windows, what makes each PC different is the database or other files and programs on that specific PC. If you have version 6.01 of Windows, it is the same as every other 6.01 version. If your switch has generic 34, it is the same as any other switch containing generic 34. Since each switch has different telephone numbers, exchange codes, and exchanges in its free calling area, the databases will be different.

The *database* will contain **parameters software**, which defines how many trunk groups, trunks in each group, registers, senders, networks, and lines are assigned in a particular SPC switch. The database will also contain **translations software**, consisting in turn of line and trunk tables. A *line translations table* defines what telephone number has been assigned to each line circuit, what features that line is permitted to use, what IEC has been PICed by the customer for that line, what ringing signal is assigned, and what type of line is being served (residence, pay station, business). A *trunk translations table* defines where the trunk circuits are located and the type of signaling used by that trunk. If the line or trunk is serving a business location, the table will also contain information on which lines or trunks are part of the hunt group for that business. A *hunt group* is a group of lines defined as belonging to one business and configured so that all lines are accessible by dialing one directory number. If the first line is busy, the call will automatically forward to the next lines in the group.

The memory allocated for program store is large enough to contain the call processing program and feature program portions of the generic program. The generic program contains many other programs that are not needed all the time, such as maintenance and administrative programs. When one of these programs is required, it is loaded by issuing the appropriate command to the CPU from a maintenance terminal. A small portion of call store memory is reserved for use by administrative/maintenance programs.

When an administrator needs to update the database, the appropriate program is loaded into this reserved memory space. For example, if a new telephone is being installed, the program for making changes to the line table is loaded. The line table program will walk the administrator through all the entries that need to be made in the table for adding, changing, or removing a telephone number to the table. If a network diagnostic program is required to help test the switching network, the necessary program is loaded by issuing the load command to the CPU. The testing program is loaded into the reserved memory space and overwrites the program currently in the reserved area. Some manufacturers call these special purpose programs *overlay programs*. Usually, only one overlay program can be loaded at a time, but as systems progress, some manufacturers are providing the ability to store more than one such program at the same time.

8.13 NO. 1, NO. 4, AND NO. 5 ELECTRONIC SWITCHING SYSTEMS

SPC switching systems were first installed around 1965, when AT&T installed a No. 1® *Electronic Switching System (ESS)* in a local class 5 exchange. These systems used reed relay networks. ESSs went through several development phases and used different types of networks. The types of networks used were XBAR, reed relay, PAM, and PCM. In the late 1970s, PCM/TDM networks begin to be the network of choice. Since the early 1980s, older electromechanical switching centers have been replaced by SPC/TDM switching centers. AT&T's PCM/TDM switching system for class 5 local exchanges are called No. 5 ESS®. The first No. 5 ESS was placed in service in 1982. It can handle 100,000 telephone lines and 650,000 calls per hour.

The No. 4® ESS was developed for use as a toll tandem exchange. It was first installed in 1976 and uses a multistage TSSSST switching network. The No. 4 ESS was initially developed to handle 53,760 trunk circuits and an Average Busy Season Busy Hour Call Attempt of 550, 000 calls per hour. The No. 4 ESS has improved since its introduction in 1976. Improvements in integrated circuit technology and microprocessor speed have enabled reductions in the physical size of the switch and increased the calls-per-hour capacity. The multiplex loop in a No. 4 ESS uses coaxial wire as a medium with 128 multiplexed time slots. Of these, 120 are used for voice circuits. The multiplex loop must be running at 8.192 Mbps to handle 8000 frames per second when each frame consists of 128 samples of 8 bits each.

8.14 DMS 1, 10, AND 100

Northern Telecom manufactures digital switches for toll and local switching. These switches are called *DMS® switches*, from the term *digital multiplexed switching system*. The toll switch is a DMS 250® and the local switch is a DMS 100®. Northern also makes computerized PBX switches using the same technology. The older PBXs are called SL1® switches, and the newer PBXs are called Meridian® switches. All these switches use a PCM/TDM switching matrix. Northern developed a DMS 10® switch that could handle the line requirements of a small office. The DMS 100® was an outgrowth of the DMS 1 and DMS 10 technology. Many small Independent telephone companies use a DMS 10 switch for a local exchange. The DMS 100 was developed for application in larger central exchanges (for example, the Bell market).

DMS architecture provides for a fail-safe switching technology by duplicating the CPU and network with hot standby mirror images. The active CPU and standby CPU are independent of each other but are both running in step with each other. The operations of each are continually checked against the other to ensure the CPU is operating properly. If the output of an operation does not match, an extensive test is launched to determine the faulty CPU and take it off line. Which CPU is active, and which is in hot standby mode, can also be changed manually. They can also be programmed to switch roles every night at midnight (or some other hour of convenience). Each CPU has its own program store and data store. Of course, the data in each is a mirror image of the other. The program store contains the generic program and can contain up to 8 million 17-bit words. The data store also contains a section used as call store. Data store contains the database for the switch. Line translations and trunk translations are part of this database. The generic program can be used in many different switches, but the database is particular and unique to one central exchange.

The CPU issues instructions to the switching network through duplicated *central message controllers*, which control a duplicated switching network known as plane 1 and plane 0. In an older DMS 100, each plane can have up to 32 network modules. Each module has 64 two-way inputs and 64 two-way outputs. These inputs and outputs are TDM loops carrying 30 voice channels and 2 control channels. The newest technology in the DMS 100 uses 512 channel fiber loops and is called *E-NET* for "Enhanced Network." Four of these 512 channel fiber loops are input to a fiber interface unit for multiplexing onto a fiber bus. Eight of the fiber interface units connect to a transmit TDM bus. Thus the transmit TDM bus can handle $512 \times 4 \times 8 = 16,384$ channels.

Remember that all voice sampling begins at the DS0 level with 8000 samples per second. If we have 16,384 samples on the TDM bus, we must handle 8000 of these each second. Therefore the signaling rate on the transmit TDM bus is 131.072 Mbps. The E-Net has 8 of these Transmit TDM buses. The E Net also has 8 receive TDM buses of 16,384 each. The transmit buses connect to the receive buses via a *Time Slot Interexchange (TSI)*. Each TDM-loop transmit bus requires 8 TSIs (one connecting to each receive bus). Since each TDM loop requires 8 TSIs, and there are 8 TDM Loops in the E-Net, there are 64 TSI modules in the E-Net. Each of the TSI modules has 16,384 storage locations (one for each channel on the multiplexed loop. Each store is 8 bits wide. With this arrangement, any input device can be connected to any output device and the E-Net is a single-stage nonblocking network. With 8 transmit TDM buses that contain 16,384 channels each, the E-NET has 131,072 digital voice channels. Each device connecting to the network requires one transmit and one receive channel. For a two-way conversation, two devices will be connected via the TDM loops, TDM bus, and TSI. With two devices, each conversation requires two time slots on the transmit bus and two slots on the receive bus. Thus, one E-NET can handle 65,536 simultaneous conversations (131,072 / 2 = 65,536).

8.15 DMS LINE CIRCUITS

The DMS 100 switch serves local exchange lines via a *Line Concentrating Module (LCM)*. Each LCM has 640 line circuit cards. Each card contains one line interface circuit for converting the analog input from a telephone into a PCM signal. The line cards in an LCM are grouped into units of 32 cards. A *Bus Interface Card (BIC)* serves 32 line cards. The BIC multiplexes (and demultiplexes) the 32 input circuits from line cards onto a TDM link operating at 2.56 Mbps. This link uses 10 bits per time slot (8 for voice and 2 for signaling) with 30 time slots for voice and 2 slots for signaling and control. Thus, $10 \times 32 \times 8000 = 2.56$ Mbps.

Since each BIC is serving 32 line cards with a maximum output of 30 voice channels, not all 32 lines can be in use at the same time. Since 32 lines are served by a BIC card and the LCM contains 640 lines, each LCM must have 20 BIC cards. The outputs of these 20 BIC cards are called *terminal links*, and they connect to a time multiplexing switch. The time multiplexing switch connects to a *Line Group Controller (LGC)* using as few as 4 or as many as 6 TDM loops called *channel links*. These channel links also operate at 2.56 Mbps and contain 30 voice channels. The 20 terminal links of 30 voice channels each contain a total of 600 voice channels. With four channel links to the LGC, the multiplexed space switch will concentrate the 600 inputs from the BIC onto the 120 possible outputs ($4 \times 30 = 120$) to the LGC.

Each LGC can serve up to 20 channel links from several LCMs. The LGC connects channel links from LCMs to the switching network. In older DMS 100s, each LGC could have up to 16 TDM loops to the network. These loops were called *network links* and also were 30–voice channel, 2–control channel multiplexed slots, operating at 2.56 Mbps. The newer DMS 100 using an E-Net has two fibers (one transmit and one receive) connecting the LGC to the E-Net. The signal on the fiber is called DS 512 because it has 512 voice signals multiplexed over it.

8.16 IMPACT OF THE MODIFIED FINAL JUDGMENT ON SWITCHING

With the equal access rulings of the Justice Department in 1984 (the Modified Final Judgment), Bell had to accelerate replacement of local central exchanges with SPC in order to provide equal access to all IECs. Being forced to accelerate replacement of local exchanges was a tremendous benefit to the BOCs. In 1984, local exchange service was regulated by state utilities organizations. These organizations set the rates Bell could charge for local service based on its capital investment and expenses. To keep rates as low as possible, the agencies would not let Bell spend wildly. Therefore, the company had to spread capital expenditures over a long time frame (usually 20 years). Bell could not just decide it wanted to replace all central offices and buy all new ones. This would drive up its capital investment, which would force the public utilities organizations into allowing it to raise local rates in order to recover its investment. In 1984, Bell was able to tell these organizations that all central exchanges must be replaced by 1987 to comply with the Justice Department's Modified Final Judgment. The state utilities organizations had to allow the replacement and had to grant the request from Bell for higher local rates.

8.17 SOME ADVANTAGES OF DIGITAL SWITCHING

Replacing older electromechanical offices with SPC/TDM switches had other benefits for Bell. Bell has been able to reduce the number of central office maintenance personnel needed. The SPC/TDM switch does not require much maintenance. The electromechanical switch it replaced had moving parts that required continual testing, lubrication, and adjustment. An electromechanical switch handling 10,000 lines usually had two to four maintenance people on site. The SPC switch for the same size office generally does not have anyone at the site. The switch is capable of running its own testing and diagnostic programs. If the switch recognizes a faulty circuit, it will take the faulty circuit out of service and issue an alarm to the remote maintenance center. Because critical circuits are duplicated in a SPC office, the switch continues to provide service using the duplicate or standby circuit. The remote maintenance center dispatches a repair person to locate and replace the faulty circuit pack at the office in trouble.

Bell has also been able to gain additional sources of revenue. SPC offices are feature rich. The implementation of features is little more than a software change. To provide features like call waiting, three-way calling, caller identification, and so on, Bell must buy the generic software program that provides these features from the switching manufacturer. Once the generic is in place to support a feature, the feature is implemented on any line by simply adding an entry in the line translations table. Bell may charge anywhere from $3 to $15 a month for a feature. Outside of the initial cost for the generic program, Bell's only cost to provide a feature is the amount of time it takes a clerk to update the line table using a remote access terminal.

Although Bell was used as an example to demonstrate the benefits it has derived from being allowed to replace its older electromechanical offices with SPC offices, all major telephone companies have benefited from the Justice Department's directives to AT&T in the MFJ of 1984. United Telephone (Sprint's local telecommunications division) has replaced 98% of its local exchanges with SPC switches. Although Sprint was not forced to do so by the MFJ (the MFJ only required equal access to be provided by the RBOCs), I am sure it cited this ruling when petitioning the public utilities organizations to allow the capital investment to be included as part of the rate base. It would not have spent the money to upgrade to SPC switching if the public agency would not grant the rate increases necessary to recapture that investment. General Telephone also agreed to abide by the MFJ and has replaced its electromechanical offices with SPC switching.

With the SPC offices in place, Bell, GTE, and Sprint's local telecommunications division have been able to make tremendous reductions in central office maintenance, engineering, and management personnel. They have also gained additional revenues from the many features available for sale in SPC switches. Although all LECs have been able to reduce maintenance and operating costs by implementing new digital switches and fiber optic transmission facilities, local rates have not gone down; they have gone up. We read every year about telephone companies reducing management and labor forces and right-sizing or downsizing. Why haven't the public utilities organizations insisted on lower rates as costs have declined? Of course, it is rare for any government agency to think of reducing taxes or rates charged for services. Why should the actions of government agencies controlling telecommunications be any exception?

8.18 DISTRIBUTED SWITCHING

Another major benefit of SPC systems is due to their modularity. The line circuits in a line bay communicate with the network and the CPU over a multiplexed medium. The line bays in a central exchange are located close to the network and CPU, but they don't have to be. They can be as far away as we desire. The multiplexed medium can be made longer to extend the distance. This allows for distributing the line equipment to different geographic locations within the exchange boundary and connecting them to the main switching system with multiplexed facilities.

When line equipment is placed in a different location from the main switch, the line equipment is called a *remote line module (RLM)*. When an exchange covers a large geographic area, RLMs can be used as a way to serve customers, at the far reaches of the exchange, without having to use thick wire and range extenders. The RLM will be placed in a location that can serve a radius of 3 mi or less. This allows the use of 26-gauge cable in the local loop (Figure 8-9).

RLMs can also be used to serve an area established as a *carrier serving area (CSA)*. This is an area of the exchange that will provide DS0 capabilities to customers. The RLM contains the same multiplexer used in a central office line module (LM). The multiplexer will combine many line circuits onto one facility. Older LMs and RLMs usually use a 32-channel multiplexer. Two channels were used for signaling. This leaves 30 channels for voice. LMs and RLMs contain 640 lines. Three multiplexed loops can be used to connect the LM or RLM to the switch. This will allow 90 of the 640 customers to use their phones at the same time. If the traffic demands of the lines require more than 90 voice channels at the same time, additional multiplexed loops can be added. Newer LMs and RLMs use fiber optic cable as the medium for multiplexers and can provide many voice channels, over one fiber, between the LM or RLM and the network at the main switch.

RLMs only contain line circuits and cannot connect one line circuit to another. The line circuit is connected to the multiplexed facility and received by the LGC at the central exchange. The LGC will connect the channels on the multiplexed

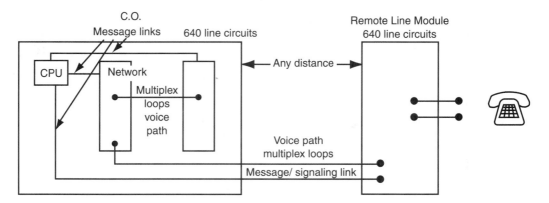

FIGURE 8-9 Remote line module.

loops to channels of the switching matrix. If one line of an RLM wishes to call another phone on the RLM, the lines are connected by the switching matrix at the central exchange. This means that calls within an RLM will tie up two channels on the multiplex loop to the central exchange. To improve the capabilities of RLMs, a switching matrix and database were added to the RLM and it was called a *remote line switch (RLS)* (Figure 8-10).

An RLS will connect calls within the remote site via its own switching matrix and does not use the channels of the multiplex loops to the central exchange for calls within the RLS. Older RLS systems still used the database and dial registers of the central exchange. When a remote line went off-hook, it was connected over the multiplexed loop to the central exchange. At the central exchange, the channel was connected to a dial register. The dial register accepted the dialed digits, and if the translator determined that the calling and called lines resided in the RLS, the CPU issued instructions to the RLS, over a data link, that the lines were to be connected together by the RLS matrix. The latest RLS systems will contain a local database, register, and switching matrix. This allows the RLS to operate completely on its own even if the data link between the RLS and central exchange goes down. Calls between stations on the RLM are still possible if an outage to the data link occurs.

Many times an RLS is used to replace an existing electromechanical exchange. Suppose the central exchange of a town is served by a SPC switch and the LEC decides to replace another exchange in the area. It can replace that exchange with an RLS and connect the RLS to the SPC exchange. The SPC office is

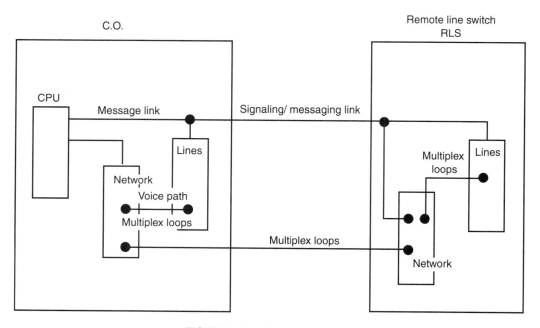

FIGURE 8-10 Remote line switch.

called the *host office*. The remote can be any distance from the host. The switching system in a small community of 500 or 2000 phones may be served by an RLS. The data link will usually use one of the channels on the multiplexed facility between the host and remote.

To ensure reliability, the data link between a host an a remote is duplicated and each link takes a different route. This route diversity ensures that the data links are in separate cables or separate multiplexing systems. If someone digs up a cable and damages the data link, the other data link is not affected. Likewise, if the carrier system containing a data link on one of its channels goes down, the other link on another carrier system is not affected. A device is placed at each end of the data links to monitor their performance. This device will switch between the links to ensure that the best link is used to connect the host to the remote. For large remotes, the multiplexed loops for the voice channels also use route diversity.

Route diversity is possible when an RLS replaces an existing exchange because an exchange usually has trunk circuits to different central offices. A data link can be placed on a carrier system directly to the host exchange. A second link can be placed on a carrier system to a third exchange, where it can be connected to a channel on the carrier system between the third exchange and the exchange serving as the host office. Route diversity is difficult when an RLS is used to serve an area of an existing exchange. Since the RLS did not replace an existing exchange, there are no carrier systems to other exchanges. In this case, the RLS only has access to a cable back to the host exchange. It takes a creative outside plant engineer to figure out a way to get a second cable route established for the second data link without having to install a second cable.

8.19 SUBSCRIBER CARRIER

SPC switching systems have an interface card that can connect directly to a cable pair serving a *digital line carrier (DLC)* such as SLC-96. This eliminates the need to have an office terminal for the SLC-96 system. The outside plant engineer can use DLC to serve a CSA. Most DLCs do not use concentration. A 96-line system uses a 96-channel TDM link between the field unit and the central office interface. Each residence is connected by its distribution cable pair to one channel unit of the DLC. The channel unit provides a 48-V battery feed to the telephone and will provide a ringing voltage as needed. The channel unit can detect dial pulses and convert them into a signal that can be transmitted over the TDM facility. The subscriber carrier central office terminal unit for an electromechanical switching system connects the TDM facility to 96-line circuits and also does conversion of signals.

When a digital subscriber carrier system connects to a digital SPC switching system, no office terminal unit is needed. The TDM facility connects to an interface unit in a trunk bay. This interface unit maps the DS0 channels of the TDM from the subscriber carrier line to the TDM highway of the SPC switch. This interface unit will have an associated database that has telephone numbers assigned to each channel. When a call is received for a line, the database instructs the SPCP which channel of the interface unit the call should be assigned to. The office channel unit sends a digital signaling message to the field unit. The digital signal is sent over a data link to the field terminal. At the field terminal, the digital signal causes the channel unit to send a ringing signal to the telephone. When the telephone answers, loop current flows from the channel unit in the field to the telephone. The field unit detects the loop current and sends a digital signal to the central office channel unit. This signal will cause the central office unit to send a message to the originating line circuit indicating that the call has been answered.

8.20 INTEROFFICE CARRIER SYSTEMS

Digital central exchanges can interface directly to interoffice digital carrier systems. Most interoffice carrier is T1, which has 24 DS0 channels. All digital central offices have multiplexed loops that carry many DS0 channels. The interface card simply maps each DS0 channel from the T1 system to one of the DS0 channels on one of its multiplexed loops. This digital trunk interface card eliminates the need for T1 central office terminals and channel units. The interface card performs the functions previously performed by these devices. Interface cards have also been developed to interface interoffice fiber to the switching systems that use fiber for their multiplexed loops.

8.21 MAINTENANCE OF SPC SWITCHING

SPC switching systems have programs that perform diagnostics on the switch. Some of these programs are loaded into memory when a system boots up and run all the time. These programs continuously test the switch by simulating the placement of calls through the switch. If one of these test calls fails, the switch activates a program to do a more through check of the network and peripherals. These programs will identify that a problem exists and will print out a report on the maintenance printer. The maintenance printer can be located with the switch or can be in a remote location. Usually, there are two maintenance printers. One printer is located at the switch. A second printer is located remotely in a centralized center that monitors many switches. The switch will perform its own diagnostics and print out the suspected cause of the problem. The maintenance person can use a maintenance terminal to run diagnostics on the switch and pinpoint the cause of a problem.

Once a defective circuit card is identified, it is replaced with a good card. The defective card is tagged with a note identifying the problem encountered and returned to the manufacturer. This process is called *repair and return*, but the manufacturer rarely repairs these cards. It is cheaper to make a new card than to repair a card. The manufacturer wants the old card returned. This provides some assurance that its technology does not fall into competitors' hands by way of the trash can. If a customer threw away defective cards, they could end up anywhere. A credit is given for returning the defective card to encourage customers to use this process. The process should be called *card exchange*. A new card is exchanged for a defective card at less than the price of a new card without exchange.

SPC systems still use visual and audible alarms and will also send these alarm signals to a remote maintenance center. Often when an alarm is generated, the switch will automatically switch affected processes to a standby unit. Maintenance personnel are sent to school where they learn how to use the maintenance terminal and how to diagnose problems. They learn the process to use for identifying which card in the system is causing a problem. Maintenance personnel repair problems by replacing cards. They do not do component-level repair. It is too costly to repair a card.

8.22 CUSTOM LOCAL AREA SIGNALING SERVICES (CLASS) AND ISDN

LECs have connected their class 5 central offices to each other and to the toll network using Signaling System 7 (SS7). This out-of-band signaling system is discussed in detail in Chapter 10. The deployment of SS7 is only possible because of the installation of SPC switches. The combination of SPC and SS7 allows the LECs to offer *Custom Local Area Signaling Services (CLASS)* and ISDN. Some CLASS features are: calling number identification, calling number identification blocking, automatic callback, automatic recall, ring again, call trace, selective call forwarding, and selective call acceptance. Equal Access Feature Group D, ISDN, and CLASS are only possible from a class 5 central office if it is a SPC switch. Customers served by an older electromechanical switch cannot access these services through the electromechanical switch; these services must be provided by a work-around solution if they are to be provided. Many small LECs will simply decide not to offer these enhanced services.

8.23 SUMMARY

The Stored Program Control Digital Multiplex Switching System is called a digital switching system and a SPC switch. The switching matrix uses PCM/TDM highways and TSIs to connect two ports on the system. The ports can be equipped with interface cards as appropriate to allow the switch to interface to analog signals or digital signals. When an analog signal is received, the interface card will convert the received analog signal into a digital signal and place it in a time slot on a TDM transmit bus. This interface card will also convert digital signals, from the assigned time slot of a receive bus of a TDM, into analog signals.

The analog signal is then sent to the external device connected to the port. It is important to remember that a digital switching system interfaces to many analog lines. Thus, when you are told that a CBX is a digital switching system, do not assume that it can interface to a digital telephone. Whether any digital switch can interface to a particular digital device will depend on whether the hardware interface device and accompanying software are present in the switch.

CBX switches usually use a single-processor architecture. Central exchange switches use distributed processing in the form of controllers so the switch can perform well during heavy calling volumes. SPCs for a central exchange also use duplicate main processors and duplicate switching matrices to provide a highly fault-tolerant and almost fail-safe switching system.

SPC switches are feature rich. Features are added to the switch by upgrading the generic program. Once the generic program is in place to support a feature, it can easily be added to any line simply by changing the line translations table for that line. SPC switches require very little maintenance and have allowed the LECs to dramatically reduce their network maintenance and network engineering personnel. Equal access was almost impossible to achieve with most electromechanical switching systems. The 1984 MFJ demanded that the Bell LECs provide equal access via Feature Group D by 1987. To accomplish this requirement of the MFJ, SPC toll tandem offices were installed and the local class 5 offices were gradually converted to SPC exchanges.

REVIEW QUESTIONS

1. What are switching systems called that are controlled by a CPU?
2. What are distributed processors called?
3. Why use distributed processing?
4. What type of switching matrix is used by a SPC switch?
5. What is an MDF?
6. What are the functions of a line circuit?
7. What is the function of a TSI?
8. How many time slots on a TDM loop?
9. How many time slots are used up by one call?
10. What type of memory does a SPC switch have?
11. What are the three types of memory in a SPC switch?
12. What is the generic program?
13. What are line translations?
14. What are the 4 types of switching networks that have been used by SPC switching systems?
15. What does AT&T call its SPC switch? What does Northern Telecom call its SPC switch?

16. How did the MFJ impact central office technology?
17. What is the difference between an RLM and an RLS?
18. How much time is required to repair a printed circuit card?
19. What are some advantages of SPC offices?
20. What is the name of the program that contains all the call processing details and also contains all the features of the switch?

GLOSSARY

Call Store A part of memory set aside to store information that is temporary in nature. The call store will keep track of each call that is in progress or that is currently up on a conversation. Call store is also used to load temporary programs used by maintenance and administration personnel.

Computerized Branch Exchange (CBX) A SPC switching system used at a private business location.

Controller A microprocessor designed to control a specific function. The controller is directed by programming in a PROM chip. The PROM chip resides on the controller circuitboard. The type of programming contained in the PROM dictates whether the controller is a line, trunk, or network controller. Controllers also receive instructions from the main processor over messaging links.

Custom Local Area Signaling Services (CLASS) Features of a SPC switch, which also require the SPC to have Signaling System 7. Some Class Features are: Caller ID, Caller ID Blocking, Automatic Callback, Automatic Recall, Selective Call Forwarding, Selective Call Rejection, and Ring Again.

Data Store That part of RAM set aside to contain the database for the switch. The database will consist of translations software (line and trunk tables) and parameters software.

Digital Multiplexed Switching System A Stored Program Controll switching system that uses a PCM/TDM switching matrix. All voice paths inside the switch are digital paths, and the SPC system switches these digital paths.

Digital Trunk Controller (DTC) A controller card that interfaces a digital trunk facility, such as a T1 carrier system line, directly to the time division multiplex loop of the SPC switch.

Distributed Processing A SPC switch architecture that uses multiple processors. Each line shelf, trunk shelf, and network shelf contains a processor to control the activities of equipment in its shelf. These processors are called controllers.

Electrically Programmable Read Only Memory (EPROM) Chip Used on controller cards. The EPROM is programmed to perform a specific function. The programming makes the controller a line, trunk, or network controller.

Generic Program The program that controls call processing in a SPC switch. The generic program contains software for all the features of the switch. To add features to a switch, the generic program must be replaced with a later version containing the desired features. The generic program is loaded into program store.

Line Card An electronic printed circuitboard that contains line circuits. The line cards of most SPC central exchanges contain one line circuit per circuit card. The first AT&T SPC switch had line cards containing 8 line circuits.

Many CBXs have line cards containing from 8 to 16 line circuits.

Line Circuit The electronic printed circuit card in a SPC switch that interfaces a local-loop telephone line to the SPC switching system. The line circuit has many functions. One function is to interface the analog signal from a telephone to the digital switching matrix of the switch.

Line Controller A circuit card in the line shelf, that contains a microprocessor and EPROM to control the activities of line circuits in that shelf.

Lined Translations Database A database table which contains the telephone number assigned to each line circuit and the features a line can access.

Line Drawer A shelf in a digital switching system that contains line circuits. A typical line drawer contains 32 line circuits.

Main Processor The CPU that controls all activities in a SPC switch.

Overlay Programs The term used to describe maintenance and administrative programs that do not permanently reside in memory but are contained on a secondary storage device. These programs are only loaded into memory as needed; when loaded, they overwrite the last overlay program contained in memory.

Parameters Software Defines a particular SPC, by indicating the number of lines, trunks, registers, and so on.

Program Store The section of mamory reserved for the generic program.

Pulse Code Modulated/Time Division Multiplex (PCM/TDM) The industry standard method used to convert an analog signal into a digital 64,000-bps signal. PCM takes a PAM signal and converts each sample (pulse) into an 8-bit code.

Random Access Memory (RAM) Highly volatile memory that will lose its contents when power to the RAM is lost. RAM is divided into three stores: program store, data store, and call store. The main processor accesses RAM to direct switching operations.

Register (Dial Register) The device in a common control switching system that receives dialed digits.

Standby Processor A central processor that runs in parallel with the main processor. If the main processor fails, the standby processor takes over.

Stored Program Control (SPC) Digital switching systems controlled by a computer using stored programs.

Time Division Multiplex (TDM) Loop The time division switching component of a SPC switch. The early TDM loops were 30-channel loops. TDM loops are now 512-channel or higher.

Time Slot Interchange (TSI) The heart of the switching matrix in a digital switching system. The TSI is a space division switching component. One TSI is needed for every channel on a TDM loop. The TSI connects a time slot on the TDM's transmit bus to a time slot on the TDM's receive bus.

Translations Software A database counting line translations tables and trunk translations tables. A database of how ports are assigned and what features are assigned to each port.

Trunk Controller A circuit card in the trunk shelf, that contains a microprocessor and EPROM, to control the activities of trunk circuits in that shelf.

9

Data Communications

Key Terms

ACK
Asymmetric Digital Subscriber Line (ADSI)
Analog Signal
Automatic Retransmission Request (ARQ)
American Standard Code for Information Interchange (ASCII)
Asynchronous Transmission
Baud
Binary Signal
Continuous ARQ
Cyclic Redundancy Checking (CRC)
Channel Service Unit (CSU)/ Data Service Unit (DSU)
Data
Data Communications Equipment (DCE)

Data Switching Exchange (DSE)
Data Terminal Equipment (DTE)
Digital Signal
Discrete ARQ
Dual Simultaneous Voice and Data (DSVD)
EIA-232 Interface
Flow Control
Full Duplex
Half Duplex
Header
Integrated Services Digital Network (ISDN)
Internet Service Provider (ISP)
Modem
NAK
Null Modem
Packet Assembler Disassembler (PAD)
Packet Network

Public Data Network (PDN)
Parallel Transmission
Protocol
Quadrature Amplitude Modulation (QAM)
RS-232
Selective ARQ
Serial Transmission
Synchronous Transmission
Trailer
Trellis Coded Modulation (TCM)
Universal Asynchronous Receiver/Transmitter (UART)
Word
X.25

The main purpose of this book is to cover voice applications and the Public Switched Telephone Network (PSTN). The PSTN is used to carry both voice and data communications. A telecommunications manager must understand both voice and data networks. Many companies will use one telecommunications manager to manage both voice and data facilities. Data communications is closely related to voice communications. Both communication systems are concerned with transmitting information over some distance.

The primary difference between voice and data lies in the signal used to convey information. Voice is converted by the transmitter of a telephone into an analog electrical signal having many different voltage levels. Data is converted into an electrical signal that has two different voltage levels. *Bi* means "two," and this two-state signal is called a **binary signal**. The two states of a binary signal are often called a 1 and a 0. Since the two binary *digits*, 1 and 0, are used to represent the two different states of a data signal, a binary signal is also called a **digital signal**. The digital signal consists of a series of binary digits (1s and 0s). Binary Digits are also called *bits*.

The PSTN started life as an analog network designed to handle analog voice signals. With the introduction of integrated circuit chips, digital communication circuitry has become cheaper and more reliable than analog circuitry. The PSTN has gradually been converted from an analog to a digital network. The telephone, the local loop, and the line circuit are the only parts of the PSTN that remain analog. The first portions of the PSTN to be converted from analog to digital were the multiplexing facilities carrying interoffice trunk circuits. The T1 carrier system developed by Bell Labs multiplexed 24 trunk circuits onto two cable pairs using digital technology. The T1 system contained 24 channel units with a codec in each channel unit. The codec was an interface between the *analog signal* from the trunk circuit of an analog switch and a TDM digital highway connecting the switch to another switch.

When digital technology was integrated into the switching matrix of class 5 SPC switching systems, the codec was moved from the channel unit of the T1 carrier system to the line circuit of the switch. The interface point between analog and digital now occurs at the line circuit of a class 5 central office. By moving the codec from the line circuit of the switch out to the telephone, we can convert the remaining analog components of the PSTN (the telephone, the local loop, and the line circuit) to digital.

Several digital technologies have been developed for voice communications that move the codec from the line circuit to the customer's premises. Chapter 11 expands on two of these technologies: *Integrated Services Digital Network (ISDN)* and *Asymmetric Digital Subscriber Line (ADSL)*. These two technologies allow us to achieve a complete end-to-end all-digital PSTN where voice and data are both transmitted using digital signals. The demand for a digital line is low because of its high cost. Most people are still being served by an analog line circuit, local loop, and analog telephone. Perhaps more people will switch to digital telephones when the price of a digital line comes down.

As mentioned earlier, a digital signal is a two-state signal. Each state can be represented by a discrete voltage level. The digital signal consist of a series of bits where each bit is either a 1 or a 0. There is no set standard for all devices as to what the voltage levels for a 1 or a 0 will be. Different devices use different voltage levels to represent a 1 or 0. One device may represent a 1 by using a +3-V signal and represent a 0 by a 0-V signal. Another device may represent a 1 by a –6-V signal and a 0 by a +6-V signal. For a digital communication circuit to work, the devices used on each end of the circuit must be using the same protocol for voltage conversion.

T1 carrier systems use alternating +3- and –3-V signals to represent a 1, and use 0 Vto represent a 0. RAM uses a very low voltage to store a 1 and no voltage to store a 0. The digital signal passed between a PC and a modem typically uses –12 V to represent a 1 and +12 V to represent a 0. This signal is called a *bipolar signal* because it has a positive and a negative state. It is also called a non-return-to-zero signal because the signal does not return to 0 V after sending either a –12-V (1) or +12-V (0) signal. The signal remains at the last voltage state until a new signal is received.

The PC belongs to the category of equipment called ***data terminal equipment (DTE)***. Data terminal equipment is used to transmit and receive data signals. The equipment used to connect a DTE to the PSTN is called ***data communications equipment (DCE)***. Some people also interpret DCE as *data circuit terminating equipment*. A DCE is used to convert the output of a DTE into a form suitable for the transmission medium. The most common form of DCE is a ***modem***. *The modem converts a digital signal into an output suitable for transmission over the analog local loop.*

The most common connection used to connect DTE to DCE is an ***EIA-232 interface*** (Figure 9-1). This interface is also called a *Recommended Standard 232 (RS-232)*. It has been approved as CCITT Standard V.24. The EIA-232 interface is the standard for serial transmission of data between two devices. The EIA-232 interface

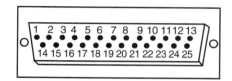

DB–25 connector

DB-9 connector

Pin number	Signal designation	Pin number	Signal designation
1	Protective ground	1	Carrier detect
2	Transmit data	2	Receive data
3	Receive data	3	Transmit data
4	Request to send	4	Data terminal ready
5	Clear to send	5	Protective ground
6	Data set relay	6	Data set ready
7	Signal ground	7	Request to send
8	Carrier detect	8	Clear to send
9	Positive DC test voltage	9	Ring indicator
10	Negative DC test voltage		
11	Unassigned		
12	Secondary carrier detect		
13	Secondary clear to send		
14	Secondary transmit data		
15	Transmit clock (DCE)		
16	Secondary receive data		
17	Receiver clock		
18	Receiver dibit clock		
19	Secondary request to send		
20	Data terminal ready		
21	Signal quality detector		
22	Ring indicator		
23	Data signal rate selector		
24	Transmit clock (DTE)		
25	Busy		

FIGURE 9-1 E1A = 232 interface.

is limited to a cable length of 50 ft between the DTE and DCE. The EIA-232 interface specifies that a signal between –3 and –15 V DC can be used to represent a logic 1, and a that signal between +3 and + 15-V DC can be used to represent a logic 0.

The EIA-232 interface cable comes in a 25-conductor and a 9-conductor cable. The 25-conductor cable has a DB-25 connector attached to each end of the cable. The 9-conductor cable uses DB-9 connectors. When the RS-232 standard was developed, it was developed with 25 leads in the interface between a DTE and a DCE. Personal computers do not use all 25 leads; they only use 9 of the 25 leads. IBM developed the DB-9 connector for its AT computer. If you look on the back of an IBM PC, you will see a male 9 pin connector. This is the EIA-232 connector of the serial port of an IBM PC. *The serial port on all DTE devices is a male DB-25 or DB-9 connector. The serial port on a DCE device will be a female connector.* It can also be either a DB-9 or DB-25 connector. A cable designed to connect a DTE to a DCE will have a female connector on one end and a male connector on the other end.

The transmit lead and receive lead in the cable get their designation from the DTE end of the cable. A PC must transmit data on the transmit lead. This lead has to connect to the receive circuit of the DCE. A PC must receive data on the receive lead. Therefore, the DCE must transmit its data over the receive lead of the cable to the PC. A DCE device has its transmit and receive leads flipped inside the DCE. The transmit circuitry of the modem connects to the EIA-232 cable receive lead so that the data from the modem arrives at the receive pin of the serial connector on the PC. The receive circuitry of the modem connects to the transmit lead from the PC. For a DB-25 connector on a PC (DTE), pin 2 is transmit and pin 3 is receive. Since the transmitter of the PC must talk to the receiver of the DCE, pin 2 of the DCE is receive. Similarly, the transmitter of the DCE connects to pin 3 of the EIA-232 so it can talk to the receiver in the PC.

A communication circuit will not work if the transmitter is talking to another transmitter. This is why two PCs cannot be connected using a standard EIA-232 cable. The two computers must be connected using modems or a null modem cable. Null means "none". *Null modem* means "no modem". DTEs have male connectors; therefore, a null modem cable will have female connectors on both ends of the cable. A null modem cable is wired so that pin 3 of the female connector at one end of the cable is wired to pin 2 of the female connector at the other end. Thus, the null modem cable reverses the transmit and receive leads between the two computers. With this reversal, the transmitter of each computer is talking to the receiver of the other computer. Null modem cables also reverse other leads (Figure 9-2).

As mentioned earlier, the telephone and local loop are analog devices. They are interfaced to the digital PSTN by a codec on the line circuit at the SPC switch. The local line circuit will only accept analog signals because of the codec in its circuitry. Thus, if the local analog telephone line is to be used for data, the data must be converted into an analog signal. The most common type of interface circuitry for interfacing data to the local loop is a modem. The word *modem* is an acronym for *mod*ulator/*dem*odulator. When a modem is used to send data over the PSTN, another modem is required at the receiving end. The modems will continuously send an analog signal to each other.

When data from a computer is fed into a modem, it modulates (changes) the analog signal being sent to the other modem. The PSTN handles the modulated signal extremely well because the signal is in the voice bandwidth of the PSTN. At

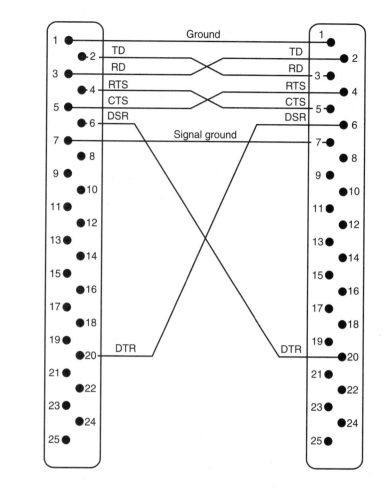

FIGURE 9-2 RS-232 Null modem cable.

the receiving modem, the signal is demodulated or changed back into a replica of the original data signal and fed into the receiving computer. Since each modem has its own modulator and demodulator, the modem can send a signal and receive a signal at the same time. A modem is a *full-duplex* device but with some protocols is capable of operating in a *half-duplex* mode.

9.1 DATA AND THE PUBLIC SWITCHED TELEPHONE NETWORK

Because the PSTN has been evolving from an analog into a digital network, it is possible to integrate data directly into the digital PSTN if the analog line circuit of an SPC switch is bypassed. *It is impossible to send data over the local telephone line without using a modem because the line circuit in today's SPC switches contains a codec.* The coder portion of a codec changes analog signals into digital signals. The coder expects to see an analog signal, not a digital signal. The codec does a great job of converting our analog voice signal into a digital DS0 signal, but it cannot do a digital-to-digital conversion. Because the local loop and line circuit are designed to handle analog signals, a modem is required to transmit data when using a local telephone number (Figure 9-3).

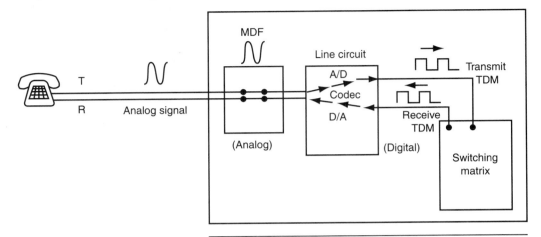

FIGURE 9-3a Voice circuit through codec.

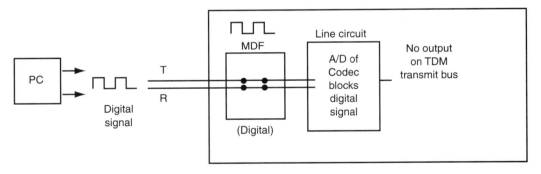

FIGURE 9-3b Data blocked by codec.

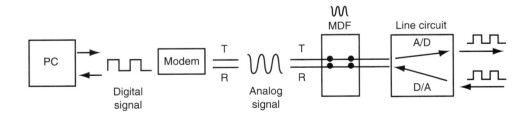

FIGURE 9-3c Modem signal passed by codec.

If a facility is to be used only for data and not for voice, a *channel service unit (CSU)/ data service unit (DSU)* is used to interface the data circuit directly into the PSTN. A CSU/DSU is a DCE. A modem is a DCE used to interface a digital circuit to an analog circuit, and a *CSU/DSU is a DCE used to interface a digital circuit to another digital circuit.* The CSU/DSU is designed to do a digital-to-digital conversion and will interface the digital speed of your computer to a different digital speed highway, such as 56 Kbps, 64 Kbps, or a 1.544-Mbps PSTN transmission facility. It is now possible to use these digital-to-digital interface devices on the customer's premises, and in a local central office because the PSTN network between central exchanges has been converted into a digital network (Figure 9-4).

It is also possible to transmit data directly over the local loop by connecting the customer's premises to the central exchange, with a Digital Subscriber Line Circuit and ISDN rather than an analog line circuit. However, an ISDN line circuit cost $100 a month from the LEC versus $20 a month for an analog line circuit. Most people prefer their analog local service line over ISDN service, because it is much cheaper than a digital ISDN line. People are also slow to convert to ISDN because services they use are not on an ISDN line circuit. It does not do me any good to convert my line to ISDN if people I transfer data to do not convert to ISDN.

The protocols used to transfer data are not the same for ISDN and a modem. *Protocol* refers to the rules of communication. Each established protocol is a definition of how data is to be communicated from one device to another. The transmitting and receiving ends must use the same communication protocol. Modems can only talk to modems using the same protocol; they cannot communicate with ISDN because the communication protocols are not the same. Remember that a receiving modem is looking for an analog signal from the PSTN. With ISDN, there is no analog signal. ISDN is an end-to-end digital network. If I wish to send data to someone using a modem, I must also use a modem. If I want to use ISDN to transfer data, an ISDN line circuit must be on the other end of the transmission facility. I am able to use ISDN

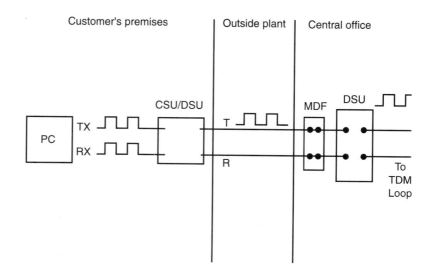

FIGURE 9-4 CSU/DSU
interface between customer's
premises and central office.

to communicate with other people via e-mail or the Internet because my *Internet service provider (ISP)* has ISDN lines. The ISP serves as an interface and protocol converter between my ISDN line and the lines of other people who are using modems to connect to the ISP.

9.2 DATA COMMUNICATIONS AND CODING

Data communications is the transmission of data from one location to another. The *data* transmitted can be numeric data or alphabetical data. When data is transmitted, some code is used to represent the various characters or numbers that will be sent over the system. Samuel Morse used a code that was named for him to send messages over his invention (the telegraph). The telegraph was our first means of long distance communications. The telegraph was a data communications network. Messages were converted into Morse code. The Morse code was sent as electrical signals over the telegraph lines. The receiving end would decode the signal back into the original message and deliver it to the intended recipient. The telephone became the preferred means of communication, and the telegraph became history. As the use of computers has skyrocketed, the need for a data communications network is essential.

The telegraph system used a code of dots and dashes to represent characters and numbers. A dot was transmitted as a short burst of electrical current and a dash was transmitted as a longer burst of electrical current. Computers were designed to code characters and numbers using 1s and 0s. The two states are represented by two different voltage levels. As we have seen, this two state condition is called a binary signal; the 1 and 0 represent the two states and are called binary digits or bits. The bit is the smallest piece of data that a computer processes and will always be a 1 or 0. Since we can only use 1s and 0s to code characters and numbers, we must use a binary numbering system. The number of different characters that can be encoded using binary numbering depends on the number of bits in the code. A personal computer (PC) uses 7 bits for coding each input from a keyboard, but each storage location is 8 bits wide to allow the use of 8-bit coding protocols when desired.

9.3 ASCII CODING

The standard binary code developed for transmission of data is the *American Standard Code for Information Interchange (ASCII)*. *ASCII is the coding scheme used by a PC to store information.* It uses 7 bits, each of which can be either a 1 or a 0. That is, each of the 7 bits can be one of two states (1 or 0). According to mathematical statistics, if we have *seven different* things (bits) that can assume *one of two* states (a 1 or 0), the *total number of different combinations* we can have is equal to 2^7. Since 2^7 is 128, ASCII provides 128 different codes from 0000000 to 1111111. These codes can be converted into decimal numbers by adding the value of each bit in a code. The value of each bit corresponds to the location of the bit in the code. The codes for decimal digits 1, 3, 65, 97, and 127 are:

2^6	2^5	2^4	2^3	2^2	2^1	2^0	=	
0	0	0	0	0	0	0	=	0
0	0	0	0	0	0	1	=	1
0	0	0	0	0	1	1	=	3
1	0	0	0	0	0	1	=	65
1	1	0	0	0	0	1	=	97
1	1	1	1	1	1	1	=	127

We can use a seven-level binary code to represent the decimal numbers from 0 to 127, or we can use the seven-level code to represent other things. ASCII uses the 128 codes that we get with the seven-level binary code to represent the letters of the alphabet and the numerals 0 to 9. See Table 9-1 for the ASCII code chart.

Look in Table 9-1 for the letter *a*. Lowercase *a* is represented by the code 1100001. We have previously shown that if this were converted to a decimal number, it would be 97. But ASCII does not use the code to convert binary numbers to decimal numbers; it uses the code to convert binary numbers to characters and the decimal numbers 1 to 9. When the lowercase letter *a* is pressed on the keyboard, it is stored as an ASCII code 1100001. Therefore when the code is read back, it is converted back into the letter *a*. One note is in order here for anyone who connects an oscilloscope to read the signal output of an RS-232 serial port: although we read an ASCII code from highest- to lowest-order bit (left to right), the bits are actually sent lowest- to highest-order.

PCs contain a hardware device between the data bus of the PC and the RS-232 serial port. This device is called a ***Universal Asynchronous Receiver/Transmitter (UART)***. The UART will send a start bit (a 0, which is +3 V or greater). It will then send the least significant data bit and each higher-order data bit until the most significant data bit is sent. It then sends the parity bit, then a stop bit (a 1, which is –3 V or less). Remember our lowercase *a*. It is 1100001 in ASCII code. When the UART sends this code out, it is sent lowest-order bit first, which is sent as 1000011. Suppose we are using even parity; when we put the start, even parity, and stop bits in place, the UART sends 0100001111. It may help to simplify what is happening by stating that although we read ASCII from left to right, the UART sends the code out from right to left. Of course, the receiving UART uses the same protocol and reconstructs the character code starting with lowest-order bit first.

9.4 EXTENDED ASCII CODING

It was not long before we needed to represent more than 128 different characters. To represent more than this number of characters, an eight-level coding system was developed called *Extended ASCII*. IBM used an eight-level coding system called *Extended Binary Coded Decimal Interchange Code (EBCDIC)* in its mainframes. EBCDIC is not the

Table 9-1 ASCII Code Chart

Back Five Bits Column #	Bit Position 2⁴2³2²2¹2⁰	Control Codes (Row 0)	Numbers Punct. (row 1)	Uppercase (row 2)	Lowercase (row 1)
Most Significant Front 2 Bits 2⁶ 2⁵		**0 0**	**0 1**	**1 0**	**1 1**
0	0 0 0 0 0	NUL	h	@	`
1	0 0 0 0 1	SOH	!	A	a
2	0 0 0 1 0	STX	"	B	b
3	0 0 0 1 1	ETX	#	C	c
4	0 0 1 0 0	EOT	$	D	d
5	0 0 1 0 1	ENQ	%	E	e
6	0 0 1 1 0	ACK	&	F	f
7	0 0 1 1 1	BEL	'	G	g
8	0 1 0 0 0	BS	(	H	h
9	0 1 0 0 1	HT	)	I	i
10	0 1 0 1 0	LF	*	J	j
11	0 1 0 1 1	VT	+	K	k
12	0 1 1 0 0	FF	,	L	l
13	0 1 1 0 1	CR	-	M	m
14	0 1 1 1 0	SO	.	N	n
15	0 1 1 1 1	SI	/	O	o
16	1 0 0 0 0	DLE	0	P	p
17	1 0 0 0 1	DC1	1	Q	q
18	1 0 0 1 0	DC2	2	R	r
19	1 0 0 1 1	DC3	3	S	s
20	1 0 1 0 0	DC4	4	T	t
21	1 0 1 0 1	NAK	5	U	u
22	1 0 1 1 0	SYN	6	V	v
23	1 0 1 1 1	ETB	7	W	w
24	1 1 0 0 0	CAN	8	X	x
25	1 1 0 0 1	EM	9	Y	y
26	1 1 0 1 0	SUB	:	Z	z
27	1 1 0 1 1	ESC	;	[	{
28	1 1 1 0 0	FS	<	\		
29	1 1 1 0 1	GS	=	]	}
30	1 1 1 1 0	RS	>	^	~
31	1 1 1 1 1	US	?	_	DEL

*a = 1100001

same coding as Extended ASCII. The advantage of using Extended ASCII is that the first 128 codes for it are the same as they are for regular seven-level ASCII. The codes from 128 to 255 are used to represent additional characters (Table 9-2).

9.5 SERIAL VERSUS PARALLEL TRANSMISSION OF DATA

Data can be transmitted one bit at a time or several bits at a time. When data is transmitted one bit at a time, it is transmitted over one wire. Each bit is transmitted one after the other onto the same wire. This type of transmission is called *serial transmission*. When data is transmitted several bits at one time, each bit is transmitted over its own wire. This type of transmission is called *parallel transmission* because the bits are transmitted over wires that are parallel. The most common form of parallel transmission is to transmit 8 bits at one time. With seven-level ASCII coding, an eighth bit was used as a parity bit to check the validity of the code. Thus, each character and number is represented by an 8-bit code (7 ASCII + 1 parity bit). By transmitting 8 bits at one time, we can transmit a character at a time instead of a bit at a time. Eight bits is called a *byte*.

In computers, an 8-bit byte is referred to as a *word*. It is not really a word. It is a character. Each character or number is represented by a unique 8-bit byte. Data inside a computer is transferred over a data bus that is a parallel bus. The first PC had eight wires for the data bus and could transfer data 8 bits (1 byte) at a time. A 486 computer uses 32 parallel leads for the data bus and can transfer data 32 bits at a time, or four 8-bit words at a time. A Pentium computer has a data bus with 64 leads and transfers data 64 bits at a time, or eight 8-bit words at a time. PCs define a word as the number of adjacent bits that can be manipulated or stored as one unit. This depends on the size of the data bus, and in a PC, the size of the data bus defines a word. In a Pentium PC a word is 64 bits, not 8 bits.

Most printers that connect to a personal computer are parallel printers transferring 8 data bits at a time. Some older printers are serial printers that connect to an RS-232 port and transfer data 1 bit at a time. It is readily apparent that parallel transmission is much faster than serial transmission, but serial transmission is most often used in data communications, because only one circuit is needed for the transmission of data. Parallel transmission would require establishing eight or more circuits between the two data devices. Parallel data transmission is used when the transmission distance is less than 25 ft. When data is sent over the PSTN, it is sent using serial transmission. It is obvious that we only have one communication circuit over the PSTN. If we only have one circuit for the transmission of data, serial transmission must be used.

9.6 ASYNCHRONOUS VERSUS SYNCHRONOUS TRANSMISSION

There are two ways to transmit data using serial transmission. One method is to use *asynchronous transmission, where each character has its own synchronizing*

TABLE 9-2 Extended ASCII Code Chart

Decimal Code	Character	Decimal Code	Character	Decimal Code	Character	Decimal Code	Character	Decimal Code	Character
0	NULL	52	4	104	h	156	£	208	–
1	J SOH	53	5	105	i	157	¥	209	—
2	1 STX	54	6	106	j	158	û	210	"
3	§ ETX	55	7	107	k	159	ƒ	211	"
4	© EOT	56	8	108	l	160	á	212	'
5	® ENQ	57	9	109	m	161	í	213	'
6	´ ACK	58	:	110	n	162	ó	214	÷
7	% BEL	59	;	111	o	163	ú	215	◊
8	BS	60	<	112	p	164	ñ	216	ÿ
9	HT	61	=	113	q	165	Ñ	217	Ÿ
10	LF	62	>	114	r	166	ª	218	/
11	VT	63	?	115	s	167	º	219	¤
12	FF	64	@	116	t	168	¿	220	‹
13	CR	65	A	117	u	169	©	221	›
14	SO	66	B	118	v	170	™	222	fi
15	} SI	67	C	119	w	171	_	223	fl
16	ÿ DLE	68	D	120	x	172	_	224	‡
17	◊ DC1	69	E	121	y	173	¡	225	·
18	◊ DC2	70	F	122	z	174	«	226	,
19	!! DC3	71	G	123	{	175	»	227	„
20	DC4	72	H	124	|	176	∞	228	‰
21	NAK	73	I	125	}	177	±	229	Â
22	SYN	74	J	126	~	178	≤	230	Ê
23	ETB	75	K	127	DEL	179	≥	231	Á
24	È CAN	76	L	128	Ç	180	¥	232	Ë
25	Í EM	77	M	129	ü	181	µ	233	È
26	Á SUB	78	N	130	é	182	∂	234	Í
27	Ë ESC	79	O	131	â	183	Σ	235	Î
28	FS	80	P	132	ä	184	∏	236	Ï
29	Ô GS	81	Q	133	à	185	π	237	Ì
30	Ò RS	82	R	134	å	186	∫	238	Ó
31	Ú US	83	S	135	ç	187	ª	239	Ô
32	SP	84	T	136	ê	188	º	240	
33	!	85	U	137	ë	189	Ω	241	Ò
34	"	86	V	138	è	190	æ	242	Ú
35	#	87	W	139	ï	191	ø	243	Û
36	$	88	X	140	î	192	¿	244	Ù
37	%	89	Y	141	ì	193	¡	245	ı
38	&	90	Z	142	Ä	194	¬	246	^
39	'	91	[	143	Å	195	√	247	~
40	(	92	\	144	É	196	ƒ	248	¯
41	)	93	]	145	æ	197	≈	249	˘
42	*	94	^	146	Æ	198	Δ	250	˙
43	+	95	_	147	ô	199	«	251	˚
44	,	96	`	148	ö	200	»	252	¸
45	-	97	a	149	ò	201	…	253	˝
46	.	98	b	150	û	202		254	
47	/	99	c	151	ù	203	À	255	ˇ
48	0	100	d	152	ÿ	204	Ã		
49	1	101	e	153	ö	205	Õ		
50	2	102	f	154	Ü	206	Œ		
51	3	103	g	155	¢	207	œ		

information. The beginning of a character is designated by a bit called the *start bit*. The start bit is a bit whose value is 0. A bit that has a value of 0 is also called a *space bit*. The end of a character is designated by a bit that is called the *stop bit*. The value of the stop bit is 1. A bit that has a value of 1 is also called a *mark bit*. Asynchronous transmission is also called *start-and-stop transmission*. The UART discussed earlier uses asynchronous communication. Therefore, the communication between the serial port of a PC and an external modem is asynchronous communication.

When mainframe and minicomputers were the only computers available, most interface to these computers was via a dumb terminal. The terminals were connected to the computer using asynchronous communication. No one can type at the speed of a computer's processor. There was no need to have the processor dedicated to one user. By using asynchronous communication protocol, many users could be connected to a central computer. When a terminal had information for the computer, each piece of data was preceded by a start bit. The use of asynchronous (start-and-stop) transmission was an efficient way to connect many devices to a central computer.

The second method of serial transmission is **synchronous transmission**. *This type of transmission sends data as a block of characters at a time.* A data block can be almost any size, but a typical data block contains 128, 256, 512, or 1024 characters. When synchronous data transmission is used, the start and stop bits are not used. A **header** is placed in front of the data sent and a trailer is placed after the data. The header will contain the destination address of the message, a synchronization signal, and control information. The trailer contains parity checking information and the address of the sender. The header usually contains about 32 bits and the trailer usually has 8 to 16 bits.

Each character sent using asynchronous transmission requires the transmission of 10 bits (start, character, stop). The start and stop bits are called *overhead bits*. If 256 characters are sent using asynchronous transmission, 2560 bits are required. If synchronous transmission is used, we must transmit the header, the 256 characters, and the trailer. Thus, 256 characters times 8 bits per character is 2048 bits. If we add 64 bits of overhead for the header and trailer, the total number of bits required is 2112. It is readily apparent that synchronous data transmission is faster than asynchronous transmission because fewer bits are needed to send the same data. Synchronous transmission is more complex than asynchronous transmission and provides more sophisticated error-detection and correction capabilities.

Synchronous transmission is usually used on a dedicated circuit connecting two high-speed computers. Asynchronous transmission is used between the PC (or DTE) and the modem (DCE) on dial-up connections through the PSTN. When we wish to transfer files of data between two PCs, we use a file transfer protocol such as X Modem, Y Modem, Z Modem, or Kermit to transfer the file. These protocols transfer a file between two PCs or between a DTE and DCE using blocks of data. For example, X Modem transfers a file 128 bytes at a time. Although the data is transferred as blocks of data, it still is an asynchronous data communications protocol. Each byte in the block of data is surrounded by a start and stop bit. The UART in the serial port dictates that the transmission of data between two serial ports will be asynchronous.

One point that should be cleared up here is that a modem has two interfaces, and each interface has its own protocol. The modem has a serial EIA232 interface to the PC and another interface to the PSTN. Remember that a modem is simply a protocol converter. It converts the binary signal received from a PC into the modulation of an analog signal to the local loop. It also converts the modulation of analog signal received from the local loop into a digital signal to the PC. Today's high-speed modems also make another conversion. They convert the asynchronous signal from the PC into a synchronous signal to the PSTN. The modem at the other end of the circuit converts the synchronous signal from the PSTN back into an asynchronous signal to its PC.

Some modems are designed to accept only asynchronous transmission from a DTE (Bell 103, Bell 202, V.21, and V.23). Some modems will only accept synchronous transmission from a DTE (Bell 201, V.26, V.27, V.29, and V.33). Several modems are designed to allow the user to select either asynchronous or synchronous transmission between the DTE and DCE (Bell 212, V.22, V.24, V.32, and V.34). The EIA-232 interface from the DTE to the DCE allows the use of either asynchronous or synchronous transmission of data.

When we talk about whether a signal between two DTEs is asynchronous or synchronous, we are talking about the signal leaving the DTE. The signal over the PSTN between the two modems' (DCEs) is a synchronous signal when high-speed modems are used. The PC will send bits in serial asynchronous transmission to the external modem. The modem contains memory and software to arrange incoming bits from the PC into a block of data for synchronous transmission over the PSTN. Since a PC is the most common form of DTE that we encounter, I will use the term *PC* instead of *DTE* throughout the remainder of this chapter.

9.7 ASYNCHRONOUS TRANSMISSION ERROR CHECKING

The simplest error checking method is parity checking. This technique is used for asynchronous transmission. ASCII uses 7 bits for coding and does not use the 8th bit. This allows the use of the 8th bit for parity checking. The transmission system will count the number of 1s sent for each character and place a 1 or 0 in the 8th-bit location as required. There are five types of parity: odd, even, mark, space, and no parity. When odd parity is used, the count of 1s for every 8 bits sent will be odd. Suppose we transmit the character *a* (1100001). This character has an odd number of 1s. For odd parity, we place a 0 in the 8th-bit location. This results in an odd number of 1s for all 8 bits. The character capital *A* is 1000001. This is an even number of 1s. When *A* is transmitted, we must send a 1 as the 8th bit to achieve an odd number of 1s for the 8 bits. With odd parity, all transmitted bytes results in an odd number of 1s. The receiver counts the 1s for each byte to make sure they are all odd.

Even parity is similar to odd parity; this error-checking protocol ensures each transmitted byte has an even number of 1s. The 8th bit is made a 1 or 0 to ensure that each byte has an even number of 1s. Mark parity is a protocol that makes every 8th bit a 1. Space parity makes every 8th bit a 0. As you can see, the 8th bit is used for the parity bit. When an eight-level code such as Extended ASCII is used for coding, the 8th bit is used as part of the code and no parity can be used. When a PC transmits data over a serial port, the protocol of the port must be set to match the protocol of the receiving device. The protocol of a serial port consists of data speed in bits per second, the number of data bits per character, parity, and the number of stop bits. The number of stop bits is usually specified as 1 bit. The parity being used by both ends of the data transmission circuit must be specified.

9.8 SYNCHRONOUS ERROR DETECTION AND CORRECTION

Parity checking provides an end-to-end check (PC to PC) on each byte transmitted between two PCs. A modem does not perform any error checking on data it receives from the transmitting PC. The distance between the PC and modem is so short that the possibility of errors is remote. Modems do perform error checking between themselves to ensure that the PSTN did not corrupt the data. The use of synchronous transmission between the two modems allows the use of sophisticated error-detection and control techniques. Modems use a synchronous error-detection technique called *Cyclic Redundancy Checking (CRC)*.

There are two types of CRC: CRC-16 and CRC-32. CRC treats a block of data as representing a large number. If the block of data contains 128 characters, it contains 1024 bits ($128 \times 8 = 1024$). Clearly, 2^{1024} is an extremely large number. If all 1024 bits were a 1, the value of this number would be 1 less than 2^{1025}. The block of data is divided by a 17-bit divisor (for CRC-16) or a 33-bit divisor (for CRC-32). The divisor is a prime number. A prime number is divisible only by itself and 1. Dividing the number represented by the block of data using the prime number divisor results in a 16-bit remainder (for CRC-16), or a 32-bit remainder (for CRC-32).

The CRC calculation is done at the transmitting modem, and the remainder is placed in the trailer behind the block of data transmitted. The receiving modem receives the block of data and trailer. The receiving modem calculates a remainder using the same CRC protocol used by the transmitting modem. It compares the remainder calculated to the remainder sent in the trailer. If they compare, the data received is okay. If they do not match, the data was corrupted. After the receiving modem has determined whether or not errors have occurred, it will notify the transmitting modem. Error conditons are handled by requesting that the originating modem retransmit the data. This technique is called *Automatic Retransmission Request (ARQ)*. There are two basic types of error correction: discrete ARQ and continuous ARQ.

Discrete ARQ is also called *stop and wait ARQ* because after sending a block of data, the transmitting end will wait for a signal from the receiving modem before sending another block of data. The signal from the receiving modem to the transmitting modem will be either a positive acknowledgment *(ACK)* that the data was good, or a negative acknowledgment *(NAK)* that the data was corrupted. For every block of data sent an acknowledgment is received. The acknowledgment will be either an ACK or a NAK. On receipt of an ACK, the next block of data is sent. If a NAK is received, the modem retransmits the last block of data from its memory.

With the use of discrete ARQ (ACK/NAK), the transmitting modem spends a lot of time waiting for the acknowledgment message. Today's PSTN transmits most data without errors. Thus, most of the received messages are ACKs and very few NAKs are received. The modem spends a lot of time waiting just to be told the data was good. *Continuous ARQ* eliminates the need for the transmitting modem to wait for an ACK or NAK. The transmitting modem identifies each block of data with a block number. The receiving modem sends an ACK or NAK to the transmitting modem for each block of data received. The transmitting modem keeps sending data without waiting as long as it is receiving ACKs every so often. If it fails to receive an ACK after a certain interval of time, the modem stops transmitting.

As long as a transmitting modem continues to receive ACKs from the receiving modem, it continues to transmit data. When the receiving modem detects an error, it sends a NAK, along with the block number affected, to the transmitting modem. Some modems use a protocol that retransmits all blocks from the NAK block forward. Some modems use a *selective ARQ* protocol. With selective ARQ, only the affected block is retransmitted. CCITT Standard V.42 is an error correction standard that allows modems to determine during handshaking whether to use MNP4 or LAP-M error-correction protocol. *Link Access Protocol for Modems (LAP-M)* provides a selective ARQ protocol.

9.9 FLOW CONTROL

Modems are continually storing and retrieving data from their memory. Some type of flow control is necessary to prevent these memory buffers from overflowing and still keep enough data in memory to allow for retransmission on receipt of a NAK. The flow control management software in the modem keeps track of data flowing into and out of the buffer memory. This software keeps track of the amount of available free memory and instructs the transmitting PC to stop sending when the buffer is almost full.

The modem uses hardware flow control to instruct the transmitting PC to stop sending. The EIA-232 interface connecting the PC to the DCE (modem) contains a request-to-send (RTS) lead on pin 4 (DB-25 connector) and a clear-to-send (CTS) lead on pin 5 (DB-25 connector). The RTS and CTS on a DB-9 connector are pins 7 and 8 respectively. When a modem has approached a full-buffer condition, it will remove the CTS signal. On losing the CTS signal, the transmitting PC stops sending. After the modem has sent enough data to free up additional buffer memory, the CTS signal will be sent to the PC and it will start sending data again. The use of RTS and CTS is called hardware *flow control*. When the PC has data to send, it sends an RTS signal to the modem. The modem returns a CTS signal if it is ready to accept data, and an RTS signal is present (Figure 9-5).

9.10 SERIAL PORTS AND UARTS

The data highway inside of a PC is a parallel data bus. The connection to most external devices is via a serial port. The UART between the parallel data bus and the EIA-232 connector converts parallel transmitted data from the computer into serial transmission. It also converts the received serial data from the EIA-232 interface into parallel-transmission for the computer. Thus, a UART is a parallel-to-serial and a serial-to-parallel converter. The maximum speed of a serial port is determined by the UART and software drivers for that port.

Microsoft Windows 3.11 allows setting a serial port to a maximum of 19,200 bps. This is because DOS 6.20 only supports 19,200 bps, which is a speed supported by older 8250 UARTs. All PCs manufactured today use a 16550 AF UART. This UART can support speeds as high as 115,200 bps. With a 16550 AF UART, the speed of data transmission between the PC and modem can be 115,200 bps.

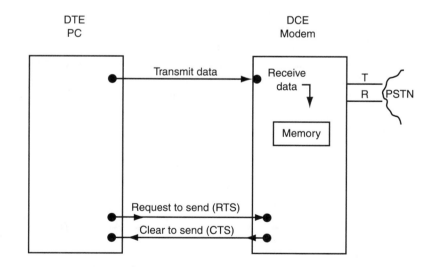

FIGURE 9-5 Hardware flow control RTS CTS.

When memory buffer approaches a full condition, CTS lead is opened and the PC stops transmitting data.

To take advantage of this higher speed, a communication software package will be needed that allows you to set the serial port of the PC to a higher speed than the DOS command "mode" allows. Windows 95 provides the software support necessary to allow setting a serial port to 115,200 bps (Figure 9-6).

Windows 95 allows setting a serial port speed as high as 921,600 bps, but the 16550 UART restricts the speed to 115,200 bps. Internal modem cards are now being sold that use a 16C650 UART. This UART chip achieves 230,000 bps. Eventually the speed of serial ports will rise as PCs, modems, and other devices equipped with this chip replace older devices.

9.11 BAUD RATE AND MODULATION OF THE CARRIER

Modems communicate with each other via audio frequency waves modulated by the digital input from a PC. The audio signal between two modems is called the *carrier signal*. The modulation (or change) to the audio signal can be accomplished by changing the amplitude, frequency, or phase of the audio signal. How fast these signal changes occur is called the *baud rate*. **Baud** refers to the number of changes occurring in a signal.

Suppose we transmit an 1800-Hz signal and change the frequency from 1800 to 1900 Hz. If we change the frequency 2400 times per second, this is 2400 baud. If each change in frequency represents 1 bit, we also have 2400 bps. Today's modems achieve higher speeds because each change in a signal represents more than 1 bit.

Baud rate is limited by the bandwidth of the PSTN voice channel. This bandwidth in the old analog PSTN was 3000 Hz (300 to 3300 Hz). By using 2400-baud signals, part of the bandwidth could be used for guardbands between signals. The bandwidth of the *digital* PSTN is 3600 Hz, and since guardbands are not necessary, modems can transmit at speeds up to 3600 baud over the digital PSTN.

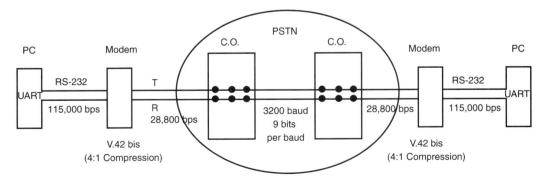

FIGURE 9-6 Personal computer-to-modem speed and modem-to-modem speed. Physical speed between modems is 28,800 bps; end-to-end speed between PCs is 115,000 bps.

The overall speed of a modem depends on the baud rate it can use and how many bits can be represented by each signal change (baud). If we wish to use different amplitudes of a signal to represent a *2-bit* combination, we need four different amplitudes. Two bits can assume any of four different combinations: 00; 01; 10; and 11. Two bits (called a *dibit*) can be represented by four different amplitudes when amplitude modulation is used:

Amplitude 1 = 00
Amplitude 2 = 01
Amplitude 3 = 11
Amplitude 4 = 10

Modems can also use *phase modulation (PM)* to modulate a carrier signal between two modems. The phase of a signal is used to describe the relationship of one signal to another. Suppose we use an 1800-Hz signal for the carrier signal of a modem. Each cycle of an 1800-Hz signal takes 0.0005555555. . . s (time = $1/f$). Each quarter of a cycle takes 0.0001388888. . . s. Each quarter of a cycle represents 90° of an AC signal. If the timing of one 1800-Hz signal is changed so that it occurs 0.0001388888. . . s later than another (reference) signal, it is 90° out of phase with the reference signal. If a signal occurs 0.00277777. . . s after the reference signal, it is 180° out of phase with the 1800-Hz reference signal.

There are three types of phase modulation: *two-phase, relative phase,* and *quadrature phase.* A two-phase signal uses 0° to represent a 1 and 180° to represent a 0. A relative phase signal uses the relationship of signals to code a 1 or 0. If the phase received is the same as the last phase received, it represents a 1. If the phase is different, it represents a 0. Thus, to transmit several 1s, the phase is changing back and forth from one phase to another. To represent a string of 0s, the phase remains the same.

A quadrature phase system uses four different phase relationships to code 2 bits at a time. If we use four different phase relationships, we can send 2 bits at a time by using the four different phases to represent the four different states a dibit can have:

Phase 1 of 0° = 00
Phase 2 of 90° = 01
Phase 3 of 180° = 11
Phase 4 0f 270° = 10

It is also possible to send 3 bits at a time by using eight different amplitudes or eight different phase shifts. Thus, $2^3 = 8$ different combinations that a tribit can have. It is also possible to send 4 bits (quadbit) at a time by using 16 different amplitudes or 16 different phase shifts. A V.29 modem achieves 9600 bps, by transmitting 4 bits at a time using 8 different phases of a signal, where each phase can be one of 2 amplitudes. There is a limit on how many different amplitudes or phase shifts a modem can use. That limit is imposed by the detecting capabilities of the receivers and the noise on the transmission facility. The more amplitude levels or phase shifts used to transmit a code, the better the receiver must be at detecting small discrete changes in the signal. The requirement to maintain a high signal-to-noise ratio also limits how many different amplitudes and phase shifts a system can use. The digital PSTN is a much quieter transmission facility than the old analog PSTN, and the digital PSTN allows for more discrete signal changes.

Because of the requirement to maintain a high signal-to-noise ratio and the need to have a change detectable by the receiver, the maximum speed at which a modem can operate often seems to have been achieved. However, chip manufacturers continue to develop new improved Signal Logic Processor chips and continue to push the speed envelope even higher. One point to keep in mind is that although newer modems such as the V.34 can achieve speeds of 28,800 bps, they can only achieve this speed with an optimal PSTN path. Most of the time, these modems will fall back to lower speeds due to an imperfect PSTN path. Modems capable of 56,000 bps are now on the market. How often this speed can be attained over the PSTN remains to be seen. It is also questionable if the technology they use will be able to transmit at high speed over the PSTN, or will they fall back to less than 28,800 bps, as V.34 modems do?

A modem also uses compression algorithms that allow it to receive data from a PC faster than it sends data over the PSTN. As data is received from a PC, the modem stores the data in memory. The compression algorithm software looks at the 1s and 0s that came from a PC and compresses the data by eliminating repeating 1s and 0s. In essence, the modem looks at several bytes of data and converts the ASCII code into a new code. The new code is basically a shorthand version of the original ASCII code. The modem transmits this shorthand code to the receiving modem. The receiving modem converts the shorthand code back into ASCII code and sends ASCII to the receiving PC. Most compression algorithms can reduce the bits required to represent data by a factor of 4. By using a 4-to-1 compression ratio (V.42 bis), the end-to-end speed of a 57,600-bps connection, can be as high as 230,400 bps. This speed cannot be realized unless the modems are interfaced to the PC via dual UARTs, or a special accelerator board, in the PC.

Quadrature Amplitude Modulation (QAM) is a technique that uses 8 different phase angles and 2 different amplitudes for each of those phases to achieve 16 detectable events. With 16 possible signals, QAM can transmit 4 bits at a time by using each one of the 16 signals to represent one of the 16 combinations a quadbit can have (2^4 = 16 combinations). If we choose 1800 Hz as our signal and change that signal 2400 times per second (2400 baud), we can transmit 9600 bps by transmitting 4 bits per baud.

Baud refers to the number of times a carrier signal changes. Older modems use 2400 baud to meet the requirements of the old analog PSTN. The older analog network used an AM/FDM carrier system to multiplex many voice channels onto one facility. This technology caused some frequency distortion as we modulated and demodulated the signal between Groups, Supergroups, and Mastergroups. These problems limited the frequency changes or baud rate to 2400 baud.

The digital carrier transmission systems used in major toll routes of the PSTN can handle frequency changes of 3600 baud. Modems designed to take advantage of the new digital PSTN are designed to V.34 standards. V.34 modems will use 2400, 2743, 2800, 3000, 3200, or 3429 baud. The baud used depends on the quality of the end-to-end PSTN circuit. If the PSTN path is a digital path and the local loop provides low loss, 3429 baud will be used. V.34 modems continually monitor the circuit and will automatically downshift to a lower baud rate if degradation of transmission occurs. V.34 modems can select from 50 different combinations of baud rate and modulation techniques. The maximum speed of 28,800 bps can be achieved by using 3200 baud and 9-bit *trellis coded modulation (TCM)*. With 9-bit TCM, we can send 9 databits per baud or 28,800 bps ($9 \times 3200 = 28,800$).

TCM is a technique that expands on QAM. The original TCM was a 5-bit code. It used a 4-bit QAM and added to it, a calculated bit to help the receiving modem decode the quadbit. Transmitting 5 bits at a time requires 32 detectable signals. The V.32 standard is the standard for TCM transmitting at 9600 bps. The 32-point constellation (or detectable events of Figure 9-7) is accomplished by using 4 different amplitudes on 8 different phases. V.32 bis was a standard used

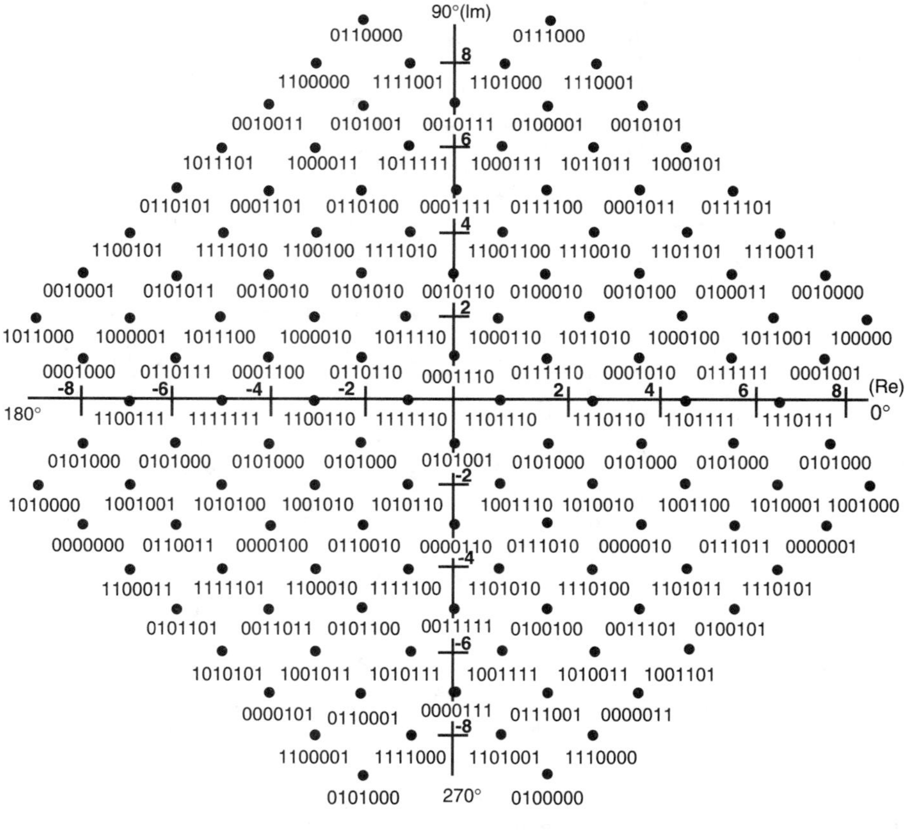

FIGURE 9-7 V.32 bis constellation pattern.

to achieve 14,400 bps. V.32 bis transmits 6 data bits plus 1 TCM bit per baud. Transmitting 7 bits per baud requires 128 detectable events. Transmitting 9 bits per baud (signal change) requires 512 detectable events, and transmitting 10 bits per signal change requires 1024 detectable events. The ITU specifications for V.34 states that the symbol rate (baud rate) for 28,800 bps is 3200. With a 3200 symbol rate, each symbol must represent 9 data bits. To transmit 9 data bits at one time requires 512 different signals (2^9 = 512). V.34 uses a 960-point constellation to provide for the 512 data bits, plus a redundant bit for trellis coding (Figure 9-8).

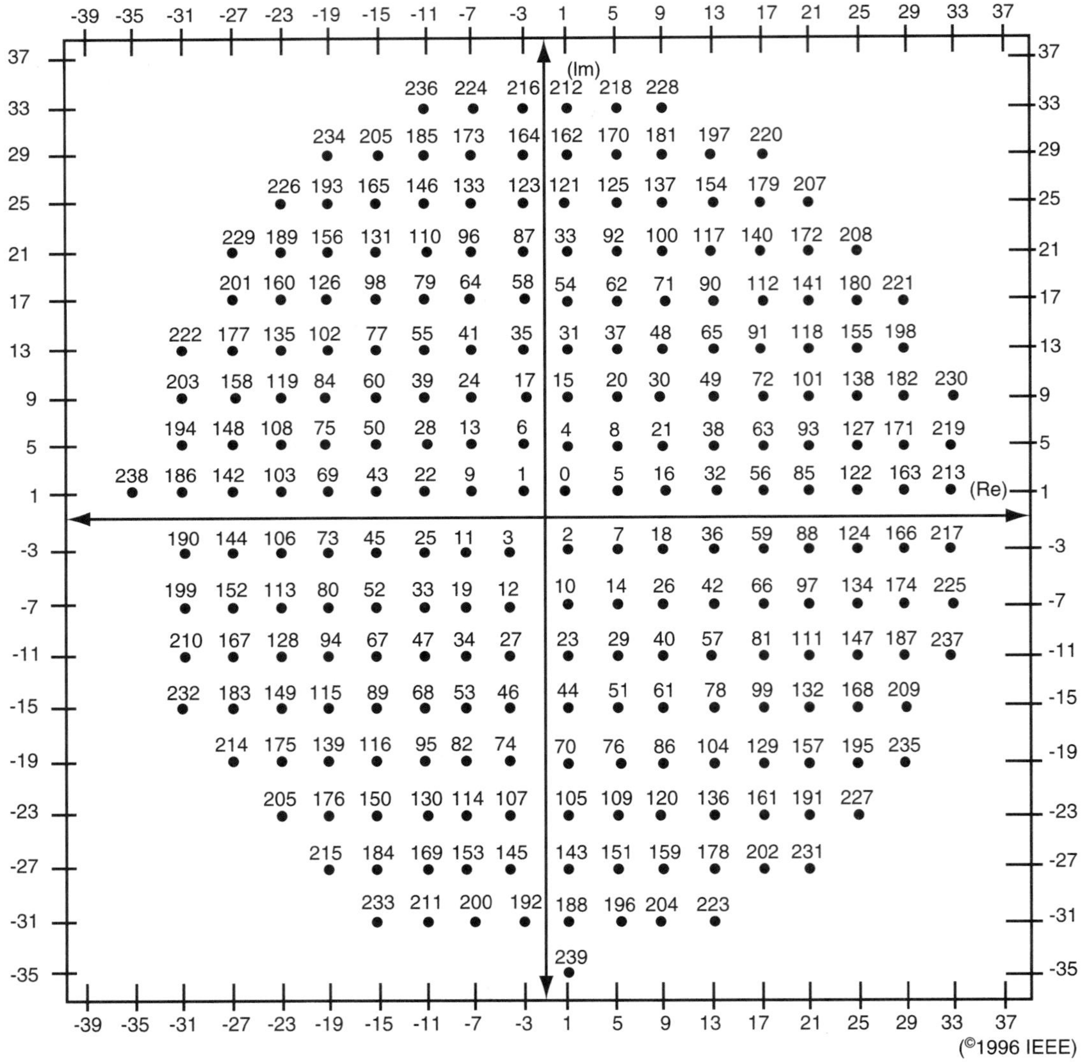

FIGURE 9-8 V.34 quarter-superconstellation with 240 signal points. The full superconstellation is obtained by rotating these points by 0°, 90°, 180°, and 270° (G. Davis Forney, Dec.1996,"The V.34 High-Speed Modem Standard," *IEEE Communications Magazine*.).

Data Communications   Chap. 9

I called some major modem manufacturers in an attempt to nail down how they are achieving 960 different signals. How many phase shifts and amplitude levels are being used to achieve a 960-point constellation? Nobody could tell me. It is done by the chip; they have no idea what goes on inside the chip. Neither do I! This is the essence of chip design. You know what goes in and what comes out, but not how it achieves what it does. A 33.6-bps speed requires the use of a 1664-point constellation.

Modems contain a microprocessor, ROM, and RAM. The ROM contains software that makes the modem very intelligent. It is advances in microprocessor technology that have allowed modems to achieve the V.34 and higher speeds. When a dial-up circuit is established, the modems send signals back and forth to set up the best possible connection between the two modems. This initial communications between modems is called *handshaking*. The V.32 bis standard established a fast retrain capability. With fast retrain, modems continually monitor the quality of the transmission facility and will automatically adjust the speed of transmission up or down. The receiving modem can determine from the incoming signal if it is receiving a data call, a fax, or voice call. The modem that handles all types of calls is called a *voice, data, fax modem*. When the PC uses software to complement the modem, the PC can automatically answer and handle all calls. The newer voice capable modems are **Digital Simultaneous Voice and Data (DSVD)**. These modems allow transmission of voice and data at the same time over the same dial-up circuit.

Modem technology has come a long way, but the fastest speed is still limited by the bandwidth of a PSTN voice channel. With a highly reliable digital PSTN, 3200 baud can be used. This allows us to transmit at 28,800 bps. Higher end-to-end speeds can be achieved by the use of file compression techniques. We are approaching a speed limit with current technology. V.42 bis data compression can achieve a 4-to-1 compression. This means data can be sent at four times the speed of the channel. At 9600 bps using V.42 bis, we can send data at 38,400 bps. At 28,800, we can send data at 115,200 bps. Modems now have a line speed of 56,000 bps. With V.42 bis, they can achieve an overall PC-to-PC speed of 224,000 bps, but the UART will not support a speed above 115,200. A special interface card is needed in the PC to support speeds up to 230,400 bps.

A 57,600-bps line speed is approaching the 64,000-bps speed of a DS0 channel. If speeds faster than 230,400 bps are desired, a digital interface higher than DS0 must be used. Digital interfaces (CSU/DSU) to the PSTN are capable of any speed desired. The CSU/DSU can be interfaced to a DS1 or DS3 signal rate. Therefore, a digital interface can achieve much higher speeds than a modem. The elimination of modems and the use of *Digital Subscriber Lines (DSL)* must be used to achieve rates higher than 230,400 bps. ISDN is discussed in Chapter 11. A Basic Rate ISDN line provides two 64,000-bps DS0 channels to the user. The user can combine these two channels to achieve a 128,000-bps channel. By using a 4-to-1 compression, a PC-to-PC speed of 712,000 bps is obtainable. As stated earlier, this speed cannot be achieved at present due to limitations within the PC.

9.12 CONNECTIONS TO THE INTERNET

Most connections to the Internet are made through a local Internet service provider (ISP). The ISP provides the customer with a local telephone number for access. If the ISP is a local business, the local telephone number connects the caller to a modem at the ISP location. The modem connects the caller to a computer at the ISP. The ISP computer connects the caller to the Internet highway. The Internet highway uses synchronous transmission. The communication from our PC to the modem is asynchronous. The communication between our modem and the modem at the ISP is synchronous. The communication between the modem at the ISP and its computer is asynchronous. The computer at the ISP converts the incoming asynchronous signal from its modem to the synchronous signal needed by the Internet. The interface standard for interface to the Internet is the *X.25* synchronous data protocol.

When your ISP is not a local company, connections to the ISP are made via (1) a local telephone number that connects via a modem to an X.25 packet network, or (2) an 800 telephone number connected to a modem at the ISP's premises. When ISPs provide a local number for access, the local telephone number connects to a value-added network (VAN) company such as Telenet. The VAN provides a packet network connection between the local telephone number and the ISP. The local telephone line connects to a modem at the VAN, and the output of the modem connects to a *Packet Assembler / Disassembler (PAD)*. The PAD converts the asynchronous signal from a customer to the synchronous X.25 signal required by the *packet network*. The customer's data is sent over the packet data network to a PAD at the ISP location. The PAD will connect the customer to the ISP's computer. The ISP's computer will connect the customer to the Internet.

9.13 PACKET SWITCHING NETWORKS

Earlier in this book, we covered circuit switching. In a circuit switched network, we establish a circuit path through the PSTN for a call. The circuit established can be used for a voice or data call. The switched circuit is dedicated to connection of one call. The call remains on this circuit for the duration of the call. The PSTN uses circuit switching because it provides a high-quality circuit for a call.

The *Public Data Network (PDN)* uses a *packet switched network (PSN)*. In the PSN, a call does not have to remain on the same path through the network. The PAD assembles data from customers into packets. These packets of data contain a header with the address of the intended destination and a *trailer* containing the address of the sender. The PAD sends the packet to a *data switching exchange (DSE)*. The DSE checks the address of the intended receiver and determines which route to take through the PDN. The route taken through the PDN can vary from one packet to the next. This is possible because each packet contains the destination address.

Each DSE reads the address and forwards the packet to an appropriate DSE or PAD. This type of switching is called a *connectionless network* because a connection is not maintained for the duration of a call. A packet is routed through the PDN but no end-to-end connection is maintained.

It is also possible to establish connection-oriented packet circuits. Switched virtual circuits and permanent virtual circuits can be established in the PDN to achieve the equivalent of a circuit switched facility. A virtual circuit allows all packets to use the same route. The connection from a sender to a receiver is made using call setup packets. All packets will use this same setup packet. The VAN carrier determines how the data will be handled. The use of a connectionless network or a connection-oriented network is invisible to the end user. Some people are using the Internet to handle voice calls. The quality of voice will be better if a connection-oriented virtual circuit is used.

X.25 is the CCITT standard that defines the interface between a DTE, or a DCE, and the PDN. The X.25 standard uses an EIA-232 interface to the DTE or DCE. A PAD using the X.25 standard will accept serial asynchronous data from the DTE or DCE and organize the data into structured frames of data for synchronous transmission over the PDN. The advantage of synchronous protocols is their extended error-detection and correction capabilities. The X.25 packet switching networks check for errors at every DSE and at the receiving PAD. Each receiver of data returns a positive acknowledgment (ACK) if the data is good and returns a negative acknowledgment (NAK) if the data contained errors. Each switch stores the data packet it sent so that it can be resent if a NAK is received. This is why the PDN is also called a store and *forward switching network*.

9.14 SIGNALING SYSTEM 7 NETWORK

The PSTN uses a packet data network for sending messages between switching centers. This network is called the *Signaling System 7 (SS7) Network*. All toll switches and many class 5 local switching centers connect to the SS7 network. SS7 can only be used by computer-controlled switching systems. The CPU in one switch can send a signal to another switch simply by sending a message over the SS7 network to the CPU of the destination switch. If a switch has an originating customer making a call to a customer on another switch, the CPU of the originating switch can send a message over the SS7 network to the CPU of destination switch informing it of the number being called. The distant CPU will test the line to see if it is idle. If the line is busy, it will return a message over SS7 informing the originating switch that the line is busy. The originating switch will attach a busy signal to the originating line. If the called line is idle, the terminating switch will inform the originating office that the line is idle. The originating office will issue a message to switching systems in the PSTN to reserve a voice path in the PSTN for the call. The terminating office will ring the called number. The originating office will connect ring-back tone to the originating line. When the called party answers, a message is transmitted over SS7 to activate the reserved path in the PSTN.

The SS7 network is necessary for ISDN and Custom Local Area Signaling Services (CLASS) such as caller ID, selective call forwarding, and selective call rejection. If a customer has caller ID and receives a call, the CPU of the receiving switch will receive the originating telephone number over the SS7 network. The CPU will then cause this number to be sent over the local loop to the receiving telephone number. Without SS7, local telephone exchanges do not receive originating telephone numbers from other exchanges. SS7 is an X.25 data network developed to complement computer-controlled switching systems. Computers can talk to each other by sending messages back and forth between them. SS7 provides a message link that connects the computers. SS7 eliminates the need for the old analog signaling used with all electromechanical switching systems.

9.15 CALLER IDENTIFICATION

Caller identification (caller ID) is provided to a local class 5 central office by the SS7 network. If the line being called has subscribed to caller ID, the line translations database will contain an entry indicating this line has caller ID. When the Stored Program Control switching system receives an inbound call message over SS7, it looks in the database to translate the dialed telephone number into a line circuit location and is informed that the called line circuit has the caller ID feature. The SS7 network passes the originating telephone number to the receiving SPC exchange. A sender is attached to the called line circuit and transmits the caller ID information. The transmission from the sender to the line circuit is in digital format over the time division switching matrix of the SPC switch.

The codec of the line circuit will convert the digital code from the sender into an audio signal and transmit this signal over the two-wire local loop. Caller ID is simplex transmission and no response is required or expected from the customer's location. The customer provides a CPE device that contains a microprocessor and a Bell 202–compatible modem. The ITU V.23 standard defines a modem that transmits/receives 1200 bps, asymmetrical (1200/75 bps), frequency shift keying (FSK) methodology. This is compatible with the Bell 202 modem. Most manufacturers of today's high-speed modems include backward compatibility, including V.23 capability. Therefore, many modems used in a PC support caller ID. If your PC is equipped with software to handle caller ID, you can have caller ID as part of your PC communication capabilities.

Most customers provide a caller ID display unit at their residence to receive the caller ID signals. This display contains a microprocessor and a Bell 202 compatible modem chip. When the modem chip detects a ring, it does not answer the call but does receive the caller ID signal from the central office. The caller ID signal is sent between the first and second ring signal from the central office.

The time frame between ring signals is 4 s. The caller ID signal will be sent between the time frame of 1/2 s after the first ring, to 1/2 s before the second ring. Thus, caller ID has a 3-s window for transmission of caller ID message to the customer's CPE.

As discussed before, the signals from a sender, located in the SPC exchange, are converted by the codec on the line card into audio signals that follow the Bell 202 protocol. That protocol results in the following signal being received by the CPE at the customer's location: for each 1 transmitted, a 1200-Hz signal is received, and for each 0 transmitted, a 2200-Hz signal is received. The signals are transmitted at 1200 baud with 1 bit per baud. This results in 1200 bps.

The caller ID protocol between the caller ID display and the class 5 central office uses asynchronous transmission and 8-bit data words with no parity check bit. Extended ASCII coding is used for messages and a checksum is transmitted for error-detection purposes. No provision is made for retransmission, but the checksum is used by the receiver to determine validity of the received message. Caller ID is transmitted in the following format: (1) channel seizure signal, (2) carrier signal, (3) message type word, (4) message length word, (5) data words, and (6) checksum word.

1. The *channel seizure signal* is 30 continuous bytes of (01010101) (this is the equivalent of 85 in decimal or 55 in hexadecimal). Thirty bytes of code 85 provides a detectable alternating function to the CPE Bell compatible 202 modem (the caller ID display).
2. The *carrier signal* consists of 130 ms of 1200-Hz tone (mark signal) to condition the receiver for data. The 130 ms of time results in 156 bits being sent.
3. The *message type word* indicates the service and capability associated with the data message. The message type word for caller ID is 04 in hexadecimal or 04 in decimal; in binary code it is 00000100.
4. The *message length word* specifies the total number of data words to follow.
5. The *data words* are encoded in ASCII and provide the following information:
   a. The first two words represent the month.
   b. The next two words represent the day of the month.
   c. The next two words represent the hour in local military time.
   d. The next two words represent the minute after the hour.
   e. The calling party's telephone number is represented by the remaining words in the data word field. If the calling party's telephone number is not available to the terminating central office, the data word field contains an ASCII *O*. If the calling party invokes the privacy capability, the data word field contains an ASCII *P*.
6. The *checksum word* contains the 2s complement of the module 256 sum of the other words in the data message (that is, message type, message length, and data words). The receiving equipment may calculate the module 256 sum of the received words and add this sum to the received checksum word. A result of 0 generally indicates that the message was correctly received. Message retransmission is not supported.

The caller ID message is composed of binary bits that represent some hexadecimal and some ASCII codes. Rather than show the actual binary bits transmitted to the caller ID display, the ASCII Code is represented by a two-digit code. The ASCII code is represented in the example below by a number containing a 10s position and a units position. The three most significant bits of the ASCII code (bits 7, 6, 5) are represented by the 10s digit. The lower 4 bits of the ASCII code (bits 4, 3, 2, and 1) are represented by the units digit. For example, the decimal digit 2 in ASCII is 0110010. The bits 7, 6, and 5 are 011 (binary code for 3), and the bits 4, 3, 2, and 1 are 0010 (binary code for 2). Thus, the decimal number 2 is ASCII code 32. Decimal 0 is ASCII code 30. The decimal numbers 0 to 9 are represented by ASCII codes 30 to 39 respectively. An example of a received caller ID message, beginning with the message type word, follows:

04 12 31 30 32 30 31 32 34 30 39 31 33 38 39 37 35 37 30 31 51

As noted earlier, some of the above two-digit codes are to be converted to hexadecimal and some to decimal. Those to be converted to hexadecimal will be noted with the letter *h*.

04h = calling number delivery information code (message type word)

12h = 18 decimal; number of data words (date, time, and calling number words). Notice that after the words 04 and 12, there are 19 words. The 19th word is the error-checking word (51). Therefore, there are 18 data words in the example given.

ASCII 31, 30 = 10 = the month of October

ASCII 32, 30 = 20 = the 20th day

ASCII 31, 32 = 12 = the 12th hour = 12:00 PM

ASCII 34, 30 = 40 = 40 minutes after the hour (that is, 12:40 PM)

ASCII 39, 31, 33, 38, 39, 37, 35, 37, 30, 31 = (913) 897-5701; this is the calling party's telephone number.

51h = checksum word

9.16 SUMMARY

Data communications is the transmission of characters, numbers, graphics, and symbols using digital signals. A digital signal is a signal composed of binary digits 1 and 0. The binary digits are represented by two different voltages. The voltage used will vary from one device to another. The EIA-232 interface standard for a serial port specifies that a 1 is represented by a voltage of –3 to –15 V, while a 0 is represented by +3 to +15 V. DTE such as a PC is used to transmit and receive digital data. DCE is used to interface a DTE to the PSTN. When a DTE is interfaced to a regular analog telephone line, the DCE device is a modem. When the DTE is interfaced to a digital facility, the DCE device is a CSU/DSU. Data is transferred within a PC using parallel transmission. A Pentium data bus contains 64 data leads and data is transferred 64 bits at a time. Data is transferred between a PC and an external modem using serial transmission. Serial transmission occurs over one transmission path, and bits are sent one behind the other over this serial path. The device in a PC that interfaces the parallel data bus of the PC to the serial path of an EIA-232 interface is a UART. Current UARTs can support speeds up to 115,000 bps. The UART uses asynchronous transmission.

The transmission between a PC and a modem is asynchronous transmission, where a start bit precedes each character and a stop bit follows each character. Each character is represented by either a seven-level ASCII code and a parity bit, or an eight-level ASCII code and no parity bit. Thus, asynchronous transmission requires the transmission of 10 bits for each character transmitted. The start bit provides synchronization for each byte transmitted. Synchronous transmission transmits blocks of data with no pauses between the bytes inside the block. This allows the use of a synchronization signal at the beginning of the block of data. Since the data is continuous, no additional sync signals are needed within the block. This eliminates the need for start and stop signals and speeds up the transfer of data. Communication between high-speed modems is done using synchronous transmission.

Modems contain sophisticated error-detection and correction techniques. Synchronous transmission allows the use of CRC-16 or CRC-32 protocols. The receiving modem notifies the transmitting modem that the blocks of data being received are okay by transmitting ACKs. When a bad block is received, a NAK along with the block number affected is sent to the transmitting modem. The transmitting modem will resend the affected block. Modems also contain a compression algorithm that converts ASCII from the PC into a shorthand code and sends the shorthand code to the distant modem, where it is converted back to ASCII. With a 4-to-1 compression, PCs can send data to the modem four times faster than the modems transmit between themselves. Using 28,800-bps modems and 4-to-1 compression allows a PC to send data to the serial port at 115,200 bps. This is the maximum speed supported by current UARTs in a PCs serial port.

Baud rate is defined as the number of times a signal changes. Baud rate is one of the components determining the physical line speed of a modem. The other component is the number of bits represented by one signal change. To transmit 9600 bps using a signal that changes 2400 times per second (2400 baud), 4 data bits must be transmitted by each baud. TCM requires an additional bit for forward error correction. To transmit 5 bits per baud requires 32 detectable signals ($2^5 = 32$). The V.34 modem requires the use of 960 different signals to transmit 9 data bits per baud. Using a baud rate of 3200, the physical data speed is 28,800 bps (9×3200). For the receiving modem to tell the difference between the 960 different signals requires a sophisticated signal logic processor in the modems, the use of sophisticated error-correction techniques, and a noise-free connection over the PSTN.

The PDN is a network of data switches that allows the public to transmit data in the form of packets. A device called the PAD is used to interface data from a user to the packet switched network. Each packet contains the destination address in its header and the address of the sender in its trailer. The interface between the PAD and a packet network is an X.25 interface. The Internet is a packet network. An Internet service provider (ISP) will accept data from a user and reformat it into packets of data for transmission over the Internet Highway.

REVIEW QUESTIONS

1. What is the difference between data communications and voice communications?
2. What type of signals are used in data communications?
3. What is the most common connection between a DTE and a DCE?
4. Why do we need a modem if our telephone line connects to a digital central office?
5. What are the two primary types of DCE?
6. What type of coding is used on most data communications?
7. What is serial transmission?
8. What is the difference between asynchronous and synchronous transmission?
9. What type of transmission is used between two high-speed modems?
10. What is the simplest error-detection technique?
11. What type of error detection is used between two high-speed modems?
12. What is the difference between discrete ARQ and continuous ARQ?
13. What type of error correction is provided by LAP-M?
14. What is the maximum speed of the latest UART (16550 AFN)?
15. How many different detectable events does QAM provide?
16. How many bits are represented by one QAM signal?
17. What is baud?
18. What is the baud rate of older modems?
19. What baud rate must be used in a V.34 modem to achieve 28,800 bps?
20. How do modems achieve an overall data speed of 115,000 bps if the baud rate is only 3600 baud?
21. What is a PAD?
22. What does the X.25 standard define?
23. What facility is used by digital central offices to send signals to other digital central offices?
24. What type of modem chip is inside the CPE equipment for caller ID?
25. What type of signaling is used between the central office and the CPE for caller ID?

GLOSSARY

ACK A positive acknowledgment (ACK) that the data was good. The receiving modem sends an ACK to the transmitting modem to indicate that the data is being received okay.

Asymmetric Digital Subscriber Line (ADSL) A digital subscriber line technology that can be used to deliver the speed of a T1 line over the local loop. An Asymmetric Digital Subscriber Line has a high bit rate in one direction and a low bit rate in the opposite direction.

Analog Signal An electrical signal that is analogous (similar) to a voice signal. The signal continuously varies in amplitude and frequency. An analog signal has an infinite number of values for voltage, current, and frequency.

Automatic Retransmission Request (ARQ) The method of error detection used between high-speed modems. ARQ is either discrete ARQ or continuous ARQ. The receiving device returns a positive acknowledgment (ACK) when it receives a good block of data, and returns a negative acknowledgment (NAK) when the block of data received contains an error.

American Standard Code for Information Interchange (ASCII) American Standard Code for Information Interchange The use of a seven level binary code to represent letters of the alphabet and the numbers 0 to 9.

Asynchronous Transmission also called start-and-stop transmission. The transmitting device sends a start bit prior to each character and sends a stop bit after each character. The receiving device will synchronize from the received start bits. Thus, synchronization occurs at the beginning of each character. Data is sent between two devices as a serial bit stream.

Baud The number of times a signal changes it state. If the amplitude of a signal changes 2400 times a second, the signal changes states 2400 times per second or at 2400 baud. If a signal changes back and forth between a frequency of 1800 Hz and 2200 Hz 3200 times per second, the baud rate is 3200. If a signal changes phase 2400 times a second, the baud rate is 2400 baud. Many people confuse baud rate and bit rate. Even some terminal emulation programs will provide an option to change Baud Rates from 9200 to 19,200. This is incorrect. They should state that you can change the bit rate from 9200 to 19,200 bps. The selection of baud rate is done automatically by modems.

Binary Signal See digital signal.

Continuous ARQ Also known as Sliding Window ARQ. Continuous ARQ eliminates the need for a transmitting device to wait for ACKs after each block of data. The device continuously transmits blocks of data and sends a block number with each block. The receiving device checks the blocks for errors and continuously returns positive ACKs to the transmitter. If an error in data occurs, the receiving device returns a NAK along with the block number affected. On receipt of a NAK, the transmitting device will retransmit the bad block of data.

Cyclic Redundancy Check (CRC) A form of error checking used between modems. A transmitting modem treats the block of data transmitted as representing a large binary number. This number is divided by a 17-bit divisor (CRC-16) or a 33-bit divisor (CRC-32). The remainder is attached in the trailer behind the block of data. The receiving modem performs the same division on the block of data received and compares its calculated remainder to the remainder sent in the trailer.

Channel Service Unit (CSU)/Data Service Unit (DSU) Also called Customer Service Unit / Data Service Unit The CSU/DSU is a

DCE used to interface a computer to a digital leased line. It is a digital-to-digital interface. It can connect a low-speed digital device to a high-speen digital highway.

Data Raw facts, characters, numbers, and so on that have little or no meaning in themselves. When data is processed, it becomes information. In telecommunications when we speak of data, we are referring to information represented by digital codes.

Data Communications Equipment (DCE) Also called Data Circuit Termination Equipment A device that interfaces Data Terminal Equipment (DTE) to the PSTN. A modem is a DCE used to interface a DTE to an analog line circuit on the PSTN. A CSU/DSU is a DCE used to interface a DTE to a leased digital line in the PSTN.

Data Switching Exchange (DSE) Also known as Packet Switching Exchange. These are the switches of a packet switched network. The DSE routes packets of information based on the packet address information found in each packet header.

Data Terminal Equipment (DTE) A DTE is used to transmit and receive data in the form of digital signals. The personal computer is the most common form of DTE.

Digital Signal An electrical signal that has two states. The two states may be represented by voltage or current. For example, the presence of voltage could represent a digital logic of 1, while the absence of voltage could represent a digital logic of 0.

Discrete ARQ Also called stop-and-wait ARQ. An error control protocol that requires an acknowledgment from the receiver after each block of data sent. The transmitting modem sends a block of data and then waits for an acknowledgment before sending the next block of data.

Dual Simultaneous Voice and Data (DSVD) A modem that can handle voice and data at the same time.

EIA-232 Interface Also called RS-232. The most common interface standard for data communications. CCITT standard V.24 is the same as EIA-232. EIA-232 defines the voltage levels needed for the various signal leads of EIA-232.

Flow Control Controlling the flow of data from one device to another usually via hardware flow control (RTS/CTS) or by software flow control (XON/XOFF). A modem contains a memory buffer to allow it to compress data before transmitting and to allow it to convert asynchronous data to synchronous data. It must be able to stop the transmitting PC when this memory buffer approaches a near-full condition. Some modems contain a very large memory buffer, which negates the need for flow control.

Full-Duplex If a transmission system allows signals to be transmitted in both directions at the same time, the system is called a full duplex transmission system.

Half Duplex A type of transmission where transmitters on each end of a medium take turns sending over the same medium.

Header A term describing the placement of control information in front of a block of data transmitted using synchronous transmission. The header contains a beginning flag (or sync signal), the destination address, and a control field.

Integrated Services Digital Network (ISDN) The use of digital line circuits to provide end-to-end digital service. The Basic Rate Interface (BRI) provides the user with two DS0 channels. The Primary Rate Interface (PRI) provides the user with 23 DS0 channels.

Internet Service Provider (ISP) A company

that provides individuals with access to the Internet.

Modem A DCE device that interfaces the digital signal from a DTE to the analog local loop and line circuit of the PSTN, by converting the digital signals into modulations of an analog signal.

NAK A negative acknowledgment transmitted by the receiving modem when it detects that an error has occurred in a block of transmitted data. If a NAK is received, the modem retransmits the last block of data from its memory.

Null Modem No modem. A null modem cable is used to connect two PCs via their serial ports, when they are connected directly without using a modem.

Packet Assembler/Disassembler (PAD) The device that interfaces data to a packet network. The PAD accepts data from a user and arranges the data into packets that can be processed by the packet switching network.

Packet Network A data network that transmits packets of data

Public Data Network (PDN) The packet data network accessible to the general public for the transmission of packet data.

Parallel Transmission The use of several transmission leads to allow the simultaneous transmission of several bits at one time. A parallel data bus with 8 leads can process data 8 bits at a time. Or 64 leads on the data bus allows information to be processed 64 bits at a time. Parallel transmission is much faster than serial transmission but requires many more transmission leads.

Protocol The rules of communication. Each protocol defines a formal procedure for how data is to be transmitted and received using that protocol.

Quadrature Amplitude Modulation (QAM) The transmission of 4 bits per baud. A V.29 modem transmits at 2400 baud with 4 bits per baud to achieve 9600 bps transmission. QAM uses 2 different amplitudes for each of 8 different phases of a 1700-Hz signal to achieve the 16 detectable events necessary to code, and decode 4 bits at a time.

RS-232 The most common interface standard for data communications. CCITT standard V.24 is the same as RS-232. RS-232 defines the voltage levels needed for the various signal leads of RS-232.

Selective ARQ With selective ARQ, only frames that have errors are retransmitted. When a transmitting modem is using selective ARQ and receives a NAK, only the frame for which the NAK was received is retransmitted.

Serial Transmission The transmission of bits, one behind the other, over one transmission medium.

Synchronous Transmission The transmission of data as blocks of bytes. Synchronization of the receiver occurs from a special bit pattern called a sync signal that is placed in front of the block of data information.

Trailer A trailer is used in synchronous communications. A trailer is data placed behind the block of information transmitted. The trailer cnotains parity checking information and the address of the sender.

Trellis Coded Modulation (TCM) TCM is a forward error-correction technique. An extra bit is added to the bits of data transmitted to help the receiver decode the data more reliably. QAM transmits 4 bits at a time. TCM added a 5th bit to the 4-bit code for error correction purposes.

Universal Asynchronous Receiver/Transmitter (UART) A piece of hardware (an integrated circuit chip) whose purpose is to interface a device using parallel transmission to a device using serial transmission. Every serial port contains a UART between the parallel data bus of the PC (or modem) and the serial port.

Word In a PC, the number of adjacent bits that can be manipulated or processed. This depends on the number of leads comprising the data bus, which is in turn dependent on the number of registers attached to the data bus. A word can be 8 bits or 1 byte in an 8088 microprocessor environment. A word can be 32 bits or 4 bytes in a 80486-based PC, or it can be 64 bits (8 bytes) in a Pentium-based PC.

X.25 The interface standard to a packet data network

10

Signaling

Key Terms

Addressing
Alerting Signals
Automatic Number
    Identification (ANI)
Caller Identification
Glare Condition

Ground Start Line
Ground Start Signaling
In-Band Signaling
Loop Signaling
Out-of-Band Signaling
Progress Signals

Reverse Battery Signaling
Signaling
Signaling System 7 (SS7)
Supervisory Signals

Signaling is anything that serves to direct, command, monitor, or inform. In telecommunications, signals are needed to inform a local switching system that someone wishes to place a call, provide directions to the PSTN on how the call is to be connected, and inform the called party that a call is waiting. Signaling is divided into four basic categories: (1) supervisory, (2) addressing, (3) alerting, and (4) progress.

10.1 CATEGORIES OF SIGNALS

Supervisory signals are used to monitor the status of a line or trunk circuit. A switching system is informed of whether a line or trunk is idle or busy by supervisory signals. Every telephone is connected by a local-loop cable pair to a line circuit in a class 5 switching center. One of the functions of a line circuit is to monitor the status of the telephone attached to it. Line circuits monitor the status of the telephone by monitoring loop current. When the phone is on-hook, there is no current in the local loop. When the telephone is taken off-hook, the switchhook of the phone closes an electrical path between the Tip and Ring of the cable pair. Negative 52 V DC is attached through the line circuit to the Ring lead.

Ground is attached through the line circuit to the Tip lead. When the phone is off-hook, the completed electrical path causes electrical current to flow from the 52-V battery, through a current-detection device on the line circuit, out over the Ring lead to the phone, and returning over the Tip lead and line circuit to ground. The current-detection device supervises the line and when loop current is detected, the line controller is notified. The line controller keeps the main CPU notified on the status of each line (Figure 10-1).

Addressing signals are used in identifying which telephone has been called. The telephone issues an address as dialed digits using dial pulses or DTMF tones. A dial register in the central office collects dialed digits and presents these to a translator. The translator actually translates dialed digits into switching instructions.

Alerting is the process of letting someone know a call is waiting. Ringing signals alert a called party that the phone should be answered. *Alerting signals* are closely tied to supervisory signals. When the called party answers a call, loop current flowing in the loop causes operation of the line relay. This causes the ringing relay to release and stop sending the ringing signal. If a called line has the call waiting feature assigned in the line translations database and is on a call, the called party will receive a tone to indicate a call is waiting. A ringing signal is usually applied to the line for 2s, and then removed. After a 4-s quiet interval, ringing is applied again for 2s, and so on.

Progress signals such as dial tone, 60-ipm busy tone, 120-ipm busy tone, ring-back tone, and recordings are used to indicate the progress of a call to the calling

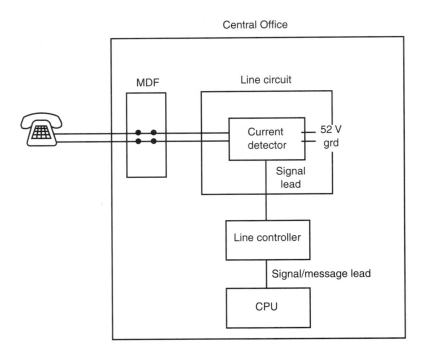

FIGURE 10-1 Line detector.

party. Dial tone is provided by the local central exchange when it connects a local telephone line to a dial register. In an electromechanical switching system, dial tone was a 350-Hz tone mixed with a 440-Hz tone. In a Stored Program Control (SPC) switching system, dial tone is stored as a digital code on a tone card. The tone card does not send a tone; it sends a digital code over the TDM switching fabric to the line circuit. The codec of the line circuit converts the digital code into the 350/440-Hz analog signal and sends it out over the loop to the telephone (Figure 10-2). Busy signal is a 480-Hz tone modulated by 620-Hz tone. Ring-back tone is 440 HZ modulated with a 480-Hz tone.

Special tones are also generated when a call is placed from a pay station to the operator. When a call from a pay station uses the same trunks to reach an operator as regular residential and business lines, some method must be used to show that the call is originating from a pay station. When the operator answers an incoming call, she receives a special tone to indicate the call is being placed from a pay station. Tones are also used to reveal the denomination of coins deposited in a pay station. These tones are received by the switching equipment to indicate how much money has been deposited. Special progress tones are also sent when an operator answers a call. These special tones tell switching equipment that the call is being sent to an operator. Because dial tone, busy tone, and other special tones all follow an industry standard, devices can use integrated circuit chips designed to recognize these tones. A modem has one of these chips to monitor for dial tone before dialing a number. When it recognizes a busy signal, the modem automatically hangs up and redials the call.

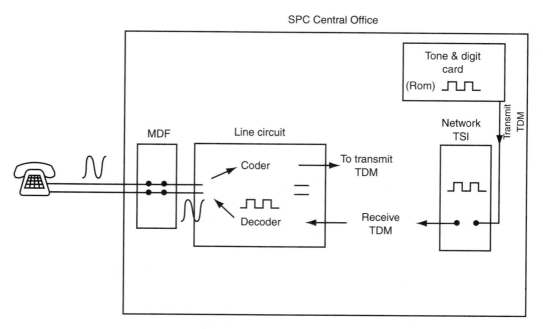

FIGURE 10-2 Tone and digit card through a codec.

10.2 LOOP SIGNALING

Activation of a signal by opening or closing the loop to complete an electrical circuit and causing current to flow is called *loop signaling*. Line circuits in a central office use loop signaling to detect when a telephone has gone off-hook. The line circuit contains a current-detection device to detect line current. This device can be a relay wired in series with the 52-V battery and Ring lead, or it can be an integrated circuit chip. Most SPC switching systems now use a solid-state electronic circuit to sense loop current. Early SPC central offices used an electromechanical line relay for the current-detection device because it could withstand abnormal voltage spikes better than an electronic chip. The outside plant comprising the local loop was subject to contact with lightning and high-voltage commercial alternating current. Central office equipment is protected from long duration exposure to these high voltages by protectors on the mainframe. Each cable pair has a protector that monitors the cable pair for high-voltage conditions. When a high voltage is present the protector opens the connection between the cable pair and the central office line circuit. The operation of a protector can take up to a quarter of a second before it opens the outside line from the central office equipment. A line relay can withstand a short duration high-voltage spike, but an integrated electronic chip circuit cannot. Therefore, relays were used for the line interface in class 5 central offices. The use of buried cable has lessened the exposure of cables to high-voltage sources, and improved protectors have been developed for the line circuit. This has led to the replacement of line relays by integrated circuit chips as current detectors.

When the telephone is taken off-hook, current flows through the line circuit and signals the central office that the line has gone off-hook. In a SPC switching system this signal is an interrupt signal to the line controller. When the line controller sees an interrupt signal, it signals the main CPU which line has gone off-hook. The CPU will request that a dial register and dial tone be connected to the line. When the telephone is put on-hook, current flow ceases and the line circuit signals the line controller that the telephone has gone on-hook. The line controller relays this information to the CPU.

10.3 DUAL-TONE MULTIFREQUENCY SIGNALING

Dual-tone multifrequency (DTMF) tones are used to send signals to the dial register of a central office, or to send signals to automated telephone systems such as answering machines and automated attendants. These systems respond to the tones from a DTMF pad to perform certain functions. The central office will respond to tones from a DTMF pad as dialed digits unless the caller sends an asterisk or star symbol (*) as the first digit. Receiving this symbol as the first digit instructs the central office that the DTMF tones following the * are to activate features such as call forwarding, recall last number, disable call waiting, and so on. Additionally, if a line circuit sees a momentary on-hook flash when a call is in progress, it recognizes this as a signal that the line circuit is about to activate a feature such as three-way calling or call transfer and will provide dial tone to the line while placing a hold on the original conversation.

10.4 USING LOOP SIGNALING FOR DIAL PULSES

Step-by-step (SXS) switching systems relied exclusively on loop signaling. Each switching stage of a SXS system was controlled by a dial at the telephone. The caller would place a finger in the dial hole corresponding to the digit desired and pull the dial off normal until the finger reached a finger stop. When the finger was pulled out of the dial hole, the dial returned to normal. On its return to normal, the dial would open the loop connecting the phone to the exchange as many times as the number dialed. If a 5 was dialed, the loop was opened five times. Each open is referred to as a *pulse*. Dials were adjusted to complete ten pulses per second. Each pulse consisted of an open loop for 60 ms and a closed loop for 40 ms. One pulse required 100 ms, and ten pulses could be completed in 1s. The SXS switching system recognized each open-loop condition as a dial pulse, and the stepping relays would step once for each pulse received.

The switching system was informed when one digit had been completed by a period of time called the *interdigital time*. This is the time frame between the last pulse of one digit and the first pulse of the next digit. When the last pulse (open-loop condition) has been sent for a digit, the loop remains closed until the next digit is dialed. Switching equipment is designed to recognize a closed-loop condition of 400 ms or more as the signal that signifies completion of the previous digit. Thus the interdigital interval must be at least 400 ms. This was accomplished by putting distance between the one hole of the dial and the finger stop. Pulling the dial around to the finger stop for any digit will require at least half a second (500 ms). SXS and other electromechanical switches had to rely on these timing relationships to identify what signal had been received. A long closed loop signified an off-hook signal. A long open loop signified an on-hook condition. Short on-hook signals (60 ms) represented dial pulses. An open loop of 750 ms or more represented an on-hook hang-up or disconnect signal.

10.5 LOOP SIGNALING LIMITATIONS

When the local loop is designed, it is designed with regard to transmission characteristics for voice, not for signaling limitations, because transmission requirements are more stringent than signaling requirements. Early switching systems had signaling requirements that were stricter than the transmission needs. Early

SXS systems had a Ring trip relay that needed 27 milliamps of current to properly recognize when a called party answered a call. Design improvements provided a Ring trip relay that would operate on 20 milliamps. This put signaling requirements at a level less critical than transmission requirements.

SPC switching systems have loop limits of 1800 Ω. This resistance results in at least 20 milliamps of current in the local loop. This is the minimum amount of current needed by the newer transmitters to provide clear transmission. Newer telephones can automatically adjust their resistance to compensate for a short loop and limit the current on a short loop to 60 milliamps. The line circuit in the central office also contains a current-limiting device to keep the current on a line circuit below 60 milliamps. High line currents cause crosstalk.

SPC line cards are designed to signal properly on loops of 1800 Ω. When the telephone is equipped with a DTMF keypad for dialing, the phone must receive at least 20 milliamps of current to properly power the DTMF signal generator. Chapter 6 provided the layout for the various tones generated when a digit is depressed on the DTMF keypad. The low-frequency tones (697 to 941 Hz) are transmitted at –6 dBm and the high tones (1209 to 1477 Hz) at –4 dBm. These amplitudes depend on the amount of loop current. On loops of 1300 Ω, the transmission loss at 1000 Hz is about 6.5 dB. An 1800 Ω loop has a transmission loss of approximately 9 dB. Since a low-frequency tone leaves the DTMF pad at –6 dBm, a loop loss of –9 dB will result in the signal reaching the central office at –15 dBm. The DTMF receiver in the dial register will recognize tones between 0 and –25 dBm. The DTMF signals will work on a local-loop designed to satisfy transmission requirements. When a loop is long enough to require loop treatment for transmission, the loop treatment device will treat the line for both voice and signaling.

10.6 ANSWER SUPERVISION: REVERSE BATTERY SIGNALING

In a SXS central office, the connector switch would signal that the called party had answered by reversing the battery feed to the calling party. This is known as *reverse battery signaling.* For loop signaling, the SXS exchange placed –50 V DC through a line circuit over the Ring lead and ground over the Tip lead when the line was idle. As a call progressed through a SXS office, each successive switching stage provided –50 V over the Ring wire and ground on the Tip wire. The last switching stage in a SXS switch was a connector that connected the calling party to the called party. When the called party answered, the connector reversed the battery and ground connections. The battery connection was placed on the Tip wire and the ground on the Ring wire. Toll trunks had a polarized relay connected to the Tip lead. When the incoming call was a toll call, this relay operated on a battery reversal and signaled the operator or the billing machine of the originating office that the called party has answered and billing could begin.

Loop start signaling is the term for providing an off-hook signal by closing the loop. When loop start is used on PABX switches, an outgoing call may inadvertently connect to an incoming call. The simultaneous seizure of a trunk from both ends is referred to as a **glare condition**. This condition occurs in electromechanical switching systems. An incoming call to a PABX will seize the trunk circuit, but a loop seizure does not mark the trunk busy to outgoing calls. When the PABX trunk receives a ringing signal, relays operate and mark the trunk as busy to outgoing calls. Ringing cycles are designed to provide a ring signal for 2s and then remove the ringing voltage for 4s. If the central office seizes a PABX trunk circuit between ringing cycles, the trunk will not mark itself busy to outgoing calls for up to 4s. During this time an outgoing call may be assigned to the trunk and glare has occurred. To prevent this condition, **ground start signaling** was introduced. With ground start signaling, the line relay at the central office is modified to remove ground from the Tip wire when the line is idle.

When a line circuit has been modified for ground start, –50-V DC battery is present on the Ring lead and the Tip lead has no connection. The trunk circuit at the PBX cannot use loop start to seize the line relay because no ground is present on the Tip lead. The PBX trunk must place a ground on the Ring lead to seize the central office line circuit. This is why it is called a *ground start trunk circuit*. When the connector switch of a SXS switching system has a call for the PABX, it applies ground to the Tip wire and battery to the ring wire. Because ground is not on the Tip lead when a circuit is idle and is placed on the Tip when an incoming call is present, this condition can be used as a signal to mark the trunk busy to outgoing calls.

With ground start signaling, we do not have to wait for a ringing signal to mark the trunk busy. Ground on the Tip wire operates a fast-acting relay in the PBX trunk circuit. Operation of this relay marks the trunk busy and prevents access by an outgoing call. If the PABX and central office are both SPC switching systems, the trunks can be loop start. Few SPC CBXs require ground start. The trunk circuits in a SPC CBX switch are designed to quickly notify the CPU when they are seized by the loop on an incoming call, and nothing is gained by using the ground start option.

SPC trunk circuits are designed to mark the circuit busy from a loop seizure condition and do not use the ringing signal to mark the circuit busy. The ground start option is provided on SPC trunk circuits in a CBX to prevent having to change existing central office line modifications when an older PABX with ground start trunks is replaced with a new CBX system. SPC systems are usually programmed to access outgoing trunks on a most idle basis. The central office is programmed to access CBX line circuits in a rotary hunt fashion from the lowest to the highest line. Calls to a CBX are assigned to the first line that is not busy. The outgoing calls from a CBX are going to be assigned on a most idle basis. The highest lines will be most idle, and outgoing calls will access the highest lines first. With incoming calls coming over the lowest lines and outgoing calls going over the highest lines, the chances of glare are significantly reduced. The trunk circuits of a SPC CBX will mark themselves busy to the CPU just as fast whether set up as loop start or ground start. No advantage is gained by using ground start. This was old technology used to solve problems encountered on old and slow electromechanical switching systems.

10.8 SIGNALING BETWEEN TRUNK CIRCUITS USING E&M SIGNALING

Early trunk circuits used analog, in-band loop signaling like line circuits. This signaling was called *loop signaling* since the signaling was done by opening or closing the loop. Another type of signaling was developed for trunk circuits and was called *E&M signaling*. When a caller was connected to a trunk circuit, the person connected using loop signaling, but the trunk circuit converted loop signals to E&M signaling using a device called a *DX circuit*.

The trunk circuit in one exchange was connected over a cable pair to a trunk circuit in the distant exchange. The DX sets were also attached to the cable pair. When a trunk circuit was seized by a caller at the originating office, the trunk circuit would place –50-V battery over an M lead to the DX circuit. This signal passed through the DX circuit and went out over the cable pair connecting the trunk circuit to the distant switching center. At the distant exchange this –50-V battery caused the DX circuit in that office to operate and put a ground on the E lead to the trunk circuit in that office.

DX circuits were polarized relays that would only operate when current flowed in one direction. If the current was flowing from the trunk circuit through the DX and out to the cable pair, the DX would not operate. If current was flowing from the cable pair through the DX to the trunk circuit, the DX would operate and send a ground over the E lead to the trunk circuit. Thus a battery signal being attached to the M lead in one office resulted in ground on the E lead at the distant office. *M* comes from the *m* in trans*m*it, and *E* comes from the *e* in rec*e*ive (Figure 10-3).

If a caller was attached to a trunk circuit and dialed the digit 5, the originating trunk circuit caused five interruptions to occur to the battery on the M lead. This resulted in five interruptions of ground on the E lead at the distant office. When cable pairs were used for these trunk circuits, the trunk circuits attached to the cable pair through a repeat coil. A repeat coil is a transformer with a primary and secondary winding. The primary winding was connected to the trunk circuit. The secondary winding was connected to the cable pair. The secondary winding was constructed so that the center point of the winding was cut into and extended outside the case. This provided two leads labeled A&B. A capacitor was placed between the A&B leads to provide continuity in the secondary winding for voice signals. The DX circuits were attached to these A&B leads. The repeat coil and DX circuits were usually part of the trunk circuit.

When carrier systems were introduced, DX circuits were not used. The E&M leads of the trunk circuit connected to the channel unit E&M leads. In FDM carrier

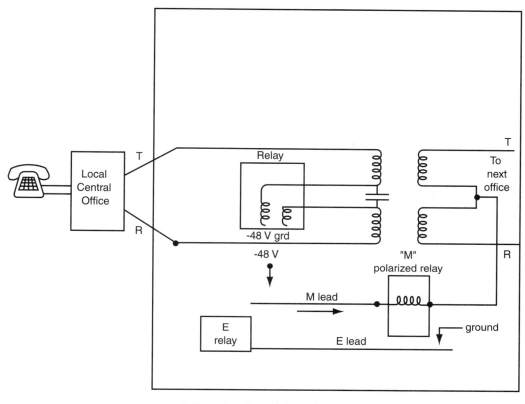

Battery placed on the M lead causes current to flow in the direction of
the arrow. Current flow in this direction will not operate the M relay.
Because current flow is in the opposite direction in the distant
(or next office), the M relay in that office operates.

FIGURE 10-3 E&M lead, repeat coil, and DX set.

systems, an idle circuit will have a 2600-Hz. tone present. In FDM channel units, M-lead signals caused removal of the 2600-Hz tone. At the receiving channel, removal of 2600-Hz tones was converted to ground on the E lead. In TDM systems, M-lead signals were converted into signaling bits to be inserted into the digitized voice signal as a replacement for a voice bit. This type of signaling is referred to as *bit-robbed signaling*. FDM carrier systems used the presence or absence of the 2600-Hz tone for supervisory signals. TDM carrier systems use bit-robbed signaling for supervisory signaling. Both signaling methods are in-band signaling techniques because the signals are carried by the voice channel.

When a T1 carrier system is used, the on-hook and off-hook signals for each channel are sent over the system by bit-robbed signaling. Since two states are needed for the signal (either on-hook or off-hook), they can be represented by a binary bit (1 or 0). Because the signaling state does not change often or fast, we do not need a signaling bit in every frame. The first signaling technique used for T1 substituted a signaling bit for each of the 24 channels in every 6th frame. This signaling bit for each channel replaced the least significant voice bit in every 6th frame on each voice channel. This causes some distortion to the voice signal, but the distortion is not noticeable. Of course, this bit-robbing technique would destroy data if the signal on the channel is a data signal. For that reason, data on a PSTN T1 system is limited to 7 bits per channel in every frame, not just every 6th frame. Seven bits per frame times 8000 frames equals 56,000 bps. Thus, a T1 system used in the PSTN can only accept a data rate of 56,000 bps on each channel. This is referred to as *56-Kbps clear channel capability*. There are some proprietary T1 systems that use a means other than bit robbing to do signaling. These systems can use all 64,000 bps of the channel for data. A T1 system used as the vehicle for ISDN does not use bit-robbed signaling. The signaling for all channels is carried by the 24th channel. This channel does not serve as a voice channel; it serves as a common channel for signals and is called the *D-channel*. Thus, a T1 system used to carry ISDN provides a clear channel capability of 64,000 bps.

Figure 10-4 shows the layout for a Superframe. A *Superframe* is composed of 12 frames. Each of the frames is 193 bits (twenty-four 8-bit voice channels plus 1 framing bit). Notice that the framing bit pattern over 12 frames is 100011011100. These framing bits allow the receiver to synchronize off the incoming bit stream. The least significant bit of every channel in the 6th frame is used for signaling and is

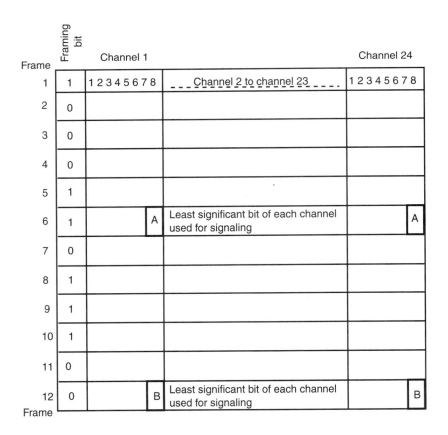

FIGURE 10-4 Superframe.

called the *A bit*. The least significant bit of each channel in the 12th frame is also used for signaling and is called the *B bit*. Another type of signaling developed for T1 signaling is called *Extended Superframe (ESF)* signaling. The ESF consists of 24 frames. Thus, there are 24 framing bits. Only 6 of the 24 framing bits are used to provide synchronization for the receiver. Another 6 of the 24 bits are used to provide a CRC error-checking protocol. The remaining 12 bits are used to provide a management channel called the *Facilities Data Link (FDL)*. The ESF provides signaling bits in every 6th frame. The signaling bit in the 6th frame is called the *A bit*. The signaling bit in the 12th frame is called the *B bit*. The signaling bit in the 18th frame is called the *C bit*, and the signaling bit in the 24th frame is called the *D bit* (Figure 10-5).

10.9 IN-BAND VERSUS OUT-OF-BAND SIGNALING

Signaling systems are classified as analog or digital and as in-band and out-of-band signaling. *In-band signals* are sent over the same circuit used for the accompanying voice signal. *Out-of-band* signals are sent over a different circuit

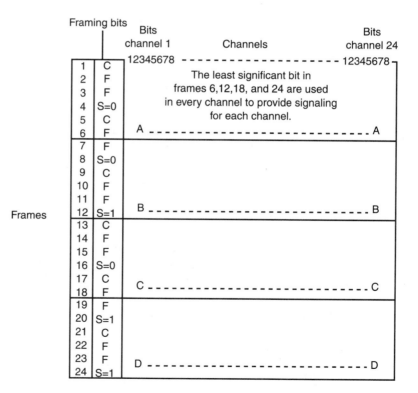

FIGURE 10-5 Extended Superframe.

ESF uses 6 bits (S) for synchronization (001001). "C" bits are used for CRC error checking and "F" bits comprise the Facilities Data Link (FDL) channel.

from the circuit used for the voice signal. The single-line telephone set uses analog, in-band signaling over the local loop. The long distance network also used analog, in-band signaling until the mid-1970s. The old analog toll network used *single frequency (SF)* signaling and *multifrequency (MF)* signaling. SF signaling used a 2600-Hz tone to signify an idle or on-hook trunk condition. Dialing over these trunks was done using MF signaling.

MF signaling was developed prior to the introduction of DTMF for telephones. The two systems cannot use the same tones for signaling. DTMF tones for touchtone telephones are significantly different from MF tones. MF signaling used combinations of tones per the following chart:

Digit Sent	Frequencies Sent
1	700 Hz + 900 Hz
2	700 Hz + 1100 Hz
3	900 Hz + 1100 Hz
4	700 Hz + 1300 Hz
5	900 Hz + 1300 Hz
6	1100 Hz + 1300 Hz
7	700 Hz + 1500 Hz
8	900 Hz + 1500 Hz
9	1100 Hz + 1500 Hz
0	1300 Hz + 1500 Hz
KP	1100 Hz + 1700 Hz
ST	1500 Hz + 1700 Hz
IDLE	2600 Hz single tone

If a trunk circuit was not in use, the trunk would place a 2600-cycle tone on the facility. A failure of the facility would cause a loss of tone, and all circuits would go busy to prevent their use on calls.

When dialing signals were transmitted, the dialed digits were preceded by *KP* and followed by *ST*—for example, KP3125551212ST would be sent over the toll network to reach telephone number 555-1212 in the 312 area code.

Toll trunks and operator positions used MF signaling over the voice circuit to set up connections through the PSTN. This type of signaling was easy to duplicate. People made MF signaling boxes referred to as *blue boxes*. They would place calls over the long distance network to nonbillable 800 telephone numbers. After initial call setup, they would use a blue box to redirect the call to any telephone number they desired. The call would not be billed because the billing machine had been told (by the original dialed digits) that the call was to an 800 number.

A tremendous number of fraudulent calls were being placed every day over the PSTN. AT&T developed out-of-band signaling to combat this fraud. With this type of signaling, the signals of many voice circuits were multiplexed together and transmitted over a separate facility or channel. With this arrangement, blue boxes were worthless. Out-of-band signaling replaced in-band MF signaling in the long distance network. MF signaling is still used in a few locations on the trunks between a class 4 toll office and the IEC office. Prior to out-of-band signaling, MF signaling was used to transmit the originating telephone number to the IEC. The originating number was automatically identified by equipment in the class 5 exchange. This was called *Automatic Number Identification (ANI)*. MF signaling transmitted the ANI from the class 5 exchange to the class 4 toll exchange while the called number was transmitted using loop signaling or E&M signaling.

AT&T's out-of-band signaling system was called *Common Channel Interoffice Signaling (CCIS)* because all signals were multiplexed over one common channel. This system was refined and improved by AT&T. As the PSTN moved from analog carrier systems to digital carrier systems, the signaling was converted into digital bit streams as well. The CCIS system also advanced as CPU technology advanced. Using CPUs to manage the signaling bits allowed the progression from an electrical signaling and tone signaling technique to a message signaling technique.

CPUs can generate, receive, and decode digital messages. Now instead of sending an electrical condition as a signal, we merely send a message from one CPU to another. Since the signaling is being done by computers talking to one another, this system can only interface to computer-controlled (SPC) switching systems. CPUs on the signaling network pass messages between the CPUs of switching systems to provide switching instructions. The CPU of the SPC switch is programmed to take specific actions for each individual message received. This Common Channel Signaling technique, using a separate network of computers to send and receive messages as signals, has been adopted by the international community as CCIS #7 and is referred to as *Signaling System 7 (SS7)*.

10.10 SIGNALING SYSTEM 7 NETWORK

SS7 is a network of its own. The network consists of *Signal Transfer Points (STPs)* and transmission facilities connecting these STPs. Signals are put into data packets and sent over the SS7 network. The SS7 network uses an X.25 signaling protocol. Prior to SS7, when a long distance call was made, the call was connected office by office through the long distance network. On calls to a busy telephone, the call would be established over the network and tie up numerous circuits in the PSTN just to return a busy signal from the far-end class 5 central office.

With SS7, when a call is placed over the long distance network, the IEC switching center will launch a query over the SS7 network to establish a call setup path and to test the called number. If the called number is busy, a busy signal is returned to the calling party *from the originating IEC switching* system. When a called line is busy, the PSTN voice network never receives the call from the originating IEC office. A busy signal is returned to the calling party from a tone generator at the originating class 5 or class 4 office.

If the called line is idle, the SS7 network issues signaling messages to each switching office that will be used to connect the call. Each office will communicate over the SS7 network and identify the trunk circuits to be used between each office for the voice circuit. These trunk circuits are reserved for the call. A signal is sent to the terminating class 5 office, instructing it to ring the called line. The caller receives ring-back tone from the originating IEC office. When the called party answers the call, the reserved trunks are then connected and the voice circuit path is completed between the calling and called parties. SS7 makes efficient use of the trunk circuits in the PSTN. SS7 was first deployed in the toll network and is now extended to many class 5 SPC offices (Figure 10-6).

10.11 SIGNALING SYSTEM 7 AND CLASS

LECs are offering services such as calling number identification, call trace, ring again, automatic recall, automatic callback, selective call acceptance, selective call forwarding, and selective call rejection. These features are referred to as *Custom Local Area Signaling Services (CLASS)*. These features are only possible if the local office is connected into the SS7 network. For example, calling number identification—better known as **caller identification** or caller ID—only works if the telephone number of the originating party is transmitted between switching centers, and the only way switching centers can pass this information between them is for both offices to be connected to SS7 network.

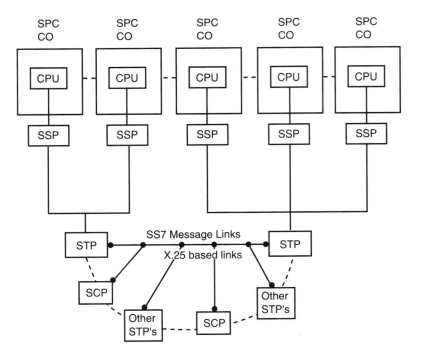

FIGURE 10-6 Signaling System 7 network.

If your office is connected into the SS7 network and you have caller ID, you can receive caller ID on most calls. Originally, this information was only passed on calls within a LATA, but the FCC has ruled that the IECs must also forward caller ID on all calls. Central offices in small rural areas may not connect into the SS7 network. Some of these exchanges are still SXS or XBAR. Calls from these offices will not forward caller ID on local calls and these offices cannot provide CLASS features. Although the IEC receives the originating caller's telephone number for billing purposes via ANI, it will not use ANI for caller ID because they are not always the same. Caller ID is a different protocol from ANI, but most of the time the number is the same. It would require a rewrite of software to provide ANI as caller ID when caller ID is not available between the LEC and IEC.

10.12 SIGNALING SYSTEM 7 AND CALLING CARD VALIDATION

IECs also use SS7 to launch quires on calling cards. All information on calling cards is contained in databases owned by the companies that issue calling cards. These databases are connected to the SS7 network. Suppose a caller wishes to use a calling card issued by one of the operating telephone companies to place a long distance call. If the caller is calling from a pay station presubscribed or PICed to Sprint, the caller is connected to a Sprint toll switching center. After dialing the area code and telephone number for the called party, the caller is prompted by a "bong-tone" signal to enter a calling card number.

After the calling card number is entered into the switching system, the switch will automatically check the number to see if it is a Sprint calling card or a LEC calling card. Suppose the switch determines the calling card is a NYNEX credit card. It will send a message over the SS7 network addressed to the NYNEX calling card database and ask if the card number is good. The NYNEX database will return a response over the SS7 network addressed to the originating IEC switching system and inform it whether the card is good or bad. If the card is validated as good, the call is completed. If the NYNEX database response indicates the card is no good, the call is routed to a Sprint operator for an alternative billing method.

10.13 SIGNALING SYSTEM 7 AND CALL SETUP

Signaling over the SS7 network is fast, and call setup is done on completion of dialing. SS7 is much faster than in-band signaling. With in-band signaling, a class 4 toll office had to wait until the caller had dialed all digits. The class 4 office would decide how the call should be routed and then resend all dialed digits to the class 3 office, which would resend all digits to the class 2 office, and so on. With SS7, all offices are connected via the SS7 network, and when the originating party completes dialing, the originating office can launch a query directly to the terminating office to check on the status of the called line. A voice path is also reserved in the PSTN. When the called party answers, the message instructing all offices to connect the reserved trunks happens faster than the called party can get a handset from the cradle of the phone up to his or her face. The calling party will hear the called party when the person answers with "hello."

SS7 has made other types of signaling on trunk circuits obsolete. Some T-carrier systems are still in use to connect local class 5 exchanges and use E&M signaling, but the toll network uses nothing but SS7. Toll switching centers are using SS7 to establish switching connections, and customers have no way to access SS7. This makes SS7 a secure system. After a voice connection is established, callers can use their DTMF pad to issue signals to computers. When someone places a long distance call using a calling card, the IEC switch is connected to the person's class 5 exchange via trunks operated under control of SS7. The IEC switching system will connect a DTMF register to the incoming trunk to collect DTMF digits from the caller. The caller will use the DTMF pad to enter the called number and a calling card number. When the IEC switch has received the calling card number and called number, it uses the SS7 network to set up voice circuits through the PSTN.

Once a connection is made through the PSTN, the caller can use the DTMF pad for signaling devices on the other end of the circuit. Because the DTMF tones fall within the voice frequencies handled by the PSTN, they are passed over the PSTN voice circuits with no problem. Since the PSTN no longer uses signals over the voice path for addressing and call setup, callers cannot redirect switching using tones over the voice circuits. Many telecommunications systems such as voice mail, answering machines, and *Automatic Call Distributors (ACDs)* ask the caller to use DTMF digits for handling of the call. A caller may be asked to press the digit 1 for the marketing department or the digit 2 for customer service. Someone calling into a voice mailbox or answering system can instruct it to play back messages by using the DTMF tones to enter a password and the action desired.

10.14 A UNIQUE PROBLEM ASSOCIATED WITH SEPARATE VOICE AND SIGNALING PATHS

SS7 was designed to place addressing and supervisory signals over a separate network from the voice network. This presents a problem not encountered with in-band signaling methods. With in-band signaling, if a facility carrying voice circuit trunks was out of service due to a cable cut or defective transmission equipment, the signaling path was also affected and all trunk circuits were made busy by the signaling system. This prevented access of trunks in trouble. Since SS7 does not use the voice path, it will issue a request for a continuity check as part of the call setup message. If no continuity is achieved, the switch takes the trunk out of service and a message is printed out on the maintenance terminal.

The switching systems and the SS7 network must be able to exchange messages and work together. This is only possible if the switching systems are SPC switching systems. The toll network uses only SPC switches. The SPC systems convert all voice and signaling into a digital binary code. SS7 was developed to meet the advanced signaling requirements of the all-digital network. Because signaling is separate from the voice channel, the full 64-Kbps voice channel can be used for voice or data. The voice channel is referred to as a *64-Kbps clear channel*. With in-band signaling, 8000 bits were robbed from the voice channel for signaling, which resulted in only 56 Kbps data capability. This clearly could not be tolerated in the end-to-end digital connectivity world.

SS7 also overcame limitations on early CCIS systems that would prevent it from handling ISDN. SS7 was developed to handle ISDN. Both system use X.25 packet circuit protocols. X.25 is an international standard defining the interface between data terminal equipment and a packet switched network. X.25 does not define the specifications for the packet switched network or what happens inside the network. X.25 merely defines the interface to the packet network. An X.25 packet switched network is nothing more than a transparent delivery service between two computers. For SS7 the packet switched network delivers messages from the CPU in one switching exchange to the CPU in another exchange or to the CPU of a service control point. SS7 is a specialized packet data network used to carry signaling information to SPC switches and messages between switches and databases (service control points).

SS7 has three major components, the *service switching point (SSP)*, the *signal transfer point (STP)*, and the *service control point (SCP)*. Each exchange using SS7 has a SSP, and the exchange will issue all requests for connection of trunk circuits to another exchange via the SSP. The SSP consists of signaling data links to a STP. Signaling between the SPC switch and the STP is done over these data links. The link consists of a data channel in each direction to provide a full-duplex channel between the two systems. The standard data rate over the link is 64 Kbps. The link has software to monitor for errors, detect errors, correct errors, control the flow of data, and transfer messages from the CPU of the SPC switch to the CPU of the STP.

STPs are installed in pairs to provide automatic switchover in case one unit fails. STPs are routing switches whose sole purpose is to connect a SSP to a SCP or to another SSP. SCPs contain a database of circuit, routing, and customer information. When the SPC switch needs to set up a call, the originating SSP at the originating SPC switch will issue a service request directly to a SCP, or the SSP may issue a request via a STP to a SCP. The SCP accepts the service request, searches its database for the appropriate information, and returns a reply to the SSP. Duplication of SCPs protects against failure of the database.

Just as there are many STPs on the SS7 network, there are many SCPs on the network also. Each SSP, STP, and SCP has its own unique address. When a message is originated by a SSP, it will specify where the message should be sent by the address field of the message, and it includes its own address in the message so the recipient knows where to send its reply. As mentioned earlier, operating telephone companies maintain databases in SCPs on all their customers. These databases can only be accessed over the SS7 network. When any IEC needs information about the legitimacy of a calling card, it accesses these databases by addressing messages to the appropriate SCP. The LEC charges IECs for each access to these databases and must provide access to any carrier.

10.16 *SIGNALING SYSTEM 7 AND THE OPEN SYSTEMS INTERCONNECTION MODEL*

SS7 uses a layered protocol similar to the *Open Systems Interconnection (OSI)* model. OSI is a standard created by the International Organization for Standardization. OSI is a seven- layered protocol that allows computers to communicate at the various levels if they are using OSI-layered communication protocols. The seven layers of OSI are: (1) physical, (2) data-link, (3) network, (4) transport, (5) session, (6) presentation, and (7) application.

SS7 is a four-layered architecture using the first four layers of the OSI model. Layered architectures allows many vendors to develop products that will work with other vendors' products. Layering is a design approach that specifies different functions and services at the different layers in a protocol stack. The protocol stack defines how communication hardware and software interoperate at the various layers. *Protocol stacks are software programs* designed for each level to define how the communication takes place between layers. Programmers only need to be concerned with the software for the layer they are working on and with how that layer interfaces to the layer above and below it. *Lower levels define physical interface* standards to a communication medium and how bits are to be transmitted over the various media.

In SS7, the first layer, or physical layer, is called the *signaling cata link layer*. It is a full-duplex connection between components of the SS7 network and normally operates at 64-Kbps. The second layer is the *signaling link layer* and is software that provides flow control, error detection, error correction, and delivery of data packets in the proper sequence. The third layer is the *signaling network layer* and is software designed to route messages from source to destination and from the signaling link layer to a user-defined fourth layer called the *signaling connection control part (SCCP)*. The SCCP is software that is application specific such as the upper-layer software used in a SCP database.

The first three layers of software are called *message transfer part (MTP)*; these three layers form an interface to conform with CCITT X.25. This MTP is a datagram service. Datagrams are packets of data that also include source and destination address information as part of the packet. Packet switches such as STPs read and process this address information to make routing decisions for the packet. The MTP provides connectionless transport service and is limited to a 14-bit signaling point address code. SCCP modifies the connectionless MTP to provide connection- or connectionless-oriented services and to provide many more addresses by a global number system.

10.17 PACKETS OF DATA

The packets of data carried over the SS7 network follow a defined structure. Queries for calling card validation have a specific location within the data packet for originating address, destination address, signaling link selection, and the message itself. The message also follows a definite structure. The calling card number is placed in a specific location. The request for validation is also in a specific location. The receiving SCP looks at the request field to determine action to be taken and looks at the field containing the calling card number to identify which card the request is for. The SCP response to the originating SSP contains a field that specifies if the card is good or bad.

Messages used for call setup are also used to indicate when billing of a call should begin. On long distance calls, when the called party answers the telephone, the terminating central office will send a message over SS7 to the IEC indicating the call was answered, and billing will begin. When a call is placed over the PSTN, the caller supplies the end office or originating IEC with the address of the called party (telephone number). The register and translator of the SPC determine how the call should be routed and the SPC sends a message over the SS7 network to establish the switching connections needed in each office. The local line circuit monitors the local line, and when the users hang up, the SPC is notified. The SPC will initiate a message over the SS7 network telling all switching centers to take the switched connections down and make all trunks idle.

10.18 OTHER SERVICES OFFERED BY THE SIGNALING SYSTEM 7 NETWORK

It is important to note that although CCIS was initially deployed as a method to achieve out-of-band signaling for setup and supervision of toll calls, SS7 was designed and implemented to achieve all this and much more. SS7 provides a communication path for computers to talk to computers. This may be the computers of a SPC switching system, the CPUs of STPs, or the computer of a database. SS7 is a suite of protocols that controls the structure and transmission of both circuit-related and non-circuit-related information.

SS7 is necessary for ISDN and Advanced Intelligent Networks (AIN). AT&T and other IECs offer a software-defined network (SDN), which gives users control over their services. With AT&T's intelligent call processing (ICP) service, customers can control the routing of inbound 800 *wide area toll service (WATS)* lines. Companies such as American Airlines can control the routing of calls coming into their customer service centers located across the United States. If problems occur in any particular center, calls can be rerouted to another center. All calls can be routed to one center at night and allow the other centers to close during this low-calling period. This can only be accomplished by extending SS7 to customers and allowing them limited access to SCPs to change routing information.

The example above also points out the reason for centralizing information into a SCP instead of providing database information as part of each SSP. If the information were located at each central office exchange, when changes in a database occurred all SSPs would need to be updated. By placing the database for many SSPs in one SCP, additions and changes to the database are easily done and controlled. It makes more sense for an operating company such as NYNEX to place all customer data in duplicated SCPs than to provide this data on an office-specific basis. Access to the data is much easier since the requester knows all information is in one database and does not need to search multiple databases. Many database searches are launched automatically by a switching system. It is much easier to write a program to do this search when the number of databases is limited. In the case of a calling card validation by an IEC, the IEC does have an on-site database to validate its own calling card. This database is searched first to see if the calling card belongs to that IEC. If it does, it can be validated immediately. If it is not the IEC's calling card, the database will inform the switch which SCP address to query over the SS7 network.

10.19 SIGNALING SYSTEM 7 AND ISDN

As mentioned in Section 10.17, SS7 was designed with ISDN in mind. ISDN could not exist without SS7. The D-channel of ISDN interfaces directly to SS7. SS7 can be extended to a customer's premises as the D-channel of an ISDN Primary Rate system. Many calling centers or service centers use this arrangement. These centers use an ACD to distribute incoming calls to agents (operators). The voice call arrives over one of the 23 voice circuits (B-channels) of Primary ISDN. The 24th channel (the D-channel) contains signaling and other information for all the B-channels.

The D-channel from the central exchange is connected to a control computer in the service center. This computer is connected in turn to the ACD and to a local area network (LAN). Attached to the LAN are computer terminals at each agent's position. When a voice call arrives at the center over one of the B-channels, the ACD connects the voice call to a headset at an agent's position and informs the control computer which agent it has sent the call to. The control computer receives information about the call via SS7 and the D-channel. Part of this information identifies who the caller is. This information is used by the control computer to search its database for additional information it may have on the caller such as past history on orders, complaints, and so on. The control computer will send this information out over the LAN to the computer terminal at the desk of the agent receiving the voice call from the ACD. The screen of the agent's terminal will be painted with the appropriate greeting to use and relevant information about the call and customer.

As the agent receives information from the customer, the keyboard of the terminal is used to access the control computer and will direct the information to the appropriate computer for processing. If the call center is one used by an IEC as an operator center, SS7 and ISDN D-channel deliver information such as the following: originating telephone number, terminating telephone number, type of service used to place the call (pay phone, residence, business, or hotel), calling card number (if used), type of access dialed (0, 0+, or 1+), and whether or not the calling card was validated by the toll switch as well as the results of that validation query. The control computer will use these information fields to determine how the agent (operator) should handle the call and will provide these instructions to the agent over the LAN.

10.20 SUMMARY

Signaling in telecommunications consists of instructions between devices in the PSTN. Signaling is divided into four basic categories: (1) supervisory (used to monitor the status of a device), (2) addressing (used to identify which device is to receive instructions), (3) alerting (used to notify an addressed party that there is a call), and (4) progress (used to keep the calling party apprised of the progress of a call). The predominant types of signaling have been loop signaling and E&M signaling, but since exchanges have been converted to SPC switches, message signaling has become more common.

SS7 is an out-of-band message signaling system. All SPC switching systems are connected via SS7. This allows an originating office to check on the status of a called telephone number before assigning voice trunks in the PSTN. If the called number is busy or does not answer, the PSTN voice path is never used. SS7 makes more efficient use of the PSTN's voice circuits. SS7 is only possible between computer-controlled switching systems.

SS7 consists of the SSP located in each SPC exchange; many STPs that connect SSPs to the SS7 network, and SCPs that contain a database. The database may contain circuit and routing information for the SS7 network, or it may be a customer-specific database and contain information for a particular company such as its calling card database. SS7 is an X.25-based packet switching network and was specifically designed to interface with the D-channel of ISDN.

REVIEW QUESTIONS

1. What is signaling?
2. What are supervisory signals?
3. How does a line circuit know when a telephone is off-hook?
4. What type of signaling is used by a regular telephone?
5. What type of signaling was required for step-by-step switching?
6. Which is more demanding on a local loop design, signaling requirements or transmission requirements?

7. What type of signaling was used by electromechanical switching systems to indicate a call had been answered?
8. Why was ground start signaling developed?
9. In the analog PSTN environment of the 1970s, what type of signaling was used by trunk circuits?
10. What is the purpose of a DX set?
11. What do *E* and *M* stand for in the term *E&M signaling*?
12. What is a progress signal?
13. What is the difference between in-band and out-of-band signaling?
14. What were some problems with in-band analog signaling?
15. What type of signaling is used between SPC switching systems?
16. What is CLASS?
17. What type of signaling does CLASS use?
18. Why was SS7 developed?
19. What are the major components of an SS7 network? Describe each.
20. How many layers of the OSI model are used by SS7?
21. What is a protocol stack?
22. Which component of the SS7 network contains a database?

GLOSSARY

Addressing The process of identifying which telephone has been called. The telephone issues an address as dialed digits using dial pulses or DTMF tones.

Alerting Signals Let someone know a call is waiting. Ringing signals alert a called party that the phone should be answered. Alerting is closely tied to supervisory signals.

Automatic Number Identification (ANI) Usually refers to the transmission of the originating telephone number (used for billing purposes) from the LEC's class 5 exchange to the IEC's intertoll switch. Billing numbers are not always the same as the telephone number for a business, but they are the same for a residence. ANI and caller ID are two different processes. The IECs will not substitute ANI for Caller ID. ANI takes place between two switches. Caller ID is an end-to-end process using SS7.

Caller Identification One of the Custom Local Area Signaling Services (CLASS) sold by the LECs. This feature requires SS7 for the transfer of the originating telephone number from the originating exchange to the terminating exchange. If the originating office is not a SPC switch, it cannot connect to the SS7 network, and the message "Not Available" will be received on the caller ID display.

Glare Condition The simultaneous seizure of a trunk from both ends. This condition occurs in electromechanical switching systems.

Ground Start Line When a line circuit has been modified for ground start, -50-V DC battery is present on the Ring lead and the Tip lead has no connection. The trunk circuit at the PBX cannot use loop start to seize the line relay because no ground is present on the Tip lead. The PBX trunk must place a ground on the Ring lead to seize the central office line circuit.

Ground Start Signaling A type of signaling used with PABX trunk circuits. The line circuit of the central office is operated by placing ground on the Ring lead.

In-Band Signaling Involves sending signals over the same circuit used for the accompanying voice signal.

Loop Signaling Activation of a signal by opening or closing the loop to complete an electrical circuit and causing current to flow.

Out-of Band Signaling Signaling that occurs over a different circuit than the circuit used for the voice signal. Many times the signals for several different voice circuits are multiplexed over one facility and called Common Channel Signaling.

Progress Signals Indicate the prograss of a call to the calling party; examples include dial tone, 60-ipm busy tone, 120-ipm busy tone, ring-back tone, and recordings.

Reverse Battery Signaling In a SXS central office, the connector switch would signal that the called party had answered by reversing the battery feed to the calling party.

Signaling Anything that serves to direct, command, monitor, or inform. In telecommu- nications, signals are needed to inform a local switching system that someone wishes to place a call, provide directions to the PSTN on how the call is to be connected, and inform the called party that a call is waiting. Signaling is divided into four basic categories: (1) supervisory, (2) addressing, (3) alerting, and (4) progress.

Signaling System 7 (SS7) A signaling network that can only interface to computer-controlled (SPC) switching systems. CPUs on the signaling network are called signal transfer points (STPs) and pass messages between the service signaling points (SSPs) of switching systems to provide switching instructions. The CPU of the SPC switch is programmed to take specific actions for each individual message received. This Common Channel Signaling technique, using a separate network of computers to send and receive messages as signals, has been adopted by the International community as CCIS #7 and is referred to as Signaling System 7 (SS7).

Supervisory Signals Used to monitor the status of a line or trunk circuit. A switching system is informed of whether a line or trunk is idle or busy by supervisory signals.

11

Integrated Services Digital Network and Asymmetric Digital Subscriber Line

Key Terms

Asymmetric Digital Subscriber Line (ADSL)
Basic Rate Interface (BRI)
B-Channel
Broadband ISDN (B-ISDN)
D-Channel
Digital Subscriber Line Circuit (DSL)
Hypertext Markup Language (HTML)

Integrated Services Digital Network (ISDN)
Internet Service Provider (ISP)
Link Access Procedure on the D-Channel (LAPD)
Narrowband ISDN
Network Termination 1 (NT1)
Network Termination 2 (NT2)
Network Termination 12 (NT12)

Primary Rate Interface (PRI)
Reference Points
Service Profile Identifier (SPID)
Terminal Adapter (TA)
Terminal Equipment 1 (TE1)
Terminal Equipment 2 (TE2)
Terminal Equipment Identifier (TEI)
2B1Q

Integrated Services Network (ISDN) is a network providing end-to-end digital connectivity and supports a wide range of services such as voice, data, and video over one facility. Prior to the introduction of ISDN, different interfaces were needed for each of these services. Each of these interfaces was tailored to provide one specific service. The Public Switched Telephone Network (PSTN) was developed to transmit analog electrical signals from one phone to another. The phone converted analog sound waves into analog electrical signals for transmission over the PSTN. In fact, anything to be transmitted over the PSTN had to be in an analog signal form. The local loop was designed for efficient transmission of analog signals between 300 and 3300 Hz. Voice signals fall in this range. The

PSTN was designed to handle voice signals, but the network has gradually evolved into a digital circuit network. Analog voice signals are converted into digital signals by a codec on the line card and carried by the digital PSTN.

11.1 ANALOG VIDEO

Video signals are also analog in nature. Video signals are higher-frequency signals than voice and occupy a much greater bandwidth. Video signals used to be transmitted from a television studio to the telephone exchange over local-loop cable pairs that were specially treated. These cable pairs were not loaded. Amplifiers and equalizers were placed at intervals along the cable route to keep the analog signal at a proper power level. At the central exchange, the cable pair from the studio was connected to a cable pair that went to the transmitter and antenna site. This pair also had amplifiers and equalizers placed at intervals along its route. Television networks such as NBC, CBS, and ABC would provide programming to local stations via the frequency division multiplexing (FDM) systems in the old analog PSTN. Special wide-bandwidth channels on the FDM systems were used to get these signals from the network (ABC, CBS, NBC, and so on.) to the class 4 toll office serving the local affiliate television station. The signal was connected over cable pairs, from the class 4 office to the local class 5 exchange, for connection to cable pairs serving the local affiliate TV control center.

Now that the PSTN has been converted to a digital network, video signals are converted into digital signals and transmitted digitally over the digital PSTN. Digitized video signals are converted back into analog signals at the transmitter site and are used to frequency modulate the carrier frequency of the transmitter. Many TV networks and TV stations no longer use the PSTN or local telephone company for transmission of their signals. Network affiliate stations receive signals from the networks via communication satellites. Some mobile units use satellite dishes to beam signals to a satellite. The signal is sent by the satellite to a receiving dish at the TV studio. Local stations also have mobile units that use microwave transmitters and dishes for local remotes. Most local stations transmit signals from their studio to their transmitter site using a microwave dish.

11.2 THE OLD ANALOG PSTN AND DATA

When the PSTN was an analog network, digital data could not be transmitted over the PSTN without first being converted into analog signals. A device called a *modem* was developed to convert the digital data signals into analog signals that were within the 300- to 3300-Hz range. This conversion process resulted in an analog signal that could be handled by the analog PSTN. The older FDM facilities used in the long distance network portion of the PSTN had a 4000-Hz bandwidth for each voice channel, but some of that channel bandwidth was used for guardbands and signaling. The effective channel bandwidth for the voice signal was 300 to 3300 Hz. Modems are low-speed devices that allow for low-speed dial-up connections between computers. The speed of dial-up modems has improved tremendously and allows for speeds of up to 56,000 bps, but the local loop, or the circuit path in the PSTN, will often limit the speed to less than 28,800 bps. Modems automatically test the circuit being used when the circuit is established and if it cannot support 56,000 bps, the modems will automatically select a lower speed that the circuit can handle.

Using an analog signal and modem for high-speed data requires a high-bandwidth and a high-frequency analog signal. When analog FDM was used to carry the signal over the PSTN, multiple FDM channels were bonded together to support the high-frequency signal. High-speed data that is being carried by a high-frequency analog signal cannot be carried over a regular dial-up line. High-speed data requires leasing an end-to-end circuit from a common carrier. These leased circuits are called *private lines* because they can not be accessed by the general public. The private line is intended for use by a few individuals, concerns, or companies. A private-line facility is required for high-speed data because high-frequency signals require special treatment of the local loop. Special conditioning of the local loop is referred to as *C conditioning*. There are five grades of C conditioning: C1, C2, C3, C4, and C5. The grade of conditioning to use depends on how high a frequency the loop needs to handle. Special conditioning of the local loop to reduce noise is called *D conditioning*.

Conditioning the local loop improves its ability to handle high-frequency signals and reduces the power loss to a signal placed over the conditioned loop. When data is carried by an analog signal, the higher the required data speed, the higher the analog frequency has to be. In some cases the local-loop facility needs treatment with amplifiers and equalizers similar to the treatment used for analog video signals. Today, high-speed data is carried as a digital signal by using multiple DS0 or DS1 facilities in the digital PSTN. If customers need to transmit high volumes of data from one location to another, they usually decide to transmit the data as a digital signal rather than an analog signal and will lease a T1 carrier facility from the LEC.

If a company requires a high-speed data line, it needs to lease a private data line over the PSTN. In the past, if an insurance company had to transmit high volumes of data between its headquarters in Chicago and a regional office in Atlanta, it had to lease special wide-bandwidth channels on the older FDM PSTN between Chicago and Atlanta from AT&T. Additionally, the LEC had to provide C1 or C2 conditioned local loops in each city for connecting the insurance computers to the AT&T toll exchanges. Now that the toll network is digital, modems are not needed. Low-speed data can connect to a digital local loop using a Customer Service Unit/ Data Service Unit (CSU/CDU) instead of a modem. The loop is conditioned to handle digital data, and the data is interfaced to the PSTN using the CSU/DSU. High-speed data lines can be connected via a CSU/DSU to a T1 TDM systems at the customer's premises. This T1 system can connect directly to leased T1 systems in the PSTN. This type of facility handles only data, and many companies use this arrangement for their data lines. Most residential customers need to use their line for both voice and data. Therefore, they order an analog voice line from the LEC. When they wish to use this line for data, they use a modem.

11.3 CONVERSION OF THE PSTN TO A DIGITAL NETWORK

In the past, modems were needed for conversion of digital signals to analog signals because the PSTN was an analog network. The PSTN is no longer an analog network but a digital network. Even though the PSTN is now digital, modems must still be used by regular phone lines because the phone lines are still analog. Engineers developed ways to convert analog signals into digital codes and found that transmitting digitized voice was more efficient and resulted in a higher-quality network than the AM/FDM network. Initially, the analog-to-digital conversion was done at the channel units of TDM carrier systems using a codec in the channel unit. Manufacturers of switching systems caught on to TDM technology and moved the codec from the trunk circuit (channel unit) to the line circuit.

Although a few analog switching systems still exist, the toll switches are 100% digital, and the majority of local switching is also digital. Most IEC networks use digital transmission systems. The only part of the PSTN that remains analog is the local loop and the line circuit. The line circuit in a digital central office has a codec to convert analog signals into digital signals at 64 Kbps (DS0). This line circuit is designed as an interface to an analog local loop. It expects to see an analog signal. The codec on the analog line circuit does not expect to see a digital signal and cannot properly code a digital input. A different line circuit (called a Digital Subscriber Line Circuit—for example, ISDN) is needed to accept a digital signal from the local loop.

Since the standard line circuit interface of a switch expects to see an analog signal input from a local loop, when a local line circuit is used for switched data, a modem is still required. A modem at the customer's premises converts the digital signal of a computer into an analog signal that can be handled by the local loop *and the codec of the line circuit*. The line circuit codec converts the analog signal back into a digital signal. To prevent all this conversion and reconversion, it would be necessary to move the codec from the line circuit to the telephone (or other station apparatus), condition the local loop to handle the digital signal, and use a line card with a digital interface to the local loop. ISDN does exactly that. The codec resides in the ISDN telephone and converts voice signals into digital signals. The telephone also provides an EIA-232 digital interface. The digital output from a personal computer (PC) can be connected to this port. The ISDN telephone can be used on a voice call while the PC is connected on a data call. Both calls will be handled by the single digital line card and local loop.

As stated above, the local loop can receive special treatment that will allow it to handle digital signals. TDM T1 carrier systems use a digital signal (DS1), and cable pairs serve as the transmission medium for the DS1 signal. The DS1 signal used by T1 is a digital signal consisting of a serial data bit stream at 1.544 Mbps. To handle this signal, the cable pair is unloaded and signal regenerators are placed at 6000-ft intervals along the cable route. Each end of the cable pair has a TDM multiplexer/demultiplexer attached to the cable pair. Cable pairs over 18,000 ft can be treated to handle either a digital signal or an analog signal but not both. Cable pairs under 18,000 ft are not loaded. Therefore, cable pairs under 18,000 ft can be used as an analog voice line or a Digital Subscriber Line without special treatment (see Chapter 5). The majority (more than 75%) of all local loops are under 18,000 ft, and these cable pairs can be used as an analog line or a digital line.

Many people are buying ISDN modems for their PC. They make arrangements with their *Internet service provider (ISP)* for connections with the ISP via ISDN. These customers purchase ISDN service from the LEC and are provided service via an ISDN line circuit in the central office. The ISP is also connected to ISDN line circuits at its serving central office. When an ISDN connection is provided between the ISDN modems at the customer's PC and the ISP's computer, data can be transferred at 128,000 bps between the two ISDN modems. The ISDN modem in a PC also has several RJ-11 jacks to allow for the connection of telephones or fax machines. Because ISDN is a digital service, the local loop must be an unloaded cable pair and no more than 18,000 ft in length. As discussed in Chapter 5, ISDN is provided to remote areas of an exchange by using Subscriber Carrier (SLC-96) to establish a Carrier Service Area that can support DS0 services such as ISDN.

11.4 IS ISDN A NECESSARY TECHNOLOGY?

The PSTN has evolved from an analog network into a digital network. Data communications professionals are not very professional when they criticize the PSTN as a voice-only network. The only portion of the PSTN that remains analog is the local loop, and it has been the exchange carriers that have been in the forefront of ISDN development. Many people have criticized the telecommunications industry as pushing ISDN, simply to extend the evolution of the PSTN into an all-digital network. They imply that ISDN was creating a solution where none was needed and that the PSTN is excellent as is, with no need to convert the local loop to digital: "If it ain't broke, don't fix it." ISDN has been called: "I Still Don't kNow what it's for," "It Still Does Nothing," and so on. If the PSTN is to handle voice only, these critics would be correct. But clearly everyone is aware of the explosion that has occurred in the data world. The PSTN must be designed to handle voice, video, and data. ISDN does offer a solution for this concept.

Opponents to ISDN point to advances in modem design that have provided very fast data speed capabilities as a reason ISDN is not needed. The speeds of new modem designs are approaching the speed ISDN offers. Most computers are now equipped with 28,800-bps modems, and many people are deploying new modems with speeds of up to 56,000 bps. These modems provide speeds that allow for fast transfers of files between the ISP and a PC. With these modems, the local loop could remain analog, as long as users wished to use the facility only for voice and connection of PCs to the network. One reason ISDN has not been accepted widely in the United States is that most people have not needed the features it provides. The Internet explosion will change that. Many people are using the Internet, and in a few years everyone will want Internet access. Internet is a packet data network. When sites on the Internet are accessed, large graphic files are transferred from the Internet site to your PC. The PC uses these graphic files to paint the screen. Fast data speeds are needed to minimize wait time between screen changes.

ISDN access to the Internet provides high-speed access. This is an excellent application for ISDN, but many ISPs do not offer ISDN access lines. Here is another example of where ISDN is a good solution but deployment of this application is hindered by high cost and the lack of support from the data community. ISDN provides faster transfer of graphic files than modems do. Opponents of ISDN will cite the explosion that is going to occur in cable modems. With deregulation of local exchange services, cable TV companies are entering the telecommunications business and Internet access. The wide bandwidth of their coaxial cables allows them to carry high-speed signals. Cable modems will attach to the cable TV facilities and will provide up to 27-Mbps download capabilities and 1-Mbps upload capabilities. This is about 60 times faster than ISDN and 300 times faster than a 28.8-Kbps modem. Cable modem service will cost about $40 a month. BRI ISDN costs about $100 a month.

The deployment of ISDN has been badly handled by everyone involved. Few applications were developed for use of ISDN and it was not widely deployed. It does not do me any good to have ISDN if people I communicate with directly cannot (or will not) get ISDN. As mentioned above, for ISDN to work, I must connect to another ISDN line. When I use the Internet, my ISP interfaces my ISDN line to the X.25 Internet highway and takes care of the protocol conversions needed. If I were to use a regular analog phone line and modem, the ISP would also do a protocol conversion for me. I can communicate with anyone on the Internet and use e-mail because the ISPs take care of protocol conversions. The speed with which *Hypertext Markup Language (HTML)* files are transferred to me depends on the speed of the slowest link. The speed of the on/off ramps to the Internet govern the speed of data transfer.

The window of opportunity for ISDN technology may have come and gone. The LECs should hope this is not the case. I predict that the general public will be demanding digital connectivity by 2005. ISDN is an international standard and appears to be the only affordable technology that LECs can use to meet this demand. The Bell LECs know the future of the local loop depends on its conversion to digital. The first step in the conversion of the local loop to digital will be to replace feeder and distribution cables with fiber.

Many local switching systems have been converted from in-band signaling to out-of-band signaling. This conversion is necessary for three reasons: (1) SS7 is needed to offer Custom Local Area Signaling Services (CLASS) such as caller ID,

(2) SS7 is necessary to offer ISDN, and (3) in-band signaling restricts the DS0 channel to 56-Kbps data. Out-of-band signaling is needed to eliminate the bit-robbing signaling technique of in-band signaling and return the DS0 to a full 64-Kbps clear channel for data. In addition to installing SS7 to all class 5 exchanges, many LECs have plans to convert from wire, as the local loop, to fiber and coaxial cable. Pacific Bell has stated that it intends to provide fiber or coaxial cable to 6.5 million homes and businesses by the year 2000.

11.5 DEVELOPMENT OF STANDARDS

ISDN was initially offered by the LECs in 1986, but because there were few applications developed that could take advantage of the technology, it did not take off. Another reason for the failure of ISDN was a lack of standardization. AT&T, Northern Telecom, Siemens, and other switching manufacturers had their own unique proprietary standard for ISDN. As a result, ISDN station equipment and CPE had to be purchased from the switching manufacturer. An ISDN device designed to work on an AT&T switch would not work on a Northern Telecom switch. A standard ISDN architecture was developed by CCITT in 1984, updated in 1988, and again in 1992. These standards are based on the OSI model. The standards are not for the network. They are *interface* standards to the ISDN network.

When a network interface standard was implemented, all switching systems were forced to use the same protocols regardless of manufacturer. Customer services are delivered over two standard interfaces, the **Basic Rate Interface (BRI)**, for basic services, and **Primary Rate Interface (PRI)**, to provide higher transmission capabilities. BRI is the lowest-level ISDN interface. BRI provides four logical two-way digital circuits over one physical cable pair between the customer's location and the central exchange. The physical cable pair is a *Digital Subscriber Line (DSL)*. The **DSL circuit** is a two-way digital circuit operating at 160 Kbps. The ISDN line circuit at the central exchange and the terminal device at the customer's location divide this 160-Kbps bit stream into four logical data channels. Three of these channels (144 Kbps) are used by the customer. The fourth channel (16 Kbps) is used by the hardware of the ISDN line circuit as a maintenance (M) channel. Two of the user channels are 64 Kbps each and are called the *bearer (B) channels*. These B-channels can be used for either voice or data. The third channel is called the *delta (D) channel* and is a 16 Kbps channel that operates only in a packet data mode.

Because BRI consists of two bearer and one delta channel, it is called 2B+D. The **D-channel** provides signaling and control for the two B-channels. The D-channel is used as a vehicle for packet data transport. When most people talk about BRI, they refer to it as 144 Kbps. This is the total user rate for the 2B+D user channels. The D-channel can be used to transport user data in a low-speed packet mode, but most often the D-channel is used only for call setup, and for communication messages between the ISDN devices on each end of the circuit. The user is left with 128 Kbps over the B-channels for use as voice or data. The **B-channels** are called bearer channels because they carry the customer's voice or data. The fourth channel is a 16-Kbps channel called the *maintenance* or *overhead channel*. The 16 Kbps are used to support performance monitoring, framing, and timing functions between the customer-premise equipment and the equipment at the central exchange. The total bit rate for all four channels is 160 Kbps.

11.6 BRI AND DSL STANDARDS

Prior to standardization, the BRI supplied by AT&T had 48 Kbps overhead and had a 192-Kbps line rate. The new standard digital line rate of 160 Kbps matches the standard developed by ANSI (T1.601-1988) for a DSL. The DSL standard provides customers with digital access to a digital local exchange using a nonloaded, two-wire, local-loop cable pair. The DSL standard provides the 144 Kbps throughput needed for Basic ISDN, plus the 16 Kbps needed as overhead for framing, synchronization, and monitoring. In 1986 ANSI chose to base its Basic ISDN standard interface to the local loop, on echo canceling technology, using the 2B1Q line coding technique. This decision was reached after more than a year of study on the properties of various line coding techniques such as AMI, MDB, 4B3T, 3B2T, and Biphase. These were standards being proposed by some of the world's leading research firms, and some of these coding techniques were already in use by different manufacturers as part of their proprietary standards. After exhaustive laboratory experiments and field trials of the various coding techniques, 2B1Q was found to provide the best service on the various loops in a typical telephone exchange territory. With 2B1Q, the line rate is actually 80 Kbaud because 2 bits are transmitted on each baud to achieve 160 Kbps. Prior to 2B1Q, the local exchange interface to the local loop for ISDN was 192 Kbps using AMI line coding, which resulted in an 96-Kbaud signal. This provided a user rate of 144 Kbps plus 48 Kbps overhead.

11.7 2B1Q LINE CODING

2B1Q coding is the optimum digital subscriber line technology. The line code is based on pulse amplitude modulation (PAM) technology, which takes 2 binary bits and converts them into a multilevel analog signal for transmission, using PAM in a time division multiplex (TDM) signal, over the analog local loop. 2B1Q is a four-level code; it codes 2 bits at a time into one of four amplitude levels:

1. If the first bit is a 1 and the second bit is a 0, transmit a +2.5-V pulse.
2. If the first bit is a 1 and the second bit is a 1, transmit a +0.833-V pulse.
3. If the first bit is a 0 and the second bit is a 1, transmit a –0.833-V pulse.
4. If the first bit is a 0 and the second bit is a 0, transmit a –2.5-V pulse.

These four codes are called *quaternary symbols* or *quats*. The first bit determines whether the transmitted pulse amplitude is positive or negative. If the first bit is a 1, the pulse has a positive amplitude. If the first bit is 0, the pulse has a negative amplitude. The second bit determines the level (or amplitude) of the pulse. If the second bit is a 0, the level is 2.5 V. If the second bit is a 1, the level of the pulse is 0.833 V. Notice that 2B1Q means 2 bits = 1 quat (Figure 11-1).

The use of 2B1Q line coding reduces the frequency of the line signaling from 160 kilobaud or Kbaud (at 1 bit per baud) to 80 Kbaud (with 2 bits per baud). Local loops are considered to act like low-pass filters and were designed to handle low-frequency signals. Attenuation of signals varies directly with the frequency of the signal transmitted. Lower-frequency signals have less attenuation and can work on longer loops. Lower frequencies also reduce crosstalk into adjacent cable pairs. The lower frequency of 2B1Q makes it an excellent choice for DSL.

The 2B1Q coding process does not result in the same number of positive and negative pulses on the transmission line, and so 2B1Q includes a scrambling algorithm to achieve the same number of positive and negative pulses on the line. This reduces line current to 0 and allows the use of a longer loop. 2B1Q allows a DSL to work on a cable pair up to 18,000 ft from the central office, if the cable pair is not loaded and has no bridge taps. A bridge tap exists when a feeder cable pair connects to two different distribution cables. The working distance of Basic ISDN can be extended beyond 18,000 ft if the DSL is treated by placing a repeater in the line. DSL will work in any cable with regular customer lines. DSL will not work properly on a cable pair that is in close proximity to a cable pair that is handling an analog subscriber carrier.

The DSL line format is based on the transmission of quats. The basic frame consist of transmitting 120 quats in 1.5 ms. In addition, 80 quats in 1 ms represents 80 Kbaud. At 80 Kbaud (the line rate for DSL) per second, we transmit 120 quats in

2B1Q Symbol Voltage Levels

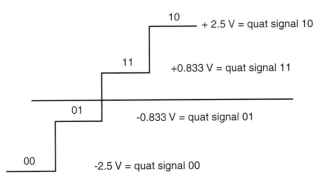

The diagram indicates how 2B1Q sends 2 bits (a dibit) with each signal voltage. Since four different signal voltage levels are used, it is called *quaternary signaling* and is represented by the letter *Q*. Thus, 2 bits = 1Q (2B1Q). At 80 Kband, we transmit 160,000 bps using 2B1Q. The 2B1Q signal is a PAM/TDM signal.

FIGURE 11-1 2B1Q Symbol voltage levels.

1.5 ms. One DSL frame requires 1.5 ms. Each frame contains 240 bits. Since each frame is 1.5 ms long (0.0015 s), we can achieve 666.66667 frames per second ($F = 1/T$: $F = 1/0.0015 = 666.66667$). With 240 bits in each frame, we achieve 160 Kbps per second (240 bits per frame × 666.66667 frames per second = 160,000 bps).

Each 240-bit DSL frame contains the following bit allocations: bits 1 to 18 for a Sync Word (SW); bits 19 to 234 for information transport of twelve 2B+D channel samples (each B sample is 8 bits and each D sample is 2 bits. The total bits needed for these twelve 2B+D samples is 216 bits ($12 \times (2 \times 8) + 2) = 12 \times 18 = 216$); bits 235 to 240 are used for 6 bits of M overhead channel. Total bits in a frame = 18 + 216 + 6 = 240 bits. Of course, the 240 bits in a frame calculation can also be arrived at by another method. Remember that the basic frame consists of 120 quats. Each quat represents 2 bits. Therefore, $120 \times 2 = 240$ bits per frame (Figure 11-2).

Time:	<------------ .0015 seconds (or 1.5 milliseconds) ------------>								
Framing	Framing	2B+D	Overhead or Maintenance bits (M1-M6)						
Quat #	1 to 9	10 to 117	118s	118m	119s	119m	120s	120m	
Bit Position	1 to 18	19 to 234	M1 235	M2 236	M3 237	M4 238	M5 239	M6 240	
Frame 1	Inverted SW	2B+D	EOC	EOC	EOC	ACT	1	1	1.5 ms
Frame 2	SW	2B+D	EOC	EOC	EOC	1	1	FEBE	1.5 ms
Frame 3	SW	2B+D	EOC	EOC	EOC	1	crc1	crc2	1.5 ms
Frame 4	SW	2B+D	EOC	EOC	EOC	1	crc3	crc4	1.5 ms
Frame 5	SW	2B+D	EOC	EOC	EOC	1	crc5	crc6	1.5 ms
Frame 6	SW	2B+D	EOC	EOC	EOC	1	crc7	crc8	1.5 ms
Frame 7	SW	2B+D	EOC	EOC	EOC	1	crc9	crc10	1.5 ms
Frame 8	SW	2B+D	EOC	EOC	EOC	1	crc11	crc12	1.5 ms

Total Time for 8 frames (i.e. 1 Superframe) = 120.0 ms

One Superframe (There are 8 superframes from the central office ISDN line circuit to the customer's NT1. They are called Superframes A, B, C, D, E, F, and G. There are also eight Superframes from the NT1 back to the ISDN line circuit. These return frames are called Superframes 1, 2, 3, 4, 5, 6, 7, and 8).

NOTE:
SW = Sync Word; EOC = Embedded Operations Channel; s = space bit; m = mark bit; CRC = Cyclic Redundancy Check for 2B+D and M4; ACT = Activation Bit; 1 = reserved bit; FEBE = far end block error bit. Note that in 1.5 ms, we acheive the transmission of 120 Quats which is 240 bits.

FIGURE 11-2 2B1Q Superframe with detail of M-channel. The chart above lays out the composition of a Superframe. Note that there are 240 bit positions in each and every frame from frame 1 to frame 8. Since each of the eight frames contains 240 bits, a Superframe (which contains eight frames) contains (8×240) or 1920 bits. Each of the eight frames in a Superframe takes 0.0015 seconds (1.5 ms) to transmit: therefore the Superframe is 8×1.5 ms) or 12 ms long. This figure represents one Digital Subscriber Line (DSL) Superframe and the bit asssingments when the DSL uses 2B1Q protocol. Note that the first 18 bits in each frame (the Sync Word) is used for physical layer synchronization between the customer's NT1 and the central office ISDN line circuit. The last 6 bits in each frame are used to establish a maintenance (M) channel. Bits 19 to 237 inclusive provides 2165 bits used to carry 2B+D information.

The Sync Word is used for physical layer synchronization and frame alignment. The M-channel is used for signaling, error detection, and maintenance messages. Since there are eight frames in a Superframe, the 6 overhead bits used in each frame for the M-channel provide a total of 48 bits per Superframe for the M-channel. Since there are 666.66667 frames per second, 6 bits per frame times 666.66667 frames per second provides 4 Kbps for the M-channel. Notice that the M-channel and Sync Word taken together comprise 24 bits per frame (18 for Sync Word + 6 for M-channel = 24 bits). This provides the 16 Kbps referred to earlier as the maintenance and synchronization channel ($24 \times 666.66667 = 16$ Kbps). Also notice that the 216 bits in each frame for 2B+D = 144 Kbps ($216 \times 666.66667 = 144,000$).

Eight frames are grouped together to form a Superframe. Since each frame is 0.0015 s long, a superframe is 0.012 s (12 ms) long ($8 \times 0.0015 = 0.012$). We can transmit 83.333 Superframes per second. Each Superframe contains $240 \times 8 = 1920$ bits and we transmit 160,000 bps ($1920 \times 83.333 = 160,000$). Superframes are transmitted over one wire of the local loop from the customer to the central office and are Superframes no.1, 2, 3,. . . , 8. Superframes are also transmitted from the central office to the customer over the second wire of local loop and are called Superframes A, B, C, . . . , H (Figure 11-3).

11.8 BRI AND NT1

Basic Rate ISDN is provided to a customer by a DSL operating at 160 Kbps. For BRI, the customer provides an interface device to the line called a *Network Termination 1*

		Group 1	Group 2 to Group 11	Group 12	M Chn	Total
No. of bits:	18	8+8+2 (18)		8+8+2 (18)	6	Number of Bits:
Frame 1	ISW	B1+B2+D	<- - - - - - - - - - - >	B1+B2+D	M	240
Frame 2	SW	B1+B2+D	<- - - - - - - - - - - >	B1+B2+D	M	240
Frame 3	SW	B1+B2+D	<- - - - - - - - - - - >	B1+B2+D	M	240
Frame 4	SW	B1+B2+D	<- - - - - - - - - - - >	B1+B2+D	M	240
Frame 5	SW	B1+B2+D	<- - - - - - - - - - - >	B1+B2+D	M	240
Frame 6	SW	B1+B2+D	<- - - - - - - - - - - >	B1+B2+D	M	240
Frame 7	SW	B1+B2+D	<- - - - - - - - - - - >	B1+B2+D	M	240
Frame 8	SW	B1+B2+D	<- - - - - - - - - - - >	B1+B2+D	M	240
Total Bits	144	144	$(10 \times 18 \times 8) = 1440$	144	48	= 1920

FIGURE 11-3 2B1Q Superframe without M-channel description. Total bits in each Superframe = 1920; Total 2B+D bits in each Superframe = 1728; Each Superframe = 12ms; Freq. = 1/t = 1/12 ms = 83.33333 Superframes per second; Total 2B+D bps = $1728 \times 83.33333 = 144,000$ bps

(NT1). The NT1 functions at the physical layer of the reference OSI model and serves as the *demarcation* or *demarc* point. Demarc, designates the point at which LEC ownership ends and CPE ownership begins. The NT1 provides for termination of the two-wire local loop and termination of the four wire S/T bus. The NT1 performs impedance matching to the digital local loop and monitors the performance of the DSL. The NT1 performs maintenance functions and ensures accurate timing, by synchronizing on signals from the ISDN line circuit of the local exchange. The NT1 performs a conversion between the AMI protocol of the four-wire S/T interface and the 2B1Q protocol of the two-wire local loop and vice versa. Of course, it is also the NT1 device that contains an integrated circuit chip that performs the DSL functions and 2B1Q line coding/decoding.

11.9 BRI INTERFACES AND REFERENCE POINTS

The 2B1Q transceiver in the NT1 has two ports. One port connects to customer terminal equipment and is called the *S/T interface*. The S/T interface is defined by CCITT Recommendation I.430. The second port connects to the twisted-wire pair of the local loop through a passive termination hybrid and a line pulse transformer. This interface point to the local loop is called the *U interface* or *U Reference Point* (ANSI Standard T1.601). Each point where two devices interface to each other are called *reference points* (Figure 11-4).

The R reference point is an interface point between a non-ISDN-compatible terminal device that is called TE2 and an ISDN adapter device called a *terminal adapter (TA)*. The EIA-232 serial output port of a PC is a non-ISDN-compliant

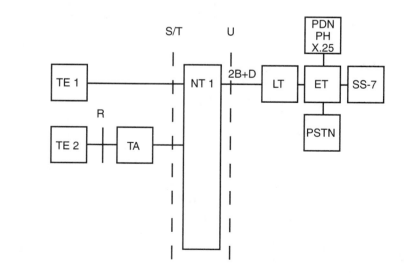

FIGURE 11-4 BRI ISDN interface points.

device. The PC serial port connects to a TA for conversion of the EIA-232 compliant bit stream into an ISDN-compliant bit stream. The *S reference point* connects an ISDN-compliant (such as TE1 and TA) device to the NT1. The *T reference point* is used to designate the interface between a **Network Termination 2 (NT2)** and an NT1 when the NT1 connects to a PRI ISDN line to the central exchange.

When the NT1 was owned by the LEC, the T reference point served as the demarc point per CCITT ISDN recommendations. The FCC does not recognize the T reference point as the demarc point. It recognizes the U reference point as the demarc point. ANSI has developed the U reference as the point of demarcation. Now that the NT1 is CPE and the demarc is the U reference point, the S and T references are often combined as the S/T BUS. For Basic Rate Access, the S and T reference points both have the same electrical specifications. When the ISDN line is a BRI ISDN (23B+D), the NT1 will be connected to an NT2 and the interface between the two is called the T reference point. The S and T reference points operate in a TDM bit stream of 192 Kbps. The S/T Interface bit stream uses pseudoternary coding for transmission. The NT1 will convert messages received over the S/T bus from the terminal equipment into 2B1Q coding over the U interface to the central office cable pair.

11.10 TERMINAL EQUIPMENT AND THE S/T BUS

Up to eight different terminal devices can be connected to the S/T bus. The S/T bus is a four-wire bus that connects the NT1 to ISDN-compliant devices such as terminal equipment (TE), *terminal adapter (TA)*, and NT2. One pair of wires is used for transmit and one pair is used for receive. Each pair is running at a bit rate of 192 Kbps. The user information (2B+D) occupies 144 Kbps. The remaining 48 Kbps are used for control, performance monitoring, and timing. These 48 Kbps of overhead enable the NT1 and TE (or TA) to recover the 2B+D channels from the TDM stream. Both the transmit and receive portion of the S/T bus group 48 bits together to form an I.430 frame, but the frame structures are different for each direction (Figure 11-5). Both frames use pseudoternary coding for transmission of bits over the S/T bus. This coding is such that a binary 1 is represented by no line signal and a binary 0 is represented by either a positive or negative 750-mV pulse. When 0s are transmitted, each subsequent 0 will have alternate polarity. By alternating the polarity of each 0, the line is DC balanced. This means that over time the net current on the line is 0. This allows the S/T bus to be longer than it would without using this coding technique. Terminal equipment (TE or TA) derives its timing from a network clock via the NT1 and S/T bus overhead bits. This timing is used to synchronize the transmitted signals (2B+D). Since station equipment (TE or TA) is being synchronized off the NT, transmission from the TE or TA is delayed by 2 bits. Transmission on the transmit portion of the S/T bus will be offset by 2 bits from transmission on the receive portion of the S/T bus (Figure 11-5).

<--------- Direction of Bits on S/T Bus from NT 1 to TE 1

Bit No.	1	2	3	4	5	6	7	8	9	10	11	12	13	14	15	16	17	18	19	20	21
Bit Designation	F	L	B1	B1	B1	B1	B1	B1	B1	B1	E	D	A	Fa	N	B2	B2	B2	B2	B2	B2

Bit No.	25	26	27	28	29	30	31	32	33	34	35	36	37	38	39	40	41	42	43	44	45
Bit Designation	D	M	B1	B1	B1	B1	B1	B1	B1	B1	E	D	S	B2	B2	B2	B2	B2	B2	B2	B2

<--------- Direction of Bits on S/T Bus from TE 1 to NT 1

Bit No.	1	2	3	4	5	6	7	8	9	10	11	12	13	14	15	16	17	18	19	20	21
Bit Designation	F	L	B1	B1	B1	B1	B1	B1	B1	B1	L	D	L	Fa	L	B2	B2	B2	B2	B2	B2

Bit No.	25	26	27	28	29	30	31	32	33	34	35	36	37	38	39	40	41	42	43	44	45
Bit Designation	D	L	B1	B1	B1	B1	B1	B1	B1	B1	L	D	L	B2	B2	B2	B2	B2	B2	B2	B2

Note: F = Framing Bit; L = DC voltage balancing bit; B1 bits for the first B-channel; E = Echo of the D bit received from the TE1 by the NT1; D = D-channel bit; A = Activation bit; Fa = auxilliary framing bit; N = balancing bit for the auxilliary framing bit; B2 bits for the 2nd B-channel; M = Multiframing bit; and S is an unspecified bit.

FIGURE 11-5 I.430 frames. The S/T bus is a four-wire bus (transmit = 2 wires and receive = 2 wires). The NT1 transmits at 144,000 bps over the transmit bus to the TE1. The NT1 willreceive data from the TE1 over the receive bus at 144,000 bps, but the bits will be delayed by 2 bits from the bits on the transmit bus. Thus, when the NT1 transmit bit 3 to the TE1, bit 1 is being received from TE1.

11.11 DATA FRAMES ON THE S/T BUS AND INFO SIGNALS

The polarity of the Framing bit (F) is always positive, and it marks the beginning of the frame. It is followed by the baLancing bit (L), which is always negative. This offsets the positive voltage of the framing bit. Balancing bits are used to keep the current on the S/T bus at zero amps. In the NT to TE direction the auxiliary framing (Fa) bit is set to 1 in every fifth frame and the Multiframing (M) bit is set to 1 in every 20th frame. The Fa and M bits are 0 in all other frames. The Fa and M bits group frames and help keep the line synchronized. The N bit is used to offset the voltage of the Fa bit and serves as the balancing bit for Fa bits. The Activation (A) bit is used by the NT to convey to the TE that the interface is active and operational. Remember that with pseudoternary signaling, a 1 bit is recognized as the absence of voltage and a 0 bit is the presence of 750 mV.

When a TE is not in use, no voltage is applied to the line. This is referred to as *INFO 0 signal*. When a TE (or TA) is activated, it transmits a positive 750 mV (a binary 0), a negative 750 mV (a binary 0), and six 1s. These 8 bits (00111111) are transmitted continuously at 192 Kbps from the TE to the NT1 and are known as the *INFO 1 signal*. The INFO 1 signal tells the NT1 that a TE (or TA) wants to activate the line (S/T bus). The NT1 acknowledges by returning 0s in all A, B, D, and E slots. This signal transmitted from the NT1 to the TE consisting of all 0s in the A, B, D, and E slots is known as an *INFO 2 signal*. The TE transceiver quickly synchronizes on the alternating voltages of the stream of 0s. The TE begins sending data over the B- and D-channels. This is known as an *INFO 3 signal*. The TE simply says okay, I have synchronized and here is some data. The NT1 recognizes the incoming data from the TE and sets the A bit to 1 as it transmits back to the TE1. This is called an *INFO 4 signal* and indicates the TE is activated and BRI is operational. Each TE has a unique identification number. When the TE sends the INFO 3 signal, it begins by sending its identification number. As the NT1 receives an ID, it echoes the ID number back to the TE using the E bit slots on the I.430 frame. The TE receives these bits and uses them to verify that it is logically attached to the NT1. Several TE devices can be in contention for connection to the NT1. As they transmit the D-channel INFO 3 signal, the NT1 will accept the information and respond back to one device. It alerts that device that it is attached by sending the ID of the attached device in the E bit locations. All other TE devices will quit sending because they did not receive an E-channel echo of their D-channel bits.

11.12 TERMINAL EQUIPMENT IDENTIFIERS AND SERVICE PROFILE IDENTIFIERS

Because the S/T bus accommodates 2 bearer channels, only two devices (out of the eight possible devices that can be connected to the S/T bus) can be active at the same time. Each device is interfaced to the NT1 by a TE or TA. To identify which of the eight devices is using one of the B channels, each TE and TA is assigned a *terminal equipment identifier (TEI)*. The TEI is set on some devices by the use of jumpers, dip switches, or thumbwheels. Some devices have a display and will prompt you to use the keyboard to enter a TEI. Still other devices will have the TEI assigned automatically by the central exchange when the device is used for the first time. The TEI is contained in the second layer's address field. It is 7 bits carried in the second octet of the address field. Seven bits allows for identifying 127 different devices ($2^7 = 128$). Address 0 is not used. Address 127 is used to broadcast a message to all devices.

When a TEI is assigned automatically, it uses messages via LAPD to get a TEI assigned. When the TE1 (or TA) is plugged into the S/T bus, it immediately requests a TEI from the network. The TE sends an Unnumbered Information Frame (I). The contents of the information field indicate this is a TEI request. The address field indicates the TEI is 00 or 127. This indicates to the NT1 and LE that the TE does not have a TEI. The network receives the TEI request and returns a response containing a TEI assignment as part of the information field. TEIs are associated with a *service profile identifier (SPID)* in the database of the LEC central exchange. Automatic initialization cannot take place if the customer has not entered a SPID in the TE. The customer receives the SPID assignment verbally from the LEC customer service center. Only after the customer has manually assigned the SPID to the TE can it initialize automatically. If only one device is attached to the S/T bus in a point-to-point (rather than a multi/point) configuration, a SPID is not needed. The device is identified by inputting the directory number in the TE. With multipoint, each device receives a directory number and SPID. SPIDs have several formats such as: area code + telephone number + a one- or two-digit suffix code + TI—for example, 913 5556666 01 00. The TI will be changed from 00 as it gets an automatic assignment upon initialization.

If only one TE1 were placed on the S/T bus, it could be up to 3300 ft from the NT1. When multiple devices are placed on the S/T bus, the length of the S/T bus from the NT1 is limited to 650 ft if using 150-Ω impedance cable, and each device must be no more than 30 ft from the S/T bus. The D-channel is informed which device is assigned to the bus, and this information is conveyed over the D-channel to the central exchange. The D-channel lets the central exchange know, which device has been assigned to the DSL and which B-channel within the DSL has been assigned to that device. Using the SPID for that device, the switch knows which service to provide. A word of caution here. It has been reported that some of Northern Telecom's DMS-100 do not support eight logical devices since they used SPIDs to identify the two B-channels. I am sure that Northern Telecom would have corrected such a problem by now. SPID configurations are dependent on whether a BRI is provisioned as a National ISDN-1 line or as a custom standard. DMS-100 switches now adhere to the National ISDN-1 standard, and AT&T is a custom standard. Additionally, DMS-100 is configured as a multipoint configuration. I believe that the Northern Telecom DMS-100 is suffering from rumors about a past problem because the DMS-100 makes use of the TI to automatically configure TEs and supports multiple terminals. The point to be made here is you should verify ISDN capabilities with your LEC and ISDN vendor before proceeding with ISDN. It may even be possible that the LEC does not offer ISDN in your exchange.

11.13 CONVERSION OF S/T BUS DATA TO U REFERENCE POINT REQUIREMENTS

As mentioned earlier, the DSL, 2B1Q line format over the local loop is using frames that contain 240 bits. Each frame on the S/T bus is 48 bits. The DSL operates at 160 Kbps and the S/T bus is operating at 192 Kbps. The difference in line rates is the difference in overhead. The NT1 has the responsibility of taking the 144 Kbps (2B+D) user information from the S/T frames and placing them in the DSL frames. The 48-bit S/T frame is composed of 16 bits from B1, 16 bits from B2,

4 bits for D, and 12 bits of overhead. Since the S/T interface is running at 192,000 bps, the time for each bit is as follows: $1/192,000 = 0.0000052083$ s. Thus, 48 bits = 0.000250 s. Since one frame of 48 bits takes 0.000250 s, we can transmit 4000 frames per second ($F = 1/T = 1/0.000250 = 4000$). Since each frame contains two 8 bit words for both B-channels, this frame rate satisfies the need to transmit 64,000 bps for each channel—that is, 4000 frames $\times$ 16 bits B1 = 64,000. ISDN may be used for analog signals by using a TA to convert the analog output of a phone into a digital bit stream. The TA will contain a codec to perform the analog-to-digital conversion. Remember that the output of a codec is 64,000 bps and the B-channels must operate at 64,000 bps to ensure that every bit generated by a codec is captured.

The DSL operates at 160 Kbps, and the time for each bit input to the 2B1Q coder/decoder is: $1/160,000 = 0.00000625$ s. Each frame is 0.0015 s ($0.00000625 \times$ 240 bits) = 0.0015 s. The DSL will convey 666.66667 frames each second ($1/0.0015$). Each frame contains 12 B1 samples of 8 bits per sample (96 bits), 12 B2 samples of 8 bits per sample (96 bits), and 12 samples of 2 bits per sample for the D-channel (24 bits). With 666.66667 frames per second, the B-channels will have 64,000 bps ($96 \times 666.6667 = 64,000$ bps) and the D-channel will have 16,000 bps ($24 \times 666.66667 = 16,000$ bps). Thus, even though the S/T and U interfaces are running at different speeds and different frame configurations, the NT1 takes care of conversions between the two interfaces to ensure correct passage of 2B+D user information from terminal equipment to the central exchange.

11.14 TERMINAL EQUIPMENT CONNECTIONS TO NT1

The terminal side of an NT1 can connect to an NT2 device via a T interface. When an NT1 interfaces to an NT2, the NT1 has a PRI to the local exchange. When an NT1 only connects to TE1s or TAs, the NT1 has a BRI to the LEC exchange. *Terminal equipment 1* or *TE1s* are terminal devices that are ISDN compliant. TE1 devices include: ISDN telephones, PCs equipped with a TA card in one of their expansion slots, and any other device designed to ISDN protocols. TE1s connect directly to the NT1 via the S/T bus. Terminal equipment not in compliance with ISDN protocols (*Terminal equipment 2* or *TE2*) must wire to a TA. TE2 devices such as telephones, personal computers, dumb terminals, and printers would wire to a TA. The TA will perform a physical interface change as well as doing conversion of analog to digital, protocol conversion, speed conversion, and multiplexing. There is not one do-it-all TA. There are several TA designs to accommodate the variety of different devices you may wish to connect to ISDN. The TA for a telephone will convert analog information coming in over two wires into a DS0 signal sent over the transmit portion of the S/T bus and will take DS0 signals received over the S/T bus, convert them to analog, and send them to the phone. The TA for a PC will accept digital data over the EIA-232 connection from the PC and map these bits into the TDM transmit of the S/T bus. It will map bits received over one of the ISDN channels to the PC. One side of the TA connects to a TE2. The other side of the TA can connect via an S interface to a NT2 or via the T interface to a NT1. Most ISDN service was initially deployed in large businesses through their PABX. When terminal equipment is used behind a PABX, the NT2 is needed as well as a 23B+D PRI NT1 for LEC interface. Why run a BRI from every TA or TE1 to the central exchange via a BRI NT1? Using a NT1 with PRI to the central exchange, the PABX distributes the 23B+D capacity of the NT1 to BRI terminal devices via the NT2. Therefore, most terminals connected via an NT2. The NT2 is placed between the NT1 and TE1, TE2, or TA of Figure 11-4 when the NT1 connects to a 23B+D PRI line.

11.15 S/T BUS DESIGN

In the residential market, there is no NT2. TE1s and TAs connect directly to the NT1 via the S/T bus. Most connections to the bus are done using an RJ-45 connector. This is an eight-pin connector that looks similar to the four-pin RJ-11 connector used on telephones. The four-wire S/T bus connects to pins 3, 4, 5, and 6. Pin 3 is + transmit and pin 6 is – transmit. Pin 4 is + receive and pin 5 is – receive. Pins 7 and 8 are used when a second power source is needed. Pins 1 and 2 are used when a third power source is needed. Power source 1 is provided over leads 3, 4, 5, and 6. from the NT1. These leads provide a phantom circuit arrangement that furnishes 40 V DC. This is a 1-W power source for all TEs. Power source 2 is used for TEs that may require up to 7 W from the NT1. Power source 3 is used to supply power from one of the TEs instead of from the NT1 (Figure 11-6).

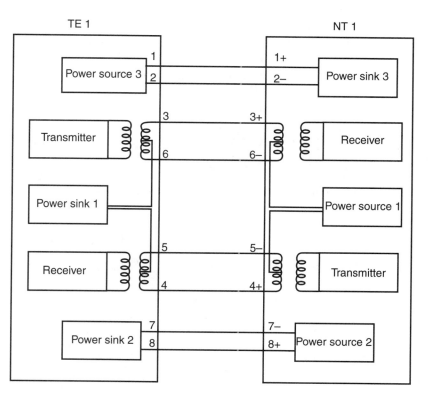

FIGURE 11-6 The S/T bus. The figure shows a drawing of the S/T bus. Power source 1 of the NT1 supplies power to the TE1 by using a phantom arrangement over the transmit and receive pairs. The TE1 can use its power supply 3 to supply power to the NT1 or other devices. The power sink for a TE1 is pins 7 & 8. Pins 7 & 8 of another TE1 can connect to pins 7 & 8 of the NT1 to receive power from the NT1's powersource 2, or the pins 7 & 8 of the second TE1 could connect to pins 1 & 2 of the S/T bus to receive its power via source number 3 from the first TE1.

11.16 LAYER 2 SIGNALING

Except for our discussion of automatic assignment of TEI, up to this point we have been involved with physical connections, interfaces, and data frame structures. All of these are similar to OSI layer 1 functions. Automatic assignment of TEI is a layer 3 function. Signaling at layer 1 has been to send various patters of on and off voltage conditions as signals between the TE1 and NT1. B-channels are strictly a level 1 layer operation when used in a circuit switched mode. One B-channel is connected over a network to a distant B-channel and they exchange digital signals over a 64 Kbps loop. The B-channels can use any protocol as long as the protocol on both ends are the same. The D-channel is part of the layer 1 physical connection but also provides layer 2 and layer 3 functions.

11.17 LINK ACCESS PROCEDURE ON THE D-CHANNEL

Instead of relying on physical signaling using voltage conditions, the D-channel uses messages for signals. The ISDN D-channel is out-of-band signaling. This signaling takes place between CPU chips and is logical rather than physical. One chip circuit tells another chip what action to take by issuing commands in the message strings between the two chips. Programming is burnt into the chip telling it what action to take, based on the message received. This type of signaling, which uses messages to direct the actions of a CPU, takes place at layer 2 and layer 3. The ISDN Data Link Layer (layer 2) protocol is called *Link Access Procedure on the D-Channel (LAPD)*. This protocol (software), defines the logical (not physical), connection between the terminal equipment (TE1 or TA) and the network (NT1 and LE). CCITT Recommendations Q.920 (I.440) and Q.921 (I.441) describe the general principles and operational procedures of LAPD. This procedure is also known as *Digital Subscriber Signaling System No. 1 (DSS1) Data Link Layer* and is a bit-oriented protocol, similar to the X.25 *Link Access Procedure Balanced (LAPB)*. The primary function of the data link layer is to provide an error-free communication link between devices and to convey information between layer 3 of these devices using the D-channel. To accomplish this, the data link layer software must do the following:

1. It must signal the beginning and end of a data packet by framing the user data.
2. It must maintain the sequential order of transmitted frames.
3. It must provide a destination and originating address.
4. It must return an acknowledgment of received frames to the transmitter.
5. It must be able to detect bit and frame sequence errors.
6. It must provide flow control to control the speed of data transfer.

One difference between the LAPB of X.25 and the LAPD of ISDN is that LAPB is a point-to-point protocol. LAPD is a point-to-multipoint protocol, which allows its use between a NT1 and multiple TEs. LAPD uses *Carrier Sense Multiple*

Access/with Collision Resolution (CSMA/CR) to allow multiple TEs on the S/T bus. I must stress again that LAPD is a protocol (software) operating at layer 2, to carry layer 2 and layer 3 messages. LAPD is a *Logical link* between CPUs in the TE and NT1 or LE. The CPUs can send and receive messages that are framed according to LAPD. A LAPD frame contains the following fields:

1. FLAG Signals the beginning and end of a frame. The flag consists of the bit pattern 01111110.
2. ADDRESS Contains the *service access point identifier (SAPI)* and TEI that is sending or is to receive the message.
3. CONTROL Identifies the type of frame.
4. INFORMATION Contains layer 2 and 3 messages.
5. FRAME CHECK SEQUENCE Uses CRC to check for errors.

The control field indicates the type of frame being transmitted. There are three types of frames in LAPD: information (I), supervisory (S), and unnumbered (U) (Figure 11-7). Information frames contain layer 2 and layer 3 messages. Supervisory frames control the flow of information frames. Unnumbered frames are used to initiate and terminate a logical link connection.

The address field contains the TEI and the SAPI. The SAPI identifies the service required (voice, switched data, packet data, and so on), and also identifies which B- or D-channel is being used to carry the customer's information. The SAPI is the logical connection between layer 2 and layer 3. Think of the SAPI as a post office box. Messages between layers are placed in the box. Each different service will have a different SAPI or different post office box. This allows each TE to tell the LE what process will take place from the TE and which channel the TE is using.

The address field consists of a TEI, which tells the LE which TE device is active, and a SAPI to tell the LE which process the TE will be using. TEI and SAPI taken together, as one address, are called the *data link control identifier (DLCI)* (Figure 11-8). Some common SAPI values are 0 for call control over circuit switching, 1 for packet mode communications over the PDN, and 16 for X.25 data over the D-channel. The SAPI address or post office box allows the layer 3 software of the TE to communicate with layer 3 software of the LE in the central exchange to provide the service desired. The Layer 3 software is embedded in the information field of the LAPD frame (Figure 11-7a).

Layer 3 software will vary depending on which SAPI we have used. Signaling is sent from layer 3 of the TE, through the NT1, to layer 3 of the LE. Each signal consists of messages containing information elements. There are about 33 different signaling messages in ISDN, such as SETUP, CONNECT, DISCONNECT, ALERTING, PROGRESS, SUSPEND, RESUME, RELEASE, INFORMATION, and so on.

If a B-channel is to be used for the duration of a call by only one TE, it is set up as a circuit switched connection. A circuit switched connection is started by sending a SETUP message over the D-channel. The TE sends the LE the desired services, the address of the called party, and the B-channel to use. If the central

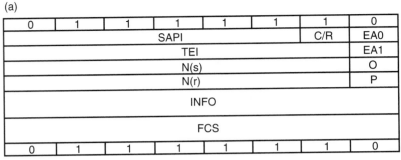

(a)

Information Frame

(b)

Supervisory Frame

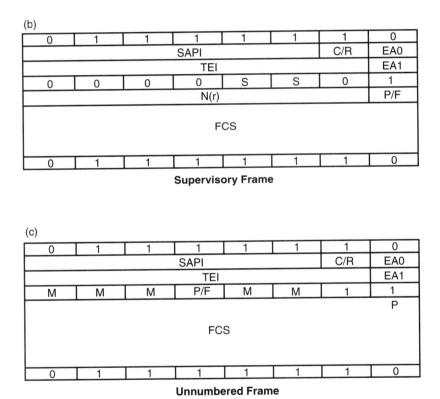

(c)

Unnumbered Frame

FIGURE 11-7 LAPD frames.

exchange is satisfied that the SETUP message is valid, it returns a CALL PRO-
CEEDING message. A SETUP message is sent to the called party, an ALERTING
message received, a CONNECT message sent, and then a CONNECT AC-
KNOWLEDGMENT sent. Layer 3 on the D-channel takes care of getting all service requests established (Figure 11-9). When both ends of the call have received
a connect message, the two TE devices can communicate and voice or data can
be passed over the B-channel. Signaling messages vary depending on the service

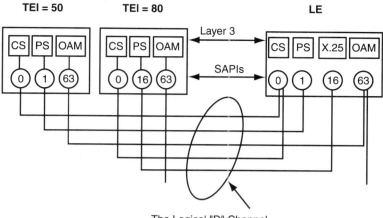

FIGURE 11-8 Layer 3 communication between the TE1 and LE.

The Logical "D" Channel

desired. Data connections can be on a dial-up basis over the Public Data Network (PDN), packet mode over the B-channel, and packet mode over the D-channel.

11.18 CALL SETUP OVER THE D-CHANNEL

The D-channel provides instructions to the network on what to do with the B-channels. It provides TEIs, SAPI, call setup instructions, and call termination instructions. The central exchange knows which services to provide based on the TEI, SPID, and SAPI. For voice calls, the TEI and SPID will instruct the switch to use the PSTN (voice) network. For data calls over one of the B-channels, the TEI and SPID will instruct the central exchange to connect the B-channels to the PSTN. The PSTN will use a 64-Kbps channel path in the PSTN to connect to the called ISDN terminal's ISDN line circuit in the terminated LEC central office. The caller can use both B-channels for data, if desired, by placing a call over each B-channel and bonding the two channels together. This is accomplished by a special command in the ISDN modem's initialization string. If you have devices that transmit packet data, the B-channel will be connected to the PDN.

The D-channel is in charge of getting end-to-end connections established. Once the originating D-channel instructs the local switching exchange of its desires, the local exchange takes over. The local exchange issues switching instructions over the SS7 network to get the voice or data call established. The far-end terminal will communicate via its D-channel to its local office. Communication between the end offices is handled by the SS7 network. Control Information on the D signaling channel is transferred over to SS7 at both end offices.

When on a data call, the B-channel will provide end-to-end 64-Kbps full duplex data over each B-channel either in a switched mode over the PSTN or in a packet switched mode over the PDN. Since the D-channel of BRI does not require all 16 Kbps

Originating TE	LE Originating CO	ISDN Network (PSTN, PDN, and SS7 Networks)	LE Terminating CO	Terminated TE
Originating Call Setup:				
SETUP- - - - - >	SETUP- - - - - >	SETUP- - - - - >	SETUP- - - - - >	SETUP - - - - - >
<- - Call Proceeding	<- - Call Proceeding	<- - Call Proceeding	<- - Call Proceeding	<- - Call Proceeding
<- -Alerting	<- - Alerting	<- - Alerting	<- - Alerting	<- - Alerting
<- -Connect	<- - Connect	<- - Connect	<- - Connect	<- - Connect
Connect Ack - ->	Connect Ack - ->	Connect Ack - ->	Connect Ack - ->	Connect Ack- ->
Transfer of Voice or Data now occurs over the B-Channel				
Disconnect call after conversation is complete or after data has been transferred over the B-Channel by issuing the following Commands over the D-Channel:				
Disconnect - - - >	Disconnect - - - >	Disconnect - - - >	Disconnect - - - >	Disconnect - - - >
< - - - Disconnect	< - - - Disconnect	< - - - Disconnect	< - - - Disconnect	< - - - Disconnect
Release - - - - >	Release - - - - >	Release - - - - >	Release - - - - >	Release - - - - >
< - - - - Release	< - - - - Release	< - - - - Release	< - - - - Release	< - - - - Release

FIGURE 11-9 Signaling message flow over the D-channel.

for control of the network, low-speed packet data can be sent over the BRI D-channel using X.25 protocol. ISDN terminals are assigned a telephone number for voice that is of the following form: area code + exchange code + terminal digits—for example, 913 + 555 + 1212. ISDN terminals may also be assigned a data address (telephone number) of this form: zone code + country code + PDN code + network terminal number—for instance, 3 + 10 + 4 + 5556660000. These 14 digits identify a unique terminal on the PDN.

11.19 LEC CENTRAL OFFICE LE AND TE

The class 5 central exchange provides the line-interface circuitry for either a BRI or PRI line. The local loop interfaces to the line equipment (LE) at the U reference point. The switching system comprises the exchange terminator (ET), packet handler, signaling processor, and switched circuit access to the PSTN and PDN. The interface point between the LE and ET is called the *V reference point*. LE is part of the class 5 switching system and LE is often considered part of the ET. The V interface is more of a logical interface than a physical interface because the LE may be switched to many different devices, depending on the service desired.

For switched packet data calls over the B-channel, the LE can be connected to a packet handler at the class 5 exchange or it can be connected to an ISDN port of the Public Switched Packet Data Network (PSPDN or PDN). If connected to a packet handler, the packet handler connects via X.75 protocol to the PSPDN. For voice calls and nonpacket data calls, the ET will connect the B-channel to the PSTN. An X.25 TE device can also connect to the D-channel to access the PDN.

The ET at the central office provides access to all services desired by a TE1. Many businesses are using the ISDN D-channel as a way to do credit card validation from point-of-sale terminals (POST). They can purchase D-channel only capability or purchase a complete BRI. With BRI they can use telephones, fax, POST credit card scanners, and alarm systems all over one BRI. The TEs will be connected to the proper service by the ET.

11.20 PRIMARY RATE ISDN AND BROADBAND ISDN

For Primary Rate ISDN (PRI) (23B+D), the customer must provide an NT1 and possibly an NT2, or a combined NT1 and NT2, which is called a **Network Termination 12 (NT12)**. NT12s are used in Europe and some PABXs. NT2's are used to interface PABXs and LANs to ISDN. PRI provides 23B+D service. As with BRI, all signaling is handled by the D-channel. The 23 B-channels can be used for voice or data. PRI appeals to businesses that use PABXs or ACDs. Each B-channel is 64 Kbps, and the D-channel is also 64 Kbps. The basic frame is 193 bits and is the same DS1 frame used for T1. The PRI line rate is 1.544 Kbps (the same rate as T1 carrier). The operational difference between PRI ISDN and T1 is that the ISDN system was designed to handle a full 64-Kbps data rate over each channel by means of out-of-band signaling. T1 uses a bit-robbed signaling technique. ISDN uses a common channel (the 24th channel) for all D-channel signals.

The European version of ISDN uses the E system line rate of 2.048 Mbps. This system provides 30B+D. Conversion of all class 5 trunk signaling to SS7 is necessary for a T1 carrier system to achieve 64-Kbps data capability. PRI B-channels can be added to achieve higher data rates. This technique is used to provide low-resolution video to small businesses and residences. The two B-channels of BRI can be combined to provide 128 Kbps for a video signal. Quite a few businesses are finding this application for ISDN useful. They can feed the signal from surveillance or security cameras to a remote location such as a security company. The LECs also offer other combinations of B-channels over a PRI to achieve higher data rates. Six B-channels can be combined to provide a data rate of 384 Kbps (6 × 64 Kbps = 384 Kbps). This service is called an *H0 channel*.

BRI and PRI services are called *narrowband ISDN* to differentiate these services, provided over wire, from the *broadband ISDN (B-ISDN)* services provided over fiber optic cable. The layer 2 protocol (LAPD) and layer 3 signaling and control protocols of BRI are the same for PRI ISDN and B-ISDN. Narrowband ISDN uses synchronous transmission, but B-ISDN is provided primarily over *asynchronous transfer mode (ATM)* facilities. ATM refers to a switching and transmission service. Data switches in the ATM network provide for the routing of data in the network, and fiber optic cable connects the ATM switches together. ATM is the underlying fabric for *Switched Multimegabit Data Services (SMDS)*. SMDS is a network service developed by Bellcore for the RBOCs. The LECs of the RBOCs use SMDS to establish *Metropolitan Area Networks (MANs)* for their large-business customers.

B-ISDN can provide bandwidth on demand services using the ATM switches and Synchronous Optical Network or SONET (fiber optic) transmission facilities. SONET is a multiple-level protocol used to transport high-speed signals. It allows multiplexers/demultiplexers to insert or drop out one DS0 signal from the high-speed line without having to demultiplex the whole facility. SONET is the vehicle that makes B-ISDN and the services it can provide possible. These services can be anything from videoconferencing and multimedia computing to medical imaging, broadcast TV, and MAN services.

Perhaps B-ISDN will have more success in the marketplace than narrowband ISDN since B-ISDN offers higher speeds (100 Mbps) and can be used to provide WAN connections or real-time video signals. These high speeds may appeal to business users, but does a residential customer need more than two 64-Kbps channels? For voice calls, the codec output is 64 Kbps. If a PC is making a circuit switched data call, the speed of the PC may be slower than the speed of a B-channel.

The UART that interfaces a PC to its serial port is usually a 8250 or 16550 UART. Although a 16550 UART is capable of 115,200 bps, its speed is limited by DOS and therefore by Windows 3.*xx* to 19,200 bps. Windows 95 can take full advantage of the speed offered by a 16550 UART because Windows 95 has the serial interface drivers required to achieve 115,200 bps and higher. If you are using Windows 3.1*x*, you will need to purchase a serial accelerator card and associated software to attain speeds higher than 19,200 bps.

The TA that interfaces a PC serial port to ISDN must use CCITT Recommendation I.465 to adapt PC speeds to the 64 Kbps of a B-channel. This protocol frames user data so that the receiving terminal knows where the data starts and ends and pays no attention to idle channel time stuffing bits or absence of bits. This protocol is used when the B-channel is used in a circuit mode. When the B-channel is used in an X.25 packet mode, the TA must convert the X.25 of the TE to ISDN protocol. A packet handler at the central exchange or at the entrance portal to the PDN will convert the ISDN protocol back to X.25 protocol.

I have tried to provide you with some insight into ISDN. You should now have a general view of what ISDN is and how a call is placed over this network. Before implementing ISDN technology as a solution to your communication needs, you need to do research on the different applications available. You should also talk with your ISDN vendors and LEC. Prior to divestiture and deregulation, you could do one-stop shopping at your LEC. The LECs were responsible for the network and all the services they sold. If you purchased a service from them, they would make it work. Today, there is no single source provider, and when trouble with a service occurs all the providers point the finger at each other.

You can no longer rely on the LEC to provide technical expertise. *You, the telecommunications manager, will be expected to be the expert. You must have a thorough technical knowledge of your telecommunications network and equipment.* The 1996 Telecommunications Reform Act allows LECs to reenter the CPE market without using a subsidiary. This act also allows LECs to manufacture equipment. Hopefully, the LECs will seize this opportunity to become a one-stop shop for all ISDN equipment and services. They have the resources and marketing expertise to design, develop, market, install, service, and provide technical assistance on ISDN products and services.

11.21 ISDN AND THE PERSONAL COMPUTER

The LECs have been slow to deploy Basic Rate ISDN (BRI). Their most aggressive marketing of BRI has been in 1996. Many people have been installing an ISDN line to achieve higher data speeds when connecting their PC to the Internet. As mentioned in Chapter 9, modem technology is developing to a point where a modem can achieve a 56,000-bps data rate and with compression can achieve a 224,000-bps end-to-end connection. This speed approaches the line rate provided by BRI ISDN (128 Kbps), but with 4-to-1 compression, the end of speed of ISDN is 512,000 bps. This speed and the fast modem speeds are usually faster than the serial port of a PC can handle. The speed is also faster than most circuit paths in the PSTN can support. Therefore, ISDN and most modem connections will fall back to a lower speed and throughput rate. The 56K modem is actually not designed for an end-to-end dial-up connection between two PCs. This modem is designed to interface with 56K technology at an ISP. The ISP has digital connections to the PSTN and does not interface via an analog line circuit. The ISP eliminates the analog-to-digital conversion of the codec in a line card. This allows it to run at a faster speed. The 56K modem on one end of a connection must be connected directly to the digital highway of the PSTN to achieve 56K speed.

ISDN can achieve a data rate of 64,000 bps on each channel. By combining both B-channels on a BRI, it is possible to achieve a data rate of 128,000 bps. Thus, even though a modem appears to be almost as fast as ISDN, the actual operating speeds are lower than ISDN. The internal ISDN card for a PC is also equipped with a compression algorithm to speed the transfer of data. With a 4-to-1 data compression, the end-to-end speed for ISDN running both B-channels in a bonded mode is 512,000 bps (4 × 128,000).

It will not be long before someone develops an internal PC ISDN card that will interface to the dynamic memory mapping of a PC in order to achieve a fast data rate transfer between the ISDN card and the data bus of the PC. I think future advances in the ISDN card for a PC will lead to the preference of ISDN modems over analog modems. ISDN will soon be faced with competition from cable modems provided by cable TV companies. The telephone companies are also beginning to deploy *Asymmetric Digital Subscriber Line (ADSL)* technology, and this technology may end up replacing ISDN.

11.22 ASYMMETRIC DIGITAL SUBSCRIBER LINE

Pacific Bell has issued press releases stating it was going to bring coaxial cable and fiber optic cable to many of its customers by the year 2000. In recent press releases it has seemed to back away from the earlier plans for coaxial and fiber cable installation to most customers. Instead it plans to embrace ADSL technology. According to the president of PAC Bell's Business Communications Services, ADSL was to begin being deployed in 1997. He stated that they were still laying fiber, but only to areas with a high concentration of customers.

Every LEC has a large investment in the twisted-pair copper wire local loop. ADSL is a technology that allows the LEC to retain that investment. Pacific Bell plans to use ADSL on copper to deliver T1 speed (1.544 Mbps) to customers. This LEC seems to be in the forefront of delivering ADSL, but other Bell LECs are also conducting field trials of this technology. GTE is currently testing ADSL technology in Texas to access the Internet.

11.23 THE LECS' ANSWER TO CABLE COMPANIES IS ADSL

ADSL was developed by Bellcore Labs of Morristown, New Jersey, for the RBOCs. It was originally created as a vehicle to carry video signals over twisted-pair copper wire. This technology would allow the LECs to provide cable TV services in direct competition with cable TV companies. Asymmetric digital data does not have the same data speed in both directions. ADSL has reached speeds of 6.14 Mbps (that's 6 million bits per second!) in the download direction and 640 Kbps in the upload direction. The download speed is 185 times faster than a 33.3-Kbps modem with no compression. If the modem is a 33.3 Kbps modem and is using 4-to-1 compression (230,000 bps), ADSL is still almost 27 times faster.

Cable modems are capable of 27 Mbps in the download direction and 1 Mbps in the upload direction, but the modem circuits will be shared circuits and overall speed will depend on the number of active users. During periods of heavy use by several users, the cable modem will not provide significantly more than the 6.14-Mbps download capability of ADSL. Thus, ADSL provides the

LECs with a vehicle that can go head to head with a cable modem. A basic comparison of the data speeds provided by the various technologies is that to transfer a file of 6 million bits:

1. A 33.3-Kbps modem requires about 6 min for the download.
2. A BRI ISDN will require about 1.5 min for the download.
3. An ADSL will require about 1 s for the download.
4. A cable modem will also require about 1 s for the download.

The goal of every Internet user is faster data speeds. The transfer of HTML files every time you select a new address takes time. The Internet user spends much of the time waiting for these file transfers. The LECs, IECs, and cable companies are beginning to compete for the Internet user's business. Some Bell companies have been cutting the prices for ISDN to entice customers to use their service. Cable TV companies have been slow to enter the Internet access market. The cable modems encountered some technical problems, but those appear to have been corrected, and we should see the more aggressive cable TV companies entering this market soon.

The terminal equipment used for ADSL is about $1000 and the service rates will be higher. For these reasons, I expect ISDN to continue attracting customers. Although the ADSL is 90 times faster than ISDN, it will take a year or two before the technology is deployed by ISPs. Remember, we can only achieve fast speeds if both ends of a circuit are deploying the same technology. The ISPs have been slow to provide ISDN access to the Internet. For that reason, many people *must use* a modem because the ISP is using modems for access.

Over the next couple of years, the cost of ADSL equipment and service charges from the LECs will go down. This will lead to wider acceptance of ADSL. It appears that sometime in the near future ADSL will bury ISDN, but for the next couple of years ISDN is the best bet for someone who needs more speed than a regular phone-line modem can provide. It remains to be seen if the LECs will seize the opportunity to aggressively market ISDN before it becomes yesterday's Model T.

11.24 ADSL AND VIDEOCONFERENCING

Real-time video uses a compression technique called MPEG2. This technique can deliver a high-quality video with 6 Mbps. Bell Atlantic ran some trials with ADSL and MPEG2 and discovered that it could only serve about 25% of its market with these technologies due to physical plant problems. Bell Atlantic has turned to wireless services as a method to deliver videoconferencing. Bell Atlantic and NYNEX have invested $100 million in CAI Wireless Cable Company with an option to invest $200,000 million more. Wireless cable (for example, Radio Wave Facilities) companies also have their eye on the Internet access market.

11.25 ANOTHER FIGHT OVER STANDARDS

One reason ISDN languished for so long was because several manufacturers insisted on using their proprietary standards. The same thing is happening to ADSL. There are two competing operating systems for ADSL: (1) *Carrierless Amplitude and Phase Modulation (CAP)*, and (2) *Discrete Multitone Technology (DMT)*. In 1996, the American National Standards Institute (ANSI) adopted an ADSL standard based on the DMT architecture. The CAP camp has been reducing prices of their equipment, and presently the trials of ADSL are using both technologies. The RBOCs have requested that ANSI recognize CAP. The standards war is on again! It remains to be seen how long it will take to settle on one standard.

CAP uses a technology similar to the *Trellis Coded Modulation (TCM)* technology used by high-speed dial-up modems. CAP uses two transmitting frequencies (one in each direction). DMT uses 256 different carrier frequencies. Each carrier has a 4-Khz bandwidth. DMT appears to be faster and works better on low-quality circuits. The Paradyne division of AT&T has chosen the CAP technology for manufacture of ADSL terminal equipment. DMT technology is being provided by smaller manufacturing firms. Therefore, I expect that CAP technology will win out in the standards war, unless AT&T or some larger player decides to jump on the DMT bandwagon

11.26 WHERE CAN ADSL BE DEPLOYED?

Remember that the loop limit for BRI ISDN without a repeater is 18,000 ft (approximately 3 mi). Longer loops are possible, but digital regenerators must be on the line. The same is true for all digital signals placed on twisted-pair copper wire. A T1 carrier system transmits 1,544,000 bps. When this signal is placed on copper wire, the signal must be regenerated every mile. ADSL will work on an 18,000-ft (approximately 3 mi) loop without a repeater, if the data rate is below 1.5 Mbps. Higher data rates will limit the length of the local loop. A speed of 6 Mbps will limit the distance from the central office to about 1.5 mi.

11.27 HOW DID THEY DO THAT?

The data speeds achieved by ADSL are staggering when you consider they are using the twisted-pair local loop for a medium. The higher data speeds that are being achieved by regular 57.6-Kbps modems and by ADSL modems are due to advances in digital signaling processors, codecs, filters, transformers, echo cancelation technology, and compression technique algorithms. Another very important factor that allows the attainment of high data rates is the vast improvements that have been made in the local-loop plant. Most outside plant is fairly new and is buried in the ground to protect it from adverse weather conditions.

The semiconductor industry will continue to improve the ADSL chipsets. This will lead to even higher data rates. ADSL uses forward error-correction techniques. The microprocessors and software used in today's modems to control their operation has provided them with tremendous capabilities. Whether it is a regular phone-line modem, an ISDN phone-line modem, or an ADSL phone-line modem, the advances in technology over the past several years have been tremendous. We should all thank Hayes for the start of smart modems.

11.28 SUMMARY

ISDN is an end-to-end digital network that expands the digital PSTN to the customer's premises. The final leg of this digital network is the DSL. The DSL uses a 2B1Q protocol to achieve a 160-Kbps transfer rate at 80 Kbaud. Since ISDN is designed to handle digital signals, digital data from a PC can be placed via an ISDN modem (NT1) over the facility. Before voice can be placed over ISDN, it is converted by a codec to a digital signal of 64,000 bps (DS0). The codec chip can reside in a TA or in the input port to an ISDN modem (NT1), which serves a telephone.

Basic Rate ISDN (BRI) is provided over the 160 Kbps DSL. Of this rate, the DSL uses 16 Kbps for synchronization and timing. The remaining 144 Kbps is used to provide to two bearer channels and one delta channel (2B+D). The D-channel is primarily used to set up calls for the B-channels, to monitor the circuit, and to provide for end-to-end message signaling. The D-channel on each end of an ISDN call is connected to the SS7 network at the exchange terminator. Thus, for ISDN to work, SS7 is mandatory.

BRI has three basic interface or reference points: U, S/T, and R. The U reference point is where the two-wire DSL from the LEC connects to the NT1 of the customer. The U reference point serves as the Demarc. The S reference point is where an ISDN-compliant device (TE1) connects to an NT1. The T reference point is where a terminal adapter (TA) connects to an NT1. The R reference point is where a non-ISDN-compliant device connects to a TA. An ISDN device (such as an ISDN modem) will often contain a NT1, a TE1, and a TA as part of its internal circuitry. The only interface we will see is the connection to the ISDN phone line from the LEC (U interface), an RJ-11 connection for a telephone (R interface), and possibly an RS-232 connection for the computer (R interface). The S/T interface will be internal to the circuit card.

When an ISDN modem is installed in a PC, the customer receives a SPID from the LEC. The customer manually assigns the SPID to the software controlling the modem when the modem is installed. When the first ISDN connection to the central office is made, the central exchange checks the SPID database and will automatically assign a TEI to the ISDN modem. The TEI is used to inform the central office, via the D-channel, what service to provide (voice, switched data, packet data, and so on) for the terminal equipment attached to the NT1.

The D-channel of ISDN operates at the lower three layers of the OSI model (physical, data link, and network). The physical layer conforms to I.430 for BRI and to I.431 for PRI. The data link layer conforms to Q.921 (link access procedure on the D-channel or LAPD). The network layer conforms to Q.931 (call setup procedure over the D-channel). The D-channel transmits signaling and information messages from the NT1 to the LE. These messages are used to set up calls for the B-channels. The D-channel provides instructions to the PSTN on what services to provide the B-channel.

BRI and PRI are called narrowband ISDN and use synchronous transmission. Broadband ISDN is asynchronous transfer mode transmission over fiber optic cables. Broadband ISDN provides speeds faster than 100 Mbps and can be used to establish WANs. Another high-speed digital technology being deployed by the LECs is ADSL. ADSL can be deployed to deliver the equivalent of a T1 facility to a customer. It can also be used at higher rates to deliver digital video over the local loop. ADSL has reached speeds of 6.14 Mbps over twisted-pair copper wire. This is 360 times faster than a 33.6 dial-up modem.

REVIEW QUESTIONS

1. How is data transmitted over the PSTN without using ISDN?
2. What part of the PSTN was the first to be converted to digital?
3. Which is faster, a cable modem or ISDN?
4. What is BRI?
5. What type of line is used for BRI?
6. What is the signaling rate on a DSL? How much of this rate is available to the customer?
7. What type of coding is used on the S/T Interface?
8. What type of coding is used on the U interface?
9. What is the baud rate of 2B1Q?
10. What is a quat?
11. If the S/T interface is running at 192,000 bps and the U interface is running at 160,000 bps, what happened to the 32,000 extra bits on the S/T interface?
12. What is a TEI?
13. What is a SPID?
14. What is a SAPI?
15. What is the difference between a TE and a TA?
16. How is signaling conveyed in ISDN?
17. What are the frame types in LAPD?
18. What channel carries all signaling and control information?
19. How does a central office know what type of service or connection to provide on an ISDN call?

20. What services are narrowband ISDN?
21. What services can be handled by ADSL?
22. How does the speed of ADSL for the residence compare to ISDN?
23. How does the speed of an ADSL compare to a cable modem?
24. What is the advantage of ADSL technology to the LEC?
25. What is the difference between DSL and ADSL?

GLOSSARY

Asymmetric Digital Subscriber Line (ADSL) A digital subscriber line technology that can be used to deliver the speed of a T1 line over the local loop. An Asymmetric Digital Subscriber Line has a high bit rate in one direction and a low bit rate in the opposite direction.

Basic Rate Interface (BRI) The basic building block of ISDN. A BRI consists of two 64-Kbps bearer channels and one 16-Kbps delta channel (2B+D). The delta (D) channel is used for signaling and sets up calls for the B-channels. The B-channels are used to carry customer voice or data.

B-Channel The channel that carries the customer's information. This information may be voice, data, or video.

Broadband ISDN (B-ISDN) Provides speeds faster than 100 Mbps and can be used to establish wide area networks. Broadband ISDN is asynchronous transfer mode transmission over fiber optic cables.

D-Channel The channel that carries signaling and control information. The D-channel is used to set up calls for the B-channels.

Digital Subscriber Line Circuit (DSL) Usually refers to a local-loop cable pair that is handling a digital signal using 2B1Q line coding.

Hypertext Markup Language (HTML) A software language used to create documents, such as home pages, for use on the Internet.

Integrated Services Digital Network (ISDN) The use of digital line circuits to provide end-to-end digital service. The Basic Rate Interface (BRI) provides the user with two DS0 channels. The Primary Rate Interface (PRI) provides the user with 23 DS0 channels.

Internet Service Provider (ISP) A company that provides individuals with access to the Internet.

Link Access Procedure on the D-Channel (LAPD) The software protocol used at the data link layer (layer 2) of the D-channel. LAPD uses Carrier Sense Multiple Access/with Collision Resolution (CSMA/CR) to allow multiple TEs on the S/T bus. LAPD is a multipoint protocol.

Narrowband ISDN Designates BRI and PRI, which use synchronous transmission.

Network Termination 1 (NT1) An interface device provided by a customer to terminate an ISDN line to the customer's premises. The NT1 interfaces the customer's ISDN equipment to the Digital Subscriber Line of the LEC.

Network Termination 2 (NT2) An interface device that interfaces devices to a PRI NT1. A NT2 is usually used in a PBX environment. NT2s will only be found in a PRI environment.

Network Termination 12 (NT12) A device combining a NT1 and a NT2.

Primary Rate Interface (PRI) Consists of twenty-three 64-Kbps B-channels and one 64-Kbps D-channel.

Reference Points The points at which various devices are connected in an ISDN circuit. The U interface serves as a demarc. It is where the cable pair from the LEC attaches to a NT1. The S/T reference is the interface where a TE1 or TA connects to a NT1. The R reference is the interface where a non-ISDN device attaches to a TA. The V reference is the interface point between the LE and ET in the central exchange.

Service Profile Identifier (SPID) Contained in a database maintained by the LEC of ISDN line equipment and the services available to each line. This database contains SPIDs. When a terminal is activated, the SPID is used to determine which services a terminal can have access to.

Terminal Adapter (TA) A device that interfaces a non-ISDN device to a NT1 in BRI and to a NT2 in a PRI environment.

Terminal Equipment 1 (TE1) Equipment designed to ISDN standards and for direct interface to a NT1.

Terminal Equipment 2 (TE2) Equipment that is non-ISDN compliant. TE2 equipment must be attached to a terminal adapter.

Terminal Equipment Identifier (TEI) Specific number assigned to each TE1 and TA. The TEI is usually assigned automatically the first time the terminal device is used. LAPD will invoke a query of the SPID database to obtain the TEI. The TEI is used by the LE and ET to determine what service to provide for a B-channel.

2B1Q The coding technique use for a digital subscriber line. This coding technique allows 2 bits to be transmitted at one time by using four distinct voltage levels. Each voltage level can represent a particular 2-bit code.

12

Private Switching Systems

Key Terms

Architecture
Average Busy Hour Peg
    Count (ABHPC)
Average Hold Time (AHT)
Blocked Calls
Busy Hour
Calling Capacity
Central Office (CO) Trunk
    Circuit
Computerized Branch
    Exchange (CBX)

Configuration
Direct Inward Dial (DID)
Direct Outward Dial (DOD)
Equipped for
Expansion Capacity
Moves, Adds, and Changes
    (MACs)
Overlay Programs
Peg Count (PC)
Peripheral Equipment
Port

Port Capacity
Printed Circuitboard (PCB)
Private Automated Branch
    Exchange (PABX)
Private Branch Exchange
    (PBX)
T1 CXR
Traffic
Wired for

PABX is the term used for *Private Automated Branch Exchange*. The more familiar acronym *PBX*, for *Private Branch Exchange*, refers to the first manual switchboards placed in a private business location. When these manual switching systems were replaced by automated switching, they were called PABXs. When the private automatic switching systems use computerized SPC switching, they are called *Computerized Branch Exchanges (CBXs)*. Now that all private switching systems are computerized (CBX), we call all private switching systems PBXs. It is understood that PBX now means a computerized, automatic switching system used in a business location. The technology used in these private switching systems is the same technology used for class 5 local switching systems. They are Stored Program Control switching systems using PCM/TDM switching techniques.

The major difference between an end office switching system and a PBX is the generic program and database. PBXs have many features not found in the local exchange. With CLASS features, central office lines can provide many PBX-type features, but some features remain unique to a PBX, such as least-cost routing and intercom services. Some PBXs are designed to handle about a hundred telephones and ten central office trunks. Other PBXs are designed to handle thousands of telephones and hundreds of trunks. A large PBX can be as large as, or larger than, some small central offices. A small business does not require a PBX. It may only require one or two telephones and is served by running two business lines to the customer. As the size of the business grows, it may be better off using a keysystem. A keysystem is the perfect solution to the communication needs of lawyers, doctors, and small retail stores like an auto parts store. If a business needs more than 100 telephones or 24 central office trunks, a PBX is the right solution.

12.1 CENTRAL OFFICE TRUNKS

A PBX allows callers on the system to call each other without having to use lines to the central office. Calls between stations are switched by the PBX. When a station needs to make a call outside the PBX, the station is provided with access to all central office trunks. Station users dial an access digit of 9, and the switching system will connect their station to an idle *central office (CO) trunk circuit*. The CO trunk circuits in the PBX connect to a local-loop cable pair. At the LEC central exchange, the local loop can connect to either a line circuit or a trunk circuit. If the local exchange switches calls to the PBX based on seven dialed digits, a line circuit is used. With this arrangement, the PBX operator answers all incoming calls.

If the PBX is equipped with *Direct Inward Dial (DID)* trunk circuits, the LEC central exchange will switch to the PBX after receiving three or four digits. The remaining digits are used to provide switching instructions in the PBX. Since the PBX needs to receive some of the dialed digits, a trunk circuit must be used in the local exchange instead of a line circuit. Trunk circuits can pass dialed digits to the local loop for use by the PBX. Line circuits cannot pass dialed digits to the local loop; they can only receive digits from the loop. Thus, central exchange lines will always connect over the local loop to a trunk circuit in the PBX, but the exchange end of the circuit can connect to a line or trunk circuit. The circuit to use in the central exchange depends on whether the incoming calls are answered by an operator or are sent directly to the desired station using DID.

12.2 PERIPHERALS

All PBX *telephones* connect to line circuits in the PBX. Calls to other switches will be connected to trunk circuits in the PBX. Line and trunk circuits are placed in peripheral shelves. Peripherals are devices attached to a computer to enhance the computer's capabilities or to provide an interface between the computer and other devices. Line circuits provide an interface between the CPU of the PBX and telephones. Trunk circuits provide an interface between the CPU and other switching systems. A PBX is designed so that it can interface to many different devices. Each device attaches to a *port*. The port has a signaling link connected to the CPU and a voice (or data) path connected to the switching matrix's TDM highway. Software is used to identify what the port will be used for. If a line circuit is inserted in the port, the port is identified as a line port in software. This arrangement allows for greater flexibility.

Peripheral equipment (PE) shelves accept printed circuit cards. The printed circuit cards usually have several circuits on a card. A line card usually has eight line circuits on it, and a trunk card usually has four circuits on it. Both cards are designed to plug into any slot on the PE shelf. Software will define what card has been plugged into a particular card slot of the PE. Each line circuit will use one port. Each trunk circuit uses two ports. Notice that there were eight line circuits on a line card but only four circuits on a trunk card. Each card slot on the PE shelf has eight ports. It must have eight ports because a line card might be inserted in any slot. When a trunk card is plugged into the card slot, four of the ports are not used and cannot be assigned (Figure 12-1).

12.3 PORT CAPACITY

Port capacity is a measure of the number of different devices that can be attached to the PBX. In calculating how many devices can be linked to the PBX, it is necessary to know how many ports are used up by each device. A standard SL1-S® PBX can have 144 ports. This is 18 card slots (8 ports per card slot) in the PE shelves. Eighteen line cards can be used to provide 144 line circuits, but then there would be no way to call outside the PBX. There would not be any PORTs or card slots left for trunk circuits or dial registers. If the PBX was configured with 10 CO trunks, 3 trunk cards would be

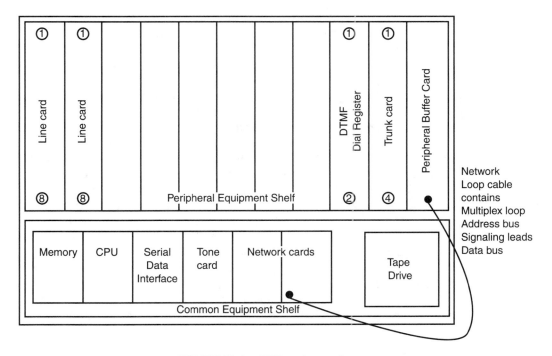

FIGURE 12-1 CBX equipment layout.

needed. Three trunk cards actually provide 12 trunk circuits, but in this case we would only use 10. The 3 trunk cards would use up 24 ports. We also need DTMF registers to receive dialed digits. These cards are also placed in the PE. If we use two DTMF register cards, this uses up 16 more ports. This will leave 104 ports available for line circuits, and 104 telephones can be attached to the PBX by inserting 13 line cards in the vacant PE card slots. Any PBX system can be configured according to the customer's needs.

12.4 CONFIGURATION

Configuration refers to the types of interfaces, number of interfaces, dialing patterns, features available, features assigned, and other database entries. Configuration is what uniquely defines a particular PBX installation. All PBXs of the same type and same manufacturer may have the same *architecture* and generic software and use the same hardware. Each installation is different because of the uniqueness of its database and the different quantities of interface cards used. Architecture refers to the way components are physically and logically organized. Each switching manufacturer uses a different (proprietary) architecture to achieve a SPC TDM switching system.

12.5 CALLING CAPACITY

Port capacity is one measure of the capacity for a PBX. Another measure is *calling capacity*. Calling capacity can be stated as the number of simultaneous conversations the switch can handle or as the number of calls per hour it can handle. These measurements are arrived at by conducting traffic studies on the switch. *Traffic* refers to telephone calls. Most PBXs can be provided with optional software to do this study. The study is usually set to run for five business days. It will log all call attempts, how long each call was, and how many calls were blocked.

Blocked calls are calls that could not be completed because no circuits were available. Traffic studies reveal how many calls can be placed at one time without incurring blockage. They will also indicate how many calls were placed in an hour. Each call placed causes a meter (or memory location) to be incremented once. The incrementation is called a *peg count (PC)*. Each call causes one peg count.

When a study is conducted, it indicates the number of calls placed each hour. The hour having the highest number of calls is termed the *busy hour (BH)*. The busy hour for each day is determined and the average figured for these five busy hours. This is called the *average busy hour peg count (ABHPC)*. Each call lasts a certain length of time. If the total amount of time for all calls is divided by the total number of calls, we get the average time of each call. This is called the *average hold time (AHT)*. The AHT of most calls is a couple of minutes. Using typical AHTs for various types of calls, an estimate can be made of the maximum ABHPC the switch can handle.

The ABHPC also needs to take memory and processing requirements into consideration. A switch may have the hardware in place to do thousands of calls per hour, but the ABHPC may be limited by processing requirements. Each function performed by a SPC switch requires processing time. Features such as call detail recording, traffic studies, and so on add to the load on a processor. Recognizing a line has gone off-hook and issuing instructions to attach a dial register and tone card to the line requires time for each step. If traffic studies on an existing switch tell you that the ABHPC is 10,000 calls per hour and you are purchasing a new switch, you must get assurance from the PBX manufacturer that the new switch will handle the ABHPC required.

12.6 EXPANSION CAPACITY

Another measure of capacity is *expansion capacity*. This measure tells you the maximum number of lines and trunks the PBX can grow to. A PBX is never purchased just to fill present needs. It must have room to grow with a business. A PBX is purchased that is wired for a certain number of lines and trunks. *Wired for* means it has the wiring in place to support lines and trunks but the printed circuit cards are not in place. When the cards are placed in the PE shelf, we say it is *equipped for* so many lines and trunks. In our scenario of port assignments, we were wired for and equipped for 12 trunks and 104 stations. We do not have room to add additional circuit cards.

We do have expansion capacity. The SL1-S system can be expanded by adding and wiring another PE shelf. A CBX usually is equipped with fewer circuits than it is wired for. As the need for additional circuits arises, the circuit cards are purchased and plugged into the PE shelf. Once a CBX reaches a condition where wired for equals equipped for, it must be expanded. The telecommunications manager must make sure that the expansion capacity of a CBX is adequate to handle future growth of the company. If the CBX is outgrown in two or three years, the manager will probably be forced to look for another job.

12.7 PRINTED CIRCUITBOARDS

CBX switches are composed of hardware and software. *Hardware* refers to the physical components of the CBX. This could be the PE shelf, common equipment (CE) shelf, the power supply, printed circuit cards, or telephones. When people talk about hardware, they are usually talking about printed circuitboards. *Printed circuitboards (PCBs)* are made from plastic and have electronic components on them. A circuit is designed by an engineer to handle a specific function, such as a line interface to a telephone. The electronic circuit will have many electronic components and chips wired together to perform the needed function.

The design of how the electronic components for a circuit are connected is arrived at using a *computer-aided design (CAD)* software program. The CAD for a circuit is fed into a *computer-aided manufacturing CAM* software program. The CAM will control an automatic component insertion machine. The components for a circuit are inserted onto a plastic board. The plastic board will have holes punched into it for mounting of the circuit components and will be acid etched, following a solder mask, to accept solder tracks. The solder tracks serve to wire the various components together. A CAM-driven machine automatically inserts the various electronic components and chips into their proper place on the board. The board is then passed over a molten bed of solder. The solder attaches itself to the board according to the design etched into the board by the solder mask. Voilá! We now have a PCB.

12.8 LINE CIRCUITS

PCBs are designed and built for all interface functions as well as the common control functions. A PCB designed for line circuits in a CBX usually has eight or sixteen line circuits on the card. When this card is replaced, you must take eight people out of service. Some central offices use this design and can only repair line circuits in the hours shortly after midnight. The DMS-100® central office switch has a superior design of only one line circuit per line card. Other switching manufacturers have adopted this design for central office switches.

Most CBXs, including Northern Telecom's SL-1, place eight line circuits on one line card. The PCBs for the CPU, memory, and network are designed to plug into a card slot on the CE shelf. The CE will also have a PCB that will interface the CE shelf to a secondary storage device such as a disk drive or tape drive. The cards in the CE shelf are connected by the backplane. The connectors that each card plug into are mounted on the backplane and are electrically connected to it. The CPU and memory cards connect over the backplane to the disk drive (or tape drive) interface card. This allows the CPU access to data on the secondary storage device.

12.9 GENERIC PROGRAM, OVERLAY PROGRAMS, AND MEMORY

As with any computer, the CPU of a PBX can only act on data stored in main memory. No computer stores all its data in main memory. A CBX stores the generic program, line translations database, and trunk translations database in memory (they are also stored on the disk or tape). This allows the CBX to do basic call processing. The generic program, which is loaded into program store memory, controls call processing and all the features of the switch. All regular station, trunk, and system features are loaded in data store memory.

If a special program needs to be activated, it must be loaded from secondary storage into memory. For example, if call detail recording is desired to record the details of each call made, the program that provides that feature must be loaded. Northern Telecom's SL-1 uses a command called Load to load these additional programs. These additional programs are called *overlay programs*. Each overlay program is assigned a two-digit code. The form of the command is LD *XX*, where *XX* is replaced by the number of the program desired. If a systems administrator wished to change the line translations database for a 2500 type of telephone line circuit, program 10 must be loaded—that is, LD 10.

12.10 SERIAL DATA INTERFACE CARD

When we issue instructions to our personal computer (PC), we use a keyboard or mouse. Both devices interface or connect to the PC through a serial port. We also communicate to the CPU of the PBX through a serial port. The CPU card usually has a serial port on its face. If we need additional serial ports, we add an additional PCB designed to provide serial ports in a CE card slot. The card is called a *serial data interface (SDI)* card. These serial ports allow us to attach a terminal or modem using an EIA-232 cable. The use of a modem allows the administrator to make all updates to the database from a remote location.

An administrator can make changes to many PABXs from a central location. This saves travel time and is the only way to go for a large company with PBXs all over the United States. The CPU will not allow anyone to LOGIN to the PBX. A user ID and password are required. The administrator will dial a telephone number assigned to the modem. When the modem answers the call, the PBX will prompt the user for input. The user types LOGIN. The PBX responds by requesting a user ID and password. After supplying these, the user can LOAD the overlay program desired. The user can end an overlay program by typing END. The user can also abort any program at any point by typing four asterisks (****).

12.11 MEMORY

The memory of a PABX is usually divided into three areas: program store, data store, and call store (Figure 12-2). A portion of call store is reserved for use by overlay programs. The size of this memory limits the number of overlay programs that can be loaded at the same time to one or two. Overlay programs use call store

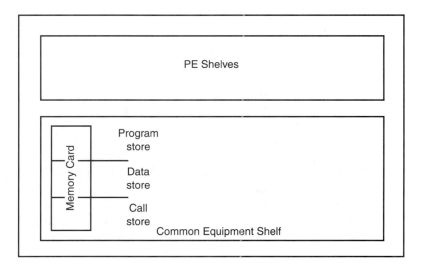

FIGURE 12-2 CBX memory layout.

to make changes in the database. The changes are made in call store and when the administrator is satisfied with the changes, they are uploaded to the data store and to secondary storage. If the PBX loses power, it can readily restore program store and data store by loading these from secondary storage. Most small PBX systems do not include a battery backup for power failures. They rely on the PBX to automatically reboot when power is restored. Rebooting and reloading from tape or disk only takes a few minutes.

Most PBX systems do include a power-fail transfer card. This PCB is wired between several telephones and their line circuits. It is also wired between several CO trunks and their local-loop cable pairs. Relays on the card are normally operated with AC power present. With these relays operated, the telephones are connected to their line circuits and the trunk circuits are connected to their cable pairs. When power fails, the relays release and connect the telephones to the CO trunk cable pairs. Incoming calls will go to these designated telephones during a power failure. These power-failure telephones can also be used to make outgoing calls. If the CO trunks are wired as ground start trunks, the power-failure telephones will need to be special phones capable of providing a ground start signal.

12.12 NONBLOCKING NETWORK

The primary function of any switching system is to connect a calling party to the called party. This is also true of a PBX switch but is often overlooked. Some administrators are overly concerned with all the features a PBX offers and forget to pay attention to this basic principle. As mentioned earlier, port capacity, calling capacity, and expansion capacity need to be addressed when purchasing a PBX. Some PBX manufacturers will tell you that their switch has a nonblocking network. A nonblocking network simply means the switching matrix has as many outlet ports as inlet ports. The matrix will not block calls offered to it, but calls may be blocked for other reasons.

If 16 people are trying to make calls outside the PBX and the PBX only has 10 CO trunks, 6 calls will be blocked. Additionally, a PBX using the 30–voice channel TDM bus for the multiplex loop can only handle 15 calls at one time. Each device assigned to the TDM bus requires a transmit and a receive bus time slot. Thus, each conversation requires 2 transmit and 2 receive time slot allocations. If the switch has one 30-channel multiplex loop, it can handle 15 simultaneous calls. If the switch has two 30-channel multiplex loops, it can handle 30 simultaneous calls. If a switch has 100 lines, 20 trunks, and 2 multiplex loops, it can only handle 30 calls because of the limitations of the multiplex loop.

The number of calls a switch can handle at the same time is limited by the number of paths available on the TDM highway. If a switch has two 30-channel TDM loops, it can handle 30 calls at one time. If 20 telephones are calling 20 other telephones within the PBX, and 10 telephones are connected to 10 CO trunks, we have 30 calls in progress. With 30 calls in progress, the system is using all 60 available speech paths on the TDM. The next phone to go off-hook will be dead. It will not get dial tone or busy tone, because there is no idle speech path on the TDM to connect the line circuit to a tone card.

The switching matrix of a switch may be a nonblocking network, but calls will be blocked if the number of calls offered exceeds the inlet/outlet capacity of the nonblocking network. Do not be misled by terms such as *nonblocking* or *virtually nonblocking*. Ask the salesperson to tell you exactly how many conversations can be held at the same time and how many calls per hour the PBX can handle. You also need to find out what the current capacity of the CBX is, and what its expansion capacity is. Current capacity and expansion capacity should be known for port capacity, calling capacity, line capacity, trunk capacity, memory capacity, and switching network capacity.

12.13 SOURCES OF BLOCKAGE

Calls can be blocked by a shortage of circuits or a shortage of paths in the switching matrix. They can also be blocked by software. Lines can be classified in software (line translations table) as low priority or high priority. This allows the system to block calls made by low-priority phones when the system is overloaded (that is, when the system is operating above engineered ABHPC capacity). When a system is offered more calls than it was engineered to handle, calls cannot be completed to their desired destination and are blocked. Blocking low-priority phones from system access during a busy period will allow high-priority phones preferred access to the PBX switching network. This type of software is usually referred to as *line load control software*. A system administrator may also use the translations table software to block telephones from getting dial tone. The translations table is set so these phones can only receive calls.

Calls may also be temporarily blocked due to a shortage of equipment such as DTMF receivers. DTMF receiver cards usually have two circuits per PCB. If the PBX has two DTMF PCBs, it has four DTMF receiver circuits. Thus, only four phones can be dialing at the same time. If a fifth person goes off-hook, no DTMF receiver is available and the phone will not get dial tone. This is why people are suppose to listen for dial tone before dialing. When one of the four circuits becomes idle, it is assigned to the next line waiting in queue. The PBX also assigns dial tone to the line. Callers hear dial tone and know they can start dialing.

12.14 TONE CARD

Dial tone, station busy tone, circuit busy tone, and ring-back tone are supplied by a PCB in the CE shelf. This card has all tones stored in ROM as a digital code. When a tone is desired, the CPU reads the appropriate ROM memory location and places the digital code into a time slot on the TDM bus. The network card connects this time slot to the line circuit. The line circuit receives the digital code, and the codec converts the digital code into an analog signal. This analog signal is passed over the local loop to the telephone, and the caller hears the tone.

When the caller dials a number using a DTMF pad, the tones are converted into digital signals by the codec of the line card. These digital codes are transmitted via the TDM bus and network card to a DTMF receiver in the PE shelf. The DTMF receiver circuit decodes these digital signals and stores each digit in call store at a specific memory address. The DTMF receiver notifies the CPU on receipt of each digit. The receiver asks the CPU to look at each digit after it has been received and provide instructions on how to proceed.

The CPU will decide whether the receiver should continue to receive digits or if the caller should be connected to another interface device. If the caller has dialed a 9 for access to a CO trunk, the CPU will instruct the network to release the connection to the register and connect the line circuit to a CO trunk circuit. The LEC's central exchange will receive all additional dialed digits. If the caller has dialed the first digit of a station number on the PBX, the register is instructed to collect more digits and come back after the collection of those digits.

12.15 2500 AND PROPRIETARY TELEPHONES

All PBXs that I am aware of support regular 500 and 2500 single-line telephones. These phones are connected to regular line circuits by one pair of wires. PBXs also offer a proprietary telephone. The proprietary phone is usually a multiline electronic telephone. Two pairs of wires are required to connect the proprietary set to a special line circuit. This special line circuit communicates with the phone over one of the pair of wires (called the *signaling pair*). The other pair of wires is used for the voice path.

The signaling path for a proprietary telephone is separate from the voice path and constitutes out-of-band signaling. This signaling works like other out-of-band signaling. A CPU in the phone and a CPU in the PBX send signals to each other by placing messages over a data link between the two. When a button is pushed on the phone, the phone sends a message over the data link informing the CPU in the PBX which button was operated. The CPU looks in the line translations table for this line circuit. The table informs the CPU of the function assigned to that button by the systems administrator. The CPU then causes that function to be performed.

With electronic telephones, the signaling processor was moved from the line circuit to the telephone. Some digital PBXs also move the codec from the line card to the telephone (digital telephone), but most of the PBXs currently in service retain the analog loop and keep the codec on the line circuit. Northern's SL series of PBXs have the codec on the line card, and their Meridian® series moves the codec to the telephone (Figure 12-3).

12.16 THE BACKPLANE

As mentioned earlier, the card slots that PCBs plug into are connected to a backplane. This backplane consists of a large PCB that has numerous solder tracks between connectors but no electronic components. The solder tracks on the backplane serve the function of a wire to connect one connector to another. The connectors that line and trunk PCBs plug into are wired to these solder tracks. The other end of the solder tracks wire to a 25-pair (50 wires) amphenol

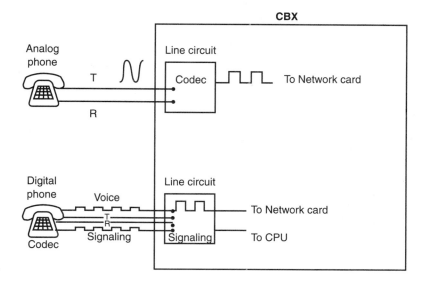

FIGURE 12-3 Analog line
circuit versus digital line circuit.

connector. The amphenol connector will accept an amphenol plug–ended cable. The other end of this cable goes to the main distribution frame (MDF) in the PBX room (Figure 12-4).

12.17 QUICK CONNECT PLUG–ENDED CABLE

The amphenol connector and cable allows for quick connection of line and trunk circuits to a MDF. Usually one end of a 50-conductor (25-pair) cable will have an amphenol connector. The cable is provided with a length long enough to reach from the PBX cabinet to the MDF. The MDF end of the cable has no connector. The wires are loose and are terminated on the MDF by punching the wires down on a quick connect block. A quick connect block does not require stripping insulation off of the wire. A special tool is used to "punch down" the wire. When the wire is punched down on the terminal, the terminal cuts through the insulation and makes electrical contact with the wire. Connections made on these types of terminals do not need to be soldered.

It is possible to buy terminal blocks for the MDF that also have an amphenol connector. With these blocks, both ends of the cable have amphenol connectors and connections are made on each end by plugging the cable into the connectors. The connectors on the backplane of the PBX switch have female connectors. The connectors on MDF blocks have male connectors. The cable between the two has a male connector on one end and a female connector on the other. This makes it easy to relocate a PBX. Most 25-pair cables are made with a male connector on one end and a female connector on the other. This allows them to be chained in series to attain any length needed. If the PBX is ever moved, another cable can be plugged in series with the existing cable to extend its length.

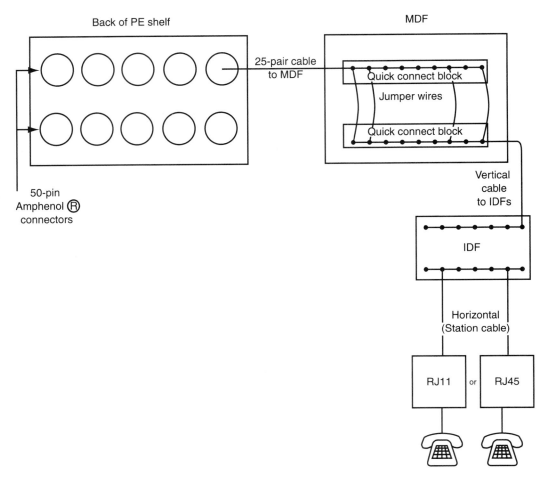

FIGURE 12-4   PBX-to-telephone wiring diagram.

12.18 PBX STATION WIRING PLAN

The MDF also has quick connect blocks with station cables attached. Sometimes the stations cable directly to this block, but most often they are cabled to *intermediate distributing frames (IDFs)*. An IDF is nothing more than a type-66 quick connect block mounted in a janitor's closet on each floor of a business. Station cables on each floor are wired to the quick connect block. These station to IDF cables are called *horizontal cables*. Cables are run from each IDF to the MDF. The cables between an IDF and the MDF are called *vertical cables* or *riser cables* (Figure 12-5).

Each telephone jack is wired with four-conductor station wiring to a quick connect block located in a wiring closet. This wiring closet is often located on the wall of a storage room or janitor's closet. Quick connect blocks are mounted on a piece of plywood attached to the wall. The cables from the MDF also terminate on these quick connect blocks. These blocks provide an intermediate wiring point.

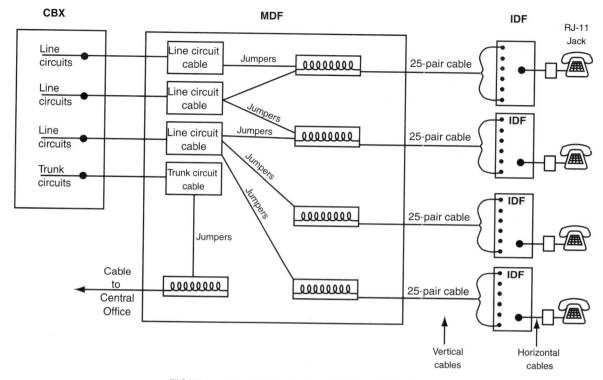

FIGURE12-5 MDF-to-IDF and MDF-to-PBX wiring.

That is why they are called *intermediate distributing frames (IDFs)*. An IDF is located on each floor of the business. In a large business such as a hospital, IDFs are placed at several locations on each floor. This reduces the length of station cable run between an IDF and each telephone jack.

The telephone jack is a standard four-pin RJ-11 jack. A four-pin jack will accommodate all 500/2500 telephones and most proprietary phones. Some proprietary phones may need six or eight wires. These phones will require eight-conductor station wire and an RJ-45 jack. If unshielded twisted-pair (UTP) wire is needed for both your voice and data network, it is advisable to run two eight-conductor station cables to each room. These can then be terminated on the type of jack required by the connecting device.

A 500 and 2500 telephone set is connected to a line circuit in the PBX via the station set line cord, the RJ-11 jack, the station wire, the IDF, the vertical cable to the MDF, jumper wires on the MDF, and the cable to the PBX line circuit. A jumper wire is a short piece of wire. It is used at the MDF to connect the IDF cable wires to PBX cable wires. This allows any line circuit to be connected to any IDF cable pair. When a telephone set is moved, it can retain the same line circuit simply by moving the jumper wire to the new IDF pair used for the new set location. When a 500/2500 set goes off-hook, loop current flows in the line circuit. This generates a signal to the signaling processor. The signaling processor notifies the CPU that this line circuit is off-hook.

A proprietary telephone must be wired to a special line circuit designed specifically for these phones. These line circuits do not contain a signaling processor; the processor is moved to the proprietary telephone. When the phone goes off-hook, the signaling processor in the phone sends a signal over the signaling wire to the line circuit. The line circuit connects the signaling wire to the CPU of the PBX. The two processors send messages to each other. If one of the line buttons has been depressed, the signaling processor of the telephone tells the CPU of the PBX which button has been depressed. The CPU looks into the line table to find the key assignment data. If it sees the depressed key is assigned as a telephone number, it attaches a dial receiver and dial tone to the voice pair. The receiver receives dialed digits over the voice path.

12.19 ACCESSING FEATURES

To activate a feature, the user of a proprietary telephone pushes a button assigned to the feature desired. The signaling processor sends a message to the CPU of the PBX informing it which button has been depressed. The CPU looks up the key assignment information in the line table to determine which feature has been assigned to the key. The feature is activated for this line circuit by the PBX. Most proprietary telephones have about ten keys. Three of these keys are assigned to telephone numbers. One key is assigned as a release key (or hang-up key). The other keys can be assigned a feature such as call transfer, conference call, call forwarding, and so on. This allows most features to be accessed by merely depressing the key assigned to the feature. If you need a feature not assigned to a key, it is accessed the same way 500 and 2500 sets access a feature.

With 500 and 2500 single-line telephones, features have to be accessed by using code dialing. A code is assigned for each feature. To access a feature, the code for that feature is dialed. To receive the dialed code, a dial register must be attached. This is done by momentarily hanging up. This momentary hang-up is done by depressing the hookswitch button briefly. The depressing of the hookswitch must be between 100 and 500 ms. This brief timed on-hook signal is referred to as *flashing the hookswitch*. If the time is too short, the switch does not recognize the signal. If it is longer than 750 ms, it thinks the signal is a hang-up signal. Because some people cannot properly flash the hookswitch, some 2500 sets designed for use with a PBX contain a flash button. This button has electronic circuitry to provide a timed hookswitch flash of 500 ms, when the button is depressed.

After the hookswitch flash, a dial register is attached to the line. The code for the feature can now be dialed. The receiver presents the code to the CPU. The CPU checks the line table to see if the desired feature is approved for this line circuit. If the line circuit is allowed to access that feature, the PBX will activate the feature for that line. A hookswitch flash notifies the CPU that the digits being received by the dial register represent the code for a feature. When a 500/2500 set goes off-hook, it is assigned a dial register for receipt of dialed digits. The CPU will used these dialed digits as the address of a telephone if it does not receive a hookswitch flash. A hookswitch flash tells the CPU that the digits that follow are not a telephone number but are a feature desired from the PBX. Once a conversation is established, a hookswitch flash causes the signaling processor to signal the CPU that the called line should be put on hold and a dial register attached to the calling line. The caller can now dial the code for a feature or dial another telephone number.

12.20 PBX AND ISDN

Most line interface circuits in a PBX will be for 2500 telephone sets and proprietary sets. Some PBXs also include interfaces for data and ISDN. The latest generation of PBXs is designed to handle both voice and data. ISDN promises to be an excellent vehicle to handle voice and data. Current PBX technology requires a four-pair station cable for voice and a separate four-pair cable for data. ISDN line circuits will only require one pair of station wires in each office. Of course, 2 four-pair station cables provides a cheaper solution than using ISDN. The ISDN software for the PBX, the ISDN line cards, and the ISDN terminal equipment adds significant cost to a basic PBX. Will your communication needs justify the added cost of ISDN?

"Can we justify the cost?" is a question that must be asked for every business decision made. The office of the future appears to be moving in the direction of integrated voice and data. Is the future already here for your firm? Only after a detailed analysis of your present and future communication needs can this question be answered. As mentioned previously, many small businesses can satisfy their needs with a keysystem and do not need a PBX.

One drawback to using a keysystem in a business with many CO trunks is inability to access all CO trunks from one station. A keysystem station accesses a CO trunk by depressing the button associated with that trunk. Once a business has more than a handful of trunks, it becomes difficult to terminate all CO trunks on each telephone. A *hybrid keysystem* was developed to overcome this problem. With a hybrid keysystem, the station user dials a 9 for CO trunk access. Trunks do not need to be terminated on the station set. Medium-sized businesses find these hybrid systems are the perfect solution to their voice communication needs. A hybrid gets its name from the fact that it is a hybrid of keysystem and PBX technologies. The major feature that tells you when a keysystem is a hybrid keysystem is when the system uses dial 9 for central office trunk access. Once a business is large enough to justify the expenditure for a PBX, a decision must be made whether the system is needed for voice only or if it should handle both voice and data.

12.21 PBX TECHNOLOGY

Many books state that PBX technology has evolved through three generations and is currently entering the fourth. Each author has a slightly different definition for the four generations, but most agree that the fourth generation is not fully defined. They all forget that the first-generation of PBXs was manual switchboards. They are all referring to four generations of PABXs. PBX manual switchboards and PABXs using electromechanical space division networks are history.

The PBXs in use today are considered third-generation systems. A third-generation PABX uses PCM/TDM switching networks and will accept either analog 2500 phones or analog proprietary telephones. The third generation has been expanded to include PABXs that also accept digital proprietary telephones. This system moves the codec to the telephone and provides for end-to-end digital connectivity.

Some of these digital PABX systems are designed for small business applications and use regular analog CO trunks over the analog local loop. In these instances, end-to-end digital connectivity is lost. The trunk circuit in the PBX must convert the digital voice received from the telephone line circuit into an analog signal for transmission over the local loop. Larger PBX systems can achieve end-to-end connectivity when the number of trunks required becomes great enough to use a *T1 CXR* between the central exchange and the PBX. Of course, a T1 system will limit any data placed on the system to 56 Kbps because of bit-robbed signaling. The limitations of a T1 system can be overcome by using a Primary Rate ISDN system. If a 23B+D ISDN system is used, data can be 64 Kbps on any B-channel.

At this point we are starting to enter what is called the fourth generation of PABXs (fifth generation of PBXs). This generation combines voice and data and allows the integration of LANs with voice. Some data communications vendors will seek to use LANs to handle both voice and data and will design software and hardware along these lines. Manufacturers in the telecommunications industry are apparently going to continue along present lines. They have a PBX capable of handling both voice and data and have developed interface devices that allow integration with LANs.

12.22 FUTURE PBX TECHNOLOGY

I believe that putting a codec in the telephone makes voice and data the same thing. Digitized voice is 64 Kbps, while data can be any speed. ISDN technology is an underlying technology that treats voice and data as the same thing. We are fast approaching the point where line circuit interface cards in a switching office should become extinct. Trunk circuit interface cards are already extinct in an all-digital world. Trunk circuit interface cards are only required when we need to provide a digital-to-analog interface. When digital trunk circuits using out-of-band signaling are interfaced to a digital switching system, only a bit-stream interface card is needed.

The bit-stream interface card maps DS0 channels of a low-speed TDM to the DS0 channels of a high-speed TDM. The same technique can be applied to line circuits. If the transmission facility is capable of carrying high bit streams, many devices can be multiplexed over this bit stream. I would expect to see this technology implemented in PBX switching systems soon. After the technology is proven in PBXs, I expect to see rapid deployment in class 5 central exchanges. PBXs can use Primary Rate ISDN to implement the technology. Central exchanges must handle a greater number of lines and would require broadband ISDN to implement this technology.

Integrated voice and data can be achieved by giving each device its own address. Telephones and PCs would include a terminal adapter as part of their hardware. This allows every device attached to the switching system to have its own unique address. Each device can be multiplexed onto a high-speed TDM highway, and the signaling channel will inform the switching system which time slot the device is on, the services required, and the destination desired. The signaling processor is part of each terminal device, and that CPU can talk to the CPU in the switching system via messages over the signaling channel. This technology will allow one high-speed data bus to be used as the transmission medium for both voice and data.

With both voice and data on the same bus, the PBX switching system may simply become a communication server on a LAN. The PABX or communication server could connect any terminal device to any other terminal device and could also connect terminal devices to other servers on the LAN, to the PSTN, or to the PDN. Computer telephony integration to this degree will require smart terminal devices. Each terminal could contain software that would convert dialed station numbers into a TE address. This would allow direct connection over the bus without the use of a switching system. A switching system would not be required on calls between stations but would be required for connection to CO trunks. The switching system could be replaced by a Primary Rate NT1 and central office line. When a central office line is desired, the TE would address the NT1 and LE/ET in the central exchange.

The next generation of PBX switching may involve replacement of the switch by intelligence in the telephone and in the transmission facility. LANs are well positioned to use ISDN technology. NT1s can interface to a NT2, which is interfaced to a LAN. This provides a portal from the LAN to the PSTN and PDN. Terminal adapters can be designed to interface a digitized voice terminal to the LAN. Software in the terminals will allow them to directly address other voice terminals on the LAN. They will use the NT1 to access the PSTN or PDN.

Do not limit yourself to telecommunications vendors when looking for a PBX. See what is available from vendors dedicated to data and LANs. The Internet is transmitting voice as packetized data over its network. Applying this technology to the present voice network (PBX and class 5 exchanges) should not be far behind. I expect to see significant advances in PBX technology over the next few years, but since they do not yet exist, they cannot be addressed in this book. As suggested above, technology may reach a point where the PBX switching system can be replaced by smart terminals, LANs, NT1, and ISDN. But this will only be possible for PBXs that interface to a digital SPC class 5 central office. The out-of-band signaling required to support this concept relies on communication between computer-controlled devices and computer-controlled switching systems.

12.23 THE RESURRECTION OF AN OLD SERVICE: CENTREX

It is also possible that the LECs will enhance their Centrex services. Centrex is a PBX service provided by a portion of the central office switch. Centrex provides all the features of a PBX and allows DID and DOD. Since Centrex is offered by a portion of the class 5 switch, customers do not have to purchase a PBX switch. Since they do not own a PBX, they have no maintenance expenses. Centrex is provided by the LEC, and customers are charged a monthly service charge. Centrex received a near fatal blow with deregulation during the 1970s. These rulings prevented ownership of CPE by a LEC. Since PBX services were deregulated and declared to be CPE, the LECs were prevented from using Centrex to compete in the deregulated PBX market. The Communication Reform Act of 1996 allows LECs to reenter the CPE market. The LECs have embraced this opportunity and have merged their spinoff CPE subsidiaries back into their regular operations. The marketing department of the LEC will be able to market PBXs, keysystems, and Centrex services.

Centrex offers all the features of a PBX without the capital outlay and operating expenses the purchase of a PBX entails. The disadvantage of Centrex is the requirement to cable all telephones to the central office. Since the switching system is at the central exchange, every telephone must be connected to the switch using a cable pair. Using Centrex to serve a large customer with hundreds of stations requires running hundreds of cable pairs from the exchange to the customer's premises. Some Centrex systems that serve large customers use TDM subscriber carrier as the medium between the exchange and the customer's location. I would expect to see the LECs and switching manufacturers design a Centrex interface that uses Primary Rate ISDN facilities to the customer. This would allow connection of LANs, telephones, and PCs to the Centrex via NT2 and NT1.

We simply do not know where the switching manufacturers are going with their technology and hesitate to predict the technology path chosen. We do know that the LECs are going to be forced to take some action that integrates voice and data into one network. The LECs own the weakest link in telecommunications (the analog local loop). The deregulation of local services has led to IECs and cable television companies entering the local services market. Using broad-bandwidth mediums, these companies can offer a wider array of services than a LEC can using the analog loop. The use of a cable modem on cable television facilities provides speeds many times faster than the fastest wire-line modems. The growth of the Internet is providing high demand for these faster data speeds. The LECs must upgrade the local loop to a broadband loop if they wish to provide more than plain old telephone service (POTS).

12.24 GROUND START CENTRAL OFFICE TRUNKS

One reason for using a PBX to serve a company's communication requirements is that the PBX allows a company to build a private network. The most common trunk found in a PBX is the central office trunk. This trunk connects the PBX to a class 5 central office and serves as the access to the Public Switched Telephone Network (PSTN). This trunk is accessed by any station wishing to access the PSTN. Access is gained by dialing a 9 in most PBXs. These trunk circuits are usually connected to the central office by a two-wire local loop. The trunk circuit may be configured as loop start or ground start. Loop start trunks use the same type of signaling used by a regular 500 telephone. Older electromechanical PBX switching systems had a problem where the same trunk could be seized on both ends at the same time. Ground start trunks were developed for the PBX to protect against this double-seizure condition. Ground start trunk circuits prevent accessing a trunk for an outgoing call when it is used by an incoming call.

Trunk circuits in an electromechanical PBX switching system did not mark themselves busy until a ringing signal was received from the central exchange. The central exchange ringing cycle is such that there is a 4-s silent interval between rings. If an incoming call was assigned to a trunk between rings, it was possible that the PBX trunk circuit could be seized by the central exchange for up to 4-s before receiving a ring signal. An incoming call could be attached to the circuit for up to 4-s before the trunk would be marked busy to outgoing calls. During this time frame, an outgoing call could see the circuit as idle and be assigned to the same trunk. This condition is called a *glare condition*. The incoming caller would be connected to the outgoing caller. They would talk to one another and realize their calls were inadvertently connected. They would have to disconnect from the call and place their call again.

Ground start was developed as a method of marking a trunk busy on an incoming call when the trunk is seized. The trunk was made busy immediately on seizure and did not need a ringing signal to mark it busy to outgoing calls. Today's SPC PBXs mark themselves busy on seizure whether using loop start or ground start. Ground start is not needed with a digital PBX. The option is made available for people who insist on ground start. When a PBX connects to the central exchange, the central exchange end of the local loop can be a line circuit or a trunk circuit. Line circuits are used when all seven digits of the telephone number are received and handled by the central exchange. Line circuits used for ground start must be modified so that the ground potential normally present on an idle loop start line is removed. When a ground start line is idle, there is no ground on the Tip lead. When an incoming call is assigned to the line, a ground is placed on the Tip lead. The PBX uses ground on the Tip lead to mark the trunk busy to outgoing calls.

When an existing PBX that used ground start trunk, is replaced with a new switch, the ground start selection is chosen because the lines at the central exchange are ground start lines. It is not difficult to remove the central office ground start modification, but it does not hurt to keep it if you prefer to use ground start operation. The choice is yours. I would recommend that you verify with a technical expert whether you PBX needs ground start. The PBX vendor should have this technical expertise available. Ask them whether, if you use loop start on the trunk circuit, it will make itself busy to outgoing calls immediately on an incoming call seizure, or must it see a ring signal before marking itself busy to outgoing calls?

12.25 DID AND DOD

When the central office interface for a PBX trunk is a line circuit, an attendant or operator at the PBX location answers all incoming calls. On being informed who the caller wishes to speak to, the attendant will forward the call to the desired station number. This is a throwback to the old manual switchboard procedure. Today, most companies using a PBX provide DID capabilities as well as *direct outward dial*

(DOD) capabilities. DOD allows a PBX user to access an outgoing trunk to the central office without the aid of an attendant (PBX operator). DOD access is usually provided by dialing the digit 9. DID allows callers to dial stations directly and bypass the attendant. The main number listed for a company will be answered by an attendant who can forward the call, but if the caller knows the station number of the called party it can be dialed direct. DID circuits can use a two-wire local loop for the connection between the central exchange and a PBX but must use a trunk circuit in both the PBX and the central exchange. The most common transmission medium used for connecting central office trunks to the exchange is a two-wire local loop, but large PBX systems will usually use a T1 carrier system or a digital line carrier (subscriber carrier) as the transmission medium.

12.26 PRIVATE NETWORKS AND WATS

Some PBXs use trunk circuits to establish private networks. The trunk circuit used is the same trunk circuit used for central office trunks. Instead of connecting to the local central exchange line and trunk circuits, they connect to special circuits such as *wide area toll service (WATS)* or connect to a *foreign exchange (FX)*. WATS was developed by AT&T as a method of providing discounts on toll to high-volume business users. Prior to WATS, all calls on the long distance facilities were billed the same for everyone. WATS lines bypass the billing machine. These lines have meters that keep track of the number of calls placed and the total conversation time for each month. Businesses could purchase flat-rate WATS or measured-service WATS. They could also purchase incoming and outgoing WATS lines. Interexchange carriers (IECs) provide WATS for out-of-state calls. Local exchange carriers (LECs) provide WATS for in-state calls. This is due to the 1984 MFJ, which states that inter-LATA calls must be handled by the IEC and intra-LATA calls must be handled by the LEC.

The 1996 Telecommunications Reform Act should make it possible for IECs and LECs to offer both interstate and intrastate calls. IECs provide six different WATS lines, which are called *bands*. A band 1 WATS can call into adjacent states. A band 2 WATS can call everywhere a band 1 line can, plus a few more states adjoining the adjacent states. Basically, out-of-state WATS bands radiate out in a circular fashion from the originating state. The circle that encompasses adjacent states is band 1. The circle that has a little longer radius is band 2. This continues with each circle having a little longer radius until a radius is long enough to cover all states. This is band 6 WATS.

For in-state WATS, the LECs divided the states up into three bands. One WATS line could not be used to dial both an in-state and an out-of-state telephone number. A WATS line was needed from the LEC for in-state calling, and one was needed from the IEC for out-of-state calls. This area is ripe for competition. Why have two separate lines. I expect the IECs will aggressively pursue an opportunity to provide both services with one WATS line. It will be more difficult for LECs to provide out-of-state WATS, but as they get into the regular long distance market, I am sure they will also offer both services over one line. Changes are presently occurring in this area. Check with your IEC and LEC to see what they offer. WATS lines are often accessed by dialing the digit 8. When PBX has more than one band, additional access codes must be used. When a company has multiple WATS lines, it often equips the PBX with the *least-cost routing (LCR)* feature. LCR determines from the called number how the call should be placed.

12.27 SOFTWARE-DEFINED NETWORKS

WATS has not gone away, but the IECs do offer attractive alternatives. They provide Software Defined Networks (SDN) and T1 systems that can be used to build private networks. When a business is large enough to have many PBXs scattered across the United States, it often ties them together using *tie trunks*. These trunks are accessed by dialing a special prefix that varies at each location to achieve a *uniform dialing plan*. One PBX may use numbers 2100 to 2900, a second location may can use 3100 to 3900, and a third location use 4100 to 4900. The first location would use access digit 3 to access trunks to the second location, and access digit 4 to access trunks to the third location. These locations would access the trunks on an outgoing basis using access digit 2. Thus, any telephone number in the company can be accessed by any phone in the company simply by dialing the four-digit telephone number. The telephone numbers at each PBX start with a different thousands digit. The thousands digit is used as a routing digit and identifies a unique PBX location on the private network.

12.28 FOREIGN EXCHANGE TRUNKS

Some PBXs may use a trunk circuit to access the line circuit in another PBX or distant central exchange. These trunks are called *foreign exchange (FX) trunks*. When the access digit is dialed at the PBX, the outgoing trunk circuit is seized. It will use a rented T1 channel from a IEC to seize the line circuit in the distant exchange or PBX. The distant exchange or PBX will return dial tone and local numbers can be dialed. If a PBX in Kansas City has an FX line to a Chicago central office, the station user in Kansas City can dial the access code for the FX trunk and then dial any number in Chicago. Additionally, the Chicago LEC would assign a local telephone number to the line circuit. This would allow customers in Chicago to dial a local Chicago telephone number and not have to pay a toll charge to reach the Kansas City PBX. Incoming calls from the Chicago telephone number are answered by the Kansas City PBX attendant. This process can be reversed to provide dial tone from the PBX to a distant location. When the PBX provides dial tone to a distant telephone rather than a distant exchange, the circuit is called an *off-premise extension (OPX)*. An OPX is used to provide extensions off of a PBX to a branch office and sometimes to a senior manager's home. The OPX can be provided with access to any feature of the PBX. An OPX connects to a line circuit in the PBX.

12.29 MANAGING THE PRIVATE NETWORK

As the size of a business grows, so does the task of managing the company's telecommunications and data communications needs. Large companies employ one manager for their voice network and another for their data network. As businesses jump onboard the "right-sizing" or "downsizing" craze, they are forcing more work onto fewer people to increase productivity. This has to result in increasing productivity. No management skill is needed to improve productivity. No pep rallies and no motivational techniques are needed, just reduce the workforce. The workload that was done by a laid-off employee will be picked up by the remaining workforce, or will be treated as unnecessary and discontinued as a cost-cutting measure. The remaining workers are forced to work 50- or 60-hour workweeks at top management's directives (explicit or implicit).

Thankfully not all companies have embraced downsizing as a substitute for managing their business. As new technologies and training lead to improved productivity and result in reduced personnel requirements, some companies look to see how they can make use of their workforce asset in other ventures. Executive-level management in these companies continually looks for future growth opportunities. Instead of laying employees off, it transfers employees to other, newly created job opportunities within the company. If top management of your company is more interested in downsizing the company than in growing it, you may be faced with a tremendous task to manage both voice and data networks. This is one of the effects of downsizing as applied to a telecommunications manager.

The use of central office trunks, WATS trunks, TIE trunks, OPX lines, and leased data lines to build a private network may be too overwhelming for the manager. You must continually monitor the traffic at all PBX locations, do a traffic analysis and a cost analysis of the various services being used, evaluate the different service offerings of many IECs, LECs, and independent carriers, and determine how the networks should be engineered. Some PBX locations may not have enough call volume to justify TIE lines or WATS lines. Continually analyzing many different network components to achieve maximization of company resources can be a tremendous undertaking. The manager must downsize this workload by placing more of it on the vendors. This can be done by opting for a *virtual private network (VPN)* also called *software-defined network (SDN)*.

VPN and SDN are names used to identify a private network for business that uses portions of the PSTN facilities to carry traffic for the private network. These virtual networks are offered by the IECs. The manager does not need to design and manage TIE lines, WATS lines, and 1+ long distance lines. These are all eliminated. The IEC will design your interoffice and long distance network into its database and provide you with this VPN/SDN. Calls between PBXs are connected by trunk circuits from the PBXs to the IEC. The IEC decides how the call should be handled and processes it accordingly. Access between the trunk circuits of the PBX and the IEC can be switched or dedicated. The LEC can provide access to the IEC or private VANs can provide dedicated access to the POP, for connection to an IEC.

12.30 SDN AND SS7

VPN and SDN are services that have been made possible with the deployment of SS7. The IEC receives the calling number and called number from the originating PBX. The calling number of a VPN/SDN customer will be in the IEC's database and identifies the originating customer as a virtual network customer. The IEC forwards this information over SS7 to a database (SCP). The database at the SCP contains entries on which telephone numbers are part of the VPN and which calls are off-net calls. The SCP translates the dialed number for a VPN customer into an area code + 7 digit number and provides the instructions for completing the call over the IEC's regular network (PSTN). Billing is applied for the call according to whether it is an on-net to on-net, or on-net to off-net, call. The VPN/SDN can also be set up to allow receiving calls from an off-network location. Company employees who travel can use special access numbers and authorization codes to access the VPN. Customers can also be included as part of your VPN.

The use of VPN/SDN has become popular. TIE lines and WATS lines are usually more expensive than paying for access lines to the IEC and paying the VPN charges. Managing the network is much easier since the IEC handles all calls. The manager gets all billing and traffic data from the IEC. This simplifies the data gathering and analysis process. The manager does not have to worry about a TIE line going down because they do not exist in a VPN. If problems occur in the IECs network, they have alternate routing capabilities to keep the VPN in service. The IECs even offer telecommunications managers limited access to the SS7 network and the SCP so they can manage the VPN. This allows the manager to reroute calls from one PBX location to another in the event a weather problem, equipment problem, or work stoppage closes a PBX location.

12.31 PBX PURPOSE AND FUNCTIONS

A PBX is a private switching system that provides switching capabilities for stations connected to the PBX. It allows the construction of a private network and provides users with a wide array of features. Most features are developed in software. The primary purpose of any PBX switching system is to connect a calling station to another station on the PBX or to connect the calling station to the desired trunk circuit. This switching function is part of the basic call processing software and is part of generic 1 software. Each time new features are developed for a PBX, they are added to the existing generic program and the number of the generic program is changed.

Generic program 31 will contain features and improvements not found in generic 30. As new features are needed, the generic is modified or added to and a new generic is released that includes the needed feature. Switching manufacturers love to see new features developed because a good portion of their revenues comes from selling existing customers the latest software release. Microsoft continually adds new bells and whistles to its Windows software for PCs. Each new release carries a different version number. Switching manufacturers do the same thing with their software.

12.32 SYSTEM FEATURES

Features developed for a PBX system are called *system features*. These features are developed to improve the efficiency of the PBX itself (such as least-cost routing and DID) or to provide enhanced administrative features (such as call detail recording). Least-cost routing is sometimes called *automatic route selection*. Each station can be assigned to a particular class of service and trunk group restriction. This will determine which trunk routes they are allowed to use. Least-cost routing will look at the trunk routes allowed by a station and place the call over the circuit that results in the lowest cost per minute. If you use a software-defined network, automatic route selection may not be required. With SDN, you usually have only two trunk routes. One group of trunks connects to the local exchange (CO trunks) for local calling. The second group of trunks connects to the IEC SDN and handles all long distance calls. The SDN will do automatic route selection instead of the PBX.

12.32.1 Direct Inward System Access

Direct inward system access (DISA) is a system feature that allows company employees to access the company's private network by dialing a special number and entering a password. The special number provides trunk access to the company's PBX. Once they are connected to the PBX, they must enter an authorization code and password. After access is approved by the system, employees can dial anywhere that a regular PBX station can dial. This allows them to access WATS lines and the SDN to place long distance calls.

Sometimes the DISA feature is accessed by dialing an 800 number. The 800 number will connect via the PSTN to a trunk circuit in the PBX. After dialing the 800 number, employees can enter the DISA code to access all the capabilities of the PBX. They can place long distance calls using the SDN rather than using a more expensive calling card. They can also access the e-mail and voice mail system to leave messages. Security of DISA is a problem. Unauthorized people have gained access to a company's private network and have used it to place long distance calls. Telecommunications managers should change authorization codes and passwords often. They should also keep a close watch on billing invoices and the call detail recording system to detect abuse.

12.32.2 Dialed Number Identification System

This system feature allows the PBX or automatic call distributor (ACD) to receive information from the central exchange about which particular station or group of stations the call should be connected to. This feature is used most of the time on an ACD system. The ACD will send the call to a particular group of ACD agents based on the *dialed number identification system (DNIS)* information. This could also be accomplished by assigning a different telephone number to each group of agents. A telemarketing group will often handle calls for many companies. A different in-WATS (800) number could be assigned to each company. When someone dials the 800 number, the call is routed to the appropriate agent group. DNIS performs a similar function and uses one group of telephone numbers for receipt of all calls. The PSTN will interpret a called number and change it into the number of the telemarketing group. The call is placed over the PSTN to the ACD, along with information identifying who the caller was calling.

12.33 STATION FEATURES

All features developed for users of the PBX are called *station features* (such as call transfer and call forward). Before station features can exist, they must be included in the system's generic load. Most PBXs have a system feature that allows more than one customer to use a PBX. This concept became popular in the 1970s. Real estate companies leased office space and telecommunications facilities to tenants. Each customer (tenant), as well as the lines belonging to that customer, was identified in the database of the PBX. Features accessible by a particular customer were also identified in the database. A station could only have access to the features identified for that specific customer. Thus, all station features had to be identifiable at the system level and the customer level before they could be added to the line translations database for that station and become a station feature.

A clear delineation cannot always be drawn between what a system feature is and what a station feature is. A feature can only exist for a station if it is a feature the system offers. Stations do not have access to some system features. System features should be thought of as features available to all station users or used by the system itself. For example, "dial-9 access for a central office trunk" is a standard feature on all PBXs and exists for all stations. This feature must be restricted on a station-by-station basis when stations are denied this system feature. A station feature can be thought of as a feature that must be specified in the line translations table for each station.

12.33.1 Call Forward

Call forward is both a system feature and a station feature. It is a system feature because the system will automatically forward unanswered calls to a voice mailbox or an attendant. It is a station feature because stations can be given this feature and allowed to program a specific location for forwarding unanswered calls. The system or the station can also provide for forwarding calls when the station is busy. "Call forward busy" can be the same forwarded location or a different location than used by "call forward no answer."

12.33.2 Call Transfer

Call transfer is a station feature. When users wish to leave their desk, they can program a location for the transfer of all calls. Call transfer will override the call forward no answer feature. Call forward no answer will forward calls after a certain number of rings (this number is flexible and is programmable). Call transfer will send calls immediately to the transferred location, and that station receives ringing immediately on receipt of a call in the system for the transferred station. As mentioned earlier, station features for 2500 sets are accessed by code dialing. A proprietary set accesses features when a button is pushed. To activate call transfer, station users must press the call transfer key on their proprietary sets, or flash the hookswitch and dial the access code for call transfer. The number of digits for call transfer is also programmable. This allows the administrator to restrict call transfer to four digits. Station users cannot transfer calls to their home telephone or any seven-digit number unless the administrator programs the line for seven-digit transfer. Four-digit transfer will allow call transfer to any station on the PBX.

12.33.3 Ring Again

Ring again, or *camp-on queuing*, allows a station user to camp on a busy station. This is a PBX feature and only works when the called number is a station on the PBX. If the called PBX station is busy, the caller activates the ring-again feature and hangs up. When the called station is finished with the present conversation and hangs up, the line is idle. The CPU will send a special ringing signal to the calling station's phone. The party knows by the special ring that it is the call camped on busy. The calling party picks up the telephone. When this happens, the PBX will ring the station that was camped on. This feature prevents callers from having to continuously place calls to a busy phone and allows them to make more productive use of their time. The ring-again feature will continuously test the line. When it becomes idle, the caller is notified.

12.33.4 Conference Call

PBXs can be ordered with *conference call* capabilities. This feature requires the addition of hardware. A PCB must be supplied that provides a conference bridge. This card is designed to connect many different time division slots. The number of slots it can connect determines how many stations can be connected in a conference. This PCB ensures that an adequate level of transmission is maintained for all connected stations. Most PBXs allow up to eight stations on one conference. The initial purchase of a conference call feature must take future needs into account. The number of conference cards to provide as well as the number of stations each PCB can connect must be considered. This feature cannot be changed without purchasing new hardware as well as software. An upgrade to handle more stations on one conference or to increase the number of conference circuits can be costly.

Establishing a conference call is easy. Flashing the hookswitch and dialing the access code or pushing the conference button on a proprietary phone provides dial tone. The user dials the station number desired. When the called station answers, the calling party announces that a conference is being established. Although most PBXs allow a private consultation with the called party at this time, I would still be careful about what is said. The original party you were talking to should not be able to hear your conversation with the station that was dialed for a conference. This feature allows you to consult privately with a third party. Once you have consulted with the third party and decide a conference is in order, you can bridge all parties together by pushing the conference button again. With 2500 sets, you would have to flash the hookswitch to link all three stations. If additional stations must be added to a conference, the same procedure is followed again: press the conference button (this puts the original party on hold and provides your station with dial tone), dial the desired station, consult or announce the conference when the called party answers, press the conference button again, and the called party is added to the existing conference.

12.33.5 Speed Dialing

PBXs have two versions of *speed dialing*; the system provides a system speed dial as well as station speed dial. *System speed dial* numbers are programmed by the attendant and can be accessed by all stations provided with the system speed dial feature in their line table. If the line table includes the *station speed dial* feature, the station can program frequently called numbers that are accessible only by that station. System speed dial numbers usually have two digits. The tens digit identifies the number as a system speed dial number. For example, the attendant could program speed dial for 1-312-555-6622 and assign this number to speed dial code 92. When any station activates the speed dial feature and dials 92, it will be connected to 1-312-555-6622. If the station activates speed dial and dials a single digit such as a 3, it will connect to the telephone number it programmed at its station as speed dial code 3.

   System speed dial is used to store numbers many station users call frequently. These may be telephone numbers for a branch office, the main office, the credit union, and so on. Station users program their individual speed dial locations for their home, a customer or vendor they deal with on a regular basis, and so forth. Most PBXs provide the capability to have ten speed dial locations for each station and ten for the system speed dial. Advanced PBXs offer the capacity for more than ten speed dial numbers. Of course it is also possible to have station sets with speed dial capabilities. This provides the user with more flexibility but is somewhat redundant and unnecessary.

12.33.6 Call Pickup Groups

Most telephones are arranged in groups. All engineering phones are assigned to one group number. Accounting is in a separate group, operations has its own group, marketing and human relations have their own group number, and so on. The group number a phone is assigned to is an entry in the line translations table (line table). When a phone is added or someone moves to a different department, the group number is one entry made in the line table. When any telephone in a group is ringing, it can be answered using the *call pickup* feature. This feature allows anyone in your group to answer a call coming in to your department. Calls cannot be picked up for other groups. If your neighbor is working for accounting and you are working for engineering, you cannot pick up calls for your neighbor unless the system administrator assigns both phones to the same group. Sometimes groups are assigned on a geographic basis rather than a departmental basis to allow for this type of pickup. This feature is helpful when people leave their desk without activating call transfer. If the system has call forward no answer activated, the incoming call will be forwarded if someone does not use call pickup to answer the call. Remember that call forward no answer forwards calls to a predetermined station after a certain number of rings.

12.33.7 Call Park

If you answer a call at your desk and need to transfer the call to another location, you could use call transfer. Sometimes this is not practical. Suppose you need to go to the basement to retrieve information the caller is asking for. There is a phone in the basement area but no one is there to answer the call. You can use *call park* in this situation. Call park is like a huge parking lot in which a call can be parked. If you do not specify a parking slot when you activate the call park feature, the call will be parked in the slot corresponding to your station number. If you wish, you can specify any parking slot desired. The parked call can be picked up at any telephone by dialing the *call park retrieve* access code. You could park your caller and then go to the basement telephone and retrieve the call from the parking lot. This feature is also used to transfer calls when you do not know the location of the desired party. On receiving a call for Joe Smith, you can park the call and use the paging system to notify Joe Smith. You would tell Joe he has a call and what parking slot the call is in—for example, "Joe Smith, please call code 64."

12.33.8 Intercom Groups

When a group of people call each other a great deal, the stations can be put in an *intercom group*. Many different intercom groups can be defined in most PBXs providing this feature. This allows dialing other stations in the group with a one- or two-digit number. Special features and restrictions can also be set up for each group.

12.33.9 Multiple Appearance Directory Numbers

When a telephone number appears on more than one telephone, it is referred to as a *multiple appearance directory number (MADN)*. This feature can allow a single-line telephone to have more than one telephone number. It also allows the use of single-line telephones to answer an incoming call. The call will cause all assigned station sets to ring unless they are on another call.

12.33.10 Executive Override

Executive-level managers are usually provided with access to all trunks and stations. Many PBXs also offer the *executive override* feature. If your station has this feature, you can override a busy signal. When you get a station busy signal, you would activate the executive override feature. Activation of this feature should cause a brief warning tone to be placed on top of the called station's conversation. You would then be connected into the existing call. Of course, this is not good phone etiquette, and since executives do not want someone doing this to them, they are provided with a *privacy* feature. This feature prevents anyone from using executive override to barge in on an executive's conversation. This feature also prevents the attendant from monitoring an executive's conversation. Although the privacy feature is assigned primarily to executive phones, it is also used on data lines. This prevents the interruption of a data transfer due to a call waiting tone. When the executive override feature is activated, the station receives a warning tone. This tone will cause an error on data handled by a modem. To prevent this from happening, the modem-equipped line is assigned the privacy feature.

A word of caution is in order here. It is unlawful to monitor someone's telephone conversation except as it is incidental to your business. You have probably noticed that when you call a large customer service or government organization, you receive a recorded announcement informing you that some calls are monitored for service improvement purposes. Your company does not have to provide you with this warning. It owns the PBX and can do as it wishes. Some security managers and executives will instruct their telecommunications managers to disable the warning tone. This allows them to use executive override to listen in on your conversations without your knowledge. You should never disable this warning tone because it is a violation of the law to monitor other people's conversations without their consent. If you do not take it on yourself to learn how to disable this tone, you can simply reply "I don't know how" if you are asked to disable it.

Your role as a telecommunications manager makes you responsible for providing technical solutions and assistance to the corporation's telecommunications needs. The corporation also expects you to ensure the integrity of its telecommunications needs. Safeguards must be in place to ensure that unauthorized people cannot gain access to your private network. If they are able to gain entry into your PBX, they can make long distance calls over your network and these calls will be billed to your company. You should also honor the trust that all employees are placing in you to ensure that their communications are private.

Almost all companies have a statement in their code of ethics emphasizing that employees should treat each other with respect. This should also mean that a company respects the rights of employees to privacy, and most companies do. The code of ethics also normally states that company property cannot be used for private use. The telephone system belongs to the company. The security department may say that an employee is using the phone system to place private calls and instruct you to activate monitoring of the employee's telephone. They are instructing you to violate the trust placed in your position. A refusal to comply may lead to your termination.

The courts have provided companies with wide latitude in the area of employee surveillance, but there is always a chance that you might be sued if you follow the directions of security. Everyone feels differently about this issue. Who is the telecommunications manager obligated to serve? You will have to follow your own conscience and personal code of ethics. I would advise that you never participate in wire tapping activity without notifying the U.S. District Attorney. Your security department should not object to this disclosure, but I am sure it will object to such notification. Why should it, if its directive is legal?

12.34 SUMMARY

Computer-controlled switching systems have replaced electromechanical switching systems in the PSTN, and this technology has also been used to replace PABXs. If a business has more than 100 stations it is probably using a CBX to provide telephone service to its employees. CBXs use a PCM/TDM switching matrix to connect calls to their desired destination. The size of the switching matrix will determine the number of calls that a CBX can support at the same time. This is referred to as the calling capacity of the switch.

The number of different devices that a CBX can support is referred to as port capacity. The ability to add more lines and/or trunks to a CBX is referred to as expansion capacity. A CBX system will be installed with the wiring in place to support a certain number of PCBs and we say the system is wired for X number of ports (lines and/or trunks). The system will also be delivered with a certain number of line circuit cards, trunk circuit cards, and DTMF receiver cards installed in the peripheral equipment shelves. This is referred to as equipped for X number of lines, trunks, and registers.

The CBX is a SPC switching system and functions much like the SPC class 5 central exchange. It is controlled by a generic program and contains a site-specific database. Although the memory of a CBX is smaller than that of a class 5 switch, it is also divided into three memory stores: program store, data store, and call store. Most CBXs have analog interface line cards and support regular 2500 telephones as well as analog proprietary telephone sets. The codecs reside on the line circuits. The latest version of CBXs will also support proprietary digital telephones (the codec resides in the telephone).

Proprietary telephones (whether analog or digital) use out-of-band signaling between a signal logic processor in the telephone and the CPU controlling the CBX. The signaling path is furnished by a second pair of wires. The voice path will be over the regular Tip and Ring pair. Thus, proprietary telephones use all four wires in the RJ-11 jack. Proprietary phones are multibutton phones and allow access to features by pushing the appropriate button. Additional telephone numbers are also assigned to several of these buttons. The line translations table for the line circuit assigned to a proprietary telephone contains a key assignment table. The systems administrator defines in this table what telephone number or feature is assigned to each button.

A PBX allows a company to construct a private network. This private network can consist of tie lines, FX lines, WATS lines, or can be constructed by using an SDN. This private network allows the company to tie all business units located across the country into one communication network. CBXs have many features to help improve communication efficiency. Using a CBX usually reduces a company's communication cost over using regular business lines from the central exchange to each station. The 1996 Telecommunications Act allows the LECs to aggressively pursue offering Centrex services. The LECs will offer both large and small Centrex. As competition in the local services area heats up, you may find Centrex a cheaper alternative than keysystems and CBXs. There is no equipment to buy. If you do not own a keysystem or CBX, no maintenance is required. When considering Centrex, make sure you know how much the LEC will charge for services such as adding a line to the system or making other changes to the database for *moves, adds, and changes (MACs)*. You will also want to know if you can be provided with access to the database to do the MACs yourself.

REVIEW QUESTIONS

1. What type of switching system are today's PBXs?
2. What is the trunk circuit between a central office and PBX called?
3. What is port capacity?
4. What is architecture?
5. What is configuration?
6. What is calling capacity?
7. What is expansion capacity?
8. What is hardware?
9. What circuit does every telephone need for connection to a switch?
10. What program contains all the features of a switch?
11. Where are features for a particular line circuit identified?
12. What is the purpose of a serial data interface to a switching system?
13. What types of memory is RAM subdivided into?
14. If a PBX does not have a battery back up system, what additional hardware should the system have?
15. What is a nonblocking network?
16. What are some sources of blocking?
17. Does the tone card in a CBX generate analog tones?
18. Does a PBX support the use of regular 2500 sets?
19. What is an MDF? An IDF?

20. What is the wiring between an RJ-11 jack and the IDF called?
21. What is the wiring between the IDF and MDF called?
22. How are PBX feature accessed when using a 2500 set?
23. How are features accessed using a proprietary set?
24. How are central office trunks accessed when using a PBX, a keysystem, and a hybrid system?
25. What is Centrex?
26. What is a glare condition?
27. What is WATS?
28. Which WATS band provides calling anywhere in the United States (except your home state)?
29. What feature do most companies want in a PBX?
30. What is an FX trunk?
31. What is VPN/SDN?
32. What is the primary purpose of a PBX?
33. What are system features?
34. Where is DNIS used?
35. What are station features?

GLOSSARY

Architecture Referes to the way components of a switching system are physically and logically organized and how these components are connected. Each manufacturer uses a proprietary design to interconnect the various components of a switching system. Thus, the architecture for CBXs will be different for each manufacturer and will even vary between the different models of switches sold by the same manufacturer. The difference in architecture makes the models different.

Average Busy Hour Peg Count (ABHPC) An average of the number of calls placed over a group of circuits during the hours when most calls occur for several days.

Average Hold Time (AHT) The average length of time a circuit is in use. It is found by dividing CCS by PC. For example, if the ABBH CCS is 500 CCS and ABBH PC is 100 PC for a five-day study, we can conclude that the average length of every call will be 5 CCs based on our sample (study).

Blocked Calls Calls that receive a fast busy signal or a recording stating all circuits are busy. When there are insufficient circuits to handle the volume of calls being placed through a switching system or over trunk circuits to a particular location, the system will block some calls.

Busy Hour The hour of a day when the most telephone usage occurs.

Calling Capacity The number of simultaneous calls that can be supported by a switching system. Sometimes this is stated as the number of calls that can be handled in one hour. Since these two approaches are not the same, I would have the salesperson state calling capacity in both ways.

Central Office (CO) Trunk Circuit A trunk circuit in a PBX/CBX that connects to the local central exchange. Most of the trunk circuits in a PBX/CBX are central office trunks, and most connect to the central office using a twisted-pair local loop.

Computerized Branch Exchange (CBX) A SPC switching system used at a private business location.

Configuration Refers to the types of interfaces, the number of interfaces, dialing patterns, station numbers, system features, station features, and other database entries that uniquely define a CBX.

Direct Inward Dial (DID) Central office trunks allowing incoming calls to bypass the switchboard operator. CBX station numbers are part of the central exchange dialing plan, and CBX stations can be reached by dialing a regular telephone number.

Direct Outward Dial (DOD) Central office trunk circuits accessed by the station user dialing a 9. The CBX operator is not needed to connect to a central office trunk.

Equipped for An installed switching system has X number of printed circuit cards installed to support X number of lines, trunks, register, and so on.

Expansion Capacity The ability to add additional equipment to a switching system. This allows your system to support your company's future communication needs.

Moves, Adds, and Changes (MACs) The changes made in the database for a SPC switch when a telephone is moved or added or has its features or number changed.

Overlay Programs The term used to describe maintenance and administrative programs that do not permanently reside in memory but are contained on a secondary storage device. These programs are only loaded into memory as needed; when loaded, they overwrite the last overlay program contained in memory.

Peg Count (PC) A term used to indicate how many calls are handled by a circuit, a group of circuits, or a switching system. This term is a holdover from electromechanical traffic registers. When the relay of the register operated, it was said to "peg the meter." Each peg of a meter made it turn a numbered dial one step.

Peripheral Equipment Equipment that exists on the periphery of a switching system matrix. Peripheral equipment interfaces external devices to the computer of a SPC switch to provide capabilities and features to these devices. Most peripheral equipment consists of line circuits to interface a telephone to the switch and trunk circuits to interface interoffice circuits to a switch. Also included in this category are dial registers, modems, and voice mail.

Port The interface point between a switching system matrix and external devices. Each switch will be equipped with a certain number of ports to handle the number of devices that will be attached tothe system. If a line circuit card is inserted in the port, it becomes a line port. If a trunk circuit card is inserted into a port, it becomes a trunk port. If a dial register card is inserted in a port, it becomes a dial register port. Line circuits usually use one prot per circuit. Some trunk circuits, dial registers, and special circuits may use two or more prots for each circuit.

Port Capacity The number of different devices that can be connected to a CBX. This definition can be misleading because some devices require more than one port. Line circuits require one port. Trunk circuits may require two ports for each circuit. DTMF register circuits may require four ports per circuit.

Printed CircuitBoard (PCB) A plastic card containing electronic circuitry, integrated circuit chips, and solder tracks connecting the circuitry.

Private Automated Branch Exchange (PABX) A term used to describe electro-mechanical switching systems at a private business location.

Private Branch Exchange (PBX) The name given to manual switchboards in a private business location. A PBX may also be small SPC switching system used by large businesses.

Traffic A term used for the volume of calls placed over a circuit, group of circuits, or a switching system. If a system is handling 10,000 calls and hour, we say the traffic is 10,000 calls an hour.

T1 CXR A TDM system used to place 24 DS0 circiuts onto one facility between switching systems. The one facility is used as a transmission medium generally consistists of a transmit cable pair and receive cable pair.

Wired for The wiring is in place to support X number of lines. A system is usually installed with "wired for" greater than "equipped for." This allows additional lines, trunks, registers, and so on. to be added to a system simply by plugging the appropriate printed circuit-board (PCB) in a vacant card slot and adding information for the PCB in the database.

13

Automatic Call Distributors

Key Terms

Automatic Call Distributor
(ACD)

Call Sequencer

Uniform Call Distributor (UCD)

Automatic Call Distributors (ACD) are switching systems originally designed to handle high volumes of incoming calls and no outgoing calls. The requirements of some telemarketing organizations have led to the design of systems that can serve as either an outbound or inbound ACDs. The original ACD was developed to handle calls to the long distance operator. In Chapter 2, we discussed the original operator's position. It was a manual switchboard with patchcords for connecting calls. The ACD replaced the cordboard with a position that had only keys and no patchcords.

All operator trunks and outgoing toll trunks were connected to the ACD. When someone dialed 0 to reach an operator, the ACD would automatically assign the incoming call to an idle operator. When the operator dialed the number desired, the ACD would attach the appropriate outgoing toll trunk to the operator trunk used by the incoming call. The operator would stay on line until the desired party answered the call. The operator would release from the call, and the ACD would signal the toll machine to keep track of conversation time for billing.

13.1 OPERATORS OR AGENTS?

The ACD has many trunk circuits connected to it and a few stations. It has many more trunks than stations. The station side of an ACD goes to the operator or agent. Operators are no longer called operators by the IECs but are called agents. This change in terminology is helpful. If an ACD is serving a telemarketing organization, a customer service department, or the operator services department of an IEC, all the people who answer the calls are called agents. The agents use headsets instead of telephones, because no ringer is needed. Calls are automatically connected to an agent on line, and no ringing signal is needed. The headset contains a transmitter and receiver.

The station wiring of an ACD system connects the line circuits of an ACD to a headset jack at each agent's position. When agents go on duty, they will sit down at their position and plug their headset into the headset jack. When the headset is plugged into the jack, line current flows from the line circuit of the ACD to provide power for the transmitter. The line circuit detects the flow of line current as the indication of a staffed position and notifies the CPU of the ACD that the position is available to accept calls. Some ACD positions also include a key in series with the headset jack to allow cutting off the headset circuit. This allows agents to make themselves unavailable for calls by operating the key. Thus, some ACD systems require the insertion of a headset and the operation of a key to signal the ACD that the position is ready to receive calls.

13.2 DISTRIBUTION OF INCOMING CALLS

The primary functions of an ACD are to answer calls, place them in a queue according to priority, and connect calls to the appropriate agent. An ACD keeps track of the amount of time that an agent has been actively engaged in handling calls. When an inbound call arrives, the ACD assigns the call to the agent that has been the least active. The ACD assigns calls on a "most idle" basis. This ensures that the workload is spread evenly across the workforce. The ACD prints out statistics on each agent. Management uses these reports to check on the productivity of each agent.

The IECs have several long distance operator centers distributed across America. I would estimate that Sprint and MCI probably have about six operator centers and AT&T probably has about ten centers. The automation of long distance calling has led to most people placing their own long distance call via the 1+ access. This has drastically reduced the number of calls handled by operators. This has allowed a reduction in the number of long distance operators and allowed for consolidation of centers.

The new SPC switching systems, SS7, high levels of multiplexing, and ACD technology have allowed a few centers to serve all operator-assisted calls. With these technologies, thousands of operator trunk circuits can be connected to one operator services center. When a call arrives at the center, the ACD automatically connects it to the most idle agent. Each center is equipped with from 50 to 200 agent positions. The center usually uses less than half of the available positions. The extra positions are available to handle emergency situations. If something happens at one center, such as equipment failure or a strike, the telecommunications manager can access the SS7 network and the SCP to redirect calls to another center. In fact, some centers shut down every night and redirect night calls to one central location.

13.3 ACDS USED BY OPERATOR SERVICES

ACD systems were designed to replace cordboards and improve the efficiency of operators in handling calls. They are designed to connect an incoming trunk circuit to the voice circuit of a staffed position. They carry out this function extremely well. With deregulation of the industry, the operator center has been called on to take care of more functions. Some operator services centers handle calls for many different operator services companies. To serve these various new functions better, the call center is equipped with an ACD, a communication server, local area network (LAN), and PCs at the agents' positions. When agents sit down at their positions to take calls, they must LOGON to the LAN.

The agents notify the ACD that the position is staffed by inserting a headset into the headset jack and pushing a key. They notify the communication server that the position is staffed by logging in to the system. The central computer notifies the ACD that the position is logged on and can accept calls. When you look at a large telemarketing or operator services center, you see many PC terminals. The PCs are not connected to the ACD. Only the headset is connected to the ACD. The PCs are connected by a LAN to a server.

The server communicates with the CPU in the ACD over an X.25 messaging link. Early ACD positions had a few key switches and a dial at each position. These were used to instruct the switching system on what to do with the call. The marriage of PCs, a server, and the ACD has eliminated keys and dials at the agent's position. Agents will provide call handling instructions via their PC to the server, which will relay these instructions to the toll switching center over the D-channel of a Primary Rate ISDN connection.

13.3.1 ACD, LAN, ISDN, and SS7 Interconnections

An operator services call center uses ACD technology married to a central computer (server) on a LAN, and PCs on the LAN. The central computer has connections to the ACD and to the serving toll office. Primary Rate ISDN is used to connect the trunks between the serving toll office and the operator center. Primary Rate ISDN is a transmission medium that carries voice on 23 channels of a T1 TDM, and all the signaling associated with those 23 channels on the 24th channel. At the operator services center the signaling channel is connected to the communications server and ACD.

When the toll office assigns a call to one of the voice channels, a signal is sent over channel 24 (the signaling channel) to notify the ACD of the voice channel assigned. The ACD will connect the incoming voice channel to an agent's position. The ACD notifies the communication server of the position assigned to the call. The server also receives information from the toll office about the call. This information is sent over the signaling channel and may include: (1) the originating telephone number, (2) the originating class of service (residence, business, pay phone, hotel), (3) the calling card number (if it has been entered at the toll center), (4) the results of any calling card validation performed by the toll center, and (5) how the customer accesses the operator center.

This information is used by the communication server to determine how the call should be handled. The server will forward information over the LAN to the PC, at the position assigned by the ACD. The X.25 messaging link between the ACD and the central computer is used to keep each system apprised of what the other is doing. It is critical that the central computer know where an incoming call was assigned by the ACD, because the central computer must send information about the call and how to handle the call to the same position.

13.3.2 *Multiple Operator Service Providers in One Center*

As mentioned earlier, some operator services centers handle calls for multiple operator services companies. If I have my own operator services company and I have a signed agreement from an IEC to handle calls from my customers, I do not need any operators. I will provide a listing of my customers to the IEC and will provide certain 800 numbers for my customers to use. Calls placed to the 800 number will be routed to the IEC's operator center. When they arrive, the signaling channel will inform the center which 800 number the calls were placed to. The server has a listing of these 800 numbers and can identify these calls as coming from an "XYZ" company customer.

The server will provide information to the PC at the position of the agent assigned by the ACD. This information results in the PC painting the screen with a custom screen for my company (XYZ) and includes a greeting. The agent will read the greeting to the customer. The greeting is something similar to "Thank you for using XYZ operator services, how may I help you?" The special access numbers used by customers to reach their operator services company are directed to an IEC center using *dialed number identification system (DNIS)* discussed in the last chapter. This allows one center to handle calls for multiple service providers. If a LEC wants to shut down its operator services centers, it can make arrangements for another operator services provider to handle its customers. The LEC could simply sign a contract with an IEC, shut down its operator center, and redirect all operator traffic to the IEC.

The signaling channel on the Primary Rate ISDN is tied into the SS7 network. The database for directing calls to operator services centers is contained in

SCPs on the SS7 network. The database can be changed to redirect calls. The SS7 network also connects to SCPs that contain information about LEC calling cards. Toll switching centers will launch queries over the SS7 network to validate LEC calling card numbers. Operator services center agents will also validate LEC calling cards over the SS7 network. Each LEC that issues calling cards maintains its own SCP calling card database. Any operator services center can access these databases. A record is kept at each database of each access and which company made the access. The owner of the calling card database charges each company for access to the database.

The calling cards issued by LECs will be honored by an IEC with which they have an agreement to provide access to the database. IECs do not allow other IECs access to their database. For this reason, IECs do not honor each other's calling cards. Using an IEC calling card requires that you use that IEC for your toll call. The telephone being used must be PICed to the IEC you desire, or you must dial the Feature Group B access code (10XXX) to reach the desired IEC.

When an agent receives a call that cannot be billed as requested, it will offer an alternative billing method. If your LEC calling card does not pass validation, or you attempted to use the calling card of another IEC, the agent will inform you of the problem and ask if you would like to use a different card or bill to a third number. The IEC does not want to lose your business and will make every attempt to find a billing method for your call.

Many IECs handle calls from prison phones. These calls are placed from special telephones at the prison. These telephones cannot receive incoming calls. The telephones do have numbers assigned to them, but the switching system they connect to will not allow incoming calls. The telephone number is assigned to provide identification of the telephone to an operator. The database in the originating switch for these telephones restrict all types of calls but will allow calls to the operator.

The prisons have a contract with a specific operator services center. All telephone calls from the prison are connected to this designated operator services center. When the call is connected to the operator services center, the server at the center verifies the originating telephone number and class of service. The server contains a database that has a listing of all prison telephone numbers assigned to the center. The database will verify that the call is originating from a prison. The agent assigned to this call is informed about the origination of the call and which billing options are allowed.

The server forwards information about the call over the LAN to the position assigned to the call by the ACD. The PC at the agent's position receives the call from the ACD and receives billing option information from the server. The screen is painted with information about the call and with the available options. Calls from prison can only be made as collect calls. The call will be placed and the called party will be asked if they will accept charges. A validation of the called party's telephone number will be made to ensure it is listed in their name and is not a public telephone or pay station. If the called phone is listed in the name of the called party and they accept charges, the call is completed.

13.4 ACDS AND TELEMARKETING CENTERS

The ACD has found tremendous application in telemarketing centers. Many companies use in-WATS (800) telephone numbers for their marketing departments. When products are advertised on television, a toll-free 800 telephone number is listed for customers to call. The advertising company can establish its own ACD center or can contract with a company providing ACD services. Each company that contracts with an ACD service provider will provide a listing of the in-WATS numbers it uses and the products it is selling.

Each 800 number coming into the ACD has its own unique connection to the ACD. This arrangement allows the ACD to identify which company the caller was calling. The call will be connected to an idle agent trained to handle customers of the called company. Some call centers use PRI ISDN and DNIS to identify both the originating customer and the company being called. A catalog services company can use this technology to better serve customers.

When a call arrives at the ACD center, ISDN provides the originating telephone number to the communication server, and if DNIS is used, it provides the telephone number dialed by the customer. The server uses this information to pull up a database on the customer. The server forwards information from the database to the PC of the assigned agent. The agent will be provided with the name, address, telephone number, billing data, and order history of the calling customer. This allows the agent to answer the call with a greeting that includes the customer's name: "Good Afternoon, Ms. Jones, how are you today?" If the customer is inquiring about a past order, the agent has all that information on the screen. If additional information is needed, the agent can use the PC to access the centralized database over the LAN.

13.5 AUTOMATED ATTENDANT

Many service centers will use an ACD with an automated attendant option. The caller is asked to provide information to assist in identifying how the call should be routed by the ACD. Agents are assigned to specific queues based on their qualifications. A certain group of agents will handle calls for service information, another group of agents will handle marketing and sales calls, and another group of agents will handle technical questions. An ACD center may have many queues, and when an agent LOGIN occurs, the system assigns them to a queue based on their qualifications. The caller will be asked to identify their needs by entering a DTMF digit on their telephone keypad. As the caller responds to the various prompts of the automated attendant, they are providing instructions to the ACD. The ACD uses this information to connect the caller to the appropriate queue and an agent in that queue.

The use of an automated attendant allows callers to leave messages when they do not need to talk to an agent. Credit card companies make use of automated attendant and *interactive voice response (IVR)* to provide account balances and information on recent transactions. The caller can access these services or can use the DTMF code that will instruct the ACD to connect an agent to the call. Some companies overuse automated attendant. Customers routed from queue to queue only to encounter extensive delays in the queues will not be pleased with the service they are receiving.

Knowledgeable customers will refuse to provide DTMF instructions and pretend that they are using a dial phone. The absence of a DTMF command will route the call into an agent queue immediately. ACDs keep track of the number of calls waiting in each queue and the average waiting time in each queue. This information is used to determine which queue a call should be placed in when several queues are qualified to handle the call. This information can also be used to notify callers of their status in the queue.

13.6 AGENT STATIONS

Many ACDs are provided with proprietary telephones that have a liquid crystal display (LCD). The display can provide some information on the call. This type of telephone with a sophisticated ACD system enables an ACD to handle calls without the use of PCs and a LAN. Many telemarketing organizations use this type of station equipment. Some ACDs will use a video display terminal instead of an LCD to enhance the amount of information the ACD can provide to the agent. What the agent uses for station apparatus will vary according to the applications handled by the center. The station apparatus will vary from a headset, or regular 2500 telephone, to a video display telephone. An ACD can provide a lot of information about a call to the agent with an LCD or video display, but when the center handles many types of various services, an adjunct LAN and PC network is required. In these cases the station apparatus of an agent is a headset connected to the ACD and a PC connected to a LAN.

The ACD has a communications port connected to the communications port of a high-powered PC server on the LAN. This server receives call detail information from the ACD and from the ISDN connection to the PSTN. The ACD, server, LAN, and PC arrangement provide a center with the capability to handle calls for many different customers. This arrangement allows for providing distinctive call handling treatment to each class of callers. The caller can be identified on entrance to the ACD, and the agent will have information about the caller that allows for personalized service. Special high-priority customers can be identified early, and the ACD will bypass all queues and connect these callers immediately to highly qualified special agents for VIP (very important person) treatment. Airline companies use ACDs with call screening capabilities to ensure that calls from a travel agent are connected immediately to an agent on the ACD.

The station apparatus to select for an agent depends on the amount of information the agent needs to process a call. If the agent needs no information, a simple 2500 telephone set will suffice. If the agent needs information about the call—such as how long the caller was held in queue, what the caller's telephone number is, what trunk group the customer came on, what the current status of the ACD system is, and what the status of calls in the system is—a video display terminal is needed. If the agent needs help in determining how the caller's wishes should be handled or needs access to a database to assist in completing the customer's request, a PC and LAN connected to the ACD are needed.

13.7 MANAGEMENT AND ADMINISTRATIVE FEATURES

An ACD system provides many administrative features to assist managers in providing the best possible service at the lowest possible cost. The three major features are reports, supervisory terminals, and scheduling assistance. Reports are generated on all facets of an ACD from the number of calls arriving over each trunk to the workload handled by each agent. Managers can use the generated reports to evaluate how the ACD system is performing and to assess the performance of each agent. The supervisory terminals provide a real-time display of current operations as well as displaying any requested report.

One supervisory terminal is usually dedicated to displaying current staffing and workload. A supervisor can glance at this screen and see information such as number of calls waiting in queues, average waiting time, the number and location of agents actively logged into the system, and the number of callers that have disconnected prematurely. The supervisor can use one of the supervisory terminals to monitor calls handled by an agent. He or she will have a visual display of the current call and how it is being handled as well as being able to monitor the conversation. This display will also provide a summary on the workload of the agent. It can also provide the time an agent logged in to the ACD, when the agent went on break and came back, when the employee went to lunch, and returned, how many calls the agent has handled during the current shift, and the average time on each call. The supervisor can use this information to determine if the agent needs additional training, supervision, or counseling.

Some call centers require the agents to do record work on each call such as processing the customer's order. This time is referred to as "wrap-up" time. The agent will usually operate a key on the terminal that notifies the ACD that the person is in wrap-up and not available for calls. Wrap-up time is also included as part of the agent information provided on a supervisor's display terminal. In addition, the supervisor's terminal allows a supervisor or manager to change the assignment of agents to various queues, to change the routing of calls, and to view a report on current performance of the center.

The ACD will provide standard reports on various functions of the ACD. Many manufacturers will also provide customized reports or the opportunity for a user to customize reports. Reports can be provided on current activities (real-time reports) and on activity over a period of time. The ACD has the capability to store information on each agent, each trunk, each queue, each trunk group, and each hour of activity.

This allows the system to generate reports for any period of time desired. The ACD will use information generated over the past month and past year to determine the average hold time on all calls, the number of calls handled in each hour, and the service levels provide at various call and staffing levels. The ACD can use this information to forecast future call volume and the number of agents that will be needed to handle that call volume. The manager of a call center can determine the level of service they wish to provide callers (that is, how long do you want callers to wait in queue?). The manager provides the ACD with the level of service desired, and it will generate a schedule for the week or month in question.

Reports can be generated for each agent over a stated period of time and printed out or displayed at the supervisor's terminal. Real-time activity and reports on an agent can also be displayed on the supervisor's terminal. Some of the indices that are important to the ACD center management are: productivity of individual agents, average answer time for the center, average work time per call for the center, average time customers are held in queues, average number of calls held in queue, percent of time all agents are busy, percent of time all trunks are busy, percent of calls abandoned, percent of time that a position is staffed, and percent of time a staffed position is available to answer calls. Reports are generated at scheduled intervals on all of these indices. These reports can also be generated on demand. If the reporting capabilities of an ACD do not satisfy your requirements, the data can be directed to an EIA-232 port. This port can be connected to a modem. This allows the ACD to send data to a remote computer. A programmer can provide programming for the remote computer that will generate customized reports or spreadsheets using the data provided by the ACD.

An ACD can provide a report on the performance of the center for any period of time. It can provide a report on the actual performance of the center as it was staffed as well as a report on what the performance would have been if staffed according to the schedule. This report will indicate the number of agents scheduled, the number of absent agents, the service level provided, and what the service level would have been if no agents had been absent. The ACD can also provide a report on each call handled. This transaction audit trail provides the time of each event on a call and will indicate how an agent handled a specific call.

13.8 STAND-ALONE AND INTEGRATED ACDS

All large ACD applications require the use of a stand-alone ACD system (Figure 13-1). This system has more processing power, contains more advanced software and features, and allows for integration to a LAN. A large service center needs the sophistication that only a stand-alone ACD can provide. The major drawback to a stand-alone ACD system is its inability to forward calls to employees not attached as agents to the ACD. If you call the 800 service number of your credit card company, you get an agent that can service your call. If you ask that agent to forward you to a manager in another department, it cannot be done. All other employees are on a PBX system, not on the

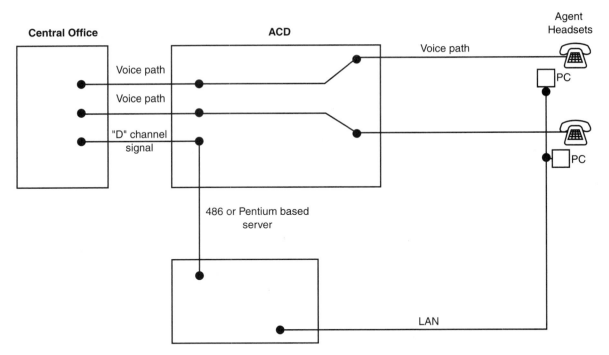

FIGURE 13-1   Automatic Call Distributor with local area network (LAN).

ACD system. Some ACD systems provide trunk circuits that can be connected to a PBX. This allows transfer of calls from agents on the ACD to stations on the PBX. Many companies do not purchase this capability because of the cost involved.

The ability to transfer calls from agents on an ACD system to station users on a PBX can be achieved easily if the ACD is integrated with a PBX system. An integrated PBX/ACD system is preferred when the system is designed to serve only one company. Customers of the company can reach the company via the PBX or ACD and can be transferred between the two systems. Companies can provide a higher quality of service to customers with the integrated system. An existing PBX can be retrofitted with an ACD. If a company is already using a PBX and is evaluating the purchase of an ACD, a low-cost option might be to add the ACD feature to the PBX. An integrated system will not offer all the features obtainable with a stand-alone ACD but may offer all the needed features at much lower cost. The integrated system provides the ability to transfer calls from agent positions on the ACD to stations on the PBX.

13.9 OUTBOUND CALLING

ACDs were originally designed as a means of distributing incoming calls to agents. Many telemarketing organizations use ACDs with outbound calling capabilities. The ACD has a program that will automatically place calls to desired customers.

The telephone numbers of potential customers are purchased from companies that offer this service. The telephone numbers are provided on a magnetic storage medium such as floppy disks and tapes. These storage mediums are accessed by the ACD and the calling program to place calls to potential customers. The ACD has voice recognition technology built into the trunks that place the outgoing calls. Using this technology, the ACD recognizes when calls have been answered and will connect an agent or a recorded message to the call.

When you receive calls from these telemarketing firms, there is a slight delay before the agent or recording is connected. I have answered the telephone and said "hello" without receiving a response. The ACD is in the process of attaching an agent to the call. Usually by the time I say a second "hello," the agent is on line and informs me of the current carpet cleaning special, insurance special, landscaping special, or whatever. The use of an ACD for outbound calls makes the agents more productive. They do not have to dial calls. Agents do not waste time on calls to busy or unanswered numbers. The ACD provides agents with the name and telephone number of the call, and they can start their pitch with a personalized greeting.

The dialing software of some ACDs is designed so that it will not place a call if all agents are busy. This ensures that an agent is available to service the call. Sometimes the software is designed to place calls based on the anticipation that an agent will be free by the end of dialing. This type of dialing software is called *predictive dialing*. The software is making a prediction on when an agent will be available, based on average holding time of calls and when current calls were assigned to agents. The system predicts how long it will be before the next agent will be available and will start dialing a call a short time before that agent is available. If the prediction is wrong, a call is placed before an agent is available. This can result in some additional delay between the time the customer answers a call and the connection to an agent, but predictive dialing will increase agent productivity. If this delay becomes too long, the called party perceives the call as a nuisance call and will hang up. You can use what you know about outbound ACDs to avoid telemarketers. If you do not receive a response after answering "hello," it is a telemarketing call and you can hang up immediately. If the call is coming from an ACD that is placing calls before agents are idle, you can usually hang up before the agent comes on line. The telemarketing organization is depending on the called number to stay on line and say "hello" several times before hanging up. Also note that these outbound calling systems do not want to get an answering machine, so they are programmed to hang up after three rings. You can avoid most telemarketing calls by not answering calls until after a fourth ring.

13.10 UNIFORM CALL DISTRIBUTORS

The precursor to ACDs was the ***uniform call distributor (UCD)***. Most PBXs offered the UCD as an option. The UCD connects calls to agents in a sequential fashion. The first agent receives the most calls and the last agent in a queue receives the fewest calls. The ACD levels the load of all agents by directing call connection on a "most idle" basis. The UCD is not equipped with the sophistication needed to achieve load leveling. The UCD will provide reports and call management features but will not provide as many options as a stand-alone ACD. The processor, memory size, and supervisory terminal of a UCD are far less sophisticated than those of an ACD and simply cannot provide all the management tools and reports provided by an ACD.

Most UCDs use a display telephone for the supervisory terminal and regular 2500 sets or headsets for the agent station equipment. Some UCDs allow the use of a video display terminal at the agent's position. A UCD cannot provide all the capabilities of an ACD, but what it provides is enough for many call centers. In fact, some centers may not even require a UCD and may use a call sequencer to handle calls. The *call sequencer* is the least sophisticated call distributor. Sequencers use proprietary agent stations with an LCD. The display indicates which incoming lines have calls in progress as well as which lines have not been answered. An indication is provided on how long calls have been waiting for an answer. Call sequencers only indicate that calls are waiting to be answered. It takes action by the agent to connect a call to the agent. The agent will determine which call has been waiting the longest and depress a button associated with that line. A call sequencer offers no management features, but they may be the perfect device to use in a small call center.

13.11 SUMMARY

Incoming call centers are diverse, ranging from small centers serving one company to large telemarketing centers and operator services centers for LECs and IECs. This wide range of diversity requires many different solutions. There is no one technology that fits all when price is a factor in the business decision. An incoming call center can use technology as simple as a keysystem to handle calls. Even the smallest keysystem can handle three to six incoming lines. If the call volume is low, a keysystem is a very satisfactory, low-cost solution. As the calling volume grows, the technology required advances to a call sequencer, UCD, and ACD. The largest centers will use a stand-alone ACD.

A call center that handles multiple companies or requires access to a database for order processing will require a LAN in addition to the ACD. Most large centers will use Primary Rate ISDN over T1 carrier systems to provide access to or from the PSTN. The arriving calls will include information about who customers were calling by using DNIS, and the customer will be identified by calling number identification. This allows the telemarketing organization to use customized greetings for each company and calling customer. Calling number identification and a LAN will allow providing the agent with information about the customer from a database. This feature will allow agents to handle customer complaints efficiently and effectively. If credit card customers calls their service bureau to inquire about a transaction or the general status of their account, it is mandatory that agents have access to the LAN and the customer database.

The ACD is capable of identifying callers (based on DNIS information) and will route them to the agent qualified to handle their call. The marriage of an outboard LAN, with a server and agent PCs, to the ACD greatly enhances the services a call center can provide. When evaluating which call distribution technology to use, you must determine if communication links are required between the ACD and stations on the company's PBX system. Integrated ACD/PBX systems offer a simple way of meeting this requirement, but stand-alone ACDs can also be provided with this capability.

The amount of management software and reporting capabilities provided by the various systems also becomes a major factor in deciding which technology to use for a call center. The reliability of the technology should be considered, too. Should battery backup or UPS be provided to protect against power outage failures? Does the system use duplicate processing technology to protect against failure of a CPU? How long does it take the system to reload after a power outage? Will the system provide call transfer to designated stations during a power failure? Is the system expandable? What is the maximum number of trunks and agents the system can support? How many separate queues can the system accommodate? What is the maximum number of trunk groups the system will support? Does the system provide DNIS capability? Will the system accommodate ISDN trunks, SS7, and X.25 interfaces? How many supervisory terminals will the system support? What are the maintenance requirements? How is maintenance performed? Will the system provide EIA-232 interfaces for report generation and remote maintenance? Each application will require different answers to these questions. Once the requirements of a call center are determined, the right technology can be chosen by asking the appropriate questions.

REVIEW QUESTIONS

1. Does an ACD have more stations or trunks?
2. Where are ACD systems found? Where were they first used?
3. How does an ACD determine which agent should be assigned to a call?
4. How does an ACD know that an agent position is staffed?
5. What benefit is gained by the marriage of a LAN to an ACD?
6. Why would you use PRI ISDN to connect an ACD to the central office?
7. What are some methods used to provide customer identification to an ACD?
8. What kind of equipment is used for agent stations?
9. What are some major differences between an ACD and a UCD?

10. What kind of information is available on a supervisor's terminal?

11. What is wrap-up time?

12. How can a manger determine staffing requirements?

13. What is the major drawback to a stand-alone ACD?

14. What is a predictive dialer?

15. How does a predictive dialer improve agent productivity?

16. Does a telemarketing center need to use an ACD?

17. Are the PC terminals connected to an ACD?

18. How can a person avoid telemarketing calls?

19. How does the SS7 network work with an ACD?

20. How is DNIS used by an ACD?

GLOSSARY

Automatic Call Distributor (ACD) A switching system that has more trunks than lines attached to it. ACDs can serve as inbound systems and will automatically connect an incoming call to an agent. They can also serve as outbound systems and will automatically dial telephone numbers. When the called party answers, an agent is connected to the call. ACDs distribute calls to agents according to which agent has been most idle (least busy).

Call Sequencer A system that uses keysystems and beehive lamp fields to indicate the presence of an incoming call to a call center. The agents answer the calls by pushing a button on the keysystem.

Uniform Call Distribitor (UCD) The predecessor of ACDs allocated calls to agents in a sequential fashion. The first agent received most calls. UCDs had limited supervisory and reporting capabilities.

14

Voice Mail
and
Automated Attendants

Early analog voice recording applications were limited to redundant applications such as weather announcement machines. Analog recordings were placed on magnetic drums and were played back to callers. These mechanical devices were large and had limited capabilities. With their limited capabilities, they had limited applications. The invention of the codec has changed the technology used for recording voice. The codec converts analog signals into digital signals. This conversion allows messages to be stored on hard disk drives in a PC. Small messages can even be stored in ROM or EEPROM. This digital technology has broadened the applications for recorded messages. This technology finds applications in answering machines, keysytems, PABXs, PCs, modems, voice mail, and many other specialized applications.

14.1 VOICE MAIL

Voice mail systems are designed that will work on a stand-alone PC or in an integrated system with a PBX. When voice mail is installed on a PC, the PC will need to be equipped with multimedia capabilities and a modem that has voice handling capabilities. Multimedia involves the installation of an audio card, speakers, a microphone, and associated multimedia software. Multimedia provides the user with the capability to record announcement messages. Several modem manufacturers sell modems that provide data, fax, and voice capabilities. These modems include a small transmitter and speaker that allow for recording and playback of messages if you do not have multimedia capabilities.

The modem manufacturer provides call management software with the modem. The modem can detect when an incoming call is a voice call by the lack of handshaking protocols used between modems. When a call is answered by a modem and receives no signal from another modem, it knows the call is a voice call. When the call management software determines that a call is a voice call, it will read the greeting message from the disk drive. The recorded greeting message is stored in digital format on the hard disk of the PC.

When a caller is answered by the modem and call management software, the greeting message is read from the disk and converted by the codec on the modem to an analog voice message. The caller hears this message and responds accordingly. A typical message might instruct the caller to key the digit 1 to leave a message for person A, key the digit 2 to leave a message for person B, and so on. The modem contains a DTMF receiver chip that decodes the digit received, and the call management software uses this digit to place a message in the appropriate mailbox. When the call management software was installed on the PC, a mailbox was established for each user. The mailbox is a directory designation on the hard drive.

Received messages are converted to digital code by the codec, and call management software places the digital message in a file of the appropriate directory (mailbox). To retrieve messages, users can LOGIN to the computer and activate the call management software and then retrieve messages via their multimedia sound card and speakers (or playback to the speaker of the modem). The combination of voice-capable modem, multimedia PC, and call management software replaces an answering machine. The latest version of call management software includes caller ID capabilities. This allows the PC to capture caller ID on all calls. The user will know who all callers were, even if they did not leave a message.

14.2 VOICE MAIL IN A PBX

In the PBX environment, PBX station users can access their voice mailbox and record a personalized greeting. This greeting is converted by a codec into a digital format and stored on a hard disk drive. Users can forward all calls to their mailbox when they are out of the office. Most PBXs are programmed to automatically forward unanswered calls to the appropriate mailbox after two rings. Calls are also automatically forwarded to the mailbox when the station is busy. When callers are transferred to the mailbox, the disk drive is accessed and the location belonging to the subject mailbox is addressed for a readout of the file at that location. When the file is read back through the codec interface, callers hear the recorded greeting.

Voice mail provides callers with an option of leaving a message or connecting to another station number. If they leave a message, the codec converts the message into a digital format. The digital message is placed in a file on the hard disk drive, and the address location on the disk drive is placed in the users' mailbox. When users call in to read messages in their mailbox, messages are retrieved from the disk drive and processed through the codec. The codec converts the digital messages to analog, and the users hear the recorded message. Voice mail relieves clerical support of the chore of taking messages. It also provides for private messages. The security department and PBX maintenance personnel can access your messages, but the secretary cannot unless you provide that individual with your password. This allows people to leave messages of a personal nature that they would not want to pass through a third party.

Voice mail has greatly reduced phone tag and allows us to be more productive. If you are busy when a call rings on your phone, you can continue what you were doing and let the call go to voice mail. When you are ready to accept new business, you can collect your messages and return calls. This feature is particularly useful when you are discussing business in your office with another employee. Letting the call go to voice mail is preferred over cutting someone off just so you can answer the phone. Voice mail can actually serve as a substitute for person-to-person conversations. A caller can leave a message asking for information. If you get the voice mail, you can return the call and leave a message containing the desired information.

14.3 VOICE MAIL'S AUTOMATED ATTENDANT

The integrated PBX and voice mail system allows a caller to reach the PBX attendant at any time by keying a 0 on a DTMF pad. Calls will be transferred automatically to voice mail when the called station is busy or if the station user has activated call forwarded to voice mail. These systems usually have an automated attendant feature, with a message that states the user is out of the office, and you can leave a message in voice mail, dial another extension number, or reach the attendant by dialing 0. This escape from voice mail feature has greatly enhanced the acceptance and use of voice mail systems by large companies.

In the past, many users were frustrated at not being able to redirect their calls after being transferred to voice mail. All they could do was leave a message. If they wanted to talk to another person, they had to hang up and dial back on a second call. This could get quite frustrating if several people they were trying to reach had their phones forwarded to voice mail. The escape capability was developed in response to early criticism of voice mail. Additional capabilities developed for voice mail systems include the following: the ability to have voice mail inform calling parties that they have reached your mailbox with a message indicating whether your phone is busy or you are out of the office; an option allowing the voice mail to transfer calls to a personal secretary instead of the attendant; messages from internal stations that include the name of the calling party; messages bearing a date and time stamp.

14.4 ACCESS TO VOICE MAIL

In a PBX system with integrated voice mail and using multiline proprietary telephones, the station user accesses voice mail by pushing a button assigned to voice mail. The light beneath this button will light when the user has mail waiting to be read. Simply by pushing the button, the station user is connected to a voice mailbox. The user can read mail or compose a new greeting message. The user will have options that allow purging or saving read messages. A central number is also provided for access to the voice mail system. This number provides access to voice mail from a standard 500/2500 station set; the number is used when a user is out of the office and needs to access a mailbox from a distant location.

The central access number is also used when someone wants to leave a message without going through your PBX station number. The user will dial the central telephone number attached to the mail system. The mail system will answer the call with a recorded message asking the caller to key in the four-digit number of the mailbox that the message is intended for. After keying in the four-digit number, the caller will hear the standard greeting message or the personalized message. The greeting includes the name of the person assigned to the mailbox. The name makes it clear whether the caller is connected to the desired mailbox. Occasions may exist when someone wishes to leave a voice mail and not go through the PBX. For example, you may be out of the office and someone else is using your desk and phone on a temporary basis. People can use the direct access to leave messages.

To retrieve messages, the person assigned to a voice mailbox pushes the button on the telephone, which is programmed by the PBX for voice mail access. Someone not at work can access a mailbox by dialing the central voice mail number. On being prompted to dial the desired mailbox, the user will key in the pound sign (#), the number of the mailbox, and a security code. The voice mail system will then allow access to any messages. The user will be able to perform any function from a remote location that could be performed from the PBX station. One handy feature of voice mail is the ability to append remarks to a message and forward the message with remarks to another mailbox. Once a person has accessed the mailbox, it is possible to perform a fast scan of all messages, scan unread messages only, save messages, delete messages, record new greetings, fast forward, reverse, pause, resume, connect to another mailbox to leave a message, and so on.

14.5 MAILBOXES FOR PBX STATIONS AND PUBLIC VOICE MAILBOXES

With an integrated voice mail system, each station on the PBX is assigned to a corresponding voice mailbox. When a mailbox needs to be assigned to someone not on the PBX system, it is called a *guest mailbox*. All guest calls are accessed via the central number for voice mail. Many companies provide voice mail to the general public by using a stand-alone voice mail system with all guest assignments. These systems usually include capabilities that will automatically call a pager, or mobile phone, to notify the guest that a message is waiting.

A disadvantage of older voice mail systems was the requirement for the caller to enter the mailbox number for the intended recipient of the message. Newer systems overcome this disadvantage by using dialed number identification system (DNIS) trunks. The voice mail company contracts with the LEC for several hundred telephone numbers. The voice mail customer is assigned a specific seven-digit telephone number by the voice mail company. The company makes arrangements with the LEC to have all called numbers forwarded to the DNIS trunks connected to the voice mail system.

When the call arrives over the DNIS trunk circuit, the called number is identified to the voice mail system. It uses this DNIS information to connect the caller to the right mailbox. Many mobile telephone systems are equipped with voice mail access as an option for their customers. If a call is not answered, it is forwarded to a voice mailbox and the caller can leave a message. A communication link is established between the mobile system and the voice mail system, which provides identification (DNIS) of the called mobile phone number to the voice mail system.

14.6 *VOICE MAILBOX IDENTIFICATION*

A public voice mail system connects to a PBX or central office via trunk ports. A voice mail service bureau will receive the identity of the mailbox desired by a caller over DNIS trunks from the central office, or by having the caller key in the four-digit mailbox number. A voice mail system integrated to a PBX will receive the identity of the desired mailbox from the microprocessor of the PBX. A communication link is installed between the processor of the voice mail system and the processor of the PBX. The ability of the microprocessors in the PBX and voice mail systems to communicate with each other provides for the integration of the systems.

The ability of the PBX to let the voice mail system know which PBX extension was called allows the system to store a message in the proper mailbox. This *party identification* feature is the feature that allows the PBX to automatically forward unanswered or busy station calls to the voice mail system. Without called party identification, the voice mail would require the calling party to enter a four-digit mailbox number before leaving a message. This leads to lost messages, because callers would not always know the mailbox numbers of people they are calling.

14.7 *VOICE MAIL CAPACITY*

How many users can be served by a voice mail system will depend on the number of access ports that connect calls to the system, the amount of disk storage provided; how long messages are; and how much the messages are compressed before storage on the disk. Disks are usually rated on how many hours of storage they can hold. The number of callers that must be connected to the system at the same time determines the number of ports needed.

The amount of disk storage needed by a voice mail system depends on the number of users, the number of messages per user, and the length of these messages. Many voice mail systems will cut off callers that try to leave a very long message. The voice mail systems administrator monitors the system to ensure that users are not placing long greeting messages on the system. The administrator also checks to ensure that users are deleting old messages from the system. Many systems will limit the amount of disk space available to each user. Most users are allocated three to five minutes of disk space for greetings and received messages. Users will not be able to receive additional messages when their disk allocation is exhausted. This limitation guarantees that users will delete old messages and keep their mailbox clean.

The voice mail system will normally monitor the use of disk space and will automatically delete old messages when the disk is full. Voice mail systems usually employ the same 8000-sample-per-second technique used in PCM, but a compression technique is used to reduce storage requirements. The high sample rate achieves good reproduction of voice on playback, and the compression technique reduces disk space storage requirements. One of the compression techniques used simply compresses the quiet intervals between words. Some voice mail systems also use compression techniques that compress the voice as well as the silent interval. Compressing the voice, can lead to alterations in the voice, and it may not be as clear and crisp as uncompressed voice.

When purchasing a voice mail system, the quality of messages on playback should be evaluated. Compression of voice lowers the cost of a system. We must evaluate cost and quality when making our purchase of a voice mail system. Cost per hour of storage should not be the only factor affecting our purchase decision. The amount of storage needed can be significantly reduced by providing only a system greeting. If your application requires allowing each user to record an individual greeting, the disk size will need to be larger. The number of ports needed should also be a consideration. Small or medium-sized companies can probably satisfy their needs with a 12-port, 12-hour voice mail system.

14.8 VOICE MAIL TECHNOLOGY AND THE INTERNET

Voice mail technology has also found applications on the Internet. Companies are beginning to offer audio applications over the Internet. Some systems provide synchronization between written text displayed on the screen and an audio file played back through the speakers of your multimedia-equipped PC. When you access a site that provides audio playback capabilities, the quality of playback depends on the speed of data transfer. Although these systems support 14.4 modems, the quality is much improved when 28.8-kbps transfer is used. Over the next few years, I anticipate improvements in voice mail applications over the Internet. The improved technology developed for Internet applications will benefit all voice mail applications.

Software has also been developed that allows a multimedia-equipped PC to send voice calls over the Internet. Sending telephone calls over the Internet is nothing more than real-time voice mail. Instead of placing a message on a disk drive for later playback, the message is transmitted over the data network and played back in real time at the distant end. The data network is a packet switched network, and the quality of packetized voice is not as good as the voice on a circuit switched network (that is, the PSTN), but the quality provided is sufficient for many people. Improvements are being made in packet and fast-packet network technologies. These improvements will lead to higher quality for voice applications placed over these data networks.

14.9 AUDIOTEXT

Audiotext is a feature of voice mail that provides a replacement for the old analog recorded announcement machines. The analog weather announcement machines have been replaced with digital announcers. These digital announcers can be used in any application that requires the distribution of information to a large audience. When people call the system with audiotext, they can select an option and the message associated with that option will be played back. This selection feature is referred to as *interactive voice response (IVR)*. With IVR, callers receive voice prompts from the system asking them to guide their call through the system using their DTMF pad to input information. This feature allows people to call a credit card company or bank and get information on an account. The IVR feature will prompt them to enter an account number and password. It will then prompt them to enter 1 for account balance, 2 for last transaction, 3 for interest paid, 0 to talk with a representative, and so on. The message played back depends on the code entered.

Audiotext has access to the company's database in order to access the information a caller is requesting. The needed information is converted from data into voice by a voice synthesizer. Audiotext provides the ability to reenter IVR prompts. This allows a caller to replay a message and to play many different messages from the audiotext. Thus, a customer can call the audiotext number and get all the information desired on an account. Some audiotext systems are integrated with voice mail and the PBX. These integrated systems allow someone to exit audiotext and leave a message for a particular person. They will also allow an individual to key in a particular PBX extension or 0 for the attendant. The PBX will connect the caller to the desired party.

14.10 AUTOMATED ATTENDANT

The integration of audiotext and IVR also provides the feature called *automated attendant*. A system with automated attendant will ask callers to identify the information they need by keying information in from their DTMF pad. Usually, they will be instructed to dial the PBX extension number of the person desired, or 0 for the attendant, or if they have a rotary dial phone, they are asked to wait and an attendant will answer their call. If an automated attendant offers too many options and provides too many queues for holding customers, callers will simply pretend to have a rotary dial phone. This causes their call to connect immediately to a person and not a recorded message.

Automated attendant can be taken to extremes. Abuse of automated attendant is frowned on by many customers. The Internal Revenue Service alienates many callers by holding them in queue for a half hour before an agent is attached to the call. The agent is not causing the problem. The problem is caused by management's failure to properly staff with the amount of agents needed to handle the call volume experienced. Although the agents do not cause the problem, they receive the brunt of customer complaints. Automated attendant is no substitute for expeditiously handling customers if you are concerned about customer relations. When customers call your company, they are either interested in buying your product or have a problem. It makes no sense to alienate the very people that are critical to the future of your company. Even government agencies that are guaranteed a future regardless of customer relations should practice prompt handling of customer inquires. By definition, government agencies have a fiduciary trust: to serve the public. They violate this trust if they abuse automated attendant.

14.11 DID VIA AUTOMATED ATTENDANT

Although an improperly designed automated attendant can lead to poor customer relations, a system that is properly designed can improve customer relations. One of the biggest advantages of automated attendant is the ability to provide DID with one central number and without expensive DID trunks. Customers dialing one centralized number will be greeted by the automated attendant and asked to input the station number desired. If they do not know the station number, they can get the attendant by keying a 0. The system can also speed up calls by providing single-digit access to department attendants. A person calling a large company can direct their call to the marketing department, the engineering department, the customer service department, the technical department, and so on. This feature is very useful for a company serving two or more products.

The design of automated attendant prompts, the number of prompts, and the levels of prompts are critical to the acceptance of automated attendant. Although automated attendant, audiotext, and interactive voice response can be implemented separately, they are usually implemented in a package that works together, and relies on the caller having a touchtone phone, in order to route the call through the system. These voice technologies enable a company to enhance its operations, reduce staffing, and increase productivity when used properly. They can also improve customer relations when used properly.

Many callers do not reach their intended party on the first call attempt. Voice mail provides a vehicle to capture these first call attempts and eliminate lost calls (and lost business). It is critical that calls answered by voice mail are responded to immediately. If customers are called back promptly, they will be receptive to leaving messages in voice mail. When people have a complaint, they usually want to speak with a person immediately. They will be hesitant to leave their complaint as a voice mail message. Some companies use voice mail as a method of eliminating complaints and will not return calls on complaints that are left. These practices reinforce the belief of callers that messages left in voice mail are a waste of time, and they demand immediate access to a live operator or attendant.

14.12 STOP THE ABUSE OF VOICE MAIL

If companies would establish methods and practices for handling voice mail, the service would receive broader acceptance. The telecommunications manager should establish practices and procedures that ensure that voice mail messages are responded to and calls returned to customers within a set time frame. The standard voice mail greeting should state something along these lines: "We are very busy and cannot answer your call. Your call is very important to us. Please leave a detailed message or request. Our standard business hours are 8:00 AM to 5:00 PM Central Time. Your call will be returned during these time frames either today or tomorrow. Please provide your telephone number and a time that is convenient for you. Thank you."

If customer relations are important to your company, they will provide the staffing necessary to return calls. Companies that have no regard for their customers and use voice mail as a dumping ground for customer complaints, with no intention of responding to these complaints, have seriously damaged the integrity of voice mail. Companies that care about the quality of services and products they provide will either provide live attendants on customer service and technical assistance calls or will establish call-back procedures for calls received by voice mail.

14.13 IMPROVEMENTS TO IVR VOICE RECOGNITION

Early IVR systems were systems that could recognize the spoken words *yes* and *no*. Calls answered by these systems would ask the caller to respond to prompt messages by answering either yes or no to the prompt. Later the DTMF register was included in the IVR system to allow the caller to respond to prompts with their touchtone phone. Today IVR systems include voice recognition as part of the system. The first use for this technology was in the mobile car telephone. The mobile phone had a voice recognition chip that could recognize short phrases such as " call home" or "call office." The phone could be programmed to dial a number associated with the voice recognition chip storage.

IECs and LECs also offer voice calling features for a monthly fee. If you subscribe to these services, you can place calls by voice commands rather than dialing. When you subscribe to these services, you establish a database of called numbers and voice codes. Upon accessing your voice command database you will be provided with an option to store commands. After selecting the store option, you will be prompted to speak the name of the person (or location) you wish to call and to enter the telephone number for that person (or location). Your voice is digitized and stored. When you use voice commands to place a call, the switching system will access the database and compare your spoken command to the commands it has on file for you. If the voice match is found, the call is placed to the destination stored for that command.

14.14 MECHANIZED OPERATOR SERVICES

Voice recognition chip technology has improved tremendously and continues to be improved. Each improvement opens the door to more applications for IVR. Several IECs are using IVR to mechanize operator services functions. Many types of calls that previously required operator assistance can be handled with IVR and an automated attendant. This technology allows any call center to mechanize calls which previously required a live agent. Customers are able to instruct automated attendants with spoken voice and are not restricted to DTMF input. This provides much greater flexibility in how calls can be handled. The software program for these IVR systems is critical. The prompt and response messages must be designed to ensure a limited number of spoken commands are received from the customer.

Multimedia-capable PCs include IVR technology as part of the system, and these computers respond to spoken commands. As the IVR technology continues to improve, the PC will be able to respond to many more spoken commands. I suspect that at some point in time a word processing program will be able to construct documents based on spoken input rather than typed input. I sure wish I had one to write this book. Voice recognition chips cannot handle fluent speech and does not recognize all words, but the technology is getting very close to that objective.

14.15 VOICE SYNTHESIS

Voice synthesis technology is also a technology that has experienced tremendous improvement. Some e-mail systems have the technology to convert text into spoken words. This allows you to access your e-mail from a telephone rather than a data connection. We should expect that it will not be to long before voice mail messages can be converted to text. When purchasing e-mail and voice mail, you will evaluate systems based on the options provided. The list of options continues to grow, and these systems can become outdated almost overnight.

Although older voice mail systems continue to provide all basic functions, you should check on the latest bells and whistles available before you purchase any new system. Over time the cost of these newer technologies will come down and will make them affordable as an option that should be purchased to improve productivity. With the newer technologies, the president of a company can issue any statement to employees simply by addressing a voice mail message to all mailboxes. A department head can communicate to all employees by issuing a voice mail, as a group mail, to his or her department. Employees can listen to the voice mail and if they wish, they could access the message from e-mail and print a copy. Caution must be exercised when using group voice mail. Flooding the system with many messages can shut the system down temporarily. Make sure group messages can be spread out over a broader time frame to eliminate this problem.

It is a responsibility of the telecommunications manager to ensure that the technology available to support communication within the company and between the company and customers is not just adequate but leads to improvements in operations, relations, and productivity. The communication system must be tailored to fit the business. Large corporations can usually benefit from the latest and greatest technology, while a mom-and-pop operation may only need a single-line instrument. Mom-and-pop operations generally do not have a telecommunications manager. Large corporations usually cannot exist without a telecommunications manager because their expertise and guidance will save the company much more than their salary. Corporations trust the telecommunications manager to select technologies that provide the most benefits at the lowest possible price.

Since large corporations can benefit the most from the more expensive technologies, I would expect to see new PBX installations at large corporations with combined voice mail and e-mail systems. Of course, some time will be required for this to occur. Many corporations are currently being served adequately by their existing communication systems and cannot justify their replacement on the basis of technology alone. The telecommunications manager must stay abreast of the changes in communication technology in order to make the right recommendations for the company. In addition to keeping informed of changes, the telecommunications manager should drive changes in technology. It is well known that small companies and large corporations respond to the demands of the marketplace. If we provide the market, they will deliver the technology.

14.16 VOICE RELAY CENTERS

Voice synthesis, voice recognition, and voice mail technologies will find tremendous applications in the relay telecommunications network. *Relay* is a telecommunications network designed to handle telephone calls for the deaf. Presently, a hearing-impaired person uses a teletype device (TTY) and modem to establish a call to the relay operator. The operator receives the call on a TTY (or PC equipped with TTY software). The person with hearing difficulties and the operator converse using the text capabilities of the TTY. The text typed at each end appears on the LCD or video display terminal (VDU) of the receiving party. The operator will make a connection over the PSTN to the individual a hearing-impaired person wishes to call.

When the relay center agent establishes a call, the agent will relay the conversation between both parties. The agent will receive text from the person who is hard of hearing and will read the text to the called party. The agent will hear what the called party has to say and convert what is heard to typed text. The hearing-impaired person reads the typed text on the screen. The relay center also receives calls from people that wish to call a hearing-impaired person. The relay center is accessed by dialing an 800 number. Sprint, MCI, AT&T, and Independent relay centers have their own unique 800 number. The relay center establishes a connection to the dataphone of a hearing-impaired person by calling that person's telephone number. When the ringing signal is received, it activates a special device, such as a light, each time the ringing signal is received. Anyone with a TTY device can call a person with hearing problems without going through the relay center. The relay center exists to handle calls for TTY devices from someone who does not have a TTY.

14.17 WHERE IS VOICE MAIL TECHNOLOGY HEADED?

Voice synthesis and voice recognition technologies should be approaching a point where these technologies could be used to automate the relay center. It appears the major hurdle is converting captured voice (that is, voice mail) to text. The deaf community is a powerful marketing force in our society. I would expect that before this book is out of print, their drive will result in such tremendous improvements to voice technology that the PC can be equipped with the technology to allow direct conversation between a hearing-impaired person, or a mute person, and anyone. The technology is not yet commercially available but should become available at an affordable price within the next ten years. ISDN would be a great vehicle for this technology.

It should be possible to augment the ISDN terminal device with a device that can replace the codec. A codec converts a digitized voice into an analog voice for connection to a receiver or speaker. The codec could be replaced with a device that converts the digitized voice into text for display on a screen. Taking this technology one step further, a device could also be added to convert outgoing text to digitized voice. A mute person could use this technology to place a call instead of using a relay center. Options could be provided, for the terminal device, to allow selecting the type of conversion needed. It could be set to convert the incoming digitized voice to an analog signal (that is, codec) for connection to a receiver; it could be set to convert outgoing analog voice signals to DS0 signals (in other words, codec); the terminal could be set to convert typed text to digitized voice signals (DS0); and it could be set to convert incoming digitized voice (DS0) to text on the screen.

As I mentioned earlier, this technology—which is a marriage of voice recognition, voice synthesis, and codec technologies—is not currently available as a commercial product. A telecommunications manager should stay on top of what is happening with technology in this area, because it could have tremendous applications in your company's business. There is always a tendency to scoff at new technology. Remember how many people laughed at Alexander Graham Bell and his invention? You must keep and open mind to the development of new technologies and push for those you think will benefit our society or your business. This technology would find tremendous applications on the Internet. I would pay close attention to what is happening to technology on the Internet, especially technologies that deal with audio or video over the Internet and the technology being used to place phone calls over the Internet.

The PSTN is a circuit switched network and provides a dedicated switched circuit between the calling and called party. It is superior to a packet data network (PDN) for handling voice calls. Since the Internet uses the PDN network, telephone calls over the Internet do not enjoy the quality of a call made over the PSTN. However, fast-packet technology, fiber optic transmission mediums, and improvements in digital voice technology are narrowing the quality gap. The circuit switching technology of the PSTN may soon give way to fast packet and asynchronous transfer mode (ATM) technologies to carry both voice and data.

REVIEW QUESTIONS

1. Where are messages stored in a voice mail system?
2. How are digital messages converted to analog voice signals?
3. What are some advantages of voice mail?
4. How does escape from automated attendant work?
5. How do calling parties know they have reached the right mailbox?
6. How do you access voice mail from a proprietary PBX telephone?
7. How do you access voice mail when not at the office?
8. How does a public voice mail system identify who a voice message is for?
9. What determines how many users can be placed on a voice mail system?
10. How is voice mail technology applied to the Internet?
11. What is audiotext?
12. What is IVR?

13. What is the integration of audiotext and IVR called?
14. What is the major problem with using automated attendant?
15. How is voice recognition used in telecommunications?
16. What is voice synthesis?
17. How is it possible to mechanize the functions of an operator?
18. How do automated attendants receive their instructions?
19. How can automated attendant be used to serve as a substitute for DID?
20. What determines how much disk storage is needed by a voice mail system?

15

Telecommunications Voice Network Design

Key Terms

All Trunks Busy (ATB)
Average Bouncing Busy Hour (ABBH)
Average Busy Hour (ABH)
Average Hold Time (AHT)
Blocked Calls Cleared (BCC)
Blocked Calls Delayed (BCD)
Blocked Calls Held (BCH)

Blocked Load
Busy Hour
Capacity
Carried Load
CCS
Demand
Erlang
Final Circuit Group

Grade of Service (GOS)
High-Usage Trunk Group
Offered Load
Peg Count (PC)
Servers
Traffic

The Public Switched Telephone Network (PSTN) was designed to achieve maximum utilization of circuits at the lowest possible cost. Each common carrier engineered its part of the PSTN to provide just the number of needed circuits and no more. The department responsible for determining how many circuits were needed was called the *network engineering department*. The function was called *network design*. In telecommunications, telephone calls being handled by a switching system are referred to as the *traffic on the system*.

Some people refer to the network design and engineering department as the traffic department. The operator services department has also been referred to as the traffic department. When the network design and engineering department was first created, it was part of the operator services department in many companies. In the 1960s, network design and engineering was spun off from the operator services department as a stand-alone department. That department is the focus of this chapter, and we will use the term *traffic* as it applies to calls placed over the PSTN facilities.

It is helpful to think of telephone calls as traffic, because then we can relate network design to highway design. The amount of traffic carried by a highway is monitored by placing counters on the highway. Many of us have seen the black rubber cords placed across a highway by the engineers from the Department of Highway and Transportation. These black rubber cords are attached to a meter. Each time a car goes over the rubber cord, the meter adds 1 to a counter. The engineer gathers this data over a certain time frame and can determine how many cars are using the highway. The engineer will use this information to determine if the highway needs to be improved or more lanes added. It is possible to make every highway in America a 12 lane highway, with 6 lanes for each direction of traffic. This would ensure enough lanes to handle traffic in most locations, but this would cost a lot of money. The engineer's job is to determine exactly how many lanes are needed for traffic on the average business day.

The job of a traffic engineer is to balance the trade-off that occurs between cost and service. If the highway engineer provides a 2-lane highway to carry traffic on an interstate highway bypass around a major city, or provides a 2-lane highway through town as part of the interstate highway, service will be terrible. The highway engineer saves money by building a 2-lane highway instead of a 12-lane highway, but the highway does not have enough lanes to carry traffic during the rush hours (busy hours of the day).

15.1 HOW MANY SERVERS DO I NEED?

In the early morning hours, just after midnight, a 2-lane interstate highway through the heart of town is probably more than enough. The engineer must engineer around the busy-hour traffic requirements. The engineer will gather data on the amount of traffic handled during the busy hours and engineer accordingly. If a major event occurs in town such as a sporting event or large convention, some traffic delays will be encountered during these times. The engineer does not engineer the highway system to handle an extraordinary event. This would result in underutilization of the highway during most of the year and would be a waste of tax dollars.

Telecommunications network design and traffic engineering is similar to highway engineering. The traffic capacity of a highway depends on the number of lanes provided between two points. The traffic capacity of a telecommunications system between two points depends on the number of circuits between the points. The job of a network design engineer is to provide just enough circuits to handle traffic during the busy hour and no more. The engineer could provide thousands of circuits between every two points in the network, but the cost of the PSTN would be astronomical with this approach.

The network design engineer is charged with keeping the cost of the network as low as possible. Again we have the trade-off between service and cost. The network design engineer must provide just enough circuits to meet the desired service level in order to meet objectives of both service and cost. To maintain the balance between service and cost, the engineer designs the network based on past traffic data. Using past history of traffic as well as design forecasting tools, the engineer predicts the number of circuits needed.

15.2 DIAL OFFICE ADMINISTRATION AND TRAFFIC STUDIES

The network design and engineering department contains a group called *dial office administration (DOA)*. This group is charged with collecting data for use by the design engineer. Some people referred to the DOA department as "dead on arrival," so this group's name has been changed in some companies. The DOA group was charged with collecting traffic measurements on every circuit. It would measure the traffic over every circuit path within a switch and would measure the traffic on all trunk circuits between switches.

Larger offices had permanently mounted traffic measuring equipment called a *traffic usage register (TUR)*. Portable equipment was taken to smaller offices when traffic data needed to be collected. Every component in the switching system was equipped with leads connected to the traffic measuring equipment. Each component was usually equipped with two leads. One lead was called a *peg count (PC)* lead and the other lead was called the *all trunks busy lead (ATB)*. These leads were cabled to the main frame. The TUR equipment was also cabled to the main frame. Jumper wires were used to make connections between the traffic leads of each circuit to the meters of the TUR.

15.3 PEG COUNT, USAGE, AND ALL TRUNKS BUSY METERS

Just as the electric company reads an electric meter to determine how much electricity you use, DOA used to read traffic meters to determine how much each piece of equipment was used. There are three meters for each piece of equipment in a switch: (1) A PC meter increments one time for each time the equipment is used. This meter is operated by a ground potential on the PC lead. Each time a device is used, it places ground on its PC lead. (2) A time duration meter is a continuously running meter like your electric meter. As long as the circuit is in use, this meter accumulates time. This device also operates off of the PC lead. (3) An ATB meter is operated by a signal on the ATB lead.

The ATB signal circuit was designed to make the ATB lead part of a chain circuit. Each trunk circuit has an ATB IN lead and an ATB OUT lead. At the first circuit of a trunk group, a ground is wired to the input of the ATB signal chain circuit. A ground is wired to the ATB IN of the first trunk circuit of a group. The ATB OUT lead of the first circuit is wired to the ATB IN lead of the next circuit, and so on. This ATB signal chain passes through each circuit in the group. If a circuit is idle, the chain is opened at that circuit. As long as one circuit in the group is idle, the ATB signal chain is open. When all circuits in a group are busy, the ATB signal chain has the link from each circuit in the group connected through to the next circuit in the chain.

When all circuits in a group are busy, the ATB signal chain forms a complete signal path from the first to the last switch in the chain. This provides a path for the ground placed on the ATB IN lead of the first circuit in the chain to the last switch in the group, and the ATB ground from the first switch will appear on the ATB OUT lead of the last circuit in the chain. The ATB meter is wired to the last circuit in the chain.

The ATB meter is operated each time it sees a ground signal on the last ATB lead. ATB meters are used predominantly on trunk circuits. When all trunk circuits in a group are busy, the **all trunks busy (ATB)** meter will be operated.

The ATB meter is incremented one time every time an *ATB condition* occurs. This type of operation occurred in electromechanical switching systems using electromechanical meters. Every time all circuits in a group were busy, the ATB meter was incremented by a count of 1. When all circuits are busy, it is possible to have more than one attempt to use the circuit group. Suppose all circuits in a group are busy for 3 min. During this time frame we might have 10, 15, 20, and so on attempts to use these circuits by other calls. No matter how many attempts were made during an ATB condition, the ATB meter would only show one count. It was also possible to have all trunks busy, and no additional calls were attempted.

The ATB meter of older switching systems counted the number of times an ATB condition occurred. The ATB meter did not count each call that was blocked. Traffic engineers did not know how many calls were blocked during an ATB condition, and they counted each ATB condition as one blocked call. This could lead to understating or overstating blocked call attempts. This situation has been corrected by SPC switching systems. With these computer-controlled switching systems, the traffic program will increment the ATB meter storage location one time every time it tries to place a call and all circuits are busy. A SPC switching system replaced all electromechanical PC, usage, and ATB meters with RAM locations. The traffic program uses these memory locations to keep track of traffic usage data. The programs also performs all calculations necessary to provide summary traffic reports.

15.4 DETERMINING BUSY HOUR

Each circuit has its own individual PC meter and time usage meter. Each group of circuits had its own ATB meter. In electromechanical switching systems, these meters were read daily by switch maintenance personnel or by an operator. The meter readings were logged on a paper form. This form was mailed to the DOA group each month. The DOA group consisted of numerous clerical personnel. They generated monthly reports from the data on the traffic log form. These reports were used by engineering to check the status of each circuit group. These reports would indicate which days of the year had the most traffic. In the following years, DOA would conduct a traffic study during these days.

During a traffic study on an older electromechanical switch, the meters were read every hour on the hour. Each hour's readings were logged on the traffic log form. At the end of a study, these forms were summarized by the clerical group in DOA and a report was generated. This report provided details on which hour had the most traffic. The hour of the day having the most traffic is called the *busy hour (BH)*. The engineering department would review these reports to determine the status of each circuit group during the busy hour.

Each year DOA will select the ten busiest hours for each circuit group from all the studies conducted during the year in an exchange. DOA will summarize these reports and take the average of the ten hours. The resulting data is referred to as the *average busy hour data (ABH)*. The busy hour usually occurs at the same time on all studies (usually 10:00 AM). In some cases, the study will just be conducted during this hour. When the busy hour is the same for all data, the average is called the *ABH*. However, some switching centers have a busy hour that varies. The busy hour may be 10:00 AM one day, 9:00 AM during another study, and 3:00 PM during a third study. When studies are summarized with the busy hour bouncing around like this, the average readings are referred to as *average bouncing busy hour (ABBH)*. When a ten-high-day summary is made, it results in an ABBH PC and ABBH time duration for each circuit as well as an ABBH ATB for each group of circuits.

15.5 CALCULATING AVERAGE HOLD TIME

The PC meter (electromechanical counter or RAM) is designed to increment the counter (or RAM) one time, at the end of equipment use. In an electromechanical switch, when a piece of equipment is activated for use, it will place a ground on the PC lead and open the ATB signal lead. When the ground is placed on the PC lead, it will activate the PC meter wired to this circuit. This energizes the PC meter, but it does not increment at this time. When the use of the circuit is over, the circuit is returned to an idle state. Returning the circuit to idle removes the ground from the PC lead and closes the loop of the ATB signal link lead of that circuit. When the ground is removed from the PC lead, the meter is deenergized. When this happens, the meter increments the count by 1. Thus, the PC meter is designed to add 1 count to the meter at the end of each use of the circuit or equipment unit.

In a SPC switch, the RAM used for PC will be incremented by the CPU of the switch every time a circuit is assigned for use by the CPU. By reading the PC meter for a piece of equipment, we know how many times the piece of equipment was used since the last reading. Many electromechanical PC meters cannot be reset to 0. The number of calls for one hour is determined by subtracting the reading at the beginning of the hour from the reading at the end of the hour. Some PC meters do have a manual reset button to reset readings to 0. At the beginning of a study, DOA will reset the meters to 0 but will not reset these meters at any time during the study.

The DOA clerical support group will perform manual calculations on the data supplied by the traffic log form to determine the hourly PCs on each piece of equipment metered. With today's SPC switching, the data for each hour is written from RAM to secondary storage and all RAM meter locations are then reset to 0. Thus, a SPC switch does not require manual readings each hour. A SPC switch also needs no clerical support for summarizing a study. Software in the switch does all summarization of data. A summary report is generated at the end of each day by pulling each hour's worth of data off the secondary storage device. The SPC traffic program will also provide a summary report at the end of a study. DOA clerks were no longer needed when all switching systems were replaced with stored program control (SPC) switches.

Since the busy hour always occurred during the day, meters usually were not read at night. Readings were taken every hour between 7:00 AM and 9:00 PM Some offices with permanently mounted traffic usage recording (TUR) meters would use a camera to take a picture of the meters. When a study was being conducted, DOA would mount a camera in front of the meters. This camera would take a picture every hour and time stamp the picture. The film was sent to the DOA's clerical group, which would read the meters using the pictures. The clerks would fill out the traffic log from the information on each picture. With a SPC switch, traffic studies can be scheduled to run at any time of the year or can be run all year long if desired.

The time duration meter was a meter designed to run continuously as long as a ground is on the PC lead. This meter was usually mounted next to the PC meter. The meter was designed much like the analog speedometer of a car. The last position on the meter would turn one time for each hour and had ten numbers on it (0–9). Each number represented one-tenth of an hour. For each revolution, it would turn the next wheel one-tenth of a turn. The second wheel had 0–9 on it to keep track of each hour. A third wheel was used to record the tens digit. This wheel was used to display hours 10 through 90. A fourth wheel was incremented a tenth of a turn for each 100 hr. The time duration meter would measure time from 0 min to 999.9 hr.

DOA could determine how many times in an hour each piece of equipment was used from the PC meter. DOA could determine the total amount of time a circuit was in use during the hour from the time duration meter. DOA could then calculate the average time of usage for each use, by dividing the time duration measurement for the hour by the PC for the hour. This average usage is called *average hold time (AHT)*. DOA would add the PCs for circuits in a group to arrive at the number of calls a group of circuits carried. The time usage for a group of circuits is called the *carried load* for the group. Carried load is equal to AHT for the group times the number of calls over the group of circuits.

15.6 DOA GENERATES SUMMARY REPORTS FROM STUDY DATA

DOA would calculate the PC, ATB, and time duration of use for each circuit by manually subtracting the previous hour's readings from the start of the next hour's readings. This was a time-consuming task, and clerks were required to calculate each hour of data. Larger telephone companies had many telephone exchanges. The BOCs and larger Independent companies such as GTE and United Telephone Company would have as many as 100 local exchanges in one state. The DOA group in these companies could conduct as many as 20 traffic studies at the same time. These DOA groups required many clerks to calculate hourly data and summarize each study.

In addition to calculating the PC, ATB, and time usage for each hour on each piece of equipment, DOA also had to supply the PC, ATB, and time usage for each hour, for each group of equipment. When the meters were installed and wired to the switching equipment, a paper form was generated that detailed which meter was assigned to each piece of equipment. This form also provided detail on which meters were part of a circuit group. The DOA clerks would add the hourly readings of these PC meters to arrive at an hourly PC for the circuit group. Likewise, the ATB and usage meter readings were added, as instructed on the form, to arrive at hourly readings for the circuit group.

When DOA clerks finished summarizing the traffic log data into a report, the network engineer was furnished with hourly data, total data for the day, total data for the week, BH data for each busy hour, ABBH data for the study, and AHT. This data was supplied for each piece of equipment (or circuit) and for each group of equipment (or circuits). DOA usually furnished the engineer with a summary report that provided the ABBH PC, ABBH ATB, ABBH AHT, and ABBH usage time for each circuit group.

The engineer was ordinarily more concerned with data on how each circuit *group* is performing than on the data about one unit of equipment. We can correlate this with highway traffic. Highway department engineers are not concerned about how many cars are using *each lane* of a interstate highway. They want to know how much usage is placed on the *total* of all *lanes* in one direction (the circuit group). Telecommunications network design engineers want to know what the *total traffic* situation is over a circuit *group*. Suppose an engineer wants to know if there are enough trunk circuits installed between two towns to handle the growing volume of calls between the two towns. DOA would conduct a traffic study on the designated trunk circuits in each office. A traffic report and summary were generated by DOA at the end of the study period. The engineer would first look at the number of ATB conditions for the study. If there were no ATBs, there had to be enough trunk circuits in place to handle current call volume.

15.7 CARRIED LOAD VERSUS OFFERED LOAD

If there were a lot of ATBs, the engineer would look at the ABBH usage time. This would provide the amount of usage during the busy hour. This actual usage is called carried traffic or carried load. The engineer could divide the carried load by the carried PC to arrive at an AHT for each carried call. Each ATB indicated the number of times a circuit group was busy. When a circuit group was busy, more than one call could attempt to use a circuit and be blocked. The SPC ATB meters would record each of these blocked attempts. Older electromechanical ATB meters did not record each blockage; they indicated how many times an ATB condition occurred. DOA considered one ATB count, or condition, as one blocked call. They also assumed that if a call was not blocked, it would have the same AHT as a call that was completed and carried by the circuit group. DOA used these two assumptions to calculate the amount of load that was not carried but was blocked. The formula is:

$$\text{ABBH ATB} \times \text{AHT} = \text{blocked load}$$

The amount of traffic offered to any circuit group is equal to the amount of traffic carried plus the amount of traffic not carried. Offered load equals carried load plus blocked load. Engineers do not design based on carried load because that can leave out a lot of traffic that wanted to use the system but could not. Engineers must engineer using offered load. The first step in circuit group design is to find out what the *offered load* is. After determining the offered load, the engineer can use traffic design tables to determine how many circuits will be needed.

Traffic design tables are composed using mathematical calculations or formulas based on probabilities and inferences from the world of statistics. The good news for an engineer that all the statistical and other mathematical computations have been done and have been reduced to a table. The basic line of reasoning behind the formulas is this: If I have X number of circuits in a group, AND I offer Y amount of traffic to the group, AND the average arrival rate of calls is Z, AND the average holding time is H, what is the probability that an offered call will be blocked? The mathematical formulas normally use ABBH AHT and ABBH PC to calculate the average arrival rate of calls.

Network design engineers do not need to know the mathematical formulas behind traffic design tables. The engineer will determine the offered load for a circuit group and will look in the traffic design table to find the section for that offered load. Underneath the load, several levels of blockage will be listed. The engineer will select the level of blockage desired for this circuit group, and listed next to the level of blockage will be the number of circuits needed. We are back to the trade-off of service versus cost. Engineering a circuit group for less blockage requires installing more circuits in the group.

Basic network design requires the engineer to know the offered load to a circuit group and the service level desired (amount of blockage) for that group. Using these two known factors, the engineer can find the unknown third factor (number of circuits required) by looking in a traffic design table. However, there are several traffic design tables. Each table is designed for specific applications. Which table to apply depends on how the telecommunications system handles blocked calls. How blocked calls are handled impacts average arrival rate and ABBH measurements. This affects the mathematical formulas underlying the tables.

15.8 HANDLING OF BLOCKED CALLS, ERLANGS, AND CCS

There are three basic ways to handle blocked calls: (1) The call can be cleared from the system and a circuit busy tone (120-ipm tone called *fast busy*) returned or a recorded announcement played stating that all circuits are busy (*ipm* stands for interruptions per minute). Calls are also cleared from a circuit group when they are routed elsewhere. This is called *alternate routing*. Handling blocked calls by one of these methods it is referred to as *blocked calls cleared (BCC)*. (2) The blocked call can be put in a holding queue until a circuit is available. This is referred to as *blocked calls delayed (BCD)*. (3) On some circuit groups, when callers are cleared from the system, they immediately redial. This is referred to as *blocked calls held (BCH)*. The network design engineer makes assumptions on how blocked calls are handled for a particular circuit group and uses the appropriate design table. The primary table used for blocked calls cleared is the Erlang B table. It is named for G. Erlang, a Danish mathematician who did much of the mathematical computations underlying the tables.

In telecommunications, traffic (or load) is also expressed in terms of Erlang. One hour of traffic is an Erlang. If a circuit is in use for 1 hr, it has been used for 1 Erlang. Individual calls usually do not last for an hour. The AHT for a call is usually in minutes and seconds. One hour has 3600 s. Roman numeral C equals 100. DOA uses a term called *CCS*, which stands for *100 call seconds*. Note that 36 CCS equals 3600 call seconds. Also note that 36 CCS equals 1 hr, and thus *36 CCS = 1 Erlang*. AHT is usually expressed in CCS. ABBH carried load may also be stated as CCS.

The Erlang B table uses Erlang as the measure for offered load. To use this table, offered load must be calculated and then stated as Erlang. Some tables use CCS instead of Erlang. The engineer will use the appropriate value (either CCS or Erlang) dictated by the table and determine the number of circuits needed to provide the desired *grade of service (GOS)* at the offered load. The probability of blocking a call is the grade of service. If the probability of blocking a call is 1 in 100, this is stated as P.01. The grade of service is stated as *P.xx* (where *xx* is replaced by the probability number. The higher the offered load and the lower the number of circuits, the higher the probability of blocking. Sometimes the probability of blocking is stated as *B.xx*. For a situation where 5% of all calls are blocked, this GOS can be stated as *P.05* or *B.05*.

15.9 NETWORK DESIGN DEPENDS ON RANDOM BEHAVIOR OF USERS

The behavior of people using the telephone system is unpredictable. When some users encounter a blocked condition, they will try immediately to place another attempt to the called number. Some users will wait and try later. Although the behavior of any individual is unpredictable, the statisticians can use the independent random actions of these individuals to predict behavior over time. The French mathematician S. D. Poisson developed a formula for the probability of call arrival. This formula provided a good estimate on the number of call attempts at one time.

Most engineers no longer use Poisson because later formulas such as Erlang B and Extended Erlang B are more accurate. Either formula can be used, but Poisson tends to overengineer. That is, Poisson will usually indicate that one or two more circuits are needed than Erlang B indicates. The traffic tables used to determine how many circuits are needed are constructed using the arrival rate probability and the average holding time for a call.

The table an engineer chooses to use is based on the engineer's best guess on the disposition of blocked calls. The Erlang B traffic table assumes blocked calls are cleared. The Poisson traffic table assumes blocked calls are held. The Erlang C traffic table assumes blocked calls are held in queue. These traffic formulas and tables are only valid when call attempts are random. When a major catastrophe such as an assassination attempt on a public figure or an earthquake occurs, everyone tries to use the telephone. Call attempts are no longer random, and the PSTN can be overloaded by the volume of call attempts.

When a traffic overload condition occurs, load control measures are enacted by DOA to reduce the call attempt volume to a manageable level. In a local exchange, line load control software will be activated. Line load control will restrict residential lines from getting dial tone. Only emergency services providers such as police, fire, hospital, government agencies, and disaster relief organizations can get dial tone. As time passes, a few residential lines at a time will be allowed access to the switching system. Gradually, the line load control is removed until all lines have service.

SPC offices use software for load control. The older electromechanical offices did not have load control. In these offices, we had to open protectors at the main frame to disable service to lines. When President John F. Kennedy was assassinated in 1963, the load was so high in my office that the main fuse blew. When a repairperson replaced the fuse, it blew up in his hand. We had to open all protectors on the main distributing frame (MDF) to disable all lines trying to place calls. With the protectors open, there was no load on the office and the main fuse could be installed. We closed the protectors on all critical service and business lines. We monitored the current at the main fuse and gradually closed the protectors on other lines. SPC load control software is more efficient and much quicker than opening protectors at the main frame. Load control software is a valuable asset when a catastrophe overloads the central office.

15.10 NETWORK ENGINEERING RELIES ON VALID TRAFFIC STUDY DATA

An engineer determines the number of circuits needed to provide service based on offered load, GOS desired, arrival rate of calls, AHT, and disposition of blocked calls. If an error is made in any of these areas, the number of circuits provided may be too few or too many. The area causing the most trouble for an engineer is calculating offered load. The engineer may not accurately calculate the blocked load portion of offered load, and the measurement of carried load may be wrong. The major reason for incorrect measurements of carried load is equipment outages. When equipment is taken out of service, it is made busy. On some systems, this causes the PC meter to step one count and also causes the usage meter to run as long as the circuit is busy. The meters count when the circuit is busy. The meter does not know if the circuit is made busy and taken out of service because of trouble or if the circuit is busy because it is handling a call. If the circuit is busy because of trouble, it distorts our traffic measurements.

During a traffic study, we only wish to measure equipment in use on a call. The traffic measuring device has no way of telling whether the busy condition is from handling a call or from a circuit outage. It is imperative that all equipment problems are fixed prior to the start of a traffic study. All equipment must be working and in service. None of the equipment can be made busy due to equipment failures. When DOA is beginning a traffic study, they will look at the meter readings for every individual piece of equipment. When they see an hourly reading indicating that the equipment has been in use for the full 60 minutes, they mark it as possible equipment trouble. A trouble ticket is generated to maintenance, and they investigate why the circuit was in use for the full 60 minutes to make sure it is not defective. If the circuit was defective and had been made manually busy to take it out of service, maintenance must repair the circuit and return it to service before the traffic study is started. DOA also looks for circuits that have no PC. Trouble tickets are generated to maintenance to investigate why these circuits are not being accessed.

15.11 STORED PROGRAM CENTRAL OFFICES

The discussions above pertained to electromechanical switching systems such as step-by-step and crossbar switching systems. These technologies have been replaced by SPC switching systems. The older traffic-measuring systems used with electromechanical switching have been supplanted by software and RAM in the SPC switches. The concepts discussed above still apply to SPC systems, but the electromechanical meters have been replaced by RAM. SPC switching systems do not use actual traffic meters. Instead of meters, they use RAM storage locations as meters. Each circuit is assigned a certain RAM location for PC and another location for time usage. A group of circuits is also assigned a RAM location for the group ATB meter.

When the CPU of the switching system assigns a circuit for use, the RAM for PC is incremented by one count. The time-usage RAM will accumulate time as long as the circuit is in use. At the end of each hour the contents of RAM being used as meters are written to a secondary memory storage location. The RAM used as meters is then reset to 0. The RAM used as meters will then track readings for the next hour. Each central office is equipped with a special disk storage device that receives data from the secondary memory storage buffer every hour. Thus, for each hour, traffic data in a SPC office flows from RAM meters to a buffer memory and then to disk storage.

DOA has a centralized computer that uses modems to call into the special disk storage devices at each SPC office once each day. The special disk storage device will transfer the previous day's traffic data to the centralized computer. The centralized computer has been programmed to perform the functions previously done by clerks. A program is written for each SPC office directing the computer to add certain memory locations to obtain data for a circuit group. The centralized computer will generate the traffic report for the office. Since this process is totally automated, the traffic can be measured and monitored for every hour of every day if so desired. Maintenance also uses reports from the automated system to find circuits that are busy all the time and to find circuits that are not being used. Once these circuits are pinpointed, maintenance will find the source of the problem and correct it.

Many SPC systems can be provided with traffic reports and maintenance reporting software as an option. The SPC system will generate actual reports instead of raw traffic meter readings. This software program eliminates the need for a centralized computer and programming to summarize studies from raw data dumps. These reports from the SPC switch can be directed to a local or remote printer or to disk storage. Maintenance and DOA can also access these reports from a remote terminal or computer. They can use the reports generated by the SPC and do not need programmers and database administrators for report generation.

When the old AT&T/Bell system shared toll revenue with the Independent LECs, DOA used the information from traffic studies to provide separations and settlement reports to the finance department. These studies were used to secure the Independent LEC's portion of toll revenue from Bell. Deregulation of the industry has made separations and settlements disappear. Automation of traffic studies has eliminated the need for clerical support. DOA is a shell of what it once was.

SPC offices are easy to engineer. Switching equipment manufacturers have packaged their hardware in modules and know the load each module will typically carry. These SPC switches are engineered by using a type of cookbook approach. You have an engineering form for the office (recipe sheet). You list the following details: originating ABBH line load and PC, terminating ABBH line load and PC, originating and terminating loads and PC for each trunk group, number of lines required, number of trunks required, and the number of dial registers and senders required. You send this form to the switching manufacturer and they engineer the office for you. Network engineering departments and DOA departments have seen dramatic reductions in the number of personnel. This is another case where the deployment of SPC switching has allowed LECs to dramatically reduce their cost of doing business.

15.12 ENGINEERING TABLES

In telecommunications, the number of circuits needed to serve a particular function is determined from traffic tables. These tables sometimes refer to circuits as *servers*. The table provides you with the number of servers needed. The word *servers* allows the use of these tables to predict quantities of servers needed in applications other than telecommunications. You can use these tables to predict how many waiters are needed in a restaurant during each hour of business. These tables are used by customer services bureaus and operator services operations to determine the number of agents needed for each hour. In telecommunications switching, servers means circuits, not people. In any environment, the number of servers needed depends on the demand placed on the serving system. This demand in turn depends on the arrival rate and the average time each server is in use.

Demand is called *load*. To be more specific, *demand is offered load*. Measured load is carried load. If the offered load is too great for the number of servers supplied, high levels of blockage occur. If too many servers are supplied for the offered load, service will be great but cost will be higher. Management usually decides on the desired service level or GOS. The GOS specifies an acceptable level of performance to management. This GOS may not be acceptable to customers. If a restaurant does not provide enough servers, customers must wait (the GOS is low). Customers may not come back. If too many servers are provided, the cost to management is higher and it loses profits. It may not stay in business. Restaurants usually provide a queue for customers during the busy hour because there are not enough tables (in this case the table is a server) available to handle the load during the busy hour. There are plenty of tables during most other hours. In fact, many tables go unused during these low-traffic periods. The owner of the business will be reluctant to build additional seating area and tables just to serve the busy-hour load without delay.

15.13 ENGINEERING VARIABLES: DEMAND, SERVERS, GRADE OF SERVICE

The basic concepts of traffic engineering are applied in many disciplines. The three basic variables used in traffic engineering are: (1) demand or load, (2) number of servers, and (3) GOS desired. GOS is stated as the amount of blockage or customers turned away. If we can live with turning away 5% of our customers, the acceptable GOS is *P*.05 (probability of blocking is 5%). The formulas that underlie a traffic table take all factors into consideration. You merely look up the offered load in the table, find the desired GOS, and next to the GOS is the number of servers required. The load used in telecommunications engineering is an average of the ten busiest hours for the year (ABBH). Since we use an average of ten hours, there will be about five hours higher than the average. There will be blockage during these hours because we do not engineer to handle the highest busy hours. Traffic on Mother's Day is typically the highest busy hour of the year. Many people will get reorder or recordings when placing calls on Mother's Day. Recordings usually inform you that all circuits are busy and will ask you to place your call later.

15.14 ENGINEERING DIRECT (HIGH USAGE) AND ALTERNATE ROUTE TRUNK GROUPS

In determining how many trunk circuits are needed between two central offices, an engineer must determine if any alternate routes are available. In a city, there may be as many as 100 or more central offices. To provide calling between all the telephones in the city, each central office must be able to connect to any other central office. Rather than provide direct trunks between all central offices, a tandem switch is used. A class 4 switch is installed in the heart of the city, and all central offices in the city will have trunks connecting to this class 4 office. This arrangement allows people to call into other central exchanges in the city. The arrangement extends their service area beyond their own local exchange. The trunks are called *extended area service (EAS) trunks*, and the class 4 office is called an *EAS tandem switch*. This switch is in tandem (or between) any two class 5 exchanges.

If your telephone service is provided by an exchange on the southeast side of town and you want to call someone served by an exchange on the north side of town, your call will be placed through the EAS tandem. Suppose your exchange code is 897-*XXXX* and the other exchange code is 941-*XXXX*. When you dial 941, the dial register collects the dialed digits and presents them to a translator. The translator in your exchange will look in a translations database to see what it should do. Since we do not have direct trunks between the two exchanges, the database has been constructed to tell the translator that any call to the 941 exchange should be placed on the trunk circuits to the EAS tandem. After the caller has dialed all seven digits, a trunk circuit to the EAS tandem switch is connected to a sender. The sender will read the digits stored in the dial register and send all seven dialed digits to the EAS tandem. The 897 exchange will then release the sender and connect the calling telephone to the trunk circuit.

In the EAS tandem switch, the dial register will receive the digits sent by the sender of the 897 exchange. The register presents the first three digits to the translator database. This database has been constructed so that when it sees 941, it provides instructions to the switch to connect the call to a trunk circuit to the 941 exchange. The EAS tandem switch will be instructed by the translator to attach a sender to the 941 trunk circuit and send the digits stored in the register associated with this call. The sender will send the last four digits (terminal digits) to the 941 exchange. A dial register in the 941 exchange receives these digits and presents them to the translator of the 941 exchange. The translator database provides a conversion for the telephone number (directory number) to an equipment number (line circuit number). The switching matrix in the 941 exchange is instructed to connect the incoming trunk circuit from the EAS tandem to the designated line circuit. The EAS tandem switch is instructed by its CPU to connect the incoming trunk from the 897 exchange to the trunk to the 941 exchange. The switching of circuits in the three exchanges results in a complete conversation path from one telephone to the other.

With the implementation of SS7 networks between class 5 exchanges, the need for senders has diminished. With SS7 when a call is placed from one exchange to another, the originating exchanges issues a message over the SS7 facility to the distant exchange to see if the called number is available. The originating office sends the called number directly to the called exchange over the SS7 facility. The EAS tandem is not involved in processing dialed digits. It is merely a switching point between all class 5 offices connected to it. Negotiation messages between the originating exchange, the terminating exchange, and the EAS tandem over the SS7 facility will determine which circuits to use between each exchange and the tandem. These circuits will be reserved for the call and when the called party answers, a connect message over SS7 will instruct the EAS tandem to connect the reserved trunk circuits from the 897 and 941 exchanges.

The use of EAS tandems is an economical way to connect all class 5 exchanges so calls can be placed between them. If the city has 20, 50, or even 100 exchanges, you do not need to install direct circuits between each exchange and every other exchange in the city. With a low volume of calls between two exchanges, it is best to route the calls through an EAS tandem. The EAS tandem will handle calls from one exchange to every other exchange in town. Since the EAS trunk group is designed to handle the sum of all traffic to other exchanges, it is a larger, more efficient trunk group and results in a lower cost than direct trunks between all exchanges.

There may be some exchanges that you do want to provide direct circuits between. It is cheaper to provide direct routes between two exchanges when the level of traffic can justify at least twelve circuits between the two. Suppose in the southeast part of town there is a 667 exchange serving an area next to the serving area of the 897 exchange. There will probably be a high volume of calls between customers served by these two exchanges. To service this high volume of calls, direct trunk circuits are established between the two exchanges. This circuit group is called a *high-usage (HU) group.*

When a call is placed from the 897 exchange to the 667 exchange, the translator will instruct the switching matrix to connect the 897 line circuit to the 667 trunk circuit. The call is established over the direct HU trunk circuits connecting the two exchanges.

If all these circuits are busy, the translator of the 897 exchange will provide instructions to connect the call to the EAS tandem trunks. The EAS tandem will connect the call to the 667 exchange trunks. This is called *alternate routing*.

The EAS trunk circuits between the local class 5 office and the EAS tandem are referred to as the *Final Group*. There is no alternate routing for a Final Group because it is the Final Group. The engineer will usually engineer a Final Group using the blocked calls held and Poisson distribution theory tables. This is because a Final Group should have the safety buffer provided by overengineering. The high-usage group is engineered using Erlang B at *P*.10 or *P*.05 GOS. By using a high blocking factor, the HU circuits maintain a high occupancy or use level. We know that blocked calls are not really blocked; they use the alternate route. The engineering of HU groups with high levels of blockage built in ensures high utilization of the circuits. High circuit utilization is achieved by providing fewer circuits. This reduces the cost of the facilities. Overall service levels will still be high because traffic overloads are handled by an alternate route.

15.15 CAPACITY DESIRED MUST BE DETERMINED FROM OFFERED LOAD

The engineer uses offered load as a prime engineering criterion. By engineering around offered load, the engineer is engineering the system with the capability to handle the offered load. The capacity of any system is its ability to handle a stated load. This can also be found by using the traffic tables. Using the two known factors of GOS and number of servers provided, you can look in the table to find the capacity (offered load) that can be handled with these two known factors. Thus capacity of a circuit group, switching system, or facility may be stated as a certain number of CCS or Erlang. This means that the present circuit group with the existing number of circuits has the capacity to handle X amount of load or traffic. If demand exceeds capacity, the amount of demand that exceeds the capacity is blocked.

In central office engineering, the capacity of a switching system is also stated as the number of calls that can be handled by a switch in 1 hr. Remember that carried load = AHT × PC. Therefore, carried PC = carried load / AHT. Traffic studies provide the engineer with the AHT of all calls handled by a switch. Traffic studies also provide the ABBH PC of all calls handled by a switch. The ABBH PC for an hour provides us with the number of calls handled in 1 hr. Since one determinant of offered load is the number of calls in an hour, load is dependent on ABBH PC.

Capacity can be stated as the number of calls that can be carried in 1 hr or as the load that can be handled in 1 hr. It is important to understand that the PC meter will score one count for every call carried on the circuit group. Thus, PC times AHT is *carried load*, not offered load. Capacity relates to the amount of load that can be carried. Demand relates to offered load. We engineer to achieve a capacity that meets our GOS objective.

We must engineer based on demand (offered load). Offered load is equal to carried load plus blocked load. Earlier I stated that carried load is equal to PC times AHT and that blocked load is equal to ATB times AHT. Thus we can state:

$$\text{Offered load} = \text{carried load} + \text{blocked load}$$
$$\text{Offered load} = (\text{PC} \times \text{AHT}) + (\text{ATB} \times \text{AHT})$$

Therefore,

$$\text{Offered load} = (\text{PC} + \text{ATB}) \times \text{AHT}$$

If a call is carried, it scores the PC meter. If the call is not carried, it will not score the PC meter but will score the ATB meter. Thus, every call will cause one of the meters to score, and the total calls attempted equals the total PCs plus the total ATBs:

$$\text{Total calls offered} = \text{PC} + \text{ATB}$$

The total calls represents demand placed on a circuit group. Total calls times AHT provides offered load (demand).

When demand on a switching system is stated as PC rather than usage, we can state how many call attempts per hour a switch can handle—for example, "this switch is engineered to handle 15,000 calls an hour." Another factor to consider is how many simultaneous calls can be handled by a circuit group or switch. In actual practice, SPC switch, which does not have distributed processing, can only handle one process at a time and one call at a time. But if we take the term *simultaneous* to mean within a few seconds, instead of at the exact same instant in time, we can state the number of simultaneous calls as a function of the number of circuit paths available.

If a switching system has 40 DTMF dial register circuits, the system can handle 40 people dialing at the same time. The number of simultaneous dialers at one time is 40. Due to the brief duration that a dial register is used, one circuit may handle as many as 60 callers over a time frame of 1 hr. At this rate, if the calls are random and spread out over the hour, 40 circuits could handle 2400 callers in an hour.

In a time division switching matrix, the capacity of the multiplex loops will be stated as the amount of load they can carry. Multiplex loops are the facilities used to connect peripherals (such as line and trunk bays) to the network. If a 30-channel loop is being used, the loop can carry 30 individual voice paths. If more than 30 voice channels are needed between a line shelf and the network, additional loops are installed. The calling party and the called party will require their own voice path. One conversation requires the assignment of two voice paths (one for each party to the call). Thus, a 30-channel multiplex loop can handle 15 conversations at the same time. The number of simultaneous conversations per equipped loop is 15. If two multiplexed loops are provided, the switching system can handle 30 conversations at the same time. In a small PBX such as an SL-1S, two 30-channel loops are the norm. A large central office is usually equipped with a 30-channel loop for every 150 telephone lines served.

Central office switches require many more voice paths than a PBX. The norm in a central office is to provide 120 voice paths per 640 line circuits. This provides a concentration ratio of just over 5 to 1. Everyone does not use their telephone at the same time, and the engineer will not provide a voice path per line circuit.

Remember, the engineer must keep cost low by providing just the number of circuits needed to handle the busy-hour load. Even during the busy hour in most offices, fewer than 1 of every 5 customers will use a phone at the same time.

The assignment of line circuits is made so that high-volume business lines and low-volume residential lines are evenly distributed across all line shelves. There are 640 line circuits in one line shelf. Some of these line circuits are assigned to businesses and some are assigned to residential customers. By controlling the assignments and mix of business to residential, DOA can control the amount of traffic offered by each line shelf. Four multiplex loops between the line shelf and the network will provide 120 voice paths.

Voice paths for trunk circuits are provided on a one-to-one basis. The engineer does not want to provide any concentration of trunks to voice circuits because the high-usage circuits must not encounter blockage within the switching network. If 20 T1 systems are terminating into the switch, 480 (20 × 24) voice paths are provided on the multiplex loops. The size of the switching network will depend on the total number of voice paths needed. The latest generation of switching systems uses fiber optic for the multiplex loops. Instead of 30-channel loops, fiber loops are 672-channel loops. Line circuits have been repackaged to get four line shelves in a bay. Thus, 4 × 640 = 2560 lines in a line bay. A single 672-channel multiplex fiber loop can serve one line bay.

15.16 NONBLOCKING SWITCHING SYSTEMS

Some switching manufacturers will state that their switching matrix is a non blocking matrix. This may be true. A nonblocking matrix simply means that for every inlet to the matrix, there is an outlet. The matrix will not result in a blocked call. Just because a switch has a nonblocking network does not mean calls will not be blocked. As stated above, the engineer purposely engineers blockage into the system to keep cost low. If a line shelf is equipped with 640 line circuits and four 30-channel multiplex loops (120 voice paths) to the network and 121 people try to use their phones in this line shelf, the 121st person will be blocked. The person is not blocked at the network but at the line bay.

If the switching system only has 40 DTMF registers, the 41st person trying to dial at the same time is blocked. These callers that are blocked get nothing. They are held in queue. If they wait until someone else hangs up, they will get dial tone. Suppose the switch has 40 toll trunks that are accessed when a caller dials 1 or 0. The 41st caller is blocked and cleared from the system. That person gets a fast busy tone. The number of simultaneous calls any part of the system can handle is directly related to the number of circuits installed. Forty trunk circuits in a circuit group means that trunk group can handle 40 simultaneous calls. The number of simultaneous calls a network matrix in a switch can handle is the number of voice circuits attached to the network divided by 2.

In the case of a 30-channel multiplexed loop, the load capacity of the loop is 30 Erlangs. Each channel can be used for 1 hr (1 Erlang) each. Both parties to a conversation are adding to the load, so we do not divide by 2. We must measure the load on each voice path. Thus, each 30-channel loop can handle 30 Erlangs in 1 hr. Converting to CCS, this is 1080 CCS. The theoretical capacity is 1080 CCS per hour for each multiplexed loop, and the number of simultaneous calls per loop is 15. Theoretical capacity requires 100% circuit utilization, and this is never achieved.

15.17 GOS, OML, AND IML

As we have stated, the engineering of circuit groups depends on offered load, arrival rate, and GOS. The GOS component is a standard provided by management. In switching systems, there are several accepted standard GOS levels. When an originating line circuit cannot be connected to a server because of the inability of the switch to find the idle server and match it to the originating line circuit, this is called originating match loss (OML). The objective is an OML of 1% or less. When an incoming trunk cannot be connected by the switching matrix to an idle line or outgoing trunk, this is called incoming match loss (IML). The IML objective is 2% or less blockage. Switching systems have an objective to provide dial tone within 3 s on 98.5% of all outgoing calls. The objective for all common control components (CPU, registers, senders, and so on) is that they will have less than 1 in a thousand (0.1%) blockage or $P.001$ GOS.

15.18 LOAD TESTING A SWITCH AT ENGINEERED ABBH LOAD

When a new switch is placed in service, it is load tested. Devices called *load boxes* are attached to line circuits scattered across all line shelves. The load box is a device that automatically places telephone calls. I hold a patent on one such device, which contains 96 telephone circuits. The load box is attached to 96 line circuits of the central office. In the load box, 48 telephone circuits are set up to originate telephone calls and 48 are set up to receive calls. The telephone numbers for the 48 receiving telephones of the load box are programmed into the central office switch for the line circuits attached to the terminating side of the load box. In the load box, the 48 originating lines are programmed to call the telephone numbers of the 48 receive lines.

The 48 originating lines will call the 48 terminating lines. The terminating lines will detect ringing and answer the call. The terminating line will send a tone to the originator, and the originating line will send a tone to the terminating line. The tone level is tested on each end to ensure that a voice path with adequate transmission was established between the two line circuits. The tones are removed and the transmission path tested in both directions for noise. The originating line will hang up, then start the test process on another call.

Each load box can generate about 12,000 call attempts each hour using 96 line circuits. If the office was designed with a capacity of 60,000 ABBH PC per hour, five load boxes are used. The engineered load is run on the switch for 24 hr. During this 24-hr time frame, the switch must provide a *P*.001 GOS, and dial tone delay must meet the criteria of no more than 1.5% delayed beyond 3 s. Current SPC switching systems with nonblocking networks eliminate OML and IML because the system will not attempt to find a path when it already knows none exists. The only measure that can be applied to these SPC systems is the *P*.001 objective for the SPC system as a whole.

15.19 SAMPLE TRAFFIC STUDY DATA

Now let us look at how the data from a traffic study or report may look. We will look at two reports for the same trunk group. The first report shows the hourly data for a five-day study conducted between 8:00 AM and 5:00 PM each day (Figure 15-1). The second report is a summary report for the ABBH data for the BH of each day (Figure 15-2). Today's SPC switching systems can provide both these reports on demand for each circuit group in the switch. The CCS column reflects the carried load. In Figure 15-1, we use the CCS column to find the hour that had the most traffic. This is the BH. Once the BH is determined, we use the PC, CCS, and ATB figures for that hour as the ABBH data for the day.

A traffic engineer or DOA clerk can look at the data from this five-day study for one circuit group and determine from the CCS values when the BHs occurred. The highest CCS on Monday is 650; therefore the BH on Monday occurred at 9:00 (more accurately, the busy hour was from 8:00 to 9:00). Thus, the ABBH CCS on Monday is 650 CCS. The ABBH PC is 270 PC.

NOTE: A higher PC occurred at 8:00 and also at 10:00, but these values are not used for the ABBH PC. ABBH PC is not necessarily the highest PC for the day. ABBH PC is the PC that occurred during the BH, and that hour is determined by the hour with the highest CCS.

Hour	Mon. PC	Mon. CCS	Mon. ATB	Tues. PC	Tues. CCS	Tues. ATB	Wed. PC	Wed. CCS	Wed. ATB	Th. PC	Th. CCS	Th. ATB	Fri. PC	Fri. CCS	Fri. ATB
8:00 AM	280	500	30	260	475	30	245	450	25	250	425	30	225	400	20
9:00 AM	270	650	45	250	475	20	250	460	25	275	475	30	225	450	25
10:00 AM	285	500	20	260	530	25	260	560	25	285	525	35	230	525	50
11:00 AM	220	450	20	285	490	30	300	450	20	250	425	20	300	450	25
12:00 AM	240	475	20	220	440	20	220	425	15	260	400	25	260	400	20
1:00 PM	250	485	30	200	400	25	250	400	15	290	400	20	250	425	15
2:00 PM	200	490	20	180	375	20	265	425	15	225	375	20	225	375	20
3:00 PM	190	500	10	185	390	10	225	375	10	200	350	10	200	350	10
4:00 PM	180	425	10	170	460	10	175	350	10	175	350	10	175	250	10

FIGURE 15-1 Five-day traffic study.

Hour	Mon. PC	CCS	ATB	Tues. PC	CCS	ATB	Wed. PC	CCS	ATB	Th. PC	CCS	ATB	Fri. PC	CCS	ATB
8:00 AM	280	500	30	260	475	30	245	450	25	250	425	30	225	400	20
9:00 AM	270	650	45	250	475	20	250	460	25	275	475	30	225	450	25
10:00 AM	285	500	20	260	530	25	260	560	25	285	525	35	230	525	50
11:00 AM	220	450	20	285	490	30	300	450	20	250	425	20	300	450	25
12:00 AM	240	475	20	220	440	20	220	425	15	260	400	25	250	425	15
1:00 PM	250	485	30	200	400	25	250	400	15	290	400	20	250	425	15
2:00 PM	200	490	20	180	375	20	265	425	15	225	375	20	225	375	20
3:00 PM	190	500	10	185	390	10	225	375	10	200	350	10	200	350	10
4:00 PM	180	425	10	170	460	10	175	350	10	175	350	10	175	250	10
Totals	2115	4475	205	2010	4035	190	2190	3895	160	2210	3725	200	2090	3625	195
Busy HR	270	650	45	260	530	25	260	560	25	285	525	35	230	525	50

ABBH CCS   558   Erlg. 15.50

ABBH PC   261   Total Traffic 5 Busy Hours = 2790

ABBH ATB   36   Total Traffic 5 days = 19,755

                                            ABH % = 0.14

AHT   2.14

Offered   635   Erlg. 17.64

FIGURE 15-2   Summary report for the ABBH data.

Once the BH is determined from the highest CCS (carried load), we use the PC, CCS, and ATB from that hour for the ABBH values. The ABBH values are determined for each day and then an average is found for each. From the above study, we get the following ABBH values:

Day	B H	ABBH CCS	ABBH PC	ABBH ATB
Monday	9:00	650	270	45
Tuesday	10:00	530	260	25
Wednesday	10:00	560	260	25
Thursday	10:00	525	285	35
Friday	10:00	525	230	50
Total for the 5 hr		2790	1305	170
ABBH		558 CCS	261 PC	36 ATB

From the above study, we determine that the ABBH load or carried traffic is 558 CCS, the ABBH PC is 261 and the ABBH ATB is 36. We can calculate the average time per call or average holding time (AHT) by diving the CCS by the PC. Thus, AHT = CCS / PC = 558 / 261 = 2.137931. Rounding up, we can state that the AHT is 2.14 CCS per call or 214 s per call. This is 3 min and 34 s per call. You must be careful when converting seconds to minutes and seconds. For example, 214 s divided by 60 equals 3.56666667. Many people mistakenly state this as 3 min and 56 s, but they are wrong! We divided by 60, not 100, so 3.5666667 in this case is 3 min and 0.5666667 min (0.5666667 of 60 s). When you multiply 0.5666667 times 60, you get 34 s. Another way to do this conversion would be to divide the AHT of 214 s by 60 s to arrive at minutes. Thus, 214 / 60 = 3.5666667. Multiply the whole minutes by 60 and subtract this from the AHT: $3 \times 60 = 180$ and $214 - 180 = 34$. We arrive again at 3 min and 34 s as the AHT.

Once we know the AHT, we can estimate the offered load. Remember that offered load equals carried load plus blocked load. Blocked Load equals ATB times AHT. In the above study, blocked load = 36×2.14 CCS = 77.04 CCS. The offered load = carried load + blocked load = 558 CCS + 77.04 CCS = 635.04 CCS. We can also state this offered load in Erlangs because 36 CCS = 1 Erlang (635.04 / 36 = 17.64 Erlangs). Another way to calculate offered load is to add PC and ATB and then multiply the total times ABBH AHT. In the above study, ABBH PC = 261 and ABBH ATB = 34. The total attempts = 261 + 36 = 297, and offered load = 297×2.14 CCS AHT = 635.5 CCS. The difference between the two calculations results from rounding up AHT. The above data was summarized manually. It can also be done automatically by entering the data into a program or spreadsheet such as Excel. The data in Figure 15-2 is calculated by Excel.

15.20 FORECASTING TRAFFIC FROM BILLING DATA USING ABH%

Note that the total figures in my calculations are only for the 8 hr of the study and not for all 24 hr of the day. An 8-hr study on a PBX captures almost all of a day's telephone usage because most businesses are only open from 8:00 AM to 5:00 PM The data generated by an actual traffic software package would provide totals for each day. A figure that is useful in traffic engineering is a figure that indicates what percentage of a day's traffic occurred during the BH. In the case of the sample study above, we are using an 8-hr day. We can calculate the ABH% for each day by dividing the BH CCS by the total CCS for the day.

We can also find the ABH% for the complete five days. First we find the total CCS for all five days by adding the total CCS for each day: 4475 + 4035 + 3895 + 3725 + 3625 = 19,755 CCS total traffic for all five days. Then we find the total for all five BHs (650 + 530 + 560 + 525 + 525 = 2790 CCS). We divide total traffic for the five BHs by the total traffic for all five days (2790 / 19,755 = 0.14123). The ABH % for the study is 14%. This means that 14% of all traffic occurred during the five BHs.

We can calculate the ABH % for each day by dividing the BH for the day by the total CCS for the day. Generally the daily numbers are close to the ABH % for the whole study, so I will not calculate the individual ABH percentages. These percentages are helpful in calculating other forecasts. Suppose we wished to calculate our WATS monthly bill and the previous study of Figure 15-1 was for our WATS lines. We can use the following formulas to convert ABBH data to monthly data: monthly = traffic / ABH % times no. of workdays in the month.

To calculate a monthly bill we will use carried CCS because we are billed for what is actually carried, not what is offered. Depending on what we need, we can use PC, CCS, ATB, and carried or offered for any of these. For forecasting our WATS bill we use: 15.5 Erlangs / 0.14 times 22 workdays next month (the number of workdays in a month varies from 20 to 24) = 2435 Erlangs or hours. I should note that for a business most of the traffic occurs between 8:00 and 5:00. Therefore, the totals of the study in Figure 15-1 would accurately reflect total traffic for the day, and ABH % would also be accurate. There is one factor that I have not included in the above calculation: setup time. Traffic measurements start on seizure of a circuit, but billing does not start until a call is answered. To arrive at a better forecast of WATS billing, we need to subtract setup time on all calls and we need to subtract time for all unanswered calls.

When we measure the time a circuit is busy, we are measuring holding time of the circuit. Holding time is comprised of two components: setup time plus conversation time. Conversation time equals billed time. We want to determine total conversation time for the month. With several traffic studies and monthly bills in hand, you should be able to establish a realistic setup time. For our purposes we will assume 20 s per call as a setup time. In our example, 20 s per call can be used to eliminate setup and unanswered call times. Let's now recalculate the monthly WATS hours:

$$\text{ABBH PC} = 261 \text{ calls}$$
$$\text{Setup time} = 261 \times .20 \text{ CCS} = 52.2 \text{ CCS}$$
$$\text{Conversation time} = \text{carried CCS} - \text{setup}$$
$$\text{CCS} = 558 - 52 = 506 \text{ CCS} = 14.05 \text{ Erlangs}$$
$$\text{Monthly Billed} = 14.05 / 0.14 \times 22 = 2209 \text{ hr.}$$

The calculation of setup time has reduced the calculation by 226 hr (2435 − 2209 = 226). We now have a pretty good handle on what our monthly WATS usage will be and can verify the accuracy of our bill. The reverse of the above procedure could be used to engineer a circuit group when no traffic study data is available. Suppose we are billed for 2209 hr (79,524 CCS) and 10,615 calls on our WATS lines. We can use this to find carried load:

$$\text{ABBH} = \text{monthly billed} + (\text{no. of calls} \times \text{setup time}) / \text{no. of workdays} \times \text{ABH \%}$$
$$= (79{,}524 + (10{,}615 \times 0.2)) / 22 \times 0.14$$
$$= (79{,}524 + 2123) / 22 \times 0.14$$
$$= 84{,}193 / 22 \times 0.14$$
$$= 535 \text{ CCS}$$

The 535 CCS calculation for carried load falls a little short of the 558 CCS gained from a traffic study because ABBH PC is not 14 % of the total PC. If you calculate ABHPC %, it is 12.3 % of the total PC. ABHPC % and ABH CCS % will not always be the same, but they are close enough that the use of the ABH CCS % for both will get us relatively close on an initial engineering attempt without any traffic data. To accurately engineer a circuit group requires accurate traffic study data, accurate assumptions on how blocked calls are handled, and application of the correct traffic table.

15.21 RELATIONSHIP OF ERLANG TO PERCENT OCCUPANCY

The Erlang can be used to obtain estimates for other numbers. As stated before, an Erlang is 1 hr. One circuit can carry 1 Erlang of traffic. Ten circuits can handle 10 Erlangs of traffic per hour. Occupancy is a number used to rate the efficiency of a trunk group or circuit. If a circuit has 75% occupancy, it is used 0.75 Erlang each hour and is idle 0.25 Erlang (1/4 hr or 15 min) each hour. If a circuit group has ten circuits and the ABBH load is 5 Erlangs, the circuit group is achieving 50% occupancy. At any point in time, the circuit group is handling approximately five calls.

On a 30-channel multiplex loop, there is a capacity of 30 Erlangs (or 30×36 CCS = 1080 CCS). Since we must measure the load of both the originating and terminating circuits, half the load (15 Erlangs) is originating load and half the load (15 Erlangs) is terminating load. For every originating circuit path, there is a terminating circuit path. Since each voice path in a switch carries 1 Erlang, the Erlang capacity of a switching matrix tells you how many circuit paths are provided. Each call requires two circuit paths (one for the originating circuit and one for the terminating circuit). Dividing the Erlang load capacity of a switching matrix by 2 will tell you how many simultaneous calls the switching matrix can handle.

15.22 ARRIVAL RATE CALCULATION

Arrival rate of calls can be calculated when the load, number of calls (PCs) in an hour, and the AHT are known. Suppose a circuit group or multiplex loop has a carried load of 30 Erlangs (1080 CCS). Thus, 1080 CCS = 1800 min (108,000 s / 60 = 1800 min). If the average holding time (AHT) per call is 2 min (1.2 CCS), we had 900 calls (PC = carried load / AHT). In other words, 1080 CCS / 1.2 CCS = 900 calls. Also, 1800 min / 2 min = 900 calls. If we had 900 calls (PC) in an hour (60 min), we must handle 900 / 60 (= 15) calls per minute. Therefore, the arrival rate is 15 calls per minute.

There is a much faster way to get this arrival rate: divide the carried load in Erlangs by the AHT in minutes (30 / 2 = 15). This short method can be seen by combining the following two formulas: ABBH PC (Calls) per 60 min = carried load per 60 min / AHT per minute; and arrival rate per minute = calls per 60 min / 60 min. This second formula simply results in the number of calls per minute. Replacing calls per 60 min in the second formula with its equivalent in the first formula, we get: arrival rate per minute = (carried load per 60 min / AHT per minute / 60 min). To get carried load per 60 min, we multiply Erlangs by 60. Arrival rate per minute = (Erlangs $\times$ 60) / AHT / 60. Instead of multiplying Erlangs by 60 and then dividing by 60 to get arrival rate, we let the multiplication and division by 60 cancel each other out. We are left with arrival rate per minute = Erlangs / AHT.

Once we know the arrival rate, we can find the mean time between arrivals. It is the reciprocal of the arrival rate. In the case above, arrival rate = 15 calls per minute. The mean time between arrivals is 1 / 15 or 0.06666667 min. One-fifteenth of a minute is 4 s. Also, 60 s (1 min) $\times$ 0.066666667 = 4 s. This makes sense. If a call arrives every 4 s, then 15 calls will arrive in 1 min.

15.23 USING THE TABLES

In the study above, we had 17.64 Erlangs (635 CCS) of offered traffic. We can use this offered traffic load, the GOS desired, and the appropriate traffic table to find the number of servers, trunks, or circuits needed. The Poisson table used for Final Trunk Groups appears in Appendix C. The Erlang B table used to engineer HU trunk groups is in Appendix D. If the Poisson table is used (Appendix C), we look up the GOS desired. Suppose we decide on a GOS of *P*.10. We look under the *P*.10 ERL column to find a number equal to or just greater than 17.64 Erlangs (or 635 CCS). The number that meets this criterion is 17.97 Erlangs. Looking all the way across to the left, we find that 24 trunks are needed to provide *P*.10 GOS to an offered load between 17.12 to 17.96 Erlangs (or 647 CCS). If we had engineered to a service level of *P*.01 GOS, we would need 29 trunk circuits. Reproduced as Table 15-1 is a portion of the Poisson table. The lookups that we just did are highlighted. Now let's look up the same offered load in the Erlang B traffic table (Appendix D). The portion of the table used by our data is included as Table 15-2.

TABLE 15-1 Poisson

| GOS | B.01 | | B.05 | | B.10 | |
Trunks	ERL	CCS	ERL	CCS	ERL	CCS
23	13.33	480	15.72	566	17.11	616
24	14.08	507	16.56	596	**17.97**	**647**
25	14.86	535	17.39	626	18.83	678
26	15.61	562	18.22	656	19.72	710
27	16.39	590	19.06	686	20.58	741
28	17.17	618	19.92	717	21.47	773
29	**17.97**	**647**	20.75	747	22.36	805
30	18.75	675	21.61	778	23.22	836

Table 15-1 Portion of Poisson Table.

TABLE 15-2 Erlang B

| GOS | B.01 | | B.05 | | B.10 | |
Trunks	ERL	CCS	ERL	CCS	ERL	CCS
20	12.03	433	15.25	549	17.61	634
21	12.84	462	16.19	583	**18.65**	**671**
22	13.65	491	17.13	617	19.69	709
23	14.47	521	18.08	651	20.74	748
24	15.29	550	19.03	685	21.78	784
25	16.12	580	19.99	720	22.83	822
26	16.96	611	20.94	754	23.88	860
27	**17.80**	**641**	21.90	788	24.94	898

Table 15-2 Portion of Erlang B Traffic Table.

In Table 15-2, we find that for an offered load of 17.64 Erlangs, we will need to provide 21 trunks to achieve a $P.10$ GOS. We would need to provide 27 trunks to achieve $P.01$ GOS. This is a difference of six trunk circuits to achieve a better GOS at $P.01$. Comparing the Poisson table to the Erlang B table, for $P.10$ GOS, we find that under Poisson 24 trunks are needed, and under Erlang B 21 trunks are needed. Poisson tells us we need 3 more trunk circuits than Erlang B does. For a $P.01$ GOS we need 29 circuits under Poisson and 27 (or 2 less) under Erlang B.

As I stated earlier in the chapter, you could engineer everything with Poisson, but in most cases it tends to overengineer. You can save circuits (and money) if you can use the Erlang B table. Remember that the choice of table depends on how blocked calls are handled. For BCH, use Poisson. For BCC, use Erlang B. As stated earlier, we will use Erlang B to determine the number of trunk circuits needed in a HU trunk group because blocked calls are cleared to the Final Group. We will use Poisson to engineer a Final Group because Poisson will provide a safety buffer by overengineering the Final Group.

15.24 SEAT-OF-THE-PANTS ENGINEERING

One final thought before we leave Erlang B and Poisson. An engineer familiar with network design, traffic tables, and the mathematics underlying traffic tables can make very good estimates on sizing circuit groups by using intuition. We know that each circuit can carry 1 Erlang (36 CCS) of traffic in 1 hr. We also know this 1 Erlang is the maximum amount of traffic that can be handled if a circuit achieves 100% utilization or efficiency. Thus, the maximum load that could theoretically be handled by any group of circuits is equal to the number of circuits. For example, a circuit group containing 35-circuits could not carry more than a 35-Erlang load.

As the load placed on a circuit group approaches the number of circuits in the group, the GOS drops dramatically, but the efficiency of the group will increase because circuits will be in use most of the time. In other words, efficiency of the circuit group and GOS of the circuit group are inversely related. This is evidenced by the traffic tables. Under the Erlang B table, 20 circuits can handle 17.6 Erlangs of offered load at GOS of $P.10$, but 27 circuits are needed if we wish to provide $P.01$ GOS. Anyone can deduce that if the same load is offered to 27 circuits that was offered to 21 circuits, the load on each individual circuit will be less when it can be spread across 27 circuits. Thus, each of the 27 circuits will carry less traffic (and is less efficient) than if 21 circuits were used.

A simple way to determine efficiency is to divide offered load by the number of circuits handling the load. In the examples above, 17.6 / 21 = 0.83 (83% efficiency) and 17.6 / 27 = 0.65 (65% efficiency). Another facet to efficiency is that efficiency is directly related to the size of a circuit group. The more circuits there are in a circuit group, the more efficient the individual circuits are.

According to the Erlang B table, at *P*.01 GOS, a circuit group consisting of 15 circuits achieves 54 % efficiency per circuit, a 27-circuit group achieves 65% efficiency, a 50-circuit group achieves 76% efficiency, and a 100-circuit group achieves 84 % efficiency. According to the Poisson table, at a *P*.01 GOS, a circuit group made up of 15 circuits achieves 50% efficiency, a 27-circuit group achieves 60% efficiency, a 50-circuit group achieves 70 % efficiency, and a 100-circuit group achieves 78% efficiency.

The experienced design engineer can use this knowledge of efficiency to guess at the number of circuits needed. The formula for efficiency can be restated to find the number of circuits required for a given load: no. of circuits = offered load / % efficiency. If the ABBH offered load is 17.6 Erlangs and we want *P*.01 GOS, the number of circuits required = 17.6 / 0.6 = 29 circuits. The % efficiency to use depends on GOS desired and the number of circuits involved. As a rule of thumb, for *P*.01 GOS use 50% for more than 15 circuits, 65% for 25 circuits, 75% for 50 circuits, and 85 % for 100 circuits. Try this on any load of your choice—for example, if the offered load = 60 Erlangs, then 60 / 0.8 = 75 circuits. Notice that I made an estimated extrapolation of 0.8, using 0.75 for 50 circuits and 0.85 for 100 circuits. For a higher grade of service such as *P*.05 or *P*.10, use a slightly higher efficiency % than you use for *P*.01. For Poisson calculations, a slightly higher efficiency % is used than for Erlang B.

These guesses are only to be used at making an initial sizing. Once a circuit group is established, a traffic study is conducted to determine the accurate sizing. If you find fault with this "seat-of-the-pants" engineering approach, let me remind you that traffic tables are based on probabilities and assumptions that are also best guesses. Tables are not without flaws. Even with flaws, the use of traffic tables does provide a more optimum solution for the provisioning of circuits. Tables help the engineer provide just the number of circuits needed for a certain GOS. The use of tables is necessary to achieve maximum circuit utilization, GOS, and lowest cost. Only when traffic data is not available do we use our seat-of-the-pants engineering approach.

One question that arises on the Erlang B table is when we look at 100 trunks and a *P*.10 GOS. You can see the table states that 107 Erlangs can be handled at *P*.10 with 100 trunk circuits. People ask how 100 trunks can handle more than 100 Erlangs. The answer is: "they can not." The 107 Erlangs is offered load, *not* carried load. With *P*.10 GOS, we accept 10% blockage, and 10% of 107 = 10.7 Erlangs. Moreover, carried load = offered load - blocked load = 107 – 10.7 = 96.3 Erlangs. Thus, 100 trunks can serve an offered load of 107 Erlangs at *P*.10 GOS by carrying 96.3 Erlangs of the offered load and blocking the remaining 10.7 Erlangs. Each of the circuits achieve 96.3 % utilization (or occupancy).

You can see from the data that greater levels of circuit efficiency are achieved at lower service levels. Note that *P*.10 is a lower service level than *P*.01. Blocking 10 calls per 100 attempts provides a lower GOS than blocking 1 call per 100 attempts. Also, the larger a circuit group, the more efficient the circuits. Since large circuit groups are more efficient, they have less capacity left to handle overloads. Smaller, less efficient groups can cope better with overloads.

If a 25% overload is placed on a small group, it will have less blockage than a 25% overload placed on a larger circuit group.

The performance of larger groups is also affected more than that of a smaller group when a circuit is removed from the group because of failure of the circuit or for maintenance of the circuit. This is a simple function of efficiency. It stands to reason that each circuit carries approximately 0.9 Erlang at 90% efficiency versus 0.5 Erlang at 50% efficiency. Thus, removing a circuit from a 90% efficiency group remove 0.9 Erlang of capacity from the group. Removing a circuit from a smaller group that is 50% efficient removes 0.5 Erlang of capacity. The engineered capacity of a circuit group does not provide extra circuits. Remember that we provide just enough circuits to handle the ABBH offered load. Because there are just enough circuits provided, it is critical that central office maintenance keep all circuits working.

15.25 SUMMARY

The PSTN is composed of many circuit switching systems and trunk circuits connecting various systems. Each portion of the PSTN has a network design engineer in charge of it. The design engineer is responsible for providing an adequate grade of service while keeping the associated cost as low as possible. Design engineers must manage and balance the trade-off between service and cost. The design of each part of the network is based on past traffic for that part. The engineer gathers traffic data such as ABBH carried load, peg count, and ATB counts. This data is used to arrive at the offered load or demand placed on that part of the network.

Various traffic tables are available to assist the engineer in determining how many circuits will be needed for a given offered load. The table to use depends on how the system handles blocked calls. If blocked calls are held, use a Poisson table. If blocked calls are cleared, use an Erlang B table. Chapter 16 discusses the tables to use when blocked calls are held in a queue. There are three basic engineering variables: (1) demand or offered load, (2) GOS desired, and (3) the number of circuits (or servers) needed. If you supply two of the variables, the traffic table will provide the third. When you enter the offered load and GOS desired. the traffic table will tell you how many circuits are needed.

REVIEW QUESTIONS

1. What is traffic?
2. How do we know how much traffic is placed on a PSTN component?

3. What is the overall responsibility of a network design engineer?
4. How does an engineer determine how many components are needed?
5. What does DOA do?
6. What are the three types of meter readings found in a traffic study?
7. What is ABBH?
8. How is AHT calculated?
9. How is carried load calculated?
10. How was it possible to replace DOA clerks with computers?
11. Is the engineer more concerned with individual circuit usage or usage for a group of circuits? Which of the two measurements would maintenance use?
12. What is the difference between carried load and offered load?
13. What is an ATB?
14. How is blocked load calculated?
15. What is the first step in circuit group design?
16. What are the three factors in a traffic table?
17. What is CCS? What is an Erlang?
18. What does $P.05$ mean? What does $B.05$ mean?
19. What happens to traffic engineering design if calls are not random?
20. How does an engineer know which traffic table to use?
21. What is the major cause of overengineering a circuit group?
22. What do SPC switching systems use for meters?
23. Who uses traffic study reports generated by DOA?
24. What is demand?
25. What is measured load?
26. What is GOS?
27. What are the three variables used in traffic engineering?
28. What is capacity?
29. Why is concentration provided in a line bay such that 640 lines may only have access to 120 outlet voice paths?
30. How many Erlangs can a 30-channel multiplex loop carry?
31. What is a load test?
32. What is the GOS objective for SPC switches?
33. Will ABBH PC always be the highest PC for the day?
34. How is the ABBH determined?
35. What is the efficiency of a trunk group consisting of (a) 15 trunks, (b) 27 trunks, (c) 50 trunks, and (d) 100 trunks?
36. Which is more efficient, a trunk group with 75 trunks or one with 50 trunks?
37. Referring to question 36, which trunk group size will be most affected by an outage of one circuit?

GLOSSARY

All Trunks Busy (ATB) Refers to a condition where all circuits in a group are busy. In SPC switching systems, a memory variable serves as an ATB meter and keeps track of how many call attempts were made to a group of circuits when all the circuits in the group were busy. An ATB meter exists for each circuit group. ATB is a count of the number of blocked calls in a SPC switching system.

Average Bouncing Busy Hour (ABBH) An average of various busy hour statistics such as PC, CCS, and ATB when the busy hour does not occur at the same time.

Average Busy Hour (ABH) If the busy hour for every day included in a study occurs at the same time (for example, if the busy hour for a five-day study occurred at 10:00 A.M. every day), the average for all hours is ABH.

Average Hold Time (AHT) The average length of time a circuit is in use. It is found by dividing CCS by PC. For example, if the ABBH CCS is 500 CCS and ABBH PC is 100 PC for a five-day study, we can conclude that the average length of every call will be 5 CCs based on our sample (study).

Blocked Calls Cleared (BCC) When insufficient circuit paths exist and blockage occurs, the blocked calls are cleared from the system by attaching them to a fast busy tone or all circuits busy recording.

Blocked Calls Delayed (BCD) When insufficient circuit paths exist and blockage occurs, the blocked calls are held in a waiting queue by the system until a circuit becomes available to serve the caller.

Blocked Calls Held (BCH) When insufficient circuit paths exist and blockage occurs, the blocked calls are not cleared or put in a queue. If callers wait, the will be served. Theis methodology occurs for dial registers.

If no dial registers are available, callers must wait or hold until one is available. This is why we are supposed to listen for dial tone. The presence of dial tone tells us our telephone has been attached to a dial register and we can dial a number.

Blocked Load The amount of offered load that does not get carried but is blocked due to insufficient circuits. blocked load is estimated by multiplying the number of attempts blocked (ATB) times AHT.

Busy Hour The hour of a day when the most telephone usage occurs.

Capacity A measure of a switching system or circuit group's ability to handle telephone calls. Capacity can be stated as either ABBH PC or ABBH CCS.

Carried Load The load carried by a circuit, group of circuits, or switching system. It is measurable. If the switch carries traffic, it can measure the traffic or load carried. Carried load equals the ABBH PC times the ABBH AHT.

CCS Centium call seconds or hundred call seconds. One CCS equals 100s; 1hr is 3600s or 36 CCS.

Demand Offered load, or the amount of traffic that users are trying to place over a circuit, group of circuits, or switching system.

Erlang A traffic usage measurement that is equal to 1hr or 36 CCS.

Final Circuit Group A group of circuits for which there is no alternate route. A final group is usually the alternate route for a high-usage group. Final groups are engineered for lower levels of efficiency to provide low levels of blockage.

Grade of Service (GOS) A probability measure of blockeage occurring. Using statistics (traffic tables), we can predict the probability that blockage will occur for a specific load placed on a specific number of servers.

High-Usage Trunk Group A group of trunk circuits purposely engineered to high levels of efficiency with a high probability of blockage. Traffic blocked is routed over an alternate route.

Offered Load The demand or amount of traffic that callers are trying to place over a system. It is equal to carried load plus blocked load.

Peg Count (PC) A term used to indicate how many calls are handled by a circuit, a group of circuits, or a switching system. This term is a holdover from electromechanical traffic registers. When the relay of the register operated, it was said to "peg the meter." Each peg of a meter made it turn a numbered dial one step.

Servers Circuits available to handle the offered load.

Traffic A term used for the volume of calls placed over a circuit, group of circuits, or a switching system. If a system is handling 10,000 calls and hour, we say the traffic is 10,000 calls an hour.

16

Telecommunications Voice Network Design: Delay and Queues

Key Terms

Delay Time
Finite Queue
First-Attempt Traffic

Load Balance
Negative Exponential
   Distribution

Queuing Theory
Retrial Methodology
Unlimited Queue

In the previous chapter, we focused on the basic principles underlying the engineering of circuits and networks, using assumptions that blocked calls were held (Poisson) and blocked calls were cleared (Erlang B). This chapter focuses on applying traffic tables used when blocked calls are held in a queue or when the customer makes subsequent attempts (retry) to complete a call. In any traffic engineering exercise, we always make several assumptions. These assumptions include the following:

1. The actions of users are independent of one another and are random.
2. The number of users is infinite.
3. The number of servers (circuits or trunks) is finite.
4. The probability of a particular user being on the system is low.
5. The number of failures is low.
6. Statistically, the system is in equilibrium.
7. We know the disposition of blocked calls.

16.1 LOAD BALANCE

DOA tries to achieve randomness in line groups by assigning both business and residential customers to the same line group. This is referred to as *load balance*. If a line group contains all business customers, it will have a higher traffic load than a

413

line group serving residential customers. With a higher load, the line group would require additional multiplex loops between the line bay and the network matrix. Load balance requires the physical distribution of business and residential customers across all line bays to achieve an optimal mix resulting in the same traffic load being placed on all line bays. DOA continually monitors the load of each line and trunk bay. DOA will make physical reassignments of lines and trunks as necessary to maintain a balanced load across all line bays and all trunk bays. This balance is necessary to meet the criterion that call attempts are random. Only when call attempts are random can the laws of probability be applied.

16.2 WHICH TABLE DO WE USE?

The laws of probability underlie all traffic formulas and tables. *How blocked calls are treated determines which traffic table to use.* Poisson is used for *blocked calls held (BCH).* The blocked calls are assumed to be held for the full duration of holding time *(average hold time or AHT).* Erlang B is used when *blocked calls are cleared (BCC).* Another table—called the *Neal-Wilkinson table*—is also used for blocked calls cleared, when call attempts exhibit a peakedness instead of randomness.

When blocked calls retry immediately, we assume BCH and can use Poisson. For calls with immediate retry (BCH), we can also use tables developed for retrials that are called *extended Erlang B tables.* When blocked calls are held in a queue (or delayed), we use an Erlang C table. Servers that provide a delay in service are dial registers, operators, and ACD systems. Erlang C makes a basic assumption that users will wait indefinitely (for an unlimited amount of time) until a server (circuit) is available. Since users will wait until a server is available, there is no overflow and carried load equals offered load.

When a call attempt is blocked, the system will reroute the call to another circuit group, place the call in queue, or return an all-circuit-busy signal. A circuit group that has calls rerouted when all circuits are busy is engineered using a Poisson or Erlang B table. The engineering for circuits that place overflow calls in a queue is done using a queue table such as Erlang C. When callers get an all-circuit-busy signal and immediately retry their call, we use an engineering methodology called *retrial methodology.*

When we engineer using retrial methodology, we use an extended Erlang B Retrial table. All callers that get blocked and receive an all-circuit-busy signal do not immediately retry their call. Some callers will try again immediately but others will wait until later to place a call. Some callers will continue to retry calls immediately and may make as many as 10, 15, or 20 attempts one after the other in an effort to get through to the desired party. Since, the behavior of callers varies when they encounter an all-circuit-busy condition, there are three basic retrial tables. One table is used if we assume that 50% of blocked callers retry immediately; another table—called a 70% table—is used when 70% of all blocked callers retry immediately; and a 100% table is used when all blocked callers retry immediately. As I will show you later, a 100% retry table is actually the same table as a regular Erlang B table.

16.3 EXTENDED ERLANG B AND THE RETRIAL TABLES

The retry tables were developed by Henry Jacobsen, an AT&T statistician. The 70% retrial table is called the *Jacobsen table* in his honor. Jacobsen made a basic assumption that if all callers would retry their calls immediately, it is almost as if they are in a queue. If everyone keeps trying their calls until they get through (100% retrial), offered load will equal carried load. Although some callers may get blocked on a first attempt, there is no final blockage because people continue trying their call until it becomes a carried call. Therefore, the offered load ends up being the carried load when 100% of callers retry immediately until they get carried. This is why a 100% retrial table is the same as an Erlang B Table. With 100% of our blocked callers making immediate retrials, we know the offered load. It is the same as the carried load. We can measure carried load. We can use the measurement as the offered load to properly size the circuit group using either the Erlang B table or the 100% retrial table.

16.4 JACOBSEN RETRIAL FORMULA

If fewer than 100% of all blocked calls retry their calls, offered load will not equal carried load, because some of the offered load was blocked and did not reenter the system. Jacobsen developed a formula to calculate the offered load based on the percentage of people who retry their call immediately. In a retrial methodology, offered load is called *first-attempt traffic*. If a user is able to complete a call on the first attempt, we would have no additional retries. If everyone gets through on the first attempt, there is no blockage, there are no retries, and all of the first-attempt load is carried. Therefore, with no retries and no blockage, carried load is equal to first-attempt load. The first-attempt formula developed by Jacobsen calculates what the first attempt, or offered load, would be based on the percent of retries:

First-attempt load = carried load × {1 − (retry % × block %)/(1 - block %)}

Remember that we can only measure carried load, but *we engineer using offered load*. With Erlang B, we arrive at offered load by adding blocked load to the carried load. With extended Erlang B, we arrive at offered load by using the Jacobsen retrial formula for first attempts. Let's use the formula for a situation when our past traffic studies tell us: the ABBH carried load is 150 CCS; the number of PCs was 72; the number of ATBs was 8; and we assume that 50% of the blocked callers made an immediate retrial. Since we have a PC of 72 and an ATB of 8, we had a total of 80 calls. Of the 80 total call attempts, 8 were blocked. Ten percent of our calls were blocked (8/80). Substituting our study information into the retrial formula:

$$
\begin{aligned}
\text{First-attempt load} &= \text{carried load} \times \{1 - (\text{retry \%} \times \text{block \%}) \, / \, 1 - \text{block \%}\} \\
&= 150 \, \text{CCS} \times \{(1 - (0.5 \times 0.1)) \, / \, 1 - 0.1\} \\
&= 150 \times \{(1 - 0.05) \, / \, 0.9\} \\
&= 150 \times \{0.95 \, / \, 0.9\} \\
&= 150 \times 1.055555\ldots \\
\text{First-attempt load} &= 158.3333\ldots
\end{aligned}
$$

We can now use the 50% retrial table from Appendix E. Notice that the table uses hours of first-attempt traffic. An hour is an Erlang. Therefore, we must convert our CCS into Erlangs before we can use the table. The formula for this conversion is: Erlang = CCS / 36. For our example, Erlang = 158.3333... / 36 = 4.398 Erlang. There is no 4.398 Erlang in the table, so we use the next higher number in the table. That number is 4.400. Looking in the 4.400 Erlang section of the 50% table, we can find the number of servers needed for various GOS levels. At the present 10% blocking level ($P.10$), we are supplying 7 circuits. If we select a GOS of 1.0%, we can see that the number of lines (servers) needed is 10. While we are in this table, note that with the first-attempt load of 4.4 Erlang at $P.10$ GOS, the hours connect (carried load) is 4.174. We stated earlier that the carried load is 150 CCS. This is 4.1666. . . Erlang (150 / 36 = 4.16. . .), and 4.16666 is very close to 4.174. Thus, this table provides both carried and offered load numbers. Carried load is hours connect and offered load is hours first attempt.

16.5 RELATIONSHIP OF BLOCKED LOAD, CARRIED LOAD, AND FIRST-ATTEMPT LOAD

Let's use the example above to look a little deeper into this retrial formula. We actually carried 150 CCS, and we were offered 158.333. . . CCS of first-attempt traffic load. In this example, 50% of our callers retried their calls and became part of the carried load. Thus, our carried traffic includes 50% of the initial blocked load. We had 8 blocked calls for a $P.10$ GOS. We know that all calls have the same AHT. If there were no blocked calls in our carried load, we could find the AHT by dividing the total CCS (150) by the total PC (72). Thus, 150/72 = 2.08333. . . CCS AHT. We had 8 ATBs, and we could calculate the blocked load (if there were no retries) by multiplying ATB×AHT (8 × 2.08333. . . = 16.667). Remember that if there are no retries we can use Erlang B methodology and the offered load = carried load plus blocked load (150 + 16.667). We calculate offered load under Erlang B and Poisson by adding the blocked load to the carried load. This can be done by stating the blocked load as a percentage of offered load. In the example above, our blocked load was 10% of our offered load. Therefore, under Poisson and Erlang B:

$$\text{Offered load} = \text{carried load} + \text{blocked load}$$

$$\text{Offered load} = \text{carried load} + 10\% \text{ of offered load.}$$

$$\text{Offered load} - 10\% \text{ of offered load} = \text{carried load.}$$

$$90\% \text{ of offered load} = \text{carried load}$$

$$0.9 \text{ offered load} = \text{carried load}$$

Now divide both sides by 0.9:

Offered load = carried load / 0.9

For our example,

Offered load = 150 / 0.9 = 166.666. . . CCS

Erlang B Offered Load

Offered load = carried load / (1 – % blocking)

If there were no retries, we would use Erlang B (or Poisson), and our offered load would have been 166.667 CCS. *But there were retries; 50% of the blocked load retried and is included in the carried load number.* Thus, we only want to add 50% of the blocked load to the carried load to arrive at offered load. Fifty percent of the blocked load (16.667) is 8.333. . . If we add this to 150, we get 158.333. . . offered load. This is exactly the same number, for first attempt, that we arrived at by using Jacobsen's formula. Jacobsen's formula is used to make adjustments to the blocked load before adding it to the carried load when retries are involved. We do not want to add all of the blocked call load to the carried load because some of the blocked load is already included in the carried load number.

16.6 100% RETRIAL TABLE

Now let's look at a 100% retry situation. If 100% of the blocked calls retried, the carried traffic already includes traffic that had been initially blocked. Therefore, with 100% retry we should not add any blocked load to the carried load to arrive at first-attempt (offered) load. Let's use the formula to check this out:

$$\text{First-attempt load} = \text{carried Load} \times \{(1 - (R\% \times B\%)) \,/\, (1 - B\%)\}$$
$$= 150 \,\{(1 - (100\% \times 10\%)) \,/\, (1 - 10\%)\}$$
$$= 150 \,\{(1 - (1.00 \times 0.1)) \,/\, (1 - 0.1)\}$$
$$= 150 \,\{\,(1 - 0.1) \,/\, (1 - 0.1)\}$$
$$= 150 \,\{1/1\}$$

First-attempt load = 150 CCS

We can see that anytime the retry % is 100%, the multiplier used against carried traffic will always be 1, because it ends up being $(1 - \%B \,/\, 1 - \%B)$. This proves our statement that *when 100 % of blocked calls retry and are carried by the system, first-attempt load = carried load.* The retry tables make a slight adjustment for the fact that one caller may make several retries. Therefore, the number of ATBs may be more than one call. Assume we had 8 ATBs that were all caused by one caller who kept trying until the 9th try was successful. This could distort our data because the caller's load is included in the carried load. If we assume 50% retry, it looks like we should add additional load to the carried load. The tables compensate for this situation. I have included the 100% retrial table and the Jacobsen 70% retrial table in Appendixes F and G. The 100% retrial table is similar to the 50% table.

16.7 *JACOBSEN 70% RETRIAL TABLE*

The Jacobsen retrial table is precalculated at 70% retry and includes both a carried load in CCS and an offered load in CCS. The reason Jacobsen used a 70% retry is that this is an average for retries for all calls in the PSTN. The advantage of this table is the fact that it already includes the calculation for first-attempt traffic. You cannot use this table unless your retry % is 70%. Let's use our previous data and assume 70% retry instead of 50% retry:

$$\text{First-attempt load} = \text{carried load} \times \{(1 - (R\% \times B\%)) / (1 - B\%)\}$$
$$= 150 \{(1 - (0.7 \times .1)) / (1 - 0.1)\}$$
$$= 150 \{(1 - (0.07)) / 0.9\}$$
$$= 150 \{0.93/0.9\}$$
$$= 150 \ 1.0333...$$
$$\text{First-attempt load} = 155 \text{ CCS}$$

The Jacobsen table is structured according to GOS at the top of the table. Let's look in the Jacobsen table (Appendix G) and find TABLE .01 (*P*.01 GOS) and a carried load of 150 CCS. The next higher load above 150 CCS is 154 CCS. Next to the carried load is the offered load number. It is also 154 CCS. Thus, at low blockage and carried traffic load, the offered traffic is equal to offered load. For an offered load at *P*.01 GOS, we need 10 circuits. This is the same number that the 50% retrial table gave us. Now let's assume that we had a carried traffic of 150 CCS *but* also had a blockage of 40%. We use the TABLE .40 and locate a carried traffic that is more than 150 CCS. We must use 170 in the carried CCS column. Note that with this carried load and % blockage, we must be serving the present circuit group with 6 circuits. Now look next to the carried traffic. The offered load is 205 CCS. We would determine the GOS desired. Suppose we desire a *P*.01 GOS. Look up an offered load of 205 CCS in the TABLE .01. We must use the next higher number of 230 CCS. For an offered load of 230 CCS at *P*.01 GOS, we need 13 circuits. Thus, to improve our GOS from *P*.40 to *P*.01 for this circuit group, we must add 7 circuits to the existing 6 in service. Let's recalculate first-attempt load using 40% blockage and a carried load of 170 CCS:

$$\text{First-attempt load} = \text{carried load} \times \{(1 - (R\% \times B\%)) / (1 - B\%)\}$$
$$= 170 \{(1 - (0.7 \times 0.4)) / (1 - 0.4)\}$$
$$= 170 \{(1 - (0.28)) / 0.6\}$$
$$= 170 \{0.72/0.6\}$$
$$= 170 \ 1.2$$
$$\text{First-attempt load} = 204 \text{ CCS}$$

This agrees with the TABLE .4 for a carried load of 170 CCS; the first-attempt load in the table is 205 CCS. Since our carried load of 150 CCS was so far below what was available in the table (170 CCS), and the offered load of 205 is very close to 203 in TABLE .01, I would use 203 CCS as the offered load, and instead of using 13 circuits, I would use 12 circuits to supply service for this circuit group.

16.8 RETRIAL METHODOLOGY VERSUS ERLANG B

Before leaving the retrial methodology, I would like to add that the retry formulas are great for helping us determine the actual offered load, but once the first-attempt load is determined, I find that the Erlang B table can also be used to determine the *number of circuits needed*. Let's assume that we are serving a carried load of 959 CCS with 30 circuits. If you look this up in the Jacobsen table, by using the number of servers, you will find that if the group size is 30 and the carried load is 959 CCS, you are close to the bottom of TABLE .20 (GOS is currently P.20). According to TABLE .20, if the carried load is 959 CCS, the offered load is 1031 CCS.

Now, let's look this 1031 CCS offered load up in the Jacobsen table and see how many servers are needed to provide a P.01 GOS. In the TABLE .01, look in the offered load column for 1031 CCS. The table stops at 1022, and since this is close to 1031, we will use this for offered load. Looking at the group size column, we find that 40 circuits would be needed to provide P.01 GOS to an offered load of 1022 CCS. Now let's convert 1022 CCS to Erlangs and use the Erlang B table. First, 1022 CCS / 36 = 28.388. . . Erlang. We do not have a 28.388-Erlang table but we can use the 30-Erlang table.

The 30-Erlang table states that for a P.01 GOS, we will need to provide 42 circuits. We can also see in the Erlang B table that if the offered load were 25 Erlang, we would need 36 circuits to provide P.01 GOS. The actual offered load of 28.388 Erlang is between 25 and 30. In fact, it is about two-thirds of the distance between 25 and 30 (3.3888 / 5 = 0.67777). The difference between 42 and 36 is 6 circuits. Two-thirds of 6 is 4. We can add 4 to 36 and we get 40 circuits required to serve a load of 28.3888 Erlang at P.01 GOS. Both tables (Jacobsen and Erlang B) tell us that 40 circuits are needed. The Jacobsen table saves us the trouble of calculating the offered load. The offered load is already calculated for 70% retry and is found in the column next to carried load.

Again, I must caution you that we can only use the retry tables that correspond to the percent of retry. If there is no retry involved and all blocked calls are cleared from the system, we must use either Poisson or Erlang B, because with no retry, we must add all of the blocked load to the carried load. This will provide a higher offered load number than the retry formula provides. For 0% retry we must calculate offered load by adding all of the blocked load to the carried load. In essence, offered load = first-attempt load when retry = 0%. As we stated earlier, if retry is 100%, then carried load = first-attempt load. The design engineer must know the disposition of blocked calls. If blocked calls are cleared from the system, the Erlang B methodology has to be applied. If blocked calls are retried, the appropriate retry table can be used.

16.9 QUEUING THEORY

If the disposition of blocked calls is to place them in a holding queue, the design engineer uses a queuing methodology and queuing table to determine the number of servers needed to handle the demand placed on the system. *The primary table used for queued traffic is the Erlang C table.* Queuing is such a common occurrence, not only in telecommunications but in a variety of business and manufacturing applications, that this body of knowledge is often called the *queuing theory* or the *waiting-line theory*. The concepts of this theory provide the basis for Erlang C. To understand Erlang C best, we should take a closer look at the three parts of a waiting-line (or a queuing) system:

1. Arrivals or inputs to the system
2. Queue (or waiting line itself)
3. Service facility

16.9.1 Arrivals:

The input source that generates arrivals (or attempts) for a system has three major characteristics:

1. The size of the population. The size can be either unlimited or finite. Examples of an unlimited population would be shoppers arriving at a store, people calling an 800 catalog number, cars entering an interstate highway, and so on. An example of a finite population would be terminals on a small LAN or people arriving for supper at your home. Most queuing models, such as Erlang C, assume an unlimited population.

2. The pattern of the arrivals. Arrivals are either random (meaning "at any time") or smooth ("scheduled"). For example, while a LAN may have a limited number of input terminals, "at any time" each terminal could request assistance of the server. By contrast, data communications can be scheduled at certain intervals via a batch process (not unlike your doctor's office, which tries to schedule one patient every 10–15 minutes). Erlang C assumes a random arrival.

3. The behavior of the arrivals. If the arrival has to wait, what does it do? Not everyone is patient. Erlang C assumes people will wait indefinitely until being "served." However, as we know, some people will drop out of a queue (known as *reneging*). Others see that the line is too long and refuse to get in line (known as *balking*). We will soon see that Erlang C does have a subset methodology (known as *equivalent queue extended Erlang B or EQEEB*) to correct for the problems of balking and reneging.

16.9.2 Queue:

The queue itself has two characteristics:

1. *The length of a queue* (or waiting line) may be unlimited or limited. An example of an unlimited queue would be a toll booth serving arriving automobiles (you can't turn around and you have to wait. . .). An example of a limited queue would be the chairs in a doctor's reception area. Erlang C assumes an unlimited queue (as do most theories); however, we will look at EQEEB, mentioned previously, that uses a limited queue length. One important note for telecommunications that the length of a queue is measured in time, not physical length!

2. The second characteristic deals with *queue discipline*, or how the requests for service are handled. Most queues employ the *first-in, first-out (FIFO)* rule. However, there are examples where requests made be handled on a priority basis, such as a hospital emergency room. Even in communications and data processing, certain requests, such as payroll processing or an important customer, may be given access sooner than others. There also is *LIFO* or *last-in, first-out*. This rule is often applied in inventory scenarios. Erlang C, like most of the queuing theories in traffic engineering, assumes the FIFO rule.

16.9.3 Servers:

The third part of a queuing system is the service facility itself and two basic parameters:

1. *The configuration* of the system is described as either single-channel or multiple-channel and either single-phase or multiple-phase.

> a. A *single-channel* system contains only one server. An example would be a drive-up window at a fast food restaurant or a trunk group in a PBX with only one trunk.
>
> b. A *multiple-channel* system contains several servers—for example, more than one trunk in a trunk group or a bank with more than one teller. While customers (or requests) may stand in a common line, any one of the tellers (servers) can handle your request; it does not matter which one you select.
>
> c. A *single-phase* system is one in which the request receives service from a server, then exits the system. From the previous bank teller example, once customers have been helped by a teller, they can exit the system. They do not need to go to another teller to complete their business.
>
> d. A *multiple-phase* system is one in which the request may have to pass through several service points (or servers) to complete the system. A good example of a multiple-phase system is school registration at the beginning of each term. To complete registration, a student has to go to several stations, not necessarily in any order.

Most communication systems operate under a multiple-channel, single-phase configuration (a general assumption of Erlang C). However, one should recognize the configuration to determine what methodology or approach should be taken. One suggestion for a multiple-phase system is to break down each phase as a single-phase system and solve each phase independently.

2. The second parameter of the service facility is the *service time distribution* or how long it takes to service the request. Most telecommunications applications assume a *random*-distribution; the time will vary widely, although averages are used. However, some systems use *constant* service time distribution; the time is always the same to service an attempt. A good example of the latter is an automatic car wash. Telecommunications examples of a constant service time could include credit verification or "midnight" batch routines.

Erlang C assumes random distribution, or to be more specific, *negative exponential probability distribution*. What this term means is that the probability of long service times is very low and the probability of short service times is very high. For example, if the average holding time is 2 min, the probability that a call is more than 2 min is less than the probability that the call is shorter. In other words, the fact that the "average" is 2 min does *not* mean 50% are less and 50% are more. (See Table 16-1 for a graphic example of this phenomenon.)

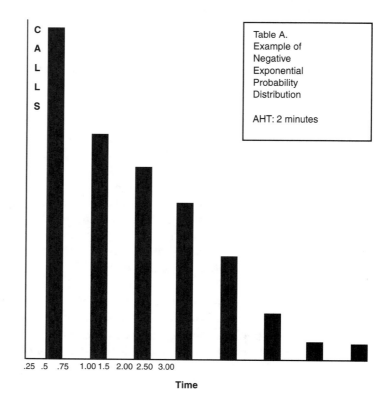

Table A.
Example of Negative Exponential Probability Distribution

AHT: 2 minutes

.25 .5 .75 1.00 1.5 2.00 2.50 3.00

Table 16-1

Time

16.10 ERLANG C ASSUMPTIONS

For most telecommunications applications that involve queuing, Erlang C is widely used. Erlang C assumes:

1. Infinite arrival population
2. Random arrival rates
3. Unlimited queue lengths (if the queue has a limited time, use EQEEB)
4. FIFO
5. Single/multiple channels
6. Single phase
7. Negative exponential probability distribution

16.11 QUEUE PERFORMANCE

In a queuing environment, performance is measured by a different standard than in a nonqueuing setting. Remember, that performance using Erlang B or extended Erlang B (nonqueuing methodologies) is measured by blockage percentage. While blockage occurs in a queuing environment (without blocking, why have a queue?), performance is measured by *delay* time. In other words, how many people are "in line" is usually not as important as how "fast" the line moves!

16.12 ERLANG C MEASUREMENTS

Besides delay time, the other measures taken in a queuing environment are:

1. Average queue length
2. Average number of units in queue
3. Probability of waiting
4. System utilization
5. Average time in the system
6. Average number of units in the system

As we first look at Erlang C and its tables, we will concentrate on delay time, the average number of units in queue, and the probability of waiting. When we continue on to EQEEB methodology, we will look at the effect of queue length. Finally, we will look at system utilization, average time in the system, and average number of units in the system when we discuss ACD applications.

16.13 ERLANG C METHODOLOGY AND TABLES

Erlang C makes the basic assumption of an *unlimited*-length queue—in other words, that attempts will wait *indefinitely* until a server is available. If that is true, there is no overflow. All blocked attempts will wait until served. Therefore, *carried traffic is equal to offered traffic*. If a traffic study shows 100 carried peg counts and 10 blocked calls, those blocked calls, which waited until a server was available, are included in the 100 carried peg counts because eventually they were carried.

The statistical concept that carried traffic is equal to offered traffic makes Erlang C easy to use in determining offered traffic. However, use of the tables becomes more interesting. The tables, like Erlang B, will calculate the percentage of blockage based on the number of servers and the offered traffic. Yet percentage of blockage is not a very important statistic in a queuing environment; what is more important is the *delay time* or the time spent in queue. (If there were no blockage, why have a queue at all?) Delay time is the time from when an attempt first encounters a busy condition to the time a server is available.

To best get acquainted with the information that can be gleamed from an Erlang C table, let's look at Table 16-2 (the complete table is in Appendix H).

Table 16-2 Unlimited Queue Table

OFFERED TRAFFIC IS .25 ERLANGS

N	P	D1	D2	Q1	Q2	P8	P4	P2	P1	PP
1	.2500	.33	1.33	.08	.33	.23	.21	.17	.12	.06
2	.0278	.02	.57	.00	.14	.02	.02	.01	.00	.00
3	.0022	.00	.36	.00	.09	.00	.00	.00	.00	.00

Traffic in Erlangs Offered Group: This is the offered traffic; in this example the offered traffic is 0.25 Erlang.

N: Number of servers. This could be trunks, circuits, or people.

P: This is the percentage of attempts that will be blocked—also referred to as the *percentage of calls delayed* or *probability of delay*. In this case, with 1 server and 0.25 offered traffic, the probability you will be delayed is 25% (or 25 out of 100 calls may be delayed/blocked). Number shown is a percentage in decimal format.

D1: This is the *average delay for all calls* (in holding times). If the average holding time is 3 minutes and you have 1 server, the average delay for all calls is 3 mins times the D1 factor of 0.33, which equals a 1-min delay. This number is basically a multiplication factor.

D2: This is the *average delay for delayed calls* (in holding times). If the average holding time is 3 minutes and you have 1 server, the average delay for delayed calls is 3 mins times the D2 factor of 1.33 which equals a 4-min delay. This number is a multiplication factor.

Q1: The *number of calls in queue at any time.* If you had 1 server, you would have 0.08 call in queue at any given time. Number is simply the number of calls (or units).

Q2: The *number of calls in queue when all servers are busy* (ATB). If you had 1 server, you would have 0.33 calls in queue when no servers are available. Number is simply the number of calls.

P8: The *probability* you will have to *wait one-eighth of a holding time.* If you had 1 server and your holding time is 3 minutes, the probability you will have to wait 22.5 s (1/8 of 3 mins) is 23%. Number is a percentage in decimal format.

P4: The *probability* you will have to *wait one-fourth of a holding time.* If you had 1 server and your holding time is 3 mins, the probability you will have to wait 45 s (1/4 of 3 minutes) is 21%. Number is a percentage in decimal format.

P2: The *probability* you will have to *wait one-half of a holding time.* If you had 1 server and your holding time is 3 mins, the probability you will have to wait 1 min and 30 s (1/2 of 3 minutes) is 17%. Number is a percentage in decimal format.

P1: The *probability* you will have to *wait one holding time.* If you had 1 server and your holding time is 3 mins, the probability you will have to wait 3 mins (1 holding time) is 12%. Number is a percentage in decimal format.

PP: The *probability* you will have to *wait two holding times.* If you had 1 server and your holding time is 3 minutes, the probability you will have to wait 6 minutes (2 holding times) is 6%. Number is a percentage in decimal format.

Table 16-3

Customer	Arrived	Time Served	Departed	Wait Time
1	10:00	10:00	10:15	0 min
2	10:30	10:30	10:45	0 min
3	10:40	10:45	11:00	5 min
4	10:45	11:00	11:15	15 min

16.14 DELAY TIME VERSUS HOLDING TIME

Note that not one column is in Erlangs! This is because the table's primary purpose is to calculate "wait" time. How long you wait depends on how long the holding time is (or it takes for service). By example, if there are ten people in front of you in a checkout line and each person takes 1 min to "get through," it will be 10 min before you get to the checkout stand. If there is only one person in front of you but that person takes 10 min, it will still take you 10 min! Again, the important concept in a queuing environment is *delay time is directly related to holding time!*

The columns D1, D2, Q1, and Q2 are at first confusing to beginning students of traffic engineering. D1 and D2 sound the same; they both calculate delay times. Q1 and Q2 also indicate number of calls in queue. The difference is in how these numbers are determined. D1 and Q1 look at the queue at all times, even when the servers are not busy, whereas D2 and Q2 look at the queue *only* at the times the servers are busy.

To illustrate this difference, consider a barbershop (with only one barber), where it takes 15 min for a haircut (holding time) (Table 16-3). During a 1-hr period, four customers arrive at different times for a haircut.

16.15 TWO TYPES OF AVERAGE TIME DELAY

What was the average time of delay? One answer would be to take the total time of waiting, 20 min, and divide by the number of customers, 4. This answer would be 5 min. But that would be misleading, particularly to customer number 4, who had to wait 15 min. The other answer would be to look only at the customers who *had* to wait because the barber was busy; customers 3 and 4 waited an average of 10 min (20 min divided by 2). Which is the more accurate description of average delay, 5 or 10 min? The first answer is an example of D1. The second is an example of D2. In general, D2 is a more accurate portrayal of delay time because it does not factor in times when the system is not busy. In most telecommunications applications, the D2 factor is considered the true reflection of delay times.

16.16 AVERAGE NUMBER OF CUSTOMERS WAITING

Similarly, what was the average number of customers waiting? From 10:00 to 10:40, there were no customers waiting. From 10:40 to 10:45, there was one customer waiting.

From 10:45 to 11:00, there was one customer waiting. Q1 would indicate that the average number of customers waiting at any time would be less than one. Q2 would only look at times when the barber is busy and would indicate that the average number of customers waiting is probably one. Which represents is the more accurate description of average customers waiting? If you are deciding how many chairs to purchase for waiting customers, you might want to purchase at least one!

Finally, D1 and Q1 are somewhat misleading numbers because they include those times when the system is *not* busy; D2 and Q2 are more reflective of how things are when the system is busy.

16.17 A TYPICAL QUEUING PROBLEM

A queuing problem:
Given a carried traffic of 72 CCS, 36 PC, and 3 servers:

1. What is the probability of delay?
2. What is the average delay of all calls?
3. What is the average delay of delayed calls?
4. What is the average number of calls in queue at any time?
5. What is the average number of calls in queue when the system is busy?
6. What is the probability a call will have to wait 50 s?

First, we know that in Erlang C, carried traffic *is* the offered traffic. Therefore, 72 CCS or 2 Erlangs is our offered traffic. We look up 2 Erlangs in the table and read across the row for 3 servers (see Table 16-4).

1. What is the probability of delay?
   Answer: $P = 0.4444$ or 44.44%
2. What is the average delay of all calls?
   Answer: 88 s or 1 min and 28 s. The average holding time (AHT) is 2 CCS (72 CCS/36 PC), and 2 CCS $\times$ 0.44 (D1) = 0.88 CCS.
3. What is the average delay of delayed calls?
   Answer: 2 CCS or 3 min and 20 s. The average holding time (AHT) is 2 CCS (72 CCS/36 PC), and 2 CCS $\times$ 1.00 (D2) = 2 CCS.

Table 16-4 Partial Erlang C Unlimited Queue Table

N	P	D1	D2	Q1	Q2	P8	P4	P2	P1	PP
3	.4444	.44	1.00	.89	2.00	.39	.35	.27	.16	.06
4	.1739	.09	.50	.17	1.00	.14	.11	.06	.02	.00
5	.0597	.02	.33	.04	.67	.04	.03	.01	.00	.00
6	.0180	.00	.25	.01	.50	.01	.01	.00	.00	.00

Offered Traffic is 2.00 Erlangs

4. What is the average number of calls in queue at any time?
   Answer: 0.89 call (Q1).
5. What is the average number of calls in queue when the system is busy?
   Answer: 2.00 calls (Q2).
6. What is the probability a call will have to wait 50 s?
   Answer: 35%. The AHT was 2 CCS. Also, 50 s is 0.5 CCS, and 0.5 CCS is one-fourth of 2 CCS. P4 represents the probability you will have to wait at least one-fourth of a holding time. The number under P4 is 0.35 or 35%.

16.18 DETERMINING NUMBER OF SERVERS REQUIRED

You solved the previous exercise by already knowing how many servers you had, yet often you need to determine the number of servers. As with Erlang B, to determine how many servers, you need to know your offered traffic and your grade of service (GOS). Whereas Erlang B GOS was based on a percentage of blockage, Erlang C GOS is based on *delay time*.

To show how this problem is solved, let's use the previous example of 72 CCS and 36 PC. Since carried traffic is equal to offered traffic (in Erlang C), the offered traffic is 72 CCS or 2 Erlangs. Suppose your GOS goal is to keep delay time to 90 s (1 1/2 min) or less. How many servers would you need to reach that goal?

Since *delay time is based on the average holding time*, you should first determine the average holding time. In this example, it is 72 CCS divided by 36 PC or 2.00 CCS (3 min and 20 s). Now look at the Erlang C table for 2 Erlangs (Table 16-5).

Which column should be used to determine how many servers would give an average delay time of 90 s or less? The P column gives us blockage or probability of delay but does not give delay times. Q1 and Q2 indicate how many calls are in queue. P8–PP show the probability of being delayed at least 1/8 to 2 holding times but do not give the average delay time. Only D1 and D2 calculate average delay times! Also note that D1 and D2 are factors used to provide the average delay time. The factor supplied in the D1 or D2 column is multiplied times AHT to get average delay.

Look down the D1 column in Table 16-6. According to the table, with 3 servers, your average delay time for all calls would be "0.44" × the average holding time. If your holding time is 2 CCS, your average delay time for all calls would be 0.44 times 2 CCS or 0.88 CCS. This 88 s is less than your goal of 90 s or less; therefore, 3 servers would be enough to achieve your goal.

Table 16-5   Erlang C Table

N	P	D1	D2	Q1	Q2	P8	P4	P2	P1	PP
3	.4444	.44	1.00	.89	2.00	.39	.35	.27	.16	.06
4	.1739	.09	.50	.17	1.00	.14	.11	.06	.02	.00
5	.0597	.02	.33	.04	.67	.04	.03	.01	.00	.00
6	.0180	.00	.25	.01	.50	.01	.01	.00	.00	.00

Offered Traffic is 2.00 Erlangs

Table 16-6 Excerpt From Erlang C Table

N	P	D1	D2	Q1	Q2	P8	P4	P2	P1	PP
3	.4444	.44	1.00	.89	2.00	.39	.35	.27	.16	.06
4	.1739	.09	.50	.17	1.00	.14	.11	.06	.02	.00
5	.0597	.02	.33	.04	.67	.04	.03	.01	.00	.00
6	.0180	.00	.25	.01	.50	.01	.01	.00	.00	.00

Offered Traffic is 2.00 Erlangs

Table 16-7 Excerpt From Erlang C Unlimited Queue Table

N	P	D1	D2	Q1	Q2	P8	P4	P2	P1	PP
3	.4444	.44	1.00	.89	2.00	.39	.35	.27	.16	.06
4	.1739	.09	.50	.17	1.00	.14	.11	.06	.02	.00
5	.0597	.02	.33	.04	.67	.04	.03	.01	.00	.00
6	.0180	.00	.25	.01	.50	.01	.01	.00	.00	.00

Offered Traffic is 2.00 Erlangs

However, if you used the D2 column to solve this problem, you arrive at a different answer (Table 16-7). According to the D2 column, with 3 servers, the average delay for *delayed* calls would be 1.00 holding times. Since the average holding time is 2.00 CCS, your average delay of delayed calls is 2.00 CCS (1.00 × 2.00) or 3 min and 20 s. This is "too long" a wait; your goal was to average 90 s or less.

Adding one more server (or 4 servers), your average delay for delayed calls would be 0.50 holding time. Now your average delay for delayed calls is 1.00 CCS (0.50 × 2.00). This is still "longer" of an average wait than your goal of 90 s.

Adding one more server (now 5 servers), your average delay for delayed calls would be 0.66 CCS (or 0.33 × 2.00). This *does* meet your goal of an average delay of 90 s or less. Therefore, you would need 5 servers. *So which answer is correct?* Do you put in 3 servers for an average delay of all calls at 88 s? Or do you put in 5 servers for an average delay of delayed calls at 66 s? *They are both correct,* because in our problem "which type" of delay was not defined! *Which type of delay is more representative?* Based on the previous discussion of the difference between D1 and D2, D1 can be misleading; D2 is more representative of what happens if a call *is* delayed. Therefore, unless you want to make things look better than they may be *or* you are asked to calculated delay for all calls, *D2 is more representative of delay times* than D1.

NOTE: The same analogy is used for when to use Q1/Q2. Q2 is more reflective of how many calls will be in queue than Q1.

16.19 DETERMINING GOS

Let's try one more scenario and determine the GOS in a queuing environment. For this problem, the carried traffic is 126 CCS and 105 PC. The carried traffic is equal to offered

Table 16-8 Excerpt From Erlang C Unlimited Queue Table

N	P	D1	D2	Q1	Q2	P8	P4	P2	P1	PP
4	.7379	1.48	2.00	5.17	7.00	.69	.65	.57	.45	.27
5	.3778	.25	.67	.88	2.33	.31	.26	.18	.08	.02
6	.1775	.07	.40	.25	1.40	.13	.09	.05	.01	.00
7	.0762	.02	.29	.08	1.00	.05	.03	.01	.00	.00

Offered Traffic is 3.50 Erlangs

traffic; therefore, the offered traffic is 126 CCS or 3.5 Erlangs. The average holding time is 1.2 CCS (126 CCS/105 PC) or 2 min. Our GOS goal is to have an average delay of *delayed* calls be 45 s or less. Look at the Erlang C (Table 16-8) under 3.5 Erlangs.

Since the problem required determining the average delay of delayed calls, you would look under the D2 column. Rather than going down this column and calculating the delay time for each number of servers, try this simple formula ("shortcut") first:

$$\text{Delay factor} = \text{delay goal} / \text{average holding time}$$

If the goal is 45 s and the AHT is 1.2 CCS, you are looking for a delay factor of 0.375. First convert delay goal and AHT into the same measurement. We will convert delay goal from 45 s to 0.45 CCS.

$$\text{Delay factor} = \text{delay goal} / \text{AHT}$$
$$= 0.45 \text{ CCS} / 1.20 \text{ CCS}$$
$$= 0.375$$

Notice the factor of 0.375 falls between "0.40" and "0.29". You would choose the smaller of the two because 0.40 of the holding time would be 0.48 CCS or 48 s (0.4 × 1.20 CSS), which is *more* than your goal of 45 s. But 0.29 of a holding time would give you 35 s (0.29 × 1.20 CCS = 0.35 CCS), and that meets your goal of having an average delay of delayed calls be 45 s or less!

This "shortcut" is a good way to quickly determine number of servers (in Erlang C), particularly when the table is very long!

16.20 FINITE QUEUE TABLE VERSUS INFINITE QUEUE TABLE

If the queue has a definite holding time, the Erlang C (unlimited queue) table cannot be used. For a definite holding time, we use the finite queue (limited queue). The Erlang C finite queue table can be used for limited queues, or you can use the equivalent queue extended Erlang B (EQEEB) table. Most engineers prefer EQEEB. Therefore, I will cover EQEEB and include an abbreviated 1-min EQEEB table is Appendix I. Because EQEEB is a methodology that assumes a definite, finite queue, many different tables are available (such as 30-s, 1-min, 5-min, 10-min, and so on.) Before looking at the EQEEB tables, let's look at an Erlang C unlimited queue table format that is similar to the format of an EQEEB table.

Let's go back to our original ABBH parameters of 72 CCS and 36 PC. Earlier, we used Tables 16-4 to 16-7 to determine how many servers are needed. The following is from an Erlang C unlimited queue table that inlcudes the amount of traffic carried by each sewer.

Hours First Attempt	Hours Conn- ect	Hours Ovfl	Delay Portion	Delay Length Call Factor	L I N E	HOURS CONNECTED TRAFFIC ON LINES					
						1	2	3	4	5	6
2.000	2.000	.000	.444	1.000	3	.889	.667	.444			
2.000	2.000	.000	.174	.500	4	.870	.609	.348	.174		
2.000	2.000	.000	.060	.333	5	.866	.597	.326	.149	.060	
2.000	2.000	.000	.018	.250	6	.865	.595	.324	.144	.054	.018
2.000	2.000	.000	.005	.200	7	.865	.594	.324	.143	.053	.017
2.000	2.000	.000	.001	.167	8	.865	.594	.323	.143	.053	.017

Each column is as follows:

Hours First Attempt: Same as offered (in Erlangs).

Hours Connect: Same as carried (in Erlangs). (Notice offered and carried are the same!)

Hours Ovfl: Overflow traffic (in Erlangs). (Why is it always 0.000?) *Answer:* If there is no blockage, there is no overflow.

Delayed Portion: Percent of calls delayed or probability of delay. In decimal format. (Same as P in Table 16-4).

Delay Length Call Factor: Average delay of delayed calls (same as the D2 column in Table 16-4). In holding times. Basically, this number times average holding time is the average delay of delayed calls.

Line: Number of servers. (Why does the number of servers always start with the first whole number greater than the offered traffic? That is, if the offered traffic is 2.2, why does the table start with 3 servers, not 1 or 2?)

Hours Connected Traffic on Lines: Amount of carried traffic per server, beginning with server 1 and ending with the number of servers in the group. In Erlangs. (Notice amount changes when number of servers in group changes!)

This table only contains N, P, and D2 information. It does not have D1, "Q1", Q2, P8–PP. However, it shows the amount of traffic carried per trunk (which Table 16-4 did not) and shows how much overflow traffic exists (which is 0.000 Erlang now, but later. . .). You will find that to solve a queuing problem you may have to use *both* tables. But back to using this table. Let's use this table to answer the following questions, given a carried traffic of 72 CCS and 36 PC and 5 servers:

1. What is the probability of delay?
2. What is the average delay of all calls?

3. What is the average delay of delayed calls?
4. What is the average number of calls in queue?
5. How much overflow traffic is there?
6. How much traffic does server 3 carry?

Since the carried load is 72 CCS, the offered load is 72 CCS (or 2 Erlangs):

Hours First Attempt	Hours Connect	Hours Ovfl	Delay Portion	Delay Length Call Factor	LINE	HOURS CONNECTED TRAFFIC ON LINES					
						1	2	3	4	5	6
2.000	2.000	.000	.444	1.000	3	.889	.667	.444			
2.000	2.000	.000	.174	.500	4	.870	.609	.348	.174		
2.000	2.000	.000	.060	.333	5	.866	.597	.326	.149	.060	
2.000	2.000	.000	.018	.250	6	.865	.595	.324	.144	.054	.018
2.000	2.000	.000	.005	.200	7	.865	.594	.324	.143	.053	.017
2.000	2.000	.000	.001	.167	8	.865	.594	.323	.143	.053	.017

1. What is the probability of delay?

   6%. Under the Delayed Portion Column, on the row with 5 servers is the number 0.060 or 6%.

2. What is the average delay of all calls?

   N/A. That is the D1 column in Table 16-4. This table does not provide that information.

3. What is the average delay of delayed calls?

   0.67 CCS. Under the Delay Length Call Factor is the number 0.333. Multiplying the average holding time of 2.00 CCS (72 CCS/36 PC) by 0.333 equals 0.666 CCS or 67 s (rounded).

4. What is the average number of calls in queue?

   N/A. That information is under the Q1 column on Table 16-4. This table does not provide that information, as well as Q2, P8, P4. . .and so on.

5. How much overflow traffic is there?

   0.000 Erlang. Since Erlang C assumes calls will wait indefinitely until a server is available, no one will leave the queue or overflow. Therefore it is always 0. (Later we will find this is *not* the reality!)

6. How much traffic does server 3 carry?

   0.328 Erlang. Read across the row for 5 servers; under the number 3 beneath the heading Hours Connect Traffic on Lines is the number 0.328. (Notice the amount of traffic for server 3 varies given how many servers there are!)

16.21 FINAL OBSERVATIONS ON ERLANG C INFINITE QUEUE THEORY

As you work the problems, notice what happens as you add or subtract servers. Basically, to reduce delay time does not require as many servers as you might think; just adding one more server can quite dramatically cut down on delay. However, adding too many servers does not make as dramatic a cut (law of diminishing returns). It is not uncommon for many queuing designs to overengineer the system—quite expensive if the servers are salaried employees!

Additionally, the percentage of blockage along with delay time is sometimes a parameter of queuing design; for example, no more than 20% probability of delay *and* no more than an average delay of 30 s is an example of a common type of request. Be prepared to solve for one, two, three, or more parameters in a queuing environment. There may be times when you cannot reach them all simultaneously. Erlang C is a widely used methodology in data and voice networking systems (particularly ACD and voice mail). It is one to practice and practice.

16.22 FINITE QUEUE THEORY (EQUIVALENT QUEUE EXTENDED ERLANG B)

The major flaw with Erlang C is the assumption of an infinite queue—that is, that calls will wait indefinitely until a server is available. That may be true in theory, but in real life, people are often in control of how long to wait in a queue. It is not uncommon for calls (people) to drop out of a queue if the wait is too long. In traffic engineering terminology, this is known as *abandoning* the queue.

When reporting traffic statistics in a queuing environment, most network systems, will show carried traffic (CCS and PC) as well as the number of *abandoned* calls. While these systems will also calculate the average delay time and number of calls in queue when all servers are busy, determining how to keep calls from abandoning the queue becomes the problem for the traffic engineer.

To compensate for the phenomenon of abandoned calls, a methodology has been developed called *equivalent queue extended Erlang B* or *EQEEB*. (This method obviously has its roots in extended Erlang B!). Despite the *Erlang B* in the title, this methodology is applied in a queuing environment. The added *extended* indicates that a retry statistic may be applied; in other words, when a person abandons the queue, he or she may "retry" later. This can lead us into a variety of scenarios of the behavior of calls in a queuing environment: 10% of the calls abandon the queue and 50% of the abandoned calls retry later; 50% abandon and 30% retry; 43% abandon and 81% retry; and so on. For this course, we will only use tables that assume *no* retry after abandoning the system.

16.23 WHAT IS A FINITE QUEUE?

But what is meant by a *finite queue?* Finite refers to length as a measurement of *time,* not distance. A finite queue is one in which, after a certain amount of delay time, a call will abandon the queue. The tables provided you are for three finite queue lengths: 1-min maximum wait, 5-min maximum wait, and 10-min maximum wait. The calculations are based on the premise that calls will abandon the queue after a maximum amount of delay.

NOTE: For purposes of this book, I will give you the offered traffic for EQEEB problems. Be aware that in real-life scenarios, your abandoned traffic (times the average holding time) needs to be added to your carried traffic to approximate the offered traffic. The best way to acquaint you with this methodology is to create a problem and then solve that problem using the EQEEB tables. These tables are provided as Appendix I.

16.24 A FINITE QUEUE SITUATION

Let's say we have an offered traffic of 72 CCS and 36 PC. Our server group is four trunks. Given a queue length of 1 min (calls will abandon the queue if the delay is greater than 1 min):

1. What is the probability of delay?
2. What is the expected carried traffic?
3. What is the expected overflow traffic?
4. What is the average delay of delayed calls?
5. How many calls will abandon the queue?

Since the problem states a 1-min queue length and the offered traffic is 2 Erlangs (72 CCS), you find 2 Erlangs on the 1-min table and find the row for 4 servers:

Hours First Attempt	Hours Connect	Hours Ovfl	Delay Portion	Delay Length Call Factor	L I N E	HOURS CONNECTED TRAFFIC ON LINES					
						1	2	3	4	5	6
2.000	1.704	.296	.271	.196	3	.695	.576	.433			
2.000	1.833	.117	.111	.084	4	.678	.551	.401	.253		
2.000	1.959	.041	.040	.030	5	.671	.540	.387	.238	.123	

1. What is the probability of delay?

   11.1%. The number 0.111 under the Delayed Portion column is the percentage of calls delayed or the probability of delay. In decimal format.

2. What is the expected carried traffic?

   1.833 Erlangs. The number 1.833 under the Hours Connect column estimates the amount of carried traffic. Under Erlang C, this number was always the same as the offered; however, since this table assumes some calls may abandon the queue before getting a server, not all of the offered traffic will be carried. In Erlangs.

3. What is the expected overflow traffic?

0.117 Erlang. This represents how much traffic, in Erlangs, will abandon the queue.

4. What is the average delay of delayed calls?

17 s. This delay length call factor is 0.084. Multiplying this by the average holding time of 2 CCS (72 CCS / 36 PC) gives you 0.168 CCS or 16.8 seconds.

NOTE: Check the delay time for the same parameters except use the Erlang C table. This table calculates a longer delay of 1.00 CCS. Why is this delay so much shorter?

Some people leave the queue. They will not wait for a server.

5. How many calls will abandon the queue?

2.1 calls. The overflow (or abandon) traffic is 0.117 Erlang or 4.212 CCS. If this number is divided by the average holding time, you will get the number of calls: 4.212 CCS / 2 CCS = 2.1!

The columns under Hours Connected Traffic on Lines" reflects how much traffic (in Erlangs) each server will carry (not unlike the previous EEB tables).

Now let's do the same problem again, changing only one parameter: the queue length is 10 min. What are the answers now to the questions? (Hopefully, you are using the "10-min" table):

Hours First Attempt	Hours Connect	Hours Ovfl	Delay Portion	Delay Length Call Factor	L I N E	HOURS CONNECTED TRAFFIC ON LINES					
						1	2	3	4	5	6
2.000	1.754	.246	.886	1.221	2	.888	.866				
2.000	1.959	.041	.359	.541	3	.749	.661	.548			
2.000	1.989	.011	.128	.195	4	.694	.576	.433	.286		
2.000	1.997	.003	.042	.065	5	.676	.547	.396	.248	.130	

1. What is the probability of delay?

12.8%. The number 0.128 under the Delayed Portion column is the percentage of calls delayed or the probability of delay. In decimal format.

2. What is the expected carried traffic?

1.989 Erlang. The number 1.833 under the Hours Connect column estimates the amount of carried traffic.

3. What is the expected overflow traffic?

0.011 Erlang. This represents how much traffic will abandon the queue. In Erlangs.

4. What is the average delay of delayed calls?

39 s. This delay length call factor is 0.195. Multiplying this by the average holding time of 2 CCS (72 CCS / 36 PC) gives you 0.39 CCS or 39 s.

5. How many calls will abandon the queue?

0.2 call. The overflow (or abandon) traffic is 0.011 Erlang or 0.396 CCS. If this number is divided by the average holding time, you will get the number of calls: 0.396 CCS / 2 CCS = 0.198 call.

Comparing these two solutions: (1 minute max to wait to 10 minutes max to wait):

    a. Why is the amount of connect traffic, delayed portion and delay time *greater* for the 10-min queue (versus the 1-minute queue)?

    b. Why is the amount of overflow traffic and the number of abandon calls *greater* for the 1-min queue (versus the 10-minute queue)?

If you can answer the above two questions, you understand the basic concept of the dynamics of queuing theory.

16.25 AUTOMATIC CALL DISTRIBUTION

Automatic call distributors (ACDs) are used by many large service organizations to answer incoming calls. All IECs use ACDs to receive calls placed to their operator services group. When you dial 0 as the first digit, the LEC connects you to its ACD. When you dial an IEC operator access code, you are connected to its ACD. If you call the reservations department for a major airline company, you are connected to its ACD. Telemarketing firms use ACDs for inbound and outbound calls. Employees that answer calls connected to an ACD are called agents. The telephone companies no longer call these employees operators. They also refer to operators as agents. The general flow of a call to an ACD is:

1. A call is placed to a number associated with an ACD
2. If a circuit to the ACD is available, the call enters the ACD
3. When a call enters an ACD, the ACD connects the call to an idle agent or places the call in a holding queue.
4. The queue will hold the call and provide the caller with directions or information. If the caller was placed in a queue because all agents are busy, the caller is informed that all agents are busy by an automated attendant recording. If the ACD uses automated attendant to answer all calls, the caller is provided with directions.
5. When the caller disconnects, the trunk between the central office and the ACD is released. The trunk circuit is now ready to accept the next call to the ACD. An ACD can only receive as many calls (at one time) as there are CO to ACD trunks.

The trunk circuits connecting an ACD to the central office are engineered according to Erlang B (or Poisson) because there is no alternative route to the ACD. If all circuits to the ACD are busy, the call is cleared from the central office system and the caller receives a fast busy signal from the central office. Thus, ACD trunk circuits are engineered using BCC methodology.

The ACD will queue calls received from the central office trunk circuits and then connect the incoming caller on the trunk circuit to an idle agent. Thus, the connection from trunk circuits to agent positions in the ACD is engineered using queuing methodology. We will employ the EQEEB methodology used in the previous section of this book. The design of traffic flow *within* the ACD is done using queuing theory and tables. Usually, an ACD is equipped with more agent positions than are needed. When the ACD was engineered initially, the design engineer inflates the expected call volume (and offered load) to provide for future growth. Large companies with many ACD centers will oversize offices in order to have sufficient agent positions to handle the call volume for several centers. If a disaster such as a flood were to close one center, the calls for that center could be redirected to other ACD centers. The number of agent positions that need to be staffed by an agent depends on the inbound calling load. Since load is directly related to the number of calls and the average holding time of each call, these numbers are ordinarily used when determining the number of needed agents each hour. An ACD system usually includes traffic engineering software within the ACD. This software program allows management to forecast how many agents will be needed each hour. They can use expected hourly call volume information and the software program to predict the number of agents needed each hour. There are two parts to designing an ACD system. The engineer must determine how many agent positions will be needed to handle the maximum number of calls that will be offered to the agents. The engineer must also determine how many trunk circuits will be needed between the central office and the ACD.

The engineer can start by engineering the ACD system. The number of agents must be sufficient to handle the incoming call volume. If we do not have enough agent positions to handle calls, it does not matter how many central office to ACD trunk circuits we have. With too few agent positions, the callers will end up being held so long in queue that they will hang up. The number of agents needed for a calling volume depends on three factors: (1) the *average conversation time (ACT)* per call, (2) the amount of records work needed on each call, and (3) agent efficiency. ACT is the average conversation time an agent spends on each call. The amount of records work needed on each call is called *wrap-up time (WUT)*. Agents for telemarketing centers or an airline company must fill out purchase orders, ticket information, credit card billing information, and so on. After each call is over, the agent must make sure all appropriate paperwork or computerized records work has been completed before making the position available for another call. Agents are not always available due to breaks, meetings, and so forth. Since they are not available 100% of the time, an efficiency number is used. The formula for calculating agent demand is:

$$\text{Agent demand} = (ACT + WUT) / \text{agent efficiency} \times \text{no. of calls/hr}$$

When doing the initial sizing of an ACD, the number of calls will be the maximum number of calls the engineer expects the ACD to handle. The engineer may use ABBH PC for this number or may use a much higher number, if the ACD is being engineered to also handle calls for other centers when those centers encounter an outage. By using the maximum number of calls expected, the engineer will be able to determine the maximum number of agent positions that the ACD must be equipped with. Once the initial sizing has been done by the engineer, the managers of an ACD center will use the number of calls forecast for each hour of operation. By using the number of calls forecast for each hour in the agent demand formula, management can determine how many agents it must have working during each hour. The managers will use this information to develop a work schedule for the agent workforce. Some ACDs have software programs that make all calculations for staffing based on past data and that will output an actual schedule for the manager.

An example can be used to illustrate how the formula is used:

$$\text{Agent demand} = (ACT + WUT) / \text{agent efficiency} \times \text{no. of calls/hr}$$
$$\text{No. of Calls} = 38$$
$$ACT = 3 \text{ CCS (5 min)}$$
$$WUT = 0.3 \text{ CCS (30 s)}$$
$$\text{Efficiency} = 91.7\%$$
$$\text{Agent demand} = (ACT + WUT) / \text{agent efficiency} \times \text{No. of calls/hr}$$
$$= (3 \text{ CCS} + 0.3 \text{ CCS}) / .917) \times 38$$
$$= 136.75 \text{ CCS (3.798 Erlang)}$$

The engineer for an ACD system uses the average time that an agent is tied up on a call (ACT + WUT) and the efficiency of the center to calculate AHT. This is different from AHT in other systems. Central office engineers state AHT as (setup time + conversation time) / no. of calls. The ACD engineer knows that the center will not be 100% efficient and also knows that the total average time per call is equal to conversation time plus WUT. An ACD engineer wants the AHT to represent the total average time per call and efficiency of the center. Let's use our example data:

$$AHT = (ACT + WUT) / \text{agent efficiency}$$
$$= (3 \text{ CCS} + 0.3 \text{ CCS}) / 0.917$$
$$= 3.6 \text{ CCS}$$

The central office engineer calculates load by multiplying ABBH AHT times ABBH PC. The ACD engineer calculates agent demand by multiplying the ACD AHT times PC. Once the AHT for an ACD center is calculated, the engineer can calculate agent demand faster by using AHT for the center. Again, we use the example data:

$$\text{Agent demand} = AHT \times \text{no. of calls/hr}$$
$$= 3.6 \text{ CCS} \times 38$$
$$= 136.8 \text{ CCS}$$
$$136.8 \text{ CCS} / 36 = 3.8 \text{ Erlangs}$$

We would use a demand of 3.8 Erlangs to determine how many agents are needed. The number of agents needed would also depend on how long we want to hold callers in queue before connecting them to an agent. We can use the 1-min table of equivalent queue extended Erlang B (EQEEB) table, to find the number of agents needed. First, we should calculate the average delay factor (D2). Remember that the D2 factor times average holding time gives us the average delay time. If we wish to have an average delay of 30 s and know the AHT, we can rearrange that formula to find D2:

$$\text{Average delay time} = D2 \times \text{AHT}$$

Therefore:

$$D2 = \text{average delay time} / \text{AHT}$$

For our example, we want an average delay no more than 35 s and the AHT (as calculated for an ACD system) is 3.6 CCS (360 s):

$$\text{Average delay time} = D2 \times \text{AHT}$$

Therefore:

$$D2 = \text{average delay time} / \text{AHT}$$
$$D2 = 35 \text{ s} / 360 \text{ s}$$
$$D2 = 0.097222\ldots$$

Now we can find out how many agents are needed to serve a load (Hours First Attempt Column) of 3.8 Erlangs, with a D2 delay factor of 0.097222... or less. We go to the 3.8-Erlang table (Table 16-9) and find 0.097222... or less in the D2 column (Delay Length Call Factor column). The number meeting this criterion is 0.094. Looking at the number of servers column (just to the right of the D2 column) in the row that contains 0.094 for D2, we find that 6 lines (servers) are needed to meet the criterion of 35 s or less delay, with a load of 3.8 Erlangs. The actual average delay will be 34 s (delay = D2 × AHT = 0.094 × 360 = 33.84 s). This table also provides additional information. If we provide 6 agents to serve a load of 3.8 Erlangs, we will have a delay of 34 s, the portion of traffic delayed is 0.124 (12.4%), the carried traffic (connect load) is 3.550 Erlangs, and the overflow is 0.250 Erlang.

Table 16-9 Excerpt from EQEEB Table

Hours First Attempt	Hours connect	Hours Ovfl	Delay Portion	Delay Length Call Factor	L I N E	HOURS CONNECTED TRAFFIC ON SERVERS						
						1	2	3	4	5	6	7
3.80	2.922	.878	.411	.287	4	.82	.77	.70	.62			
3.80	3.311	.489	.237	.174	5	.81	.75	.68	.59	.48		
3.80	3.550	.250	.124	.094	6	.80	.74	.66	.56	.45	.33	
3.80	3.682	.118	.060	.046	7	.79	.73	.65	.55	.43	.31	.21
3.80	3.748	.052	.026	.020	8	.79	.73	.64	.54	.42	.30	.20

Once we have decided how many agents will be used to serve an offered load, we will know how many calls will be delayed and the average delay of calls delayed. This information can then be used to engineer the trunk circuit group connecting the central office to the ACD. The steps used to engineer the CO-to-ACD trunk group are:

1. Determine the number of calls for the ABBH.
2. Multiply these calls by the ACT.
3. Determine how many calls will be delayed.
4. Determine the ADT: ADT = D2 × AHT.
5. Multiply number of delayed calls by the Average Delay Time (ADT)
6. Add the answers for steps 2 and 5. This will be the load offered to the CO-to-ACD trunk group. This trunk group is engineered using the Erlang B Table (Appendix D).

We use the maximum loads expected, or we use the ABBH data, to initially size the ACD agent positions and the number of CO-to-ACD trunk circuits. If we assume that the example above represented data from the ABBH, we can use that example to illustrate how we determine the quantity of ACD trunks needed:

1. Number of ABBH calls = 38.
2. Number of calls × ACT (38 × 3 CCS) = 114 CCS.
3. Determine how many calls will be delayed: 12.4% of 38 = 4.71 calls delayed.
4. Determine ADT (D2 × AHT = 0.094 × 3.6 CCS = 0.34 CCS).
5. Multiply delayed calls (4.71) times ADT (0.34 CCS) = 1.6 CCS.
6. ACD trunks carried load = 114 CCS + 1.6 CCS = 115.6 CCS. And 115.6 CCS / 36 = 3.21 Erlangs. Checking the Erlang B Table: with an offered load of 3.21 Erlangs and a $P.01$ GOS, we need 9 CO-to-ACD trunk circuits.

Have you noticed that the ACD we have just engineered needs more incoming CO-to-ACD trunks (9) than agent positions (6)? This is because calls are being held in queue by the ACD. This evens the flow of calls to the agents and makes them more productive. An ACD will always be engineered with more incoming trunk circuits than agent positions. At any point in time, some trunks will be connected to agents and some will not be connected to agents but will be connected to the holding queue. The sum of calls held in queue plus calls connected to agents will be the number of CO-to-ACD trunks needed.

16.27 SUMMARY

When blocked calls are placed in a holding queue, three basic engineering methodologies are used to determine the number of servers needed to handle the demand. The predominant method is unlimited queue Erlang C. It assumes that calls arrive randomly from an unlimited population and that callers are willing to wait forever for a server. The second most common method is finite queue Erlang C. It assumes callers will not wait in queue forever but are willing to wait a specified amount of time for a server. The third method is equivalent queue extended Erlang B. It assumes that callers have a limit on how long they will wait and uses a finite queue table.

Closely related to queuing theory is retrial methodology. It assumes that if callers are blocked they will retry immediately. The switching system is not putting callers in a queue, but their retry behavior is almost as if they were in a queue. The retry tables adjust blocked load before adding it to carried load, because if all of the blocked load is added to carried load in order to arrive at offered load, the demand will be overstated. In retrial methodolgy, carried load aready includes some of the load that was blocked on its first attempt, and if no adjustment is made to blocked load, some of it is counted twice. It is counted once as blocked load and once as carried load. In retrial methodology, demand is referred to as first-attempt traffic instead of offered load.

If 100% of callers retry their calls until they get carried, there is no blocked load and carried load equals first-attempt load. If 0% of callers retry their calls, we must add all of the blocked load to carried load to arrive at first-attempt load, and first-attempt load equals carried load plus blocked load. Thus 0% retry is calculated the same way as in Erlang B methodology. If 50% of all callers retry until they are carried, we must add only 50% of blocked load to carried load to arrive at first-attempt load. If 70% of callers retry, 70% of the blocked load got carried and we must add only 30% of the blocked load to carried load to arrive at first-attempt load.

REVIEW QUESTIONS

1. What assumptions are made by a network design engineer when designing a circuit group?
2. What is load balance?
3. When is the Erlang C table used?
4. What are the different ways that a switching system can handle blocked calls?
5. What is the premise behind a retrial table?
6. What is first-attempt load?
7. What will the first-attempt load be with 100% retry?
8. What basic assumptions are made when using an Erlang C table?
9. What is meant by negative exponential probability distribution?
10. If callers do not retry their call immediately when they receive an all-circuit-busy condition, which traffic table would you use?
11. Why do many engineers use the Jacobsen retrial table?
12. What assumption is made when using Erlang C about how long a caller will stay in queue?

13. What is the important concept in a queuing environment?
14. How is grade of service measured in a queuing environment?
15. What is the major flaw in Erlang C theory?

GLOSSARY

Delay Time Refers to the length of time that a call for a customer is held in queue before being served.

Finite Queue A queue that has a devinite waiting period.

First-Attempt Traffic The offered load in retrial traffic engineering methodology.

Load Balance The physical distribution of residential and business telephones across all line circuits in such a manner that the same percentage of business to residential occurs in all line groups. This is done to prevent a heavy concentration of business lines in one group.

Negative Exponential Distribution Assumes the probability of long service times in a queuing environment is very low and the probability of short service times is very high.

Queuing Theory Assumes that people are willing to wait in the queue or holding line for service and that based on the length of the queue, the behavior of people in the queue, and how the queue handles new arrivals, the number of servers needed can be forecast.

Retrial Methodology Assumes people will redial their call upon being blocked. It makes an adjustment to blocked load before adding it to carried load to get offered load. It makes this adjustment because this methodology assumes some of the blocked calls were eventually carried due to retry by the caller. Blocked load that was carried should not be added to carried load because it will make offered load seem higher than it actually was.

Unlimited Queue Assumes customers in a queue are willing to wait forever to get service.

17

Personal Communication Systems

Key Terms

Advanced Mobile Phone
System (AMPS)
Base Station Controller (BSC)
Base Transceiver Station
Basic Trading Area (BTA)
Cellular Radio Advanced
Mobile Phone System
(AMPS)
Code Division Multiple Access
(CDMA)

Code Excite Linear Predictive
Coding (CLIP)
Digital Advanced Mobile
Telephone System (DAMPS)
Improved Mobile Telephone
System (IMTS)
Major Trading Area (MTA)
Mobile Telephone Serving
Office (MTSO)
Mobile Telephone System (MTS)

Personal Communications
System (PCS)
Personal Communications
System 1900 (PCS 1900)
Roaming
Time Division Multiple Access
(TDMA)

The newest wave of technology to hit telecommunications is the technology behind the *personal communications system (PCS)*. This technology takes advantages of the technology advances made in SPC switching systems, the SS7 network, advanced intelligent networks (connected by SS7), and mobile telephone systems. PCS itself is not so much a technology as it is a concept. The concept of PCS is to assign someone a *personal telephone number (PTN)*. This PTN is stored in a database on the SS7 network. That database keeps track of where a person can be reached. When a call is placed for that person, the *artificial intelligence network (AIN)* of the SS7 determines where the call should be directed. To provide mobility to the PTN, it was decided that the primary location for most PTNs should be reached by using radio waves rather than copper wire for a transmission medium. The PCS is provided by radio frequencies in the 1900-MHz range, and the service is often referred to as *PCS 1900*. Before discussing PCS, I think it appropriate to briefly review the developments that have occurred in mobile telephone systems.

Mobile telephone service begin in 1946 and was called *MTS*. These systems used the radio frequencies between 35 and 45 MHz. Although MTS stands for mobile telephone service, this acronym also meant "manual telephone system." All calls had to be handled by an operator. These early systems used one frequency for the mobile phone and the base station. The handset had a pushbutton. The mobile party used a push-to-talk protocol. When the button was depressed, the transmitter of the mobile unit was activated. This would cause a signal to be transmitted to the base station. On receiving the signal, the base station would light a light above a jack at the operator's position. The operator would answer the call by plugging a patch cord from her position into the jack connected to the base station transmitter. She would ask the calling mobile party for information on whom they wished to call. The operator would use the other end of the patch cord to connect the mobile telephone caller to the desired PSTN destination.

The PSTN telephone was connected to the base station's transmitter through the patch cord at the operator's position. A conversation could now take place with the caller and called party taking turns. The MTS systems usually had only one radio frequency for both the base and mobile phones. Therefore, only one phone call at a time could be placed over the MTS. Because all mobile phones used the same frequency, the MTS was like a party line. All MTS mobile phones could hear any conversation on the system. To place a mobile-to-mobile phone call, the mobile caller would reach the operator by pressing the push-to-talk button. The person would then ask the operator to ring the other mobile phone. Once the operator rang the mobile phone, she was no longer needed. The base station was not needed for a mobile-to-mobile call because the transmitters and receivers of both phones were tuned to the same frequency.

The early MTS phones were connected to a transmitter/receiver that was placed in the trunk of the car. This transceiver was a tube-type device and placed a heavy drain on a car's battery. The transceiver was about 24 in. long, 18 in. wide, and 6 in. high. The receiver of the MTS mobile unit contained a mechanical decoder device. The tube-type MTS mobile units were replaced by transistor-type mobile units around 1970. Solid-state logic circuits in these units served as the decoder circuit. When the operator placed a call to a mobile phone, the dial pulses from her dial were converted into pulses of signaling tone. Let's assume that a 40-MHz signal is being used by a mobile telephone system. If a 5 was dialed, the base station would modulate the 40-MHz carrier signal 5 times with a tone of 2400 Hz.

All receivers were tuned to accept the 40-MHz signal and would see the 5 pulses of tone. The output of the receiver went to a decoder as well as to the receiver of the handset. The input of the decoder included a filter that would only pass the frequency of the signaling tone. The decoder would count the pulses and store the digit. The decoder contained a chain circuit. This chain circuit was wired at the time of installation as the phone number of the mobile phone. The chain circuit of the decoder for each telephone was wired so that if the first digit dialed matched the first digit of the mobile phone number, the decoder remained active and looked at the second stream of tones sent for the second digit of the phone number. If the second digit matched the wired number, the decoder looked at the next digit, and so on. This process was done for each digit. If at any digit, a match was not made between the received number of tones and the number wired in the decoder for that digit, the decoder released and reset itself. If, however, each digit was matched as it was received, on receiving the last digit, the decoder recognized that the received number matched the number assigned to this mobile unit and would ring the phone. It also sent a signal out on a wire that could be connected to the horn relay of the vehicle, if desired. This would cause the horn to blow one time for about 2 s. On hearing the ringer or horn, the mobile customer would pick up the handset and press the talk button to answer the call.

MTS mobile phone numbers had no relationship to the telephone numbers of the PSTN. MTS numbers were distinct and separate from PSTN numbers. The Bell Company of each state kept a registry of numbers assigned to the MTS systems in that state. Each MTS service provider would get a number assignment from the Bell Company records department when a new MTS phone was installed. These mobile phone numbers were usually five-digit numbers. They could not be accessed directly over the PSTN and were not part of the PSTN number plan. To reach a MTS mobile, callers had to know what serving area the mobile was in. Callers would have their local operator connect them to the operator of that serving area. They would then ask the operator of that area to ring the desired mobile number.

For the MTS system that I maintained, we had a base station transmitter, receiver, and antenna on top of a large hill. The transmitter had a range of about 50 mi. I cannot remember the exact power output, but most of these systems had a 100- to 200-W output. The transmitters in the mobile units had a much lower level of power output (about 20 W). The mobile transmitter had a range of about 15 mi. To provide coverage for the 50-mi radius from the base transmitter, we had receivers placed in the north, east, south, and west directions, about 20 mi from the base transmitter/receiver site. We had a total of five receiver sites. One of these (the central site) was also the site for the base transmitter. All sites were connected by various land transmission mediums to the toll office. The toll office had a mobile communication bay. This bay received the inputs from all receivers and the base transmitter. The equipment in this bay would determine which receiver was providing the strongest signal and would connect the signal from that receiver to a jack at the operator's position. This jack was labeled MTS. The operator knew that a call coming in on this jack was a mobile phone call (Figure 17-1).

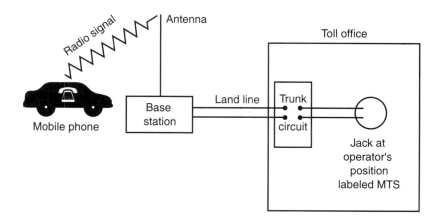

FIGURE 17-1 Mobile telephone system.

17.2 IMPROVED MOBILE TELEPHONE SYSTEMS

Improved Mobile Telephone System (IMTS) was introduced around 1964 but was not implemented where I worked until around 1973. The major benefit of IMTS was that it used several different frequencies and could support many different conversations at the same time. IMTS was also connected to the local class 5 office instead of the toll office. IMTS was connected to regular telephone numbers and line circuits. Unlike MTS, the IMTS was assigned a regular PSTN telephone number. Anyone could reach an IMTS phone by dialing the PSTN number assigned to the mobile phone. This eliminated the need for operators to handle mobile phone calls. When the IMTS mobile unit picked up their handset, they were connected to a line circuit in the central office and received dial tone. They placed their call just as they would from a regular telephone. The frequency spectrum used by IMTS was 454 to 512 MHz.

When a call was placed to a mobile IMTS phone, the caller dialed a number in the PSTN that was assigned to the mobile telephone. The PSTN telephone number was connected to a mobile radio terminal in the central office. The ringing signal for the called number activated a relay in the radio system's terminal equipment. The terminal equipment in the central office communicated with the controller of the transmitter at the antenna/transmitter site. The control system would automatically select an idle, available transmitting frequency and would modulate that frequency with signaling tones representing the dialed telephone number. The decoder of the mobile IMTS units would decode the signaling tones. If the decoded number matched the number programmed into the mobile unit, it would ring.

17.3 ADVANCED MOBILE PHONE SYSTEMS

MTS and IMTS used base station transmitters with 100 to 200 W of power and mobile transmitters with 5 to 25 W of power. These high power levels allowed the MTS and IMTS systems to cover a wide area using one base transmitter for the area. The next phase in mobile phone systems was the introduction of cellular radio. Cellular radio is also known as *advanced mobile phone systems (AMPS)*. AT&T had been testing cellular radio technology in the late 1970s and early 1980s. This technology took a novel approach. It uses many low-powered base transmitters (7 W) spread out over a geographic area (multiple base transmitter sites) and uses computer technology to decide which transmitter should be used. All the base station receivers and low-powered base transmitters are connected to a centrally located computer, which controls all operations. When a mobile phone is turned on, it communicates with several base site receivers. The computer determines which receiver is closest to the mobile unit on the strength of the received signal. The mobile unit transmits its identity (its telephone number) to the central computer. The central computer always knows the location of the mobile unit and which receiver/transmitter is closest to the mobile unit. When a call comes in for a mobile unit, the computer will use the transmitter closet to the mobile unit and send an alerting signal to the mobile unit.

On October 13, 1983, Illinois Bell implemented the first cellular network. The large metropolitan area of Chicago was divided into smaller geographic areas called *cells*. Each cell site is equipped with its own transmitting and receiving base station.

Each station was assigned several transmitting frequencies and several different receiving frequencies. By using a low-powered transmitter in the base station and mobile units (7 W), the frequencies could be reused in many cells around the city. The frequencies were not reused in adjacent cells but were in cells several miles away. Basically, cells were arranged in a group of seven cells that formed a distinct pattern. This pattern of cells that keeps repeating across the metropolitan area. Frequencies could not be assigned to more than one cell in the pattern.

With seven cells in the pattern, you have seven times the number of frequencies assigned in the cell. If each cell site had 20 frequencies, you would have a total of 140 (7×20) frequencies for the pattern. The 140 frequencies for the pattern of cells would continually repeat as the pattern repeated. No adjacent cells would be using the same frequencies. Thus, this arrangement results in our theoretical city having a total of 140 different radio frequencies. Different companies use different numbers of cells in the cell pattern and different total number of frequencies in each cell. The quantity used is dictated by how many calls the system must handle at the same time. By using low-powered transmitters, we can reuse frequencies many more times than a high-powered transmitter permits. This allows us to provide many more communication channels without having to use many more frequencies (Figure 17-2).

The example above resulted in 140 frequencies. There are 400 frequencies available for cellular radio. A cellular service provider may use more than 7 cells in a pattern and may use as many as 128 of the 400 frequencies at each cell site. One of the problems with radio communication is we continually run out of frequency allocations.

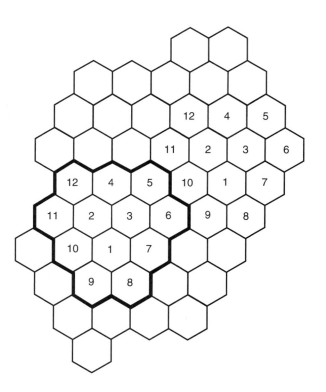

FIGURE 17-2 Twelve-cell reuse pattern.

The cellular technique allows us to use a set of frequencies over and over. If we use 100 high-powered transmitters at different frequencies in Chicago, we could only have 100 mobile conversations at the same time. If we use 100 low-powered transmitters, which can be reassigned 10 times, we can have 1000 simultaneous mobile conversations (assuming they are spread out across Chicago). The key to cell technology is the centralized computer at the *mobile telephone serving office (MTSO)* and the constant communication between a mobile unit, the base station cell site, and the computer of the MTSO. The computer will always know which cell site it should use to connect a call over.

Cellular radio uses frequencies from 825 to 845 MHz and from 870 to 890 MHz. The lower frequency band (825 to 845) is assigned as the transmit frequencies of mobile units. Base station receivers are tuned to these frequencies. The upper frequency band (870 to 890) is assigned to the base station transmitters. Mobile receivers are tuned to receive on these frequencies. The use of two frequencies for each call provides a full-duplex communication medium. The FCC controls the allocation of these frequencies and will provide a license for frequencies in a certain geographic area called a *cellular geographic serving area (CGSA)*. The CGSA is designed to fit the borders of a *standard metropolitan statistical area (SMSA)*. A SMSA defines geographic areas used by marketing agencies. The SMSA was also used by the 1984 MFJ to define most *local access transport areas (LATAs)*. Therefore, most CGSAs should correspond to the area covered by a LATA.

The cellular radio telephone system contains many cell sites, a central switching system, controllers at each cell site, and a central controller at the switching system. Voice channels from each cell site are connected to the MTSO. MTSO also stands for *mobile telephone switching office*. In the PCS network, the *mobile switching center* is also called the *MSC*. Each cell site controller is connected to the central controller of the MTSO by messaging links. The central controller also has messaging links connected to the central switching system. A cell site can contain up to 128 different transmitters and receivers. The cell site controllers keep track of the signal strength received from a mobile unit and report this information to the central controller. The central controller uses this information to decide which cell site should handle a call. The central controller keeps a list of all active mobile units and their cell site location.

The central controller and switching system for cellular radio is called the MTSO. The MTSO can be a stand-alone switch, owned by a private cellular company, or it can be integrated into a local switch if the LEC is the cellular service provider. The MTSO is similar to a class 4 switching center but is much smaller. It is like a class 4 switching center because the MTSO does not have line circuits. The MTSO only has trunk circuits. It has trunk circuits connecting the MTSO to the transmitter site and has trunk circuits connecting it to a switch in the PSTN. Many MTSOs are connected to the SS7 network. This allows a cellular phone to roam. The mobile telephone is continuously reporting to the closest cell site. As you drive across country and pass from one service provider to another, the central controller of the MTSO owned by other service providers will report your location over the SS7 network to your home base.

Suppose you have cellular service from Southwestern Bell in Kansas City, Missouri, and you are traveling to Atlanta. Anyone can call the telephone number in Kansas City assigned to your mobile phone and the call will be automatically forwarded over the PSTN to the MTSO presently serving your mobile phone. If your mobile phone is turned on, it has been communicating with other MTSOs. These MTSOs have been informing the MTSO in Kansas City of your location using the SS7 network. Suppose you are just outside St. Louis and someone calls your number in Kansas City. The Kansas City MTSO will contact the St. Louis MTSO over the SS7 network and ask it to ring your mobile phone. The Kansas City MTSO will also issue instructions to IEC switches in the PSTN and the MTSO in St. Louis to reserve a voice path between the Kansas City and St. Louis MTSOs. If the mobile phone is answered, the reserved voice path is established and the person that called you will be connected to your mobile phone. They do not know where you are. They probably assume you are in Kansas City, since that is the PSTN location for the number they dialed (Figure 17-3).

17.4 DIGITAL ADVANCED MOBILE PHONE SYSTEMS

The major difference between AMPS and *the digital advanced mobile telephone system (DAMPS)* is that AMPS uses *analog* radio signals and DAMPS uses *digital*

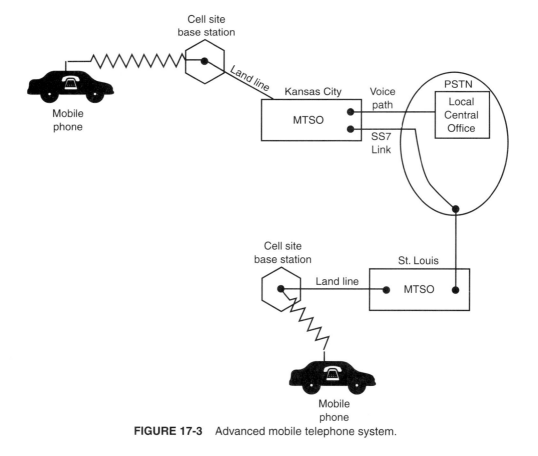

FIGURE 17-3 Advanced mobile telephone system.

radio signals. DAMPS is a technique that places multiple calls over one radio frequency using pulse code modulation (PCM) and time division multiplexing (TDM). The PCM signal is not the standard PCM signal found in the PSTN. The PSTN converts an analog signal into 64,000 bps. DAMPS technology converts the analog signal into 16,000 bps. DAMPS was introduced in 1992 for the existing AMPS. DAMPS allowed cellular operators to carry four times as many calls as a regular AMPS system. DAMPS is backward compatible with analog AMPS. The system will handle either a regular analog call per channel or multiple calls per channel using DAMPS. The manufacturer of mobile telephone sets provides a switch on the set that allows the user to select between AMPS and DAMPS. Cellular service providers could gradually change their systems from AMPS to DAMPS. This technology was used in the 825- to 895-MHz frequency band for AMPS. Cellular service providers will also use DAMPS in the higher frequency band of 1850 to 1990 MHz to provide PCS.

17.5 PERSONAL COMMUNICATION SYSTEM

The *personal communication system (PCS)* is quite similar to the technique used to provide DAMPS. PCS uses cellular technology. PCS also uses digital technology (CELP at 7.95 Kbps) and TDM. PCS utilizes lower-wattage transmitters than AMPS in the base and mobile stations. PCS also employs higher radio frequencies than AMPS. PCS uses radio frequencies between 1850 and 1910 MHz for mobile unit transmitters and between 1930 and 1990 MHz for base station transmitters. The channels are numbered from 512 (1850.2 MHz and 1930.2 MHz) to 810 (1909.8 MHz and 1989.8 MHz). This provides 299 carrier slots (512 to 810 inclusive). Channel 512 is a transmitting frequency of 1850.2 MHz by the mobile unit transmitter. This of course is also the receiving frequency in the base station unit for channel 512. The base station transmits at 1930.2 MHz, and the mobile unit also receives at this frequency for channel 512. Each channel is assigned a frequency for the base transmitter, and a different frequency for the mobile unit, to provide a full-duplex communication circuit.

The base unit is located at a *base transceiver station (BTS)*. The BTS site will consist of a small building to house the base station transmitters and receivers, and an antenna tower approximately 150 ft high. The antenna tower will contain 9 antennas arranged in a triangle array. Each side of the triangle will contain one transmitting antenna and two receiving antennas. Looking at one side of this triangle array, the transmitting antenna is located at the center point of the side of the triangle. The receiving antennas are located on each side of the transmitting antenna. The receiving antennas are placed about 3 ft on each side of the transmitting antenna. With a triangle arrangement, we have three separate antenna arrays that are pointing in three directions. If a circle is drawn around this triangle, you would see that each side of the triangle covers 120° of the total coverage area for the site. The triangle array provides a full 360° of coverage for the site.

17.6 THE FCC AUCTION, MTAS, AND BTAS

The FCC held an auction in 1995 to sell the licenses for PCS frequency allocations. This auction raised $7 billion. The United States was divided into 51 segments. Each segment was termed a *major trading area (MTA)*. Each MTA was subdivided into a *basic trading area (BTA)*. There are 492 BTAs. The auction sold two licenses per MTA and four per BTA, for a total of 2070 licenses. Several companies were sold licenses on different frequencies in the same BTA. These licenses were for broadband PCS, which provides a bandwidth of 30 KHz. The bandwidth of broadband PCS is necessary for voice communication. Narrowband PCS is a service used for paging service providers. The narrowband frequencies were auctioned off in 1994 and the broadband PCS frequencies the following year. Many of these licenses were bought by IECs and LECs. They intend to use PCS technology to provide local phone service. By using radio waves to provide local service, the IECs can get into the local services market without having to lease local loops from the LECs. Sprint purchased licenses for 29 MTAs. AT&T purchased licenses for 21 MTAs. Prime Communications (NYNEX, Bell Atlantic, US West, and Air Touch) purchased 11 licenses. Bell South, Pacific Telesis, and Southwestern Bell also purchased licenses for BTAs in their area of LEC operations. Existing cellular service providers were not allowed to bid for an MTA license.

17.7 DAMPS VERSUS GSM VERSUS CDMA

As stated earlier, many cellular service providers will use DAMPS (IS-136), which is a time division multiple access (TDMA) technology, to provide PCS. AT&T Wireless (the old McCaw Cellular) will use DAMPS. This allows it to use its current equipment. It merely changes the frequencies of the transmitters and changes some software in its controllers. Other companies entering the PCS market have chosen to use the *Global System for Mobile communications (GSM)* 1900 standard for PCS. GSM 1900 also uses TDMA technology. The method used to change an analog voice signal into a digital code varies between systems. Early DAMPS systems used a process called *adaptive pulse code modulation* to achieve digital voice at 16 Kbps. The PCS 1900 system uses a process called *code excited linear predictive (CLIP) coding* to achieve digital voice with 7.95 Kbps. CLIP is also available in a 16-Kbps chip. Another method of providing PCS will use an evolving technology called *code division multiple access (CDMA)* technology.

Development of the GSM standard began within CEPT, the Council of European PTTs, in 1982. The goal of the CEPT was to develop a standard for the digital cellular network that would allow international *roaming*. The first GSM networks entered service in 1992. By the middle of 1995, more than 75 networks were in service in 45 countries. The GSM standard is the most successful digital cellular standard in the world. More than 20 million customers use the GSM standard, with growth of around 600,000 new customers per month.

GSM has spread far beyond the European community. GSM standards have been adopted by South Africa, Saudi Arabia, Egypt, Australia, China, Hong Kong, Singapore, New Zealand, and many other countries. More than 120 operators in 80 countries have signed a Memorandum of Understanding on GSM with plans to implement GSM standard networks. GSM standards were based on the frequency spectrum around 900 MHz. In 1992, a variant on GSM called *DCS 1800* was also specified as a standard by the European Commission. DCS 1800 uses the same network technology as GSM but operates at frequencies in the 1800-MHz band. DCS 1800 was used to provide enhanced services to the MTS community. The first DCS 1800 networks were installed in 1993 in France, the United Kingdom, and Germany. GSM 1900 is simply another evolution of the original GSM standard. The frequencies of operation were just changed to the 1900-MHz range.

CDMA has been adopted as IS-95 Standard for PCS Networks. IS-95 Q-CDMA was codified as a standard in 1993. CDMA can provide ten times the capacity of DAMPS. CDMA is not backward compatible with analog AMPS in the same way that D-AMPS is. CDMA can not reuse DAMP technology. To implement CDMA, the service provider must install a new network. This is a problem for existing cellular service providers but not for new entrants to the mobile communications market. Sprint is not a new entrant to cellular communications, but in order to bid on PCS licenses, it was forced to sell its old cellular markets. It spun off the cellular division, which is no longer part of Sprint. Thus, for all intents and purposes, Sprint had to start with no technology in place. This made it easy for it to decide on CDMA technology. With CDMA technology, each active mobile transmitter is assigned a code by the central controller of the MTSO. Each transmission includes the code assigned to a particular call. Many mobile units can be transmitting over the same frequency and at the same time, but each transmission has a unique code appended to the transmission. The receivers will determine which signal is meant for them by checking the code assigned.

In the AMPS, the different mobile phone call were allocated to a specific radio frequency. Each radio frequency was 30 KHz wide. Thus, AMPS employs frequency division multiplexing (FDM). DAMPS also used FDM and the same 30-KHz bandwidth per frequency, but each call does not occupy the 30-KHz spectrum for the full duration of time. With DAMPS, each call will occupy a certain portion of time on a specific frequency. CDMA uses a wider bandwidth than AMPS and DAMPS. CDMA is referred to as a *spread spectrum technology* because the signal is spread out over a wider bandwidth. The bandwidth of each frequency is 1.25 MHz. Multiple signals will be placed using the same frequency and all will be transmitted at the same time. Since they are all transmitted at the same time, they will interfere with each other, and it is important that transmit power levels are very low. CDMA terminals transmit at less than 1W. In PCS the terminals transmit at 0.85 W. Each voice signal is assigned a pseudorandom number (PN) code pattern that will only be recognized by the receiver assigned to recognize that PN (Figure 17-4).

CDMA is another example of the advances made in signal processing technology and Super Very Large Scale Integrated circuit design. The advances in these technologies make CDMA possible. The receiver must be smart enough to analyze a captured signal, despread the signal, and identify which signal should be reconstituted and then code that signal into a 16-Kbps code. CDMA technology allows the reuse of transmitter frequencies in adjacent cells and allows the cell site coverage to be larger. The signal processor provides a higher gain for the received signal, and so the larger cell radius is due to receiver improvements, not higher transmitter power. As stated before, CDMA also allows lower transmitter power. In fact, the use of CDMA requires lower transmitter power. Since a wider coverage area per cell

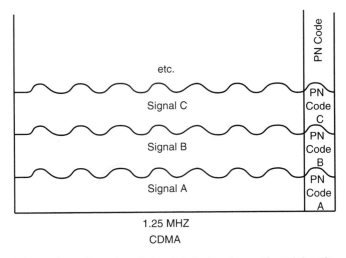

1.25 MHZ
CDMA

FIGURE 17-4 CDMA showing several signals and PN code. The receiver will receive all signals but will only amplify and decode the signal for the PN code assigned to the signal that matches the PN code assigned to the receiver.

reduces the number of BTSs needed and lowers the cost of the PCS system, many PCS providers will elect to use CDMA if they are implementing new systems.

17.8 PCS NETWORK

The PCS network is similar to the DAMPS network. In PCS, the MTSO contains a home location register (HLR) database, a visitor location register (VLR) database, an equipment identification registry (EIR) database, an authentication center (AUC), and a short messages services service center (SMS-SC). The MTSO connects to a base station subsystem (BSS). The BSS consists of a TCU (which is located at the MTSO), a *base station controller (BSC)*, and several base transceiver stations (BTSs) (Figure 17-5).

One TCU is required for each PCM link between the MTSO and the BSC. The TCU provides an administrative and maintenance interface to the BSC and performs a digital rate conversion between the BSC and the MTSO. The DS1 link connecting a BSC to the TCU will contain twenty-four 64-Kbps (DS0) channels. Each 64-Kbps channel contains 4 calls. Remember that the mobile units are using coders that convert voice to 16 Kbps. The TCU will demultiplex the four 16-Kbps voice channels and convert each of these voice channels to 64 Kbps. This conversion is necessary because the codecs (coders/decoders) of the PSTN use 64-Kbps coding. We cannot pass the 16-Kbps signal out over the PSTN because when it arrives at the distant decoder, it will not be decoded properly. The decoder at the end of a PSTN facility is looking for 64 Kbps to decode. Likewise the TCU will accept 64-Kbps voice signals from the PSTN, convert them to 16 Kbps, and multiplex this signal onto the appropriate DS0 channel to the DSC. In PCS, each DS0 between the DSC and the MTSO contains four digitized voice signals.

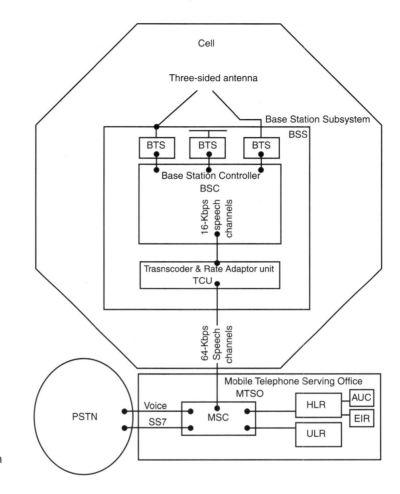

FIGURE 17-5 Block design for PCS1900.

The BSC connects to all the transmitter sites. The transmitter sites are called BTS sites. The BTS provides an interface between the mobile phone and the BSC. The BTS is controlled by the BSC. Communications are constantly taking place between the BSC and the BTS. The BSC tells the BTS which radio channel to use for communication with a particular mobile station. The BTS contains the transmitters, receivers, antenna connection equipment, and antennas. The BSC will receive information from all BTSs connected to it. This allows the BSC to make a decision on which BTS should handle a call. The BSC is informed of the signal strength received by each BTS. The BSC will assign the call to the BTS receiving the strongest signal. As a mobile moves from one cell to another, the BSC will recognize the changing signal strength and will reassign the call from one BTS to another BTS.

17.9 INITIATING A PCS CALL FROM THE MOBILE STATION

When mobile users initiate a call, their equipment will search for a local BTS. Each BTS has at least one of its radio channels assigned to carry control signals in addition to traffic. The BTS will inform the BSC of each call attempt. The BSC will allocate a dedicated bidirectional signaling channel and will set up a route to the MTSO. The BSC is responsible for the management of the radio resource within a region. Its main functions are to allocate and control traffic channels, control frequency hopping, undertake handovers (except to cells outside its region), and provide radio performance measurements. When a mobile unit initiates a call it must provide its *International Mobile Subscriber Identity number (IMSI)*. This is a unique number that will allow the PCS system to initiate a process to confirm that the mobile customer is allowed to access it. Each PCS phone has an entry in the HLR of its home MTSO. The HLR contains information about the services the subscriber is allowed to access. The HLR also contains a unique authentication key and associated challenge/response generators. The AUC will verify that the customer is legitimate and allow the call to proceed.

Whenever a mobile is switched on, and at intervals thereafter, it will register with the system. This allows the location of the caller in the network to be established and its location area to be updated in the HLR. A location area is a geographically defined group of cells. When the first registration attempt occurs, the MTSO will use the IMSI to interrogate the customer's home-base HLR and will add the customer data to its associated VLR. The VLR now contains the address of the customer's HLR, and the authentication request is routed back through the HLR to the customer's AUC. This generates a challenge/response pair that is used by the local network to challenge the mobile. In addition, the PCS system will check the mobile equipment against an EIR in order to control stolen, fraudulent, or faulty equipment.

The authentication process is powerful and is based on advanced cryptographic principles. It especially protects the network operators from fraudulent use of their services. The TDMA nature of GSM coupled with its frequency-hopping facility make it difficult for an eavesdropper to lock onto the correct signal and monitor a conversation. CDMA makes it impossible for someone to eavesdrop on your conversation since the voice is a digital signal with a pseudorandom number (PN) code pattern. Customers desiring a secure conversation over facilities using GSM TDSA should use encryption.

Once the user and the user's equipment are accepted by the network, the mobile must define the type of service it requires (voice, data, supplementary services and so on) and the destination number. At this point, a traffic channel with the relevant capacity will be allocated and the MTSO will route the call to its desired destination. The MTSO will use the SS7 network to establish call setup over the PSTN. If the called number is busy, the local switching system will return a busy signal.

17.10 RECEIVING A PCS CALL

When a PCS number is dialed, the MTSO will look in its HLR to identify the location of the PCS number. If it is informed that the location is in a remote MTSO, it will launch a query over SS7 to find the mobile phone. The MTSO will send out a paging signal to page the mobile unit. If the mobile unit is in the area, it will respond to the page. The serving MTSO will connect the caller to the mobile via its SSB if the PCS phone is still in the local serving area of the MTSO. If the serving MTSO finds the PCS phone in a remote MTSO, the MTSO will connect to the remote MTSO over the PSTN and will instruct the remote MTSO to connect the call via the remote's SSB.

17.11 SMART CARD

The *Smart Card* or *Subscriber Identification Module (SIM)* card contains a microprocessor and a small amount of memory. With a SIM card, a customer can use any PCS phone that has a card reader to make a call. The SIM card—the size of a credit card—is inserted into the PCS phone to activate the phone. In effect, while the card is in the phone, the PCS phone becomes personalized and becomes the user's personal phone. All the customer's personal data, personal identification number (PIN), services subscribed to, authentication key, IMSI, speed dialing lists, and so forth. are stored in the SIM card.

17.12 SUMMARY

This is the last chapter in the book because this is the latest technology to hit the telecommunications industry. I am sure that the technology behind PCS and the services offered by PCS will advance rapidly. Check out the Internet and check the websites of the LECs and IECs to see what is happening currently with PCS. Future revisions of this book will incorporate future advances. I have tried to provide as much information as possible on this emerging technology, but as with any new technology, changes occur almost overnight. I think I have given you a good start on PCS. As the technology grows, you will need to keep abreast of the changes. Although many different voice coding techniques are used in wireless communication, you do not need to understand how they work. You simply need to know that they exist so that you will know the type of coding your terminal equipment requires. Your equipment must utilize the same coding technique that the wireless radio system is using. If the wireless system uses CDMA, you must purchase CDMA terminal equipment. If the system is using 16-Kbps CLIP, your terminal equipment must also be 16-Kbps CLIP.

Since this is the last chapter of the book, its time to thank you for the many hours you have spent reading it. If you read this book as part of a degree program to prepare for a management position in telecommunications, I hope it has served you well. I trust it has laid a solid foundation on which you can continue to build. If you are a telecommunications manager and read the book to broaden your knowledge base on the varied technologies presented, I hope it has fulfilled your needs. I have attempted to cover all the major technical points but have tried to avoid presenting the material in the format of another dry, technical, jargon-filled book. I hope my writing sytle made the book an easy read for you.

The telecommunications industry has a bright future and offers many challenging opportunities. I wish you much success in your future within that industry. I encourage everyone to drop me an e-mail at DeVry Technology Institute of Kansas City (mcole@kc.devry.edu). All suggestions on improvements for the next edition of the book are most welcome. Let me know what you think of content and style. Thanks again for spending time with me, and I hope you have a happy, successful career.

REVIEW QUESTIONS

1. What is PCS?
2. What is PCS 1900?

3. What is MTS?
4. What is AMPS?
5. What is DAMPS?
6. What is TDMA?
7. What is CDMA?
8. What is BSC?
9. What is BTS?
10. What is MTSO?
11. What is HLR?
12. What is MTA?
13. What is BTA?
14. What is the difference between PCM in the PSTN and in DAMPS?
15. How does a company get a license to sell mobile telephone service?

GLOSSARY

Advanced Mobile Phone System (AMPS) Also called cellular radio. The serving area for AMPS is broken up into cells. Low-powered transmitters are used for AMPS. Radio frequencies can be reused in nonadjacent cells. Cellular radio uses frequencies from 825 to 845 MHz and between 870 to 890 MHz.

Base Station Controller (BSC) Connects to all the base transceiver station (BTS) sites serving a BTA. The BTS provides an interface between the mobile phone and the BSC. The BTS is controlled by the BSC. Communication is constantly taking place between the BSC and the BTS. The BSC tells the BTS which radio channel to use for communication with a particular mobile station.

Base Transceiver Station (BTS) The BTS site for a PCS 1900 system will consist of a small building to house the base station transmitters and receivers and an antenna tower approximately 150 ft high. The antenna tower will contain 9 antennas arranged in a triangle array. Each side of the triangle will contain one transmitting antenna and two receiving antennas.

Basic Trading Area (BTA) Each of the 51 major trading areas (MTAs) was subdivided into a BTA. There are 492 BTAs. PCS licenses were sold for each BTA.

Cellular Radio Advanced Mobile Phone System (AMPS). The serving area for AMPS is broken up into cells. Low-powered transmitters are used for AMPS. Radio frequencies can be reused in nonadjacent cells. Cellular radio uses frequencies from 825 to 845 MHz and between 870 to 890 MHz.

Code Division Multiple Access (CDMA) has been adopted as IS-95 Standard for PCS Networks. IS-95 Q-CDMA was codified as a standard in 1993. CDMA can provide 10 times the capacity of DAMPS. CDMA is not backward-compatible with analogue AMPS in the same way that D-AMPS is. CDMA can not reuse DAMP technology. To implement CDMA, the service provider must install a new network. CDMA is spread spectrum technology. Many conversations are multiplexed over one frequency. Each conversation is assigned a code. The receiver strips out conversations individually by using the code.

Code Excited Linear Predictive Coding (CLIP) Also known as vector-sum excited linear predictive (VSELP) coding. This speech coding algorithm is EIA Standard IS-54. This is the speech coding technique recommended for TDMA cellular radio systems. The bit rate is 7.95 Kbps. VSELP is also available in a 16-Kbps chip. AT&T makes a digital signal processor using this technology; it is called a DSP-1616.

Digital Advanced Mobile Telephone System (DAMPS) is a technique which places multiple calls over one radio frequency using Pulse Code Modulation (PCM) and Time Division Multiplexing (TDM). The PCM signal is not the standard 64,000bps PCM signal found in the PSTN but is a 16,000 bps signal. DAMPS was introduced in 1992 for the exixting AMPS and is backwar compatible with anolog AMPS. The TDM feature allows DAMPS to carry four times as many calls as AMPS.

Improved Mobile Telephone System (IMTS) IMTS was introduced around 1964. The major benefit of IMTS was that it used several different frequencies and could support many different conversations at the same time. It was also connected to the local class 5 office instead of the toll office. IMTS was connected to regular telephone numbers and line circuits. An IMTS phone was assigned a regular PSTN telephone number. Anyone could reach an IMTS phone by dialing the PSTN number assigned to the mobile phone. This eliminated the need for operators to handle mobile phone calls.

Major Trading Area (MTA) The United States is divided into 51 areas for mobile telephone licensing purposes. Each of these segments is called a major trading area (MTA). Each MTA was subdivided into a basic trading area (BTA). There are 492 BTAs.

Mobile Telephone Serving Office (MTSO) The central controller and switching system for cellular radio. The MTSO can be a stand-alone switch owned by a private cellular company, or it can be integrated into a local switch if the LEC is the cellular service provider.

Mobile Telephone System (MTS) Mobile telephone service began in 1946 and was called MTS. These systems used the radio frequencies between 35 and 45 MHz. Although MTS stands for "mobile telephone service," it also meant "manual telephone system". All calls had to be handled by an operator.

Personal Communications System (PCS) PCS is not so much a technology as a concept. The concept of PCS is to assign someone a personal telephone number (PTN). This PTN is stored in a database on the SS7 network. That database keeps track of where a person can be reached. When a call is placed for that person, the artificial intelligence network (AIN) of the SS7 determines where the call should be directed.

Personal Communications System 1900 (PCS 1900) PCS is provided by radio frequencies in the 1900-MHz range. PCS 1900 is the latest evolution in mobile communication. PCS 1900 can be provided by the time division multiple access (TDMA) technology of DAMPS or can use code division multiple access (CDMA) technology.

Roaming The mobile telephone of AMPS or DAMPS is continuously reporting its location to the closest cell site. As you drive across country and pass from one service provider to another, the central controller of the MTSO owned by other service providers will report your location over the SS7 network to your home base.

Time Division Multiple Access (TDMA) TDMA technology is used by DAMPS. With TDMA, each conversation occupies a time slot on a common transmission medium. In DAMPS, each conversation occupies a time slot on a particular radio frequency.

Appendix A

Answers to Chapter Review Questions

CHAPTER ONE

1. The telephone was invented by Alexander Graham Bell and patented in 1876.
2. The first BOC was established in 1877.
3. The first Independent was created in 1893.
4. AT&T entered into the Kingsbury Commitment as a way to get the government to drop its pursuit of antitrust action against AT&T.
5. The Communications Act of 1934 effectively eliminated the provisions of the Graham Act.
6. A tariff is the document that provides details on the service a common carrier wishes to provide, or is currently providing, and the rates the common carrier wishes to have approved for that service.
7. The Federal Communications Commission (FCC) regulates interstate telecommunications service.
8. The Public Utilities Commission (PUC) in each state regulates the telecommunications within its state (intrastate).
9. Congress enacted legislation in 1949 that extended coverage of the 1936 Rural Electrification Act to include coverage of rural telecommunications. The amended act provided low-cost government loans for the establishment of telecommunications in rural America. State legislation guaranteed a positive rate of return and franchised territory to carriers.
10. Companies providing telecommunications service to the general public are referred to as common carriers.
11. The antitrust lawsuit filed in 1949 was settled seven years later by the 1956 Consent Decree.
12. The Carterphone decision of 1968 led to deregulation of station equipment.
13. Privately owned equipment is called customer-provided equipment (CPE).
14. CPE is sold by interconnect companies. As a result of the Communications Reform Act of 1996, CPE can also be sold by LECs.

15. The MCI ruling of 1969 by the FCC and settlement of the lawsuit brought against AT&T by MCI led to deregulation of long distance service.

16. No! Deregulation of telecommunications has actually led to improved service and a plethora of service offerings. Deregulation of long distance service meant a loss of toll sharing revenue to the LECs, and local phone rates were raised to offset the loss of this revenue.

17. The Modified Final Judgment of 1984 forced AT&T to divest itself of the local BOCs. Now that the BOCs are separate companies, AT&T can no longer force them to buy solely from Western Electric; they buy from manufacturers providing equivalent equipment at the lowest price. Although the 1956 Consent Decree stopped some of Western Electric's pricing schemes, it did not prohibit BOCs from doing business with Western Electric. As long as Western Electric and local BOCs were part of AT&T, the "Baby Bells" would continue to use Western Electric as their favored supplier.

18. The two broad categories of action taken by the MFJ of 1984 were divestiture and deregulation. The MFJ forced AT&T to divest itself of the BOCs (Baby Bells). The MFJ deregulated the long distance services and forced the new BOCs to provide equal access to all IECs.

19. The toll revenue sharing arrangement was referred to as separations and settlements. The 1984 MFJ led to the elimination of revenue sharing.

20. No! Although the FCC made rulings favorable to Carter Electronics and MCI, it was the pursuit of lawsuits by these companies and the involvement of the Department of Justice (DOJ) that led to the MFJ.

21. Common carriers providing local telephone service are called local exchange carriers (LECs).

22. Common carriers providing long distance services were initially referred to as specialized common carriers but today are referred to as Interexchange carriers (IECs).

23. The PSTN is composed of the networks of all common carriers, and services on the PSTN are available to the general public. A private network is privately owned and managed for the benefit of its owners.

24. The Carterphone decision of 1968 led to deregulation of station equipment.

25. The major provisions of the MFJ were: (1) the breakup of the Bell system by forcing AT&T to divest itself of the local BOCs. (2) deregulation of the long distance services and the requirement that the new BOCs provide all IEC customers with equal access to their IEC (this requirement led to the establishment of POPs); (3) the establishment of LATAs and the ruling that calls within a LATA must be handled by a LEC, while calls between LATAs must be handled by a IEC.

26. A PIC is preferred interexchange carrier. Telephone customers inform the LEC who they want to handle their long distance calls (whom they prefer) and the LEC programs its central office to assign the preferred IEC to that customer's telephone number.

27. Any PIC can be overridden by using the Feature Group B access code for the IEC a caller wishes to use for the current call.

28. LATAs are local access transport areas. A LATA defines the area within which calls must be handled by a LEC. There are 184 LATAs. LATAs may or may not follow the boundaries of established area codes.

29. A Point of Presence (POP) office is the central office location that connects the circuits of the LEC to the circuits of the IEC.

30. The development of fiber optic transmission systems with their high levels of multiplexing and the installation of computer-controlled electronic switching systems have led to flatter network hierarchies.

31. The Telecommunications Reform Act of 1996 (Public Law 104-104) replaces the 1984 MFJ. The LECs are allowed to sell CPE; the IECs are allowed to provide local service. Any qualified common carrier can provide local service. If the Bell LECs do not hinder competition in the local services arena, they can provide Inter-LATA toll services after February 1997. The Telecommunications Reform Act of 1996 officially deregulates the local service market and also allows the BOCs to enter the long distance market.

CHAPTER TWO

1. The three parts to a communications system are: (1) transmitter, (2) medium, and (3) receiver

2. The receiver must accept a signal from the transmission medium, decode the signal, and convert it into a form readily understood by the receiving device.

3. Half-duplex systems transmit messages in both directions over one medium by the transmitters on each end of the medium taking turns sending messages. Full-duplex systems are capable of transmitting in both directions at the same time.

4. Wire is used as the transmission medium by LECs because it is the perfect medium to provide power from the central office to the telephone. Wire will conduct the low-power electrical signals of telephones extremely well.

5. The carbon transmitter was invented by Thomas Edison.

6. DC power (power supplied by a battery) provides voltage for the transmitter of the telephone.

7. AC power provides the power to ring a phone.

8. The central wiring center is called the central office, central exchange, exchange, or local exchange.

9. The first central exchange was installed in 1878, almost immediately after the first phones were installed.

10. Two wires are used to connect the telephone to a central exchange.

11. The two wires are called Tip and Ring.

12. A switchboard has ten pairs of patch cords and can handle ten calls at a time.

13. A line is the name used to refer to a pair of wires connecting the telephone to a central exchange.

14. A trunk is the name given to a pair of wires connecting one exchange to another exchange.

15. Exchange names were needed for calls between offices. If a number was given without an exchange name, the operator assumed the call was for a local line. When an exchange name was given to the operator, the call was placed over a trunk to that exchange for completion by the operator at that exchange.

16. Calls were completed between lines appearing on different switchboards by the use of interposition jacks.

17. The advantages of a central exchange were: (1) a customer needed only one phone and it could be connected to any other phone by the operator; (2) the central exchange furnished power for the phone and eliminated the need for a local battery supply at the phone; (3) all lines were rung by the operator, eliminating the need for a hand-crank generator at the phone.

18. The American National Standards Institute (ANSI) provides telecommunications standards for North America.

19. The CCITT of the ITU provides international standards.

20. AT&Ts Bell Labs developed most standards that have become de facto standards for the telecommunications industry. Many of these de facto standards will be found in the Local Switching Systems Generic Requirements (LSSGR).

21. The major factor that always delays new technology is the development of and agreement on a universal standard.

22. The first automated switching system was invented in 1892 by Almon B. Strowger.

23. Computer-controlled switching systems are called Stored Program Control (SPC) switching systems.

24. The major factor delaying the implementation of time division multiplexing was the lack of an underlying technology. TDM was invented before the development of integrated semiconductor chip circuits. TDM could not be implemented effectively using vacuum tube technology.

25. Plastic insulated cable is called PIC.

CHAPTER THREE

1. The power for a telephone is provided from the central office by the battery at the central office. This power is a DC voltage.

2. The type of voltage used to ring a telephone is AC voltage

3. DC voltage is produced by a chemical reaction.

4. A cell refers to a chemical cell that produces a DC voltage by a chemical reaction within the cell. A battery refers to a battery of cells. A battery of cells is the connection of two or more chemical cells in series to produce a higher voltage.

5. The first central exchange used a 24-V battery.

6. A molecule is a material composed of two or more atoms of elements. A molecule is the smallest piece of matter that will still possess the characteristics of the larger mass. If the matter is broken down smaller than a molecule, it is broken down into atoms of the elements comprising the molecule.

7. An element is a material in which all atoms are the same. There are approximately 100 natural elements such as oxygen, hydrogen, carbon, copper, gold, zinc, and so on. There are also several human-made elements.

8. An atom is the smallest piece of an element that will still retain the characteristics of the element. An atom is composed of a nucleus containing positive charges called protons (and possible neutrons, which have no charge) as well as negative charges called electrons surrounding the nucleus.

9. An electron is a negatively charged particle that exists in the shells surrounding the nucleus of an atom.

10. Electric current flow is the movement of electrons in an orderly fashion.

11. Resistance is opposition to the flow of electric current.

12. Voltage is a force that tends to cause electric current flow. Voltage occurs when there is a difference between the number of electrons a material has and the number it normally has.

13. The amount of electrons that flow in a circuit depends on the voltage and resistance in the circuit. This is stated in a formula: current = resistance / voltage.

14. A dry cell develops 1.5 V per cell.

15. A wet cell develops 2.17 V per cell.

16. The central office battery is composed of 24 wet cells connected in series. This battery provides 52 V but is often called a 48-V battery.

17. The physical size of a battery determines how long it can provide a voltage before it becomes discharged.

18. The power supplied by the power company is AC voltage.

19. AC is a voltage that constantly changes state. It will change from positive to negative and will also change from 0 V to its maximum value.

20. DC voltage was selected to power the phone because the phone was invented before commercial AC power was available. Today an added benefit of using DC from batteries is continuance of service in the event of a commercial AC power failure.

21. An RMS value for AC allows comparison to DC

22. Five different ringing signals were used in rural exchanges.

23. A tuned ringer was a ringer that would only ring on one frequency of ringing signal.

24. A straight-line ringer will ring on any and all ringing frequencies.

25. Four 100 Ω resistors connected in series provides 400 Ω total resistance.

26. Four 100 Ω resistors connected in parallel provides 25 Ω total resistance.

27. The dBm is used to measure electrical voice signals.

28. The dBrnC0 is used to measure noise signals.

29. 30 dBrnC0 = −60 dBm.

30. At the tuned frequency for an *LC* network, *XL* = *XC*.

CHAPTER FOUR

1. Multiplexing is combining two or more signals and placing them over one transmission facility.

2. Space division multiplexing occurs when each signal has its own path within the medium. If a cable is considered a medium, each wire pair in the cable can handle a signal, and since each wire pair occupies space, this is space division multiplexing.

3. Frequency division multiplexing occurs when each signal has its own distinct frequency over the transmission medium. The signals can be combined for transmission and then separated at the distant end by using a receiver tuned to each frequency.

4. Time division multiplexing occurs when each signal occupies a shared transmission medium for a brief interval of time.

5. The purpose of a channel unit is to change a signal so that it may be multiplexed. The channel unit also contains a receiver designed to pull out only one signal from the multiplexed signal.

6. A four-wire system is a transmission system that contains separate paths for the transmission and reception of signals.

7. Amplitude modulation is a technique used by a channel unit in an FDM system to convert the incoming voice signal (0–4000 Hz) into a higher-frequency signal. The amplitude of the voice signal causes the amplitude of the carrier signal for the channel to vary in direct relationship to the amplitude of the voice wave. Each channel unit has a different carrier frequency. This allows combining the output of the channel units into a FDM signal.

8. Frequency modulation is a modulation technique where the frequency of the carrier signal is changed in direct relationship to the amplitude of the input voice signal.

9. There is a big difference between FM and FDM. You must remember that FM is a modulation technique to prepare a signal so that it may be multiplexed with other signals using FDM. However, FM is rarely used. AM is used in most carrier systems to prepare signals for FDM.

10. The L-600 carrier system has 600 voice channels.

11. The L5 carrier system has 10,800 voice channels, and the L5E has 13,200 voice channels.

12. The L carrier systems use single-sideband amplitude modulation.

13. PAM transmits the amplitude of a small portion of a signal. PCM takes the PAM signal, converts it into a digital code, and transmits the digital code that represents the signal.

14. The industry standard method for converting an analog signal into a digital signal is pulse code modulation (PCM).

15. The sampling rate in the PCM is 8000 samples per second.

16. The sampling rate was derived from studies that demonstrated that a signal could be faithfully reproduced by taking samples at twice the highest frequency. The highest voice frequency in the PSTN is 4000 Hz. Therefore, $2 \times 4000 = 8000$ for a sampling rate.

17. Since each amplitude sample taken can assume any value, the number of different levels is infinity. With an 8-bit coding technique, PCM can only code 256 signal levels. Quantization is the process of converting an input sample's amplitude level to the nearest of the 256 levels that can be coded.

18. The codec is used to perform a PCM process.

19. The DS0 signal rate is 64,000 bps. The DS1 signal is 1,544,000 bps.

20. The T1 carrier uses the DS1 signal rate of 1,544,000 bps. To do that many bits in one second, each bit must be no longer than 1 s/ 1,544,000. This is 647.6684 ns.

21. There are 193 bits in each DS1 frame.

22. The DS1 has 8000 frames per second. Since we sample voices at 8000 samples per second at the input to a digital TDM system, every multiplexing device in the system must have 8000 frames per second. The higher the signal rate, the more bits in each of the frames.

23. When a T1 system uses cable pairs for a transmission medium, repeaters must be placed in each central office and at 6000-ft intervals along the cable route.

24. The basic building block for SONET is OC-1.

25. OC-48 contains 32,256 channels.

26. Fiber optic cable is used as the transmission medium for OC-48.

27. The major advantages of SONET standards are that different vendors' equipment can work together and lower-level signals can be stripped out of higher-level signals without having to demultiplex the entire signal.

CHAPTER FIVE

1. Copper wire will continue to be the facility of choice for the local loop because it is already in place, it costs less than other mediums, and the telephone receives its power over the local loop. Wire is a medium that conducts electricity and can provide power to the phone from the central office.

2. Fiber optic cable is the medium of choice for use in the toll network because of its ability to carry 32,256 conversations over one fiber pair. This high capacity is achieved at a lower cost with fiber than with wire, coaxial, or microwave facilities.

3. The device that governs design of the local loop is the carbon transmitter found in most telephones. This transmitter requires a minimum of 20 milliamps of current to work properly.

4. Outside plant is designed according to the resistance design concept.

5. The two wires of the local loop are called Tip and Ring.

6. *PIC* stands for plastic insulated cable.

7. The telephone receives its power from the central exchange battery via the local loop.

8. The central exchange applies –52V DC to the Ring wire and ground to the Tip wire.

9. A telephone can be located about 7 1/2 mi from the exchange before loop treatment is needed.

10. *SLC-96* stands for subscriber line carrier–96.

11. Subscriber carrier is used to serve a group of customers when there is not enough cable from the exchange. When a new subdivision is built, it may be built in an area where cable pairs are in short supply. Subscriber carrier is a cheaper alternative to placing a new cable over a long distance.

12. Mutual capacitance is the capacitance that exists between the Tip and Ring wires of the same cable pair.

13. Mutual capacitance is offset by placing load coils on the cable pairs.

14. D spacing is placing load coils 6000 ft apart.

15. The local loop is loaded if the loop will be more than 3 mi. The first load coil is placed 3000 ft from the office. The spacing for additional load coils is every 6000 ft.

16. An OC-48 system can support 32,256 simultaneous conversations.

17. *SONET* stands for Synchronous Optical Network. This is the standard for fiber optic transmission facilities.

18. A wire pair is twisted to reduce the possibility of crosstalk.

19. The major source of noise in the local loop is from deterioration of the plastic insulation on wires due to water entering the cable. The outer covering of the cable may be damaged by excavation activity or by animals eating the covering. When the outer covering is damaged, moisture is sucked into the cable.

20. Fiber uses light signals, which are not affected by electrical disturbances.

CHAPTER SIX

1. Alexander Graham Bell was awarded the patent for the telephone in 1876.

2. A key element of both designs was a variable resistance transmitter.

3. Thomas Edison improved the carbon granule transmitter found in most telephones.

4. A carbon granule transmitter needs at least 23 milliamps of current to operate properly. Newer transmitters require 20 ma of current.

5. The standard single-line telephone set requires DC voltage to power the transmitter and DTMF pad.

6. There is no standard for the telephone. It must be registered with the FCC and must work with central offices. Therefore, designers must take these requirements into consideration, but there is no standard.

7. Station equipment includes telephones, computer terminals, and modems.

8. Station equipment must be registered with the FCC to ensure it will not cause harm to the PSTN.

9. A multiline phone allows the connection of up to four lines to one telephone set with pushbutton access to each line.

10. A keysystem allows the connection of more than four central office trunks to the system. It provides intercom paths, features such as speed call lists, and optional automated attendant functions.

11. A hybrid keysystem requires dialing a 9 to access a central office trunk.

12. A hands-free phone will enable the user to hear the progress of a call but the person called will not be able to hear you talking. The handset must be used for two-way conversation. A speakerphone allows the called person to hear you without having to pick up the handset.

13. The old 1A2 keysystem required 25-pair cables for each extension because all central office trunks had to be wired to each phone. There was no switching matrix. 1A2 switching was done at the phone. The pushbuttons for each line connected the handset to the line for that button. EKTS use four-conductor station wire. Switching is done by a space division switching matrix in the KSU. One voice pair and one signaling pair are needed between the KSU and the phone. The signaling lead will tell the KSU which central office trunk it is to connect to the voice pair.

14. An EKTS uses an electronic crosspoint switching matrix.

15. System features are programmed using the attendant's telephone.

16. Station features are programmed by station users with their telephone.

17. The purpose of a hybrid circuit in the telephone is to convert the two-way, full-duplex, two-wire, local loop to the four-wire circuit of the telephone (transmitter and receiver).

18. Sidetone is feedback between the transmitter portion of a hybrid and the receive portion of the hybrid. This couples some of the transmitted signal to the receiver so users can hear themselves talking in the ear covered up by the receiver.

19. When a DTMF button is depressed, two tones are sent out over the local loop.

20. A varistor is used to limit the amount of current flowing in the transmitter and receiver.

CHAPTER SEVEN

1. Significant improvements in switching and multiplexing technologies have been the greatest telecommunications developments.

2. Almon B. Strowger invented the SXS switching system 1892.

3. XBAR is cheaper, requires less maintenance, and can select alternate routes if a primary route is busy.

4. The PSTN is the Public Switched Telephone Network. This is a circuit switched network that allows the connection of any telephone in the United States to any other telephone in the country.

5. The number of telephone exchanges had grown so much in some area codes that it was necessary to add new area codes in order to have more exchange codes.

6. In 1948, AT&T introduced the no. 5 XBAR.

7. SXS cannot work on DTMF tones, and a converter is required between the linefinder and first selector to convert DTMF into rotary-type dial pulses.

8. A dial register receives digits sent to the switch by a telephone.

9. A stored program control (SPC) switch is the name of this type of switching system.

10. On January 15, 1995, the first *NXX* area code was added.

11. A SXS system uses direct control. Each switching step is controlled directly by the dial of the telephone. XBAR and SPC switches use common control. Dialed digits do not directly control the switching process in a common control system.

12. The primary common control components are marker, register, and translator.

13. The marker is the brain of the XBAR switch.

14. The primary purpose of a translator is to translate the dialed telephone number into an equipment address location.

15. The DTMF pad sends two tones out over the Ring lead when a button is depressed.

CHAPTER EIGHT

1. Switching systems controlled by a computer are called Stored Program Control (SPC) switching systems.

2. Distributed processors are called controllers.

3. Distributed processing is used to take redundant mundane chores away from the main processor. This allows the system to operate faster and more efficiently.

4. The SPC uses a pulse code modulated–time division multiplexing network.

5. The MDF is the main distributing frame. All local-loop cable pairs and all interoffice cable pairs terminate on the MDF. All equipment in the central office also terminates on the MDF (either directly or via an IDF). This allows the connection of any equipment to any cable pair by running a jumper wire between the two.

6. The functions of a line circuit are to do the following: provide power to the telephone; provide a ringing signal to the telephone on incoming calls; detect when the telephone goes off-hook and when it hangs up; provide a test access for the testboard; offer overvoltage protection; provide a hybrid network to interface the two-wire cable pair to the four-wire TDM network;

provide a codec to interface the analog signal of the telephone to the digital signal path of the TDM network; and provide gate circuits to connect the codec to the TDM bus.

7. The time slot interchange (TSI) is the space division switching component of the digital switch. The TSI connects one time slot from a transmit TDM to one time slot of a receive TDM.

8. A TDM loop can have different time slots. The older loops had 30 voice time slots per TDM loop. The new networks have 512 time slots per TDM loop.

9. Two time slots are required for each call.

10. A SPC has RAM memory.

11. The RAM is divided into three types of memory storage: program store, data store, and call store.

12. The generic program is the program that contains all the call processing instructions for the switch and all the features for the switch.

13. Line translations are contained in the database. They identify which telephone number and what features are assigned to each line circuit.

14. The four types of switching networks that have been used by SPC are: crossbar, reed relay, PAM/TDM, and PCM/TDM.

15. AT&T calls its SPC switch for local service an ESS No. 5. Northern Telecom calls its large local digital switch a DMS100 and smaller switches a DMS10.

16. The MFJ ruling on equal access allowed the LECs to accelerate the replacement of older switches with new SPC switches.

17. An RLM cannot switch calls between lines connected to it, but an RLS can.

18. None! Nobody repairs them. Not even the manufacturer. It is cheaper to make new cards than to repair an existing card. End users do not have the knowledge or parts required to repair a printed circuitboard.

19. SPC offices require almost no maintenance because there are no moving electromechanical parts. This has allowed LECs to drastically reduce central office maintenance personnel and managers. They are easier to maintain. They are easier to engineer. They are easier to install, and the time frame to install an SPC switch is much less than the older electromechanical switch. An SPC can offer many features not available from an electromechanical switch. The features are easy to add.

20. The generic program contains all the call processing details and all the features of the switch.

CHAPTER NINE

1. A data communications system handles data signals that are binary digital electrical signals. A voice communications system handles voice signals that are analog electrical signals. Because today's voice communications systems converts the analog voice signal into a digital signal, the distinction between the two systems is diminishing.

2. Data communications uses a binary electrical signal.

3. The most common connection between a DTE and a DCE is an EIA-232 interface.

4. All telephone offices are designed to interface to analog signals over the local loop. In a digital central office, the line card contains a codec to convert that analog signal into a digital signal. A codec can not do anything with a digital signal on the local loop.

5. The two primary types of DCE are a modem and a CSU/DSU.

6. ASCII is the coding technique used for most data signals.

7. Serial transmission is sending bits one behind the other over a transmission medium.

8. Asynchronous transmission places a start bit and stop bit around each byte transmitted. Synchronous transmission transmits data as blocks of bytes using a header and trailer.

9. Synchronous transmission is used between two high-speed modems. The computers attached to the modem may use asynchronous transmission between the computer and the modem. A PC to a modem is asynchronous, but between the modems, synchronous transmission is used.

10. The simplest error-detection technique is the use of parity bits.

11. Two high-speed modems use forward error correction, usually LAP-M with continuous ARQ.

12. Discrete ARQ is a stop-and-wait protocol. After sending a block of data, the sending modem waits for an ACK or NAK. With discrete ARQ, the modem continues to send as long as it is getting ACKs. It does not wait for the ACK. It will continue sending blocks of data until a NAK or no ACKs are received.

13. LAP-M provides continuous ARQ.

14. The 16550 AFN UART has a speed of 115,000 bps.

15. Standard QAM achieves 16 different detectable events using 8 phases and 2 amplitudes.

16. One QAM signal represents 4 bits.

17. A baud is the rate that a signal changes. If a 1000-cycle signal changes back and forth between a 900- and a 1000-cycle signal 3000 times a second, it is a 3000-baud signal.

18. The baud rate of most modems is 2400 baud. Today's high-speed modems use 3200 baud.

19. A V.34 modem uses a baud rate of 3200. The signal changes 3200 times a second. Each signal represents 9 bits. Thus, $3200 \times 9 = 28,800$ bps.

20. The overall speed of 115,200 bps is achieved by using data compression algorithms in the modems. A 4 to −1 compression allows 115,200 bps to be compressed into 28,800 bps.

21. A PAD is a packet assembler / disassembler. The PAD packages data into packets for transmission over a packet switching network. It also receives packets of data and depacketizes the data.

22. X.25 *defines an interface* to a packet network.
23. SPC offices use the Signaling System 7 (SS7) Network to communicate with each other. Signals are sent in the form of word commands from one office to another.
24. Caller ID equipment uses a 1200-baud Western Electric 202 compatible modem.
25. The signaling used between the central office and the caller ID equipment at the customer's location is a 1200- and 2200-Hz signal in asynchronous 1200-baud mode.

CHAPTER TEN

1. Signaling is anything that serves to direct, command, monitor, or inform. In telecommunications, signals are needed to inform a local switching system that someone wishes to place a call, provide directions to the PSTN on how the call is to be connected, and inform the called party that a call is waiting. Signaling is divided into four basic categories: (1) supervisory, (2) addressing, (3) alerting, and (4) progress.
2. Supervisory signals are used to monitor the status of a line or trunk circuit. A switching system is informed of whether a line or trunk is idle or busy by supervisory signals.
3. When the phone is off-hook, the contacts on the hookswitch are closed. This completes an electrical path and causes electrical current to flow from the 52-V battery of the central exchange through a current-detection device on the line circuit out over the Ring lead to the phone, and returning over the Tip lead and line circuit to ground. The current-detection device supervises the line; when loop current is detected, the line controller is notified.
4. Telephones designed for residences are called single-line telephones, and they use loop signaling.
5. Step-by-step switches require DC loop signaling
6. Transmission requirements are more stringent than signaling requirements.
7. Reverse battery signaling was used to indicate a call had been answered.
8. Ground start signaling was developed to reduce "glare" conditions.
9. Analog trunk circuits mostly used E&M signaling but sometimes used loop signaling.
10. A DX set was used to derive E&M signal leads. The DX set interfaced E&M signaling devices to loop signaling devices.
11. *E* stands for receive and *M* stands for transmit (*E* for "ear" and *M* for "mouth").
12. Progress signals such as dial tone, 60-ipm busy tone, 120-ipm busy tone, ring-back tone, and recordings are used to indicate the progress of a call to the calling party.
13. With in-band signaling, signals are transmitted over the same transmission path used for voice. Out-of-band signaling uses a different path for signaling than is used by voice.

14. In-band signaling tones could be produced by anyone, and people could use these tones to bypass billing for calls.
15. SPC switching systems use message signaling over the SS7 network
16. CLASS defines a group of Custom Local Area Signaling Services such as caller ID, selective call rejection, and so on. offered by LECs from SPC switches.
17. CLASS depends on SS7; it cannot exist without SS7.
18. SS7 was developed to support ISDN
19. SS7 has three major components: the service switching point (SSP), the signal transfer point (STP), and the service control point (SCP). The SSP resides in a SPC office and is the interface between the CPU of the SPC switch and the CPU of the STP. Signaling data links connect the SSP to the STP. Signaling between the SPC switch and the STP is done over these data links. The link consists of a data channel in each direction to provide a full-duplex channel between the two systems. The standard data rate over the link is 64 Kbps. STPs are routing switches whose sole purpose is to connect a SSP to a SCP or to another SSP. SCPs contain a database of circuit, routing, and customer information. When the SPC switch needs to set up a call, the originating SSP at the originating SPC switch will issue a service request directly to a SCP, or the SSP may issue a request via a STP to a SCP. The SCP accepts the service request, searches its database for the appropriate information, and returns a reply to the SSP.
20. SS7 uses the first four layers of the OSI protocol.
21. The protocol stack defines how communication hardware and software interoperate at the various layers. *Protocol stacks are software programs* designed for each level to define how the communication takes place between layers.
22. The SCP contains a database.

CHAPTER ELEVEN

1. Data transmitted over the PSTN without using ISDN by using a modem or data service unit.
2. The first parts of the PSTN converted to digital were the carrier systems used to connect central offices together.
3. Cable modems are approximately 60 times faster than ISDN.
4. BRI is Basic Rate ISDN. BRI provides the user with two 64,000-bps DSL) channels for voice or data and a 16-Kbps signaling channel.
5. The physical cable pair used by BRI is a Digital Subscriber Line (DSL). The DSL circuit is one two-way digital circuit operating at 160 Kbps. The ISDN line circuit at the central exchange and the terminal device at the customer's location divide this 160-Kbps bit stream into four logical data channels (B1, B2, D, and M channels).
6. The signaling rate on a DSL is 160,000 bps. The customer has access to 144,000 bps.
7. The S/T interface uses AMI line coding

8. The U Interface uses 2B1Q coding.

9. The baud rate of 2B1Q is 80 Kbaud.

10. A quat is a quaternary symbol. There are four signals in 2B1Q coding. Each signal represents a dibit.

11. The 32,000 bits are part of the overhead and control sent over the S/T Interface to the NT1. The NT1 uses this information to identify which device is active. The S/T bus uses 48 Kbps for control, performance monitoring, and timing. The 2B1Q protocol only needs 16 Kbps for framing and control.

12. To identify which of the eight devices is using one of the B-channels, each TE and TA is assigned a terminal equipment identifier (TEI). The TEI is set on some devices by the use of jumpers, dip switches, or thumbwheels. Some devices have a display and will prompt you to use the keyboard to enter a TEI. Still other devices will have the TEI assigned automatically by the central exchange when the device is used for the first time.

13. SPID stands for service profile identifier. The SPID is in the database of the LEC central exchange and defines which services can be accessed by an ISDN line.

14. The SAPI identifies the service required (voice, switched data, packet data, and so on), and also identifies which B- or D-channel is being used to carry the customer's information. The SAPI is the logical connection between layer 2 and layer 3.

15. TE is ISDN compliant terminal equipment and a TA is an interface device used to connect a non-ISDN device such as a regular telephone to an NT**1.**

16. Instead of relying on physical signaling using voltage conditions, the D channel uses messages for signals. The ISDN D-channel is out-of-band signaling. This signaling takes place between CPU chips and is logical rather than physical. One chip circuit tells another chip what action to take by issuing commands in the message strings between the two chips. Programming is burnt into the chip telling it what action to take, based on the message received. This type of signaling, which uses messages to direct the actions of a CPU, takes place at layer 2 and layer 3. The ISDN data link layer (layer 2) protocol is called Link Access Procedure on the D-channel (LAPD). This protocol (software) defines the logical (not physical), connection between the terminal equipment (TE1 or TA) and the network (NT1 and LE). CCITT Recommendations Q.920 (I.440) and Q.921 (I.441) describe the general principles and operational procedures of LAPD. This procedure is also known as Digital Subscriber Signaling System No. 1 (DSS1) Data Link Layer and is a bit-oriented protocol.

17. A LAPD frame contains the following fields: (1) the FLAG field, which signals the beginning and end of a frame. The flag consists of the bit pattern 01111110; (2) the ADDRESS field contains the SAPI and TEI that is sending or is to receive the message; (3) the CONTROL field identifies the type of frame; (4) the INFORMATION field contains layer 2 and layer 3 messages; and (5) the FRAME CHECK SEQUENCE field uses CRC to check for errors.

18. The D-channel carries all signaling and control information.

19. The central office knows what type of service or connection to provide on an ISDN call by the TEI and SAPI.

20. BRI and PRI services are called narrowband ISDN to differentiate these services, provided over wire, from the broadband ISDN (B-ISDN) services provided over fiber optic cable.

21. ADSL was originally developed as a vehicle to carry video signals over twisted-pair copper wire. This technology would allow the LECs to provide cable TV services in direct competition against cable TV companies.

22. In the download direction ADSL is about 90 times faster than BRI.

23. The speed of ADSL is about the same as that of a cable modem.

24. ADSL technology allows the LEC to extend the life of its investment in the twisted copper wire pair local-loop plant.

25. DSL has the same speed in both directions. ADSL has a much faster download speed and a much slower upload speed.

CHAPTER TWELVE

1. Today's PBXs used stored program control with digital PCM time division switching matrixes. These computerized PBXs are called CBXs.

2. The trunk circuit is a central office (CO) trunk.

3. Port capacity is the number of different devices that can be attached to a CBX.

4. Architecture defines the way the components of a switch are physically and logically organized.

5. Configuration refers to the types of interfaces, number of interfaces, dialing patterns, features available, features assigned, and other database entries. Configuration is what uniquely defines a particular PBX installation.

6. Calling capacity is the number of calls that a switch can handle at the same time.

7. Expansion capacity refers to the ability to add more equipment to a switching system.

8. Hardware includes the physical components of the CBX. This could be the peripheral equipment (PE) shelf, the common equipment (CE) shelf, the power supply, the printed circuitboards (PCBs), or the telephones. When someone talks about hardware, they are usually talking about PCBs.

9. Every telephone connects to a switch via a line circuit.

10. The generic program contains all the features of a switch.

11. The line translations table identifies which features a particular line can have access to.

12. A serial data interface provides an interface that allows us to attach a terminal to the switching system. This interface allows us to communicate with the CPU for maintenance or administrative purposes.

13. RAM memory is subdivided into program store, data store, and call store.

14. A PBX without battery backup should contain a power-fail transfer card.

15. A nonblocking network contains as many outlets as inlets.

16. Some sources of blocking are insufficient paths in the switching fabric, insufficient trunk circuits, circuit outages, and software control.

17. No. The tone card generates digital codes, which are converted to analog signals by the codec on the line circuit.

18. Yes, a PBX supports the use of regular 2500 sets.

19. An MDF is the main distributing frame. In the PBX environment, all central office exchange cables, all station equipment cables, and all cables from the PBX terminate on the MDF. An IDF (which stands for intermediate distributing frame) provides an intermediate connection point for connecting horizontal station cables to vertical cables.

20. The wiring between an RJ-11 jack and the IDF is called horizontal wiring or station wiring.

21. The wiring between the IDF and MDF is called vertical cabling

22. 2500 sets access PBX features by dialing a code for the feature desired.

23. A proprietary set accesses features by pushing the button assigned for a feature. If a feature is not assigned to a button, it can be accessed by code dialing.

24. Central office trunks are accessed from a keysystem by depressing the button assigned to the CO line. Trunks are accessed from a PBX and hybrid keysystem by dialing the digit 9.

25. Centrex is PBX-type service provided by the LEC using a portion of the central exchange switching system.

26. Glare refers to the simultaneous seizure, from both ends, of a central office to PBX trunk circuit

27. WATS is wide area toll service. WATS trunks provide users with discounted toll.

28. Band 6 WATS allows calling anywhere in the United States but not within your home state.

29. Most companies want least-cost routing as a PBX feature.

30. An FX or a foreign exchange trunk is a trunk that connects to any class 5 central office via facilities in the PSTN to provide you with dial tone from a distant exchange.

31. VPN stands fot virtual private network; SDN is software-defined network. VPN and SDN use facilities in the PSTN to provide a logical private network to a customer.

32. The primary purpose of a PBX (and any other switching system) is to provide a means of connecting any incoming circuit to any outgoing circuit.

33. System features are features that make a PBX system operate more efficiently.

34. Dialed number identification system or DNIS trunk circuits are used to connect the PSTN to ACD and voice mail systems.

35. Station features are features that improve the efficiency of the end user.

CHAPTER THIRTEEN

1. An ACD has many more trunk circuits than line circuits.

2. ACD systems were first used to automate operator services centers for the IECs. ACDs are found in all major service and telemarketing organizations.

3. An ACD assigns the next call it receives to the agent that has been most idle.

4. The agent plugs in a headset and turns a key to notify the ACD that a position is staffed.

5. The server on the LAN and the PC at the agent's position can furnish an agent with volumes of information on a call and the caller as well as instructions on how to handle the call.

6. PRI ISDN uses the 24th channel to connect the server and ACD to the SS7 network. This arrangement provides caller ID, called number ID, and more information about each call.

7. Some methods used to provide caller identification are: have the caller enter information about themselves when the call is answered by an automated attendant with IVR capabilities, connect the ACD to the PSTN with DNIS trunks or connect using PRI ISDN.

8. Agent station equipment runs the gamut from a simple headset or telephone to a proprietary telephone or an integrated telephone and personal computer.

9. ACDs route calls to the most idle agent. UCDs route calls to the first available agent in the queue. ACDs provide more sophisticated management and supervisory reports.

10. The supervisory terminal can provide information on each agent and the total load being offered to the ACD systems. The supervisory terminal will indicate how many calls each agent has handled, how long the person has been logged on the system, and when the employee took breaks.

11. Wrap-up time is the amount of time an agent spends after each call filling out paperwork or records associated with a call.

12. The ACD will provide a report indicating how many agents are needed for each hour of every day.

13. The major drawback to a stand-alone ACD is the inability to transfer calls to a stand-alone PBX that serves other employees.

14. A predictive dialer is a software package in an ACD that predicts when an agent will be available to handle a call and places a call before the agent is idle.

15. Predictive dialers improve agent productivity because the agent does not spend time dialing numbers and does not lose time on unanswered or busy numbers.

16. Small telemarketing organizations do not need an ACD; they can use a call sequencer and keysystem. But large telemarketing call centers need an ACD to improve productivity for the center.

17. PC terminals are not connected to an ACD. They are connected via a LAN to a Server. The server connects to the ACD.

18. A person can avoid most telemarketing calls by not answering calls until after the third ring.

19. SS7 connects via the D-channel of PRI ISDN to an ACD and server at the ACD site. SS7 can be used to convey much information about a call. SS7 can provide caller ID and called number ID and can indicate where the call is originating from—a residence, business, hotel, pay station, and so on.

20. An ACD uses DNIS to identify what telephone number the customer dialed. It uses this number to route the call to the appropriate group of agents.

CHAPTER FOURTEEN

1. Messages are stored on a hard disk drive.

2. When digital messages are retrieved from the disk drive storage media, they will be sent through a codec. The codec converts the digital message to an analog message.

3. Voice mail can improve productivity by allowing people to leave messages for each other.

4. Escape from automated attendant provides the caller with an option to reach an attendant, usually by dialing 0.

5. You hear a greeting message from the party called.

6. You can access voice mail with a proprietary telephone by pushing the button on the telephone that is assigned to voice mail.

7. Voice mail systems have a central telephone number. You can dial the telephone number and provide an authorization code and account number to reach your voice mailbox.

8. Public voice mail uses DNIS trunks to identify which mailbox a call is for.

9. The number of ports and the disk storage space determine how many users can be on a voice mail system.

10. Sending telephone calls over the Internet involves nothing more than real-time voice mail. Instead of placing a message on a disk drive for later playback, the message is transmitted over the data network and played back in real time at the distant end.

11. Audiotext is a feature of voice mail that provides a replacement for the old analog recorded announcement machines. These digital announcers can be used in any application that requires the distribution of information to a large audience.

12. With IVR, you receive voice prompts from the system asking you to guide your call through the system using your DTMF pad to input information. This feature allows you to call your credit card company or bank and get information on your account. The IVR feature will prompt you to enter your account number and password. It will then prompt you to enter a DTMF digit for information desired.

13. The integration of audiotext and IVR is called automated attendant.

14. The major problem with using automated attendant is that it tends to alienate customers of the business when used improperly.
15. Voice recognition is being used to provide automated dialing, or other call handling instructions, based on voice commands.
16. Voice synthesis is technology used to convert text files into spoken words.
17. The functions performed by an operator have been automated for several types of calls by using automated attendant and voice synthesis technology.
18. Most automated attendants receive instructions from the digit pressed on a DTMF dial pad. Some automated attendants have voice recognition technology and can respond to spoken commands.
19. Automated attendants can serve as replacement for DID by asking callers to input the station number or name of the person they wish.
20. The length of greeting messages and how often mailboxes are cleaned out will determine disk storage requirements.

CHAPTER FIFTEEN

1. In telecommunications, traffic is telephone calls.
2. Every switching system has meters to measure the number and duration of calls that each piece of equipment handles.
3. The network design engineer must provide just enough circuits to meet the desired service level in order to meet objectives of both service and cost.
4. A network design engineer determine how many components are needed by gathering information from past traffic studies.
5. Dial office administration (DOA) is the department that conducts traffic studies and summarizes the data from a study.
6. The three types of meter readings found in a traffic study are peg count (PC); time duration; and all trunks busy (ATB).
7. ABBH is average bouncing busy hour. It is the average of several busy hours of data. To find ABBH PC, the peg counts for the busy hour, for each day of a traffic study, are added and then divided by the number of days in the study.
8. Average hold time (AHT) = average carried load divided by average carried peg count
9. Carried load = average hold time × average carried peg count.
10. DOA clerks summarized studies by adding hourly data together and then performing other mathematical calculations to arrive at averages or totals for a group of circuits. One thing computers do very well is math. Thus, mechanization of the DOA clerical function was a natural.
11. The network design engineer is more concerned about the total traffic for the total circuits in a group. Maintenance personnel are more concerned about individual circuit measurements. They can use individual circuit measurements to identify circuits in trouble. DOA also uses individual readings to make sure the traffic study data has not been corrupted by circuits in trouble.

12. Carried load is load that was carried and measured by the switch. Offered load is traffic that was offered to a switch. Most of the offered load will be carried, but some of this offered load may be blocked and not carried. Thus, offered load = carried load plus blocked load.

13. ATB is all trunks busy. Although the word *trunk* is used, it means all circuits busy.

14. In Poisson and Erlang B, blocked load = AHT × ABBH ATB.

15. The first step in circuit group design is to find out what the *offered load* is.

16. The three factors in a traffic table are demand; GOS, and number of servers.

17. CCS stands for 100 call seconds. Roman numeral C equals 100; 3600 s = 36 CCS; 3600 s (36 CCS) = 1 hr. One hour of traffic is an Erlang; 36 CCS = 1 Erlang.

18. *P*.05 and *B*.05 mean the same thing. They mean the *Probability* of *Blocking* is 5%.

19. Traffic engineering design is based on the premise that traffic is random. When traffic is not random due to a major catastrophe, traffic design tables and principles cannot be used.

20. The traffic table to use is determined by how blocked calls are handled.

21. The area causing the most trouble for an engineer is calculating offered load. If the traffic study data is in error due to equipment outages occurring during the study, the carried load and blocked load will be misstated.

22. SPC switching systems use RAM memory for meters. Each hour, data from these RAM meters is stored on a disk drive.

23. The traffic study reports generated by DOA are used by network design engineers and network maintenance. Prior to 1984, reports were also generated for the toll separations and settlements department.

24. Demand is offered load.

25. Measured load is carried load.

26. GOS is grade of service or the amount of blockage encountered for a group of circuits.

27. The three variables used in traffic engineering are demand, grade of service, and number of servers.

28. The capacity of a circuit group is its ability to handle an offered load.

29. Concentration is provided because we know that all 640 lines will not be in use at the same time (assuming random behavior).

30. A 30-channel multiplex loop can carry 30 Erlangs of traffic. There are 30 circuits. If each circuit is in use for an hour, the loop carries 30 hr of traffic; 30 hr = 30 Erlangs. This is also equal to 1080 CCS (36 CCS × 30 = 1080 CCS).

31. A load test is a test of the switching system that places the engineered ABBH call volume over the switch for 24 hr.

32. The GOS objective for SPC switches is *P*.001.

33. No! ABBH is based on carried load. The PC during this hour may or may not be the highest PC for the day.

34. Average bouncing busy hour (ABBH) is based on the hour of a day that has the highest carried load.

35. According to the Erlang B table, at $P.01$ GOS, a circuit group consisting of (a) 15-circuits achieves 54 % efficiency per circuit, (b) a 27-circuit group achieves 65% efficiency, (c) a 50-circuit group achieves 76% efficiency, and (d) a 100-circuit group achieves 84 % efficiency. According to the Poisson table, at a $P.01$ GOS, (a) a circuit group composed of 15 circuits achieves 50% efficiency, (b) a 27-circuit group achieves 60% efficiency, (c) a 50-circuit group achieves 70 % efficiency, and (d) a 100-circuit group achieves 78% efficiency.

36. The higher the number of circuits in a group, the more efficient each circuit will be. Thus, a 75-circuit group attains more efficient use of each circuit than a 50-circuit group.

37. The more efficient a circuit group is, the more it is impacted by a circuit outage. Thus, the 75-circuit group will be impacted more by a circuit outage.

CHAPTER SIXTEEN

1. In any traffic engineering exercise, we always make several assumptions. These assumptions are:
   a. The actions of users are independent of one another and are random.
   b. The number of users are infinite.
   c. The number of servers (circuits or trunks) is finite.
   d. The probability of a particular user being on the system is low.
   e. The number of failures is low.
   f. Statistically, the system is in equilibrium.
   g. We know the disposition of blocked calls.

2. Load balance requires the physical distribution of business and residential customers across all line bays to achieve an optimal mix resulting in the same traffic load being placed on all line bays.

3. When blocked calls are held in a queue (or delayed), we use an Erlang C table.

4. When a call attempt is blocked, the system will reroute the call to another circuit group, place the call in queue, or return an all-circuit-busy signal.

5. The basic premise behind a retrial table is that some users will immediately re-dial their call when they encounter an all-circuit-busy condition. There are three retrial tables: (1) 50% assumes 50% of callers retry, (2) 70% assumes 70% of callers retry, and (3) 100% assumes all callers will retry when they encounter an all-circuit-busy condition.

6. First attempt is offered load when using retrial methodology. First-attempt traffic is based on the percent of retries:

First-attempt load = carried load × {1 − (retry % × block %) / (1 − block %)}

7. With 100% retry, everyone eventually gets their call completed because they keep trying. Since everyone gets through, all traffic is carried. Therefore, at 100 % retry, first-attempt load = carried load.

8. For most telecommunications applications that involve queuing, Erlang C is widely used. Erlang C assumes:
   a. Infinite arrival population
   b. Random arrival rates
   c. Unlimited queue lengths
   d. FIFO
   e. Single/multiple channels
   f. Single phase
   g. Negative exponential probability distribution
9. Negative exponential probability distribution means the probability of long service times is very low and the probability of short service times is very high. For example, if the average holding time is 2 min, the probability that a call is more than 2 min is less than the probability that the call is shorter. In other words, that fact that the "average" is 2 min does *not* mean 50% are less and 50% are more.
10. If there is no retry involved and all blocked calls are cleared from the system, we must use either Poisson or Erlang B.
11. The Jacobsen retrial table is a 70% retrial table. Studies have provided documentation that 70% of all callers will retry a call after being blocked.
12. Erlang C makes the basic assumption of an *unlimited*-length queue—in other words, that attempts will wait *indefinitely* until a server is available. If that is true, there is no overflow. All blocked attempts will wait until served. Therefore, *carried traffic is equal to offered traffic.*
13. The important concept in a queuing environment is that *delay time is directly related to holding time*!
14. When using Erlang C, grade of service is based on *delay time*. How long does a person wait for service?
15. The major flaw in Erlang C is the assumption of an infinite queue—that is, that callers will wait indefinitely until a server is available. That may be true in theory, but in real life, people are often in control of how long to wait in a queue. It is not uncommon for calls (people) to drop out of a queue if the wait is too long.

CHAPTER SEVENTEEN

1. PCS is a concept. The concept of PCS is to assign someone a personal telephone number (PTN). This PTN is stored in a database on the SS7 network. That database keeps track of where a person can be reached. When a call is placed for that person, the artificial intelligence network (AIN) of the SS7 determines where the call should be directed.
2. PCS 1900 is provided by radio frequencies in the 1900-MHz range. PCS 1900 is the latest evolution in mobile communications; it can be provided by the time division multiple access (TDMA) technology of DAMPS or can use

Code Division Multiple Access (CDMA) technology. PCS is cellular radio technology; it uses digital technology (PCM at 16 Kbps). PCS utilizes lower-wattage transmitters than AMPS in the base and mobile stations. PCS also uses higher radio frequencies than AMPS. PCS uses radio frequencies between 1850 and 1910 MHz for mobile unit transmitters and between 1930 and 1990 MHz for base station transmitters. The channels are numbered from 512 (1850.2 MHz and 1930.2 Mhz) to 810 (1909.8 MHz and 1989.8 MHz). This provides 299 carrier slots (512 to 810 inclusive). Each channel is assigned a frequency for the base transmitter and a different frequency for the mobile unit to provide a full duplex communication circuit.

3. MTS was the first mobile telephone system installed (1946). These systems used radio frequencies between 35 and 45 MHz. Although MTS stands for "mobile telephone service," it also meant "manual telephone system." All calls had to be handled by an operator.

4. Advanced mobile phone systems (AMPS) is also known as cellular radio. Cellular radio uses many low-powered base transmitters (7 W) spread out over a geographic area (multiple base transmitter sites) and uses computer technology to decide which transmitter should be used. When a mobile phone is turned on, it communicates with several base site receivers. The computer determines which receiver is closest to the mobile unit on the strength of the received signal. The mobile unit transmits its identity (its telephone number) to the central computer. The central computer always knows the location of the mobile unit and which receiver/transmitter is closet to the mobile unit. When a call comes in for a mobile unit, the computer will use the transmitter closest to the mobile unit and send an alerting signal to the mobile unit.

5. Digital advanced mobile telephone system (DAMPS) is a technique that places multiple calls over one radio frequency using code excited linear predictive coding (7.95 Kbps) and time division multiplexing (TDM).

6. TDMA technology is used by DAMPS. With TDMA, each conversation occupies a time slot on a common transmission medium. In DAMPS, each conversation occupies a time slot on a particular radio frequency.

7. Code division multiple access or CDMA has been adopted as IS-95 standard for PCS networks. CDMA can provide ten times the capacity of DAMPS. CDMA is not backward compatible with analog AMPS in the same way that D-AMPS is. CDMA cannot reuse DAMP technology. To implement CDMA, the service provider must install a new network. CDMA is spread spectrum technology. Many conversations are multiplexed over one frequency. Each conversation is assigned a code. The receiver strips out conversations individually by using the code.

8. BSC is the base station controller. The BSC connects to and controls all base transceiver station (BTS) sites serving a BTA. The BSC tells the BTS which radio channel to use for communication with a particular mobile station.

9. The base transceiver site or BTS for a PCS 1900 system will consist of a small building to house the base station transmitters and receivers as well

as an antenna tower approximately 150 ft high. The antenna tower will contain 9 antennas arranged in a triangle array. Each side of the triangle will contain one transmitting antenna and two receiving antennas.

10. The mobile telephone serving office or MTSO is the central controller and switching system for cellular radio. The MTSO can be a stand-alone switch owned by a private cellular company or can be integrated into a local switch if the LEC is the cellular service provider.

11. The home location registry or HLR contains information about the services the PCS subscriber is allowed to access. The HLR also contains a unique authentication key and associated challenge/response generators. The AUC will verify that the customer is legitimate and allow the call to proceed.

12. The United States is divided into 51 areas. Each of these areas is called a major trading area (MTA). Each MTA is subdivided into basic trading areas (BTAs). There are 492 BTAs. The FCC sold two licenses per MTA and four per BTA for PCS.

13. A basic trading area or BTA is a subset of an MTA.

14. PCM in the PSTN is 64 Kbps and in DAMPS is 8 Kbps.

15. Companies had to bid at auctions held by the FCC.

Appendix B

The Library of Congress *Summary* of Senate Bill 652

S.652

PUBLIC LAW: 104-104 , (became law 02/08/96)

SPONSOR: Sen Pressler, (introduced 03/30/95)

TABLE OF CONTENTS:

Telecommunications Act of 1996 - Title I: Telecommunication Services - Subtitle A: Telecommunications Services - Amends the Communications Act of 1934 (the Act) to establish a general duty of telecommunications carriers (carriers): (1) to interconnect directly or indirectly with the facilities and equipment of other carriers; and (2) not to install network features, functions, or capabilities that do not comply with specified guidelines and standards.

Sets forth the obligations of local exchange carriers (LECs), including the duty: (1) not to prohibit resale of their services; (2) to provide number portability; (3) to provide dialing parity; (4) to afford access to poles, ducts, conduits, and rights-of-way consistent with pole attachment provisions of the Act; and (5) to reestablish reciprocal compensation arrangements for the transport and termination of telecommunications.

Imposes additional obligations on incumbent LECs (incumbent LEC requirements), including the duty to: (1) negotiate in good faith the terms and conditions of agreements; (2) provide interconnection at any technically feasible point of the same quality they provide to themselves, on just, reasonable, and nondiscriminatory terms and conditions; (3) provide access to network elements on an unbundled basis; (4) offer resale of their telecommunications services at wholesale rates; (5) provide reasonable public notice of changes to their networks; and (6) provide physical collocation, or virtual collocation if physical collocation is impractical.

Directs the Federal Communications Commission (FCC) to complete, within six months, all actions necessary to establish regulations to implement such requirements. States that nothing precludes the enforcement of State regulations that are consistent with those requirements.

Requires the FCC to create or designate one or more impartial entities to administer telecommunications numbering and to make such numbers available on an equitable basis. Directs that the cost of numbering administration and number portability be borne by all carriers on a competitively neutral basis.

Exempts a rural telephone company from incumbent LEC requirements until such company has received a bona fide request from interconnection, services, or network elements and the State commission determines that such request is not unduly economically burdensome, is technically feasible, and is consistent with universal service provisions, except the public interest determination. Sets forth provisions regarding: (1) State termination of the exemption and the establishment of an implementation schedule; and (2) limits on the exemption.

Authorizes an LEC with fewer than two percent of the subscriber lines installed in the aggregate nationwide to petition for a suspension or modification of specified requirements for the telephone exchange service facilities specified in the petition. Directs the State commission to grant such petition to the extent that it is necessary to avoid significant adverse economic impacts on users of telecommunications services or to avoid imposing an undue economic burden or a technically infeasible requirement, where such suspension or modification is in the public interest.

Provides for the continued enforcement of exchange access and interconnection requirements. Authorizes an incumbent LEC to voluntarily negotiate and enter into a binding agreement with a requesting carrier without meeting incumbent LEC requirements. Directs that such agreement: (1) include a detailed schedule of itemized charges for interconnection and each service or network element included in the agreement; and (2) be submitted to the State commission. Permits any party negotiating such an agreement to ask a State commission to participate in the negotiation and to mediate any differences arising in the course of the negotiation.

Authorizes the carrier or any other party to the negotiation, from the 135th through the 160th day after the date on which an incumbent LEC receives a request for negotiation, to petition a State commission to arbitrate any open issues. Sets forth provisions regarding the duty of the petitioner, opportunity to respond, action by the State commission, refusal to negotiate, standards for arbitration, and pricing standards. Requires any interconnection agreement adopted by negotiation or arbitration to be submitted for approval to the State commission. Sets forth provisions regarding grounds for rejection, preservation of authority by the State commission, the schedule for decision, failure of the State commission to act, and review of State commission actions.

Authorizes a Bell operating company (BOC) to prepare and file with a State commission a statement of the terms and conditions that such company generally offers within that State to comply with incumbent LEC requirements and applicable regulations and standards. Sets forth provisions regarding State commission review, the schedule for review, and authority to continue review. Specifies that submission or approval of the statement shall not relieve a BOC of its duty to negotiate the terms and conditions of an agreement regarding interconnection.

Sets forth provisions regarding: (1) consolidation of State proceedings; (2) a required filing by the State commission; and (3) availability of any interconnection, service, or network element provided under an approved agreement to which the LEC is a party to any other requesting carrier on the same terms and conditions as those provided in the agreement.

Preempts any State and local statutes, regulations, or requirements that prohibit or have the effect of prohibiting any entity from providing interstate or intrastate telecommunications services. Preserves a State's authority to impose, on a competitively neutral basis and consistent with universal service provisions, requirements necessary to preserve and advance universal service, protect the public safety and welfare, ensure the continued quality of telecommunications services, and safeguard the rights of consumers.

Authorizes a State, without violating the prohibition on barriers to entry, to require a competitor seeking to provide service in a rural market to meet the requirements for designation as an eligible carrier. Makes this provision inapplicable to: (1) a service area served by a rural telephone company that has obtained an exemption, suspension, or modification that effectively prevents a competitor from meeting such requirements; and (2) a provider of commercial mobile services.

Requires: (1) the FCC to institute and refer to a Federal-State Joint Board a proceeding to recommend changes to any of its regulations to implement specified requirements, including the definition of the services that are supported by Federal universal service support mechanisms and a specific timetable for completion of such recommendations; (2) one member of the Board to be a State- appointed utility consumer advocate nominated by a national organization of State utility consumer advocates; and (3) the Board, after notice and opportunity for public comment, to make its recommendations to the FCC within nine months.

Directs the Board and the FCC to base policies for the preservation and advancement of universal service on: (1) availability of quality services at just, reasonable, and affordable rates; (2) access to advanced telecommunications and information services to all regions of the nation; (3) access and costs in rural and high cost areas that are reasonably comparable to that provided in urban areas; (4) equitable and nondiscriminatory contribution by all telecommunications services providers; (5) specific and predictable support mechanisms; (6) access to advanced telecommunications services for schools, health care, and libraries; and (7) such other principles as the Board and the FCC determine are in the public interest.

Defines "universal service" as an evolving level of telecommunications services that the FCC shall establish periodically, taking into account advances in telecommunications and information technologies and services.

Requires all carriers providing interstate telecommunications services to contribute to the preservation and advancement of universal service. Authorizes the FCC to exempt a carrier or class of carriers if their contribution would be "de minimis."

Provides that only designated eligible carriers shall be eligible to receive specific Federal universal service support.

Grants States authority to adopt regulations not inconsistent with the FCC's rules. Requires all providers of intrastate telecommunications to contribute to universal service within a State in an equitable and nondiscriminatory manner, as determined by the State. Permits a State to adopt additional requirements with respect to universal service in that State as long as such requirements do not rely upon or burden Federal universal service support mechanisms.

Directs: (1) the FCC, within six months, to adopt rules to require that the rates charged by providers of interexchange telecommunications services to subscribers in rural and high cost areas shall be no higher than those charged by each such provider to its subscribers in urban areas; and (2) such rules to require that a provider of interstate interexchange telecommunications services provide such services to its subscribers in each State at rates no higher than those charged to its subscribers in any other State.

Requires a carrier, upon receiving a bona fide request, to provide telecommunications services: (1) which are necessary for the provision of health care services in a State, including instruction relating to such services, to any public or nonprofit health care provider that serves persons who reside in rural areas in that State at rates that are reasonably comparable to those charged for similar services in urban areas in that State; and (2) for educational purposes included in the definition of universal service for elementary and secondary schools and libraries at rates that are less than the amounts charged for similar services to other parties, as necessary to ensure affordable access to and use of such services. Permits a carrier providing such service to have an amount equal to the amount of the discount treated as an offset to its obligation to contribute to the mechanisms, or receive reimbursement utilizing the support mechanisms, to preserve and advance universal service.

Directs the FCC to establish competitively neutral rules to: (1) enhance access to advanced telecommunications and information services for all public and nonprofit elementary and secondary school classrooms, health care providers, and libraries; and (2) define the circumstances under which a carrier may be required to connect its network to such public institutional telecommunications users.

Specifies that: (1) telecommunications services and network capacity provided to health care providers, schools, and libraries may not be resold or transferred for monetary gain; and (2) for-profit businesses, elementary and secondary schools with endowments of more than $50 million, and libraries that are not eligible to participate in State-based plans for funds under the Library Services and Construction Act are ineligible to receive discounted rates.

Requires the FCC and the States to ensure that universal service is available at rates that are just, reasonable, and affordable.

Prohibits a carrier from using services that are not competitive to subsidize those that are subject to competition. Requires the FCC, with respect to interstate services, and the States, for intrastate services, to establish any necessary cost allocation rules, accounting safeguards, and guidelines to ensure that services included in the definition of universal service bear no more than a reasonable share of the joint and common costs of facilities used to provide those services.

Requires that: (1) if readily achievable, manufacturers of telecommunications and customer premises equipment ensure that equipment is designed, developed, and fabricated to be, and providers of telecommunications services ensure that service is, accessible and usable by individuals with disabilities; and (2) whenever such requirements are not readily achievable, such a manufacturer or provider shall ensure that the equipment or service is compatible with existing peripheral devices or specialized customer premises equipment commonly used by such individuals to achieve access, if readily achievable.

Directs the Architectural and Transportation Barriers Compliance Board to develop guidelines for accessibility of telecommunications and customer premises equipment in conjunction with the FCC and to review and update the guidelines periodically.

Requires the FCC to establish procedures for its oversight of coordinated network planning by carriers and other providers of telecommunications service for the effective and efficient interconnection of public telecommunications networks used to provide such service. Authorizes the FCC to participate in the development by industry standards-setting organizations of public telecommunications network interconnectivity standards that promote access to public telecommunications networks used to provide service, network capabilities and services by individuals with disabilities, and information services by subscribers of rural telephone companies.

Directs the FCC to: (1) complete a proceeding for the purpose of identifying and eliminating market entry barriers for entrepreneurs and other small businesses in the provision and ownership of telecommunications and information services, or in the provision of parts or services to providers of such services; (2) seek to promote the policies and purposes of the Act favoring diversity of media voices, vigorous economic competition, technological advancement, and promotion of the public interest; and (3) periodically review and report to the Congress on any regulations prescribed to eliminate such barriers and the statutory barriers that it recommends be eliminated, consistent with the public interest.

Prohibits a carrier from submitting or executing a change in a subscriber's selection of a provider of telephone exchange service or telephone toll service except in accordance with such verification procedures as the FCC shall prescribe. Makes any carrier that violates such procedures and collects charges for such a service from a subscriber liable to the carrier previously selected by the subscriber in an amount equal to all charges paid by such subscriber after such violation.

Directs the FCC to prescribe regulations that require incumbent LECs to share network facilities, technology, and information with qualifying carriers where the qualifying carrier requests such sharing for the purpose of providing telecommunications services or access to information services in areas where the carrier is designated as an essential carrier. Establishes the terms and conditions of such regulations. Requires LECs sharing infrastructure to provide information to sharing parties about deployment of services and equipment, including software.

Prohibits any LEC subject to interconnection requirements under this Act from: (1) subsidizing its telemessaging service directly or indirectly from its telephone exchange service or its exchange access; and (2) preferring or discriminating in favor of its telemessaging service operations in its provision of telecommunications services. Directs the FCC to establish procedures or regulations thereunder for the expedited receipt and review of complaints alleging violations that result in material financial harm to providers of telemessaging services.

(Sec. 102) Specifies that a common carrier designated as an "eligible telecommunications carrier" shall: (1) be eligible to receive universal service support; and (2) throughout the service area for which the designation is received, offer the services that are supported by Federal universal service support mechanisms either using its own facilities or a combination of its own facilities and resale of another carrier's services, and advertise the availability of such services and the charges therefor using media of general distribution.

Requires a State commission to designate such a carrier for the service area. Authorizes (in the case of an area served by a rural telephone company) or requires (in the case of all other areas) the State commission to designate more than one common carrier as an eligible carrier for a service area designated by the State commission, as long as each additional requesting carrier meets the requirements of this section and such designation is in the public interest.

Sets forth provisions regarding: (1) designation of eligible carriers for unserved areas; and (2) relinquishment of universal service (in areas served by more than one eligible carrier).

(Sec. 103) Amends the Public Utility Holding Company Act of 1935 (PUHCA) to allow registered holding companies to diversify into telecommunications, information, and related services and products where the Securities and Exchange Commission (SEC) determines that a registered holding company is providing telecommunications, information, and other related services through a single purpose subsidiary, designated an "exempt telecommunications company" (ETC). Requires prior State approval before any utility that is associated with a registered holding company may sell to an ETC any asset in the retail rates of that utility as of December 19, 1995.

Specifies that the ownership of ETCs by registered holding companies shall not be subject to prior approval or other restriction by the SEC, but the relationship between an ETC and a registered holding company shall remain subject to SEC jurisdiction, with exceptions.

Requires any registered holding company or subsidiary thereof that acquires or holds the securities, or an interest in the business, of an ETC to file with the SEC such information as the SEC may prescribe concerning: (1) investments and activities by the registered holding company, or any subsidiary thereof, with respect to ETCs; and (2) any activities of an ETC within the holding company system that are reasonably likely to have a material impact on the financial or operational condition of the holding company system.

Prohibits public utility companies from assuming the liabilities of an ETC and from pledging or mortgaging the assets of a utility for the benefit of an ETC.

Sets forth provisions regarding: (1) protection against abusive affiliate transactions; and (2) non-preemption of rate authority.

Prohibits reciprocal arrangements to avoid the provisions of this section among companies that are not affiliates or associate companies of each other.

Authorizes State commissions to: (1) examine the books and records of the ETC and any public utility company, associate company, or affiliate in the registered holding company system as they relate to the activities of the ETC; and (2) order an audit of a public utility company that is an associate of an ETC.

(Sec. 104) Amends the Act to specify that a purpose of the Act is to make available service to all the people of the United States without discrimination on the basis of race, color, religion, national origin, or sex.

Subtitle B: Special Provisions Concerning Bell Operating Companies - Requires a BOC to obtain FCC authorization prior to offering "interLATA" (i.e., long-distance; "LATA" means "local access and transport area") service within its region unless those services are previously authorized or incidental to the provision of another service, in which case interLATA service may be offered after the date of this Act's enactment. Permits a BOC to offer out-of-region services immediately after such date.

Sets forth requirements for a BOC's provision of interLATA services originating in an in-region State, including: (1) the presence of a facilities-based competitor or competitors (but the presence of a competitor offering exchange access, telephone exchange service offered exclusively through the resale of the BOC's telephone exchange service, and cellular service does not meet such requirement); or (2) the failure of a facilities-based competitor to request access or interconnection.

Establishes specific interconnection requirements, including a competitive checklist that a BOC must satisfy as part of its entry test (e.g., interconnection in accordance with specified requirements, nondiscriminatory access to 911 services, and reciprocal compensation arrangements).

Sets forth administrative provisions regarding applications for BOC entry. Authorizes the Attorney General to provide to the FCC an evaluation of an application using any standard the Attorney General deems appropriate. Sets forth provisions regarding FCC determinations, limits on FCC actions, publication of determinations, and enforcement of conditions required for approval. Directs the FCC to establish procedures for the review of complaints concerning failures by BOCs to meet such conditions.

Prohibits joint marketing of local services obtained from the BOC and long distance service within a State by carriers with more than five percent of the nation's presubscribed access line for three years after the date of enactment, or until a BOC is authorized to offer interLATA services within that State, whichever is earlier.

Requires any BOC authorized to offer interLATA services to provide intraLATA toll dialing parity coincident with its exercise of that interLATA authority. Bars States from ordering a BOC to implement toll dialing parity prior to its entry into interLATA service. Provides that any single-LATA State or any State that has issued an order by December 19, 1995, requiring a BOC to implement intraLATA toll dialing parity is grandfathered under this Act, with the prohibition against "non-grandfathered" States expiring three years after this Act's enactment date.

Sets forth "incidental" interLATA activities that the BOCs are permitted to provide upon the date of enactment.

Prohibits a BOC (including any affiliate) which is an LEC from providing specified services (including manufacturing activities, origination of interLATA telecommunications services other than incidental interLATA services, out-of-region services, or previously authorized activities and interLATA information services other than electronic publishing and alarm monitoring services) unless it does so through an entity that is separate from any entities that provide telephone exchange service.

Delineates structural and transactional requirements that apply to the separate subsidiary, including operating independently from the BOC, maintaining separate books and records, having separate officers, not obtaining credit under any arrangement that would permit a creditor upon default to have recourse to the BOC's assets, and conducting transactions with the BOC on an arm's length basis.

Sets forth provisions regarding: (1) non-discrimination safeguards; (2) biennial audit requirements; (3) sunset of provisions of this section; and (4) joint marketing.

Permits a BOC to: (1) engage in manufacturing after the FCC authorizes the company to provide interLATA services in any in-region State; (2) collaborate with a manufacturer of customer premises or telecommunications equipment during the design and development of hardware or software; and (3) engage in research activities relating to manufacturing and enter into royalty agreements with manufacturers of telecommunications equipment.

Requires each BOC to maintain and file with the FCC information on protocols and technical requirements for connection with and use of its telephone exchange service facilities.

Sets forth provisions regarding: (1) manufacturing limitations for standard-setting organizations; (2) alternate dispute resolution; (3) BOC equipment procurement and sales; and (4) FCC enforcement authority.

Prohibits a BOC or any affiliate from engaging in the provision of electronic publishing that is disseminated by means of such BOC's or any of its affiliates' basic telephone service, but allows a separated affiliate or electronic publishing joint venture (EPJV) operated in accordance with this section to engage in electronic publishing.

Requires a separated affiliate or EPJV to be operated independently from the BOC and to maintain separate books and records. Prohibits the affiliate from incurring debt in a manner that would permit a creditor upon default to have recourse to the BOC's assets. Sets forth provisions governing the manner in which transactions by the affiliate must be carried out (to ensure that they are fully auditable) and governing the valuation of assets transferred to the affiliate (to prevent cross subsidies). Prohibits the affiliate and the BOC from having corporate officers or property in common.

Prohibits the separate affiliate or EPJV from marketing the name, trademarks, or service marks of an existing BOC except for those that are owned by the entity that owns or controls the BOC.

Prohibits a BOC from engaging in joint marketing of any promotion, marketing, sales, or advertising with its affiliate, except that a BOC may: (1) provide inbound telemarketing or referral services related to the provision of electronic publishing if the BOC provides the same service on the same terms, conditions, and prices to non-affiliates as to its affiliates; (2) engage in non-discriminatory teaming or business arrangements; and (3) participate in EPJVs, provided that the BOC or affiliate has not more than a 50 percent (or, for small publishers, 80 percent) direct or indirect equity interest in the publishing joint venture.

Requires a BOC that enters the electronic publishing business through a separated affiliate or EPJV to provide network access and interconnection to electronic publishers at just and reasonable rates that are not higher on a per-unit basis than those charged to any other electronic publisher or any separated affiliate engaged in electronic publishing.

Entitles a person claiming a violation of this section to file a complaint with the FCC or to bring suit as provided in the Act.

Prohibits a BOC or affiliate thereof from engaging in the provision of alarm monitoring services before five years after the date of this Act's enactment, except for such services by a BOC that was engaged in providing such services as of November 30, 1995, directly or through an affiliate (but such BOC may not acquire an equity interest in or obtain financial control of any unaffiliated alarm monitoring services entities from November 30, 1995, until five years after the enactment date).

Provides that an incumbent LEC engaged in the provision of alarm monitoring services shall: (1) provide nonaffiliated entities, upon reasonable request, with the network services it provides to its own alarm monitoring operations on non-discriminatory terms and conditions; and (2) not subsidize its alarm monitoring services directly or indirectly from telephone exchange service operations. Requires the FCC to establish procedures for the receipt and review of complaints concerning violations of such provision or the regulations thereunder that result in material financial harm to a provider of alarm monitoring service. Bars an LEC from recording or using in any fashion the occurrence or contents of calls received by providers of alarm monitoring services for purposes of marketing such services on behalf of such LEC or any other entity.

Directs the FCC to adopt rules that eliminate discrimination between BOC and independent payphones and subsidies or cost recovery for BOC payphones from regulated interstate or intrastate exchange or exchange access revenue. Authorizes the FCC, if it determines that it is in the public interest, to allow the BOC's to have the same rights as independent payphone providers in negotiating with the interLATA carriers for their payphones. Grants the location provider the ultimate decision-making authority in determining interLATA services in connection with the choice of payphone providers.

Title II: Broadcast Services - Requires the FCC, if it determines that it will issue additional licenses for advanced TV services, to: (1) limit the initial eligibility for such licenses to persons who are licensed to operate a TV broadcast station, who hold a permit to construct such a station, or both; and (2) adopt regulations that allow such licensees or permittees to offer such ancillary or supplementary services on designated frequencies as may be consistent with the public interest, convenience, and necessity. Provides for the: (1) recovery for FCC reallocation or reassignment of the original or additional license of a person licensed to operate a TV broadcast station; and (2) charging and collection of fees from licensees by the FCC for the authorized use of designated frequencies. Requires a report from the FCC to the Congress on the implementation of this provision. Requires the FCC, within ten years after the first issuance of additional licenses, to conduct an evaluation of the advanced TV services program.

(Sec. 202) Directs the FCC to modify its multiple ownership rules to eliminate its limitation on the number of radio stations which may be owned or controlled nationally. Limits the number of radio stations an entity may own, operate, or control in a local market, with an exception when the FCC determines that such ownership, operation, or control will increase the number of radio broadcast stations in operation. Directs the FCC to: (1) eliminate its limitation on the number of TV stations which may be owned or controlled nationally; (2) increase to 35 percent the national audience reach limitations for TV stations; and (3) conduct a rulemaking proceeding to determine whether its rules restricting ownership of more than one TV station in a local market should be retained, modified, or eliminated.

Directs the FCC to extend its waiver policy with respect to its one-to-a-market ownership rules to any of the top 50 markets. Directs the FCC to permit a TV station to affiliate with an entity that maintains two or more networks unless such networks are composed of: (1) two or more of the four existing networks (ABC, CBS, NBC, FOX); or (2) any of the four existing networks and one of the two emerging networks (WBTN, UPN). Directs the FCC to: (1) permit an entity to own or control a network of broadcast stations and a cable system; and (2) revise ownership regulations if necessary to ensure carriage, channel positioning, and nondiscriminatory treatment of nonaffiliated broadcast stations by a cable system. Requires the FCC to revise all such rules biennially. Repeals current restrictions on broadcast- cable crossownership under the Communications Act.

(Sec. 203) Provides an eight-year license term for both TV and radio broadcast licenses.

(Sec. 204) Revises provisions regarding renewal procedures for the operation of TV broadcast stations. Includes standards for both renewal and denial of an application. Requires each renewal applicant to attach to such application a summary of comments and suggestions from the public regarding violent programming. Makes such amendment effective with respect to applications filed after May 1, 1995.

(Sec. 205) Extends to direct broadcast services current protections against signal piracy. Empowers the FCC with exclusive jurisdiction to regulate direct-to-home satellite services.

(Sec. 206) Provides that any ship documented under U.S. laws operating under the Global Maritime Distress and Safety System provisions of the Safety of Life at Sea Convention shall not be required to be equipped with a radio telegraphy station operated by one or more radio officers or operators.

(Sec. 207) Directs the FCC to promulgate regulations to prohibit restrictions that impair a viewer's ability to receive video programming services through devices designed for over-the-air reception of TV broadcast signals, multichannel multipoint distribution service, or direct broadcast satellite services.

Title III: Cable Services - Revises the definitions of "cable service" and "cable system" for purposes of the Act. Directs the FCC to: (1) review any complaint submitted by a franchising authority after the date of enactment of this Act concerning an increase in rates for cable programming services; and (2) issue a final order within 90 days, unless the parties agree to extend the review period. Terminates such review authority for cable programming services provided after March 31, 1999. Makes such provision inapplicable with respect to: (1) operators providing video programming services in areas subject to effective competition (as defined); or (2) any video programming offered on a per channel or per program basis. Exempts from certain cable rate regulation provisions small cable operators (serving fewer than one percent of all cable subscribers in the United States, serving no more than 50,000 subscribers, and not affiliated with any entity whose gross annual revenues exceed $250 million). Revises provisions with respect to cable TV market determinations, requiring an expedited decisionmaking process. Prohibits any State or franchising authority from restricting in any way a cable system's use of any type of subscriber equipment or transmission technology. Sets forth provisions with respect to: (1) cable equipment compatibility; and (2) subscriber notice (allowing any reasonable means at the cable operator's discretion). Repeals anti-trafficking restriction provisions of the Act. Directs the FCC to allow cable operators to aggregate equipment costs into broad categories, regardless of the function levels of such equipment within such categories. Provides for the treatment of prior-year losses of a cable system.

(Sec. 302) Subjects common carriers providing video programming to subscribers using radio communications to the requirements of title III and to the ownership and joint venture restrictions set forth in the following paragraph, but not to other requirements of title VI of the Act. States that such carriers providing such programming on a common carrier basis shall be subject to such requirements and restrictions, but not to other requirements of title VI. Allows such carrier to elect to provide such programming by means of an open video system, stating that such a provider need not make capacity available on a nondiscriminatory basis to any other person for the provision of cable service directly to subscribers.

Prohibits any LEC or affiliate from purchasing or otherwise acquiring more than a ten percent financial interest, or any management interest, in any LEC providing telephone exchange service within such cable operator's franchise area. Prohibits an LEC and a local cable operator from entering into a joint venture to provide video programming directly to subscribers or to provide telecommunications services within such market. Provides exceptions, including exceptions for joint ventures in rural areas, joint use of transmission facilities in limited circumstances, acquisitions made in competitive markets, exempt cable systems (cable systems serving less than 17,000 subscribers, with other restrictions), and small cable systems located in nonurban areas. Authorizes the FCC to waive such financial interest or joint venture restrictions in cases of undue economic distress, economic viability, anticompetitive effects of such restrictions, or when the local franchising authority approves such waiver.

Authorizes an LEC to provide cable service to its subscribers through an open video system that complies with this section. Outlines, with respect to the provision of such service through such system, provisions concerning: (1) certificates of compliance; (2) dispute resolution; (3) FCC regulations; (4) consumer access; (5) reduced regulatory burdens for such systems; and (6) FCC implementation of appropriate rules and regulations within six months after the enactment of this Act. States that an operator of an open video system may be subject to the payment of fees based on gross revenues in lieu of cable TV franchising fees.

(Sec. 303) Sets forth provisions regarding preemption of franchising authority regulation of telecommunications services. Prohibits a franchising authority from ordering a cable operator to discontinue the provision of a telecommunications service or a cable system to the extent it is used to provide a telecommunications service by reason of the failure of the cable operator to obtain a franchise or franchise renewal for the provision of such service. Prohibits a franchising authority from requiring a cable operator to provide any telecommunications service or facilities, other than institutional networks, as a condition of the initial grant of a franchise, franchise renewal or franchise transfer.

(Sec. 304) Directs the FCC to adopt regulations to ensure the commercial availability of convertor boxes, interactive equipment, and related equipment used to access multichannel video programming (MVP) from manufacturers, retailers, or other vendors not affiliated with any MVP distributor. Ensures the continued system security of MVP services. Provides FCC waiver authority with respect to provisions adopted under this section.

(Sec. 305) Directs the FCC, within 180 days after the enactment of this Act, to complete an inquiry to ascertain the level at which video programming is closed captioned. Provides closed captioning accountability criteria and requires a schedule of deadlines for the provision of such service. Provides exemptions from such requirements in cases of economic burden, inconsistency with current contracts, or undue burden of a significant difficulty or expense (with specified factors). Directs the FCC to: (1) commence an inquiry to examine the use of video descriptions on video programming in order to ensure the accessibility of such programming to persons with visual impairments; and (2) report to the Congress on its findings.

Title IV: Regulatory Reform - Directs the FCC to forbear from applying any regulation or provision of the Act to a telecommunications carrier or service if it determines that: (1) enforcement is not necessary to ensure that charges, practices, and classifications are just and reasonable and not discriminatory; (2) enforcement is not necessary for the protection of consumers; and (3) forbearance is consistent with the public interest. Directs the FCC to consider whether such forbearance will promote competitive market conditions. Allows any carrier to petition for such forbearance, requiring an FCC ruling within one year of such petition. Prohibits State enforcement of a regulation or provision after FCC-granted forbearance.

(Sec. 402) Directs the FCC, in every even-numbered year beginning with 1998, to: (1) review all regulations issued under the Act that apply to the operations or activities of a provider of telecommunications services; and (2) determine whether such regulation is no longer necessary in the public interest. Requires the FCC to repeal or modify any regulation so determined. Provides procedures for streamlining such repeals or modifications.

(Sec. 403) Eliminates or reduces specified FCC regulations, functions, and authority with respect to: (1) amateur radio examination procedures; (2) the designation of inspection entities; (3) instructional TV fixed service processing; (4) the setting of depreciation rates; (5) the use of independent auditors; (6) the delegation to private laboratories of equipment testing and certification; (7) the uniformity of license modifications; (8) jurisdiction over Government-owned ship radio stations; (9) the operation of domestic ship and aircraft radios without licenses; (10) fixed microwave service licensing; (11) foreign directors; (12) limitations on silent station authorizations; (13) construction permit requirements; (14) inspections of broadcast station equipment and apparatus; and (15) inspections by entities other than the FCC.

Title V: Obscenity and Violence - Subtitle A: Obscene, Harassing, and Wrongful Utilization of Telecommunication Facilities - Communications Decency Act of 1996 - Revises provisions of the Communications Act prohibiting obscene or harassing telephone calls and conversation to apply to obscene or harassing use of a telecommunications facility and communication. Increases the penalties for violations. Prohibits using a telecommunications device to: (1) make or initiate any communication which is obscene, lewd, lascivious, filthy, or indecent with intent to annoy, abuse, threaten, or harass another person; (2) make or make available obscene communication; (3) make or make available an indecent communication to minors.

Provides that no person shall be held to have violated such prohibition solely for providing access or connection to a telecommunications facility, system, or network not under such person's control. Provides employers with a defense for actions by employees unless the employee's conduct is within the scope of employment and is known, authorized, or ratified by the employer. Establishes as a defense to prohibited communications that a person has taken, in good faith, reasonable, effective, and appropriate actions to prevent access by minors or has restricted access by requiring use of a verified credit card, debit account, or adult access code or personal identification number.

(Sec. 504) Requires cable operators, upon request, to fully scramble or block programming to which the subscriber does not subscribe.

(Sec. 505) Requires a multichannel videoprogramming distributor: (1) to fully scramble or block sexually explicit adult programming so that nonsubscribers do not receive it; and (2) until it complies with such requirement, to not provide such programming during the hours of the day when a significant number of children are likely to view it.

(Sec. 506) Allows cable operators to refuse to transmit any public access or leased access program which contains obscenity, indecency, or nudity.

(Sec. 507) Amends the Federal criminal code to specify that current obscenity statutes prohibit using a computer to import or transport in interstate or foreign commerce, for sale or distribution, obscene material, including material designed, adapted, or intended for producing abortion or for any indecent or immoral use.

(Sec. 508) Prohibits using any facility or means of interstate or foreign commerce to persuade, induce, entice, or coerce a minor to engage in prostitution or any sexual act for which any person may be criminally prosecuted.

(Sec. 509) Provides that no provider or user of an interactive computer service shall be held liable for any voluntary action taken to restrict access to, or to enable information content providers to restrict access to, material that the user or provider considers to be objectionable, whether or not such material is constitutionally protected.

Subtitle B: Violence - Directs the FCC, if it determines that video programming distributors have not, within one year, voluntarily established rules for rating programming that contains sexual, violent, or other indecent material about which parents should be informed before it is displayed to children and voluntarily agreed to broadcast signals that contain such ratings, to: (1) establish an advisory committee to recommend guidelines and procedures for rating such programming; (2) prescribe such guidelines and procedures; and (3) prescribe rules requiring programming distributors to transmit such rating to permit parents to block inappropriate programming. Directs the FCC, not less than two years after enactment of this Act, to require apparatus designed to receive TV signals that are shipped in interstate commerce or manufactured in the United States and that have a picture screen of 13 inches or greater (measured diagonally) to be equipped with a feature designed to enable viewers to block display of all programs with a common rating. Authorizes the FCC to allow apparatus manufacturers to comply with such requirement using alternative technology that meets certain standards of cost, effectiveness, and ease of use.

(Sec. 552) Encourages broadcast television, cable, satellite, syndication, and other video programming distributors to establish a technology fund to encourage electronics equipment manufacturers to facilitate the development of technology which would empower parents to block programming deemed inappropriate for children and to encourage availability of such technology to low income parents.

Subtitle C: Judicial Review - Provides for the expedited review of any civil action challenging the constitutionality of this title by a district court of three judges and by direct appeal to the Supreme Court.

Title VI: Effect on Other Laws - Provides that any conduct or activity that was, before the enactment of this Act, subject to any restriction or obligation imposed by the AT&T Consent Decree, the GTE Consent Decree, or the McCaw Consent Decree shall, after enactment of this Act, be subject to the restrictions and obligations imposed by the Communications Act as amended by this Act.

Provides that nothing in this Act shall be construed to modify, impair, or supersede: (1) the applicability of the antitrust laws; or (2) any State or local law pertaining to taxation, except with respect to fees for open video systems. Repeals a provision of the Communications Act permitting the FCC to render a proposed merger of competing local telephone companies exempt from any Act of Congress making the transaction unlawful.

(Sec. 602) Exempts any provider of direct-to-home satellite service from the collection or remittance of any local tax or fee on such service.

Title VII: Miscellaneous Provisions - Prohibits a party calling a toll-free telephone number from being assessed a charge by virtue of being asked to connect or otherwise transfer to a pay-per-call service. Prohibits the calling party from being charged for information conveyed during a call to a toll-free (800) number unless the calling party: (1) has a written agreement specifying the material terms and conditions under which the information is offered and which includes the rate at which charges are assessed and certain identifying information; or (2) is charged for the information only after the information provider includes an introductory disclosure message regarding the charge, rate, and means of billing for the call and the calling party is charged by means of a credit, prepaid, debit, charge, or calling card. Outlines provisions concerning: (1) billing arrangements; (2) required use of a personal identification number by the subscriber to obtain access to the information provided; (3) exceptions to the written agreement requirement; and (4) termination of service if a telecommunications carrier reasonably determines that a complaint against an information provider is valid.

Amends the Telephone Disclosure and Dispute Resolution Act to authorize the FCC to extend the definition of "pay-per-call services" under such Act to other services that the FCC determines are susceptible to the unfair and deceptive billing practices addressed by such Act.

(Sec. 702) Makes it the duty of every telecommunications carrier to protect the confidentiality of proprietary information of other carriers, equipment manufacturers, and customers. Permits a carrier that receives proprietary information from another carrier or a customer for purposes of providing any telecommunications service to use such information only for such purpose. Directs a carrier to disclose customer proprietary network information upon the customer's request. Permits a carrier to use, disclose, or permit access to aggregate customer information for other purposes. Requires a carrier that provides telephone exchange service to provide subscriber list information to any person upon request for the purpose of publishing directories in any format.

(Sec. 703) Directs the FCC to prescribe regulations to: (1) govern the charges for pole attachments used by telecommunications carriers to provide telecommunications services, when the parties fail to resolve a dispute over such charges; and (2) ensure that utilities charge just, reasonable, and nondiscriminatory rates for the pole attachments. Requires a utility to apportion the cost of providing space on a pole based on the number of attaching entities. Requires any increase in the rates for pole attachments to be phased in over a five-year period. Requires a utility to provide a cable television system or any telecommunications carrier with nondiscriminatory access to any pole or right-of-way owned by it. Allows a utility company providing electric service to deny a cable television system or telecommunications carrier access to such poles when there is insufficient capacity and for reasons of safety, reliability, and generally applicable engineering purposes. Requires utilities that engage in the provision of telecommunications services or cable services to impute to its costs of providing such service an equal amount to the pole attachment rate for which such company would be liable. Requires utilities to provide written notification to attaching entities of any plans to modify or alter its poles or other rights-of-way. Requires any attaching entity that modifies its own attachments to bear a proportionate share of the costs of such modifications. Prevents a utility from imposing the cost of rearrangements to other attaching entities if done solely for the benefit of the utility.

(Sec. 704) Preserves State or local authority over decisions regarding the placement, construction, and modification of personal wireless service facilities, but prohibits

State or local regulation thereof from: (1) unreasonably discriminating among providers of functionally equivalent services; or (2) prohibiting the provision of personal wireless services. Requires State or local action on requests regarding such facilities to occur within a reasonable time, with denials of requests to be in writing and supported by substantial evidence in a written record. Prohibits State or local regulation of such facilities on the basis of environmental effects of radio frequency emissions to the extent such facilities comply with FCC regulations. Provides for expedited judicial review and petitions of the FCC for relief from adverse State or local actions.

Directs the President to prescribe procedures by which Federal agencies may make available property and rights-of-way for the placement of new telecommunications services that are dependent upon the utilization of Federal spectrum rights.

(Sec. 705) Prohibits a commercial mobile services provider from being required to provide equal access to common carriers for the provision of telephone toll services. Directs the FCC, if it determines that subscribers to such services are denied access to the provider of telephone toll services of the subscribers' choice, contrary to the public interest, to prescribe regulations to afford subscribers unblocked access to the provider of telephone toll services of the subscribers' choice through the use of a carrier identification code assigned to such provider or other mechanism. Provides that such regulations shall not apply to mobile satellite services unless the FCC finds it to be in the public interest.

(Sec. 706) Requires the FCC and each State telecommunications commission to encourage the deployment of advanced telecommunications capability to all Americans by utilizing price cap regulation, regulatory forbearance, measures that promote competition, or other regulating methods that remove barriers to infrastructure investment. Requires the FCC to regularly initiate a notice of inquiry concerning such availability and, if it determines it to be necessary, to take action to accelerate deployment of such capability by removing barriers to infrastructure investment and by promoting competition in the telecommunications market.

(Sec. 707) Establishes the Telecommunications Development Fund as a corporate body in the District of Columbia to promote access to capital for small businesses in order to enhance competition in the telecommunications industry, to stimulate new technology development, to promote employment and training, and to support universal service. Directs the Fund to: (1) make loans, investments, or other extensions of credit and provide financial advice to eligible small businesses; and (2) prepare research, studies, or financial analyses.

(Sec. 708) Recognizes the National Education Technology Funding Corporation as a nonprofit corporation independent of the Federal Government and operating under the laws of the District of Columbia. Authorizes the Corporation to receive discretionary grants, contracts, gifts, contributions, or technical assistance from any Federal department or agency.

Requires audits of the Corporation by independent certified public accountants. Provides reporting and recordkeeping requirements. Requires the accessibility of

Corporation books for audit and examination. Directs the Corporation to report annually to the President and the Congress on operations and activities of the previous fiscal year. Requires Corporation members to be available to testify before the Congress concerning such operations and activities.

(Sec. 709) Directs the Assistant Secretary of Commerce for Communications and Information to report annually to specified congressional committees concerning the activities of the Joint Working Group on Telemedicine, together with any findings in the studies and demonstrations on telemedicine funded by the Public Health Service or other Federal agencies. Specifies that such reports shall examine questions related to patient safety, the efficacy and quality of the services provided, and other legal, medical, and economic issues related to the utilization of advanced telecommunications services for medical purposes.

(Sec. 710) Authorizes appropriations.

Appendix C

Poisson Table

GOS	B.01		B.05		B.10	
Trunks	ERL	CCS	ERL	CCS	ERL	CCS
1	.01	.36	.05	1.8	.11	3.8
2	.15	5.4	.36	12.9	.54	19.4
3	.44	15.7	.82	29.4	1.1	39.6
4	.82	29.6	1.37	49.1	1.75	63
5	1.28	46.1	1.97	70.9	2.44	88
6	1.79	64.4	2.61	94.1	3.14	113
7	2.33	83.9	3.28	118	3.89	140
8	2.91	105	3.97	143	4.67	168
9	3.50	126	4.69	169	5.42	195
10	4.14	149	5.42	195	6.22	224
11	4.78	172	6.17	222	7.03	253
12	5.42	195	6.92	249	7.83	282
13	6.11	220	7.69	277	8.64	311
14	6.78	244	8.47	305	9.47	341
15	7.47	269	9.25	333	10.28	370
16	8.18	294	10.06	362	11.14	401
17	8.89	320	10.83	390	11.97	431
18	9.61	346	11.64	419	12.83	462
19	10.36	373	12.44	448	13.67	492
20	11.08	399	13.25	477	14.53	523
21	11.83	426	14.08	507	15.39	554
22	12.58	453	14.89	536	16.25	585
23	13.33	480	15.72	566	17.11	616
24	14.08	507	16.56	596	17.97	647
25	14.86	535	17.39	626	18.83	678
26	15.61	562	18.22	656	19.72	710
27	16.39	590	19.06	686	20.58	741
28	17.17	618	19.92	717	21.47	773
29	17.97	647	20.75	747	22.36	805
30	18.75	675	21.61	778	23.22	836
31	19.53	703	22.47	809	24.11	868
32	20.33	732	23.33	840	25.00	900
33	21.11	760	24.19	871	25.89	932
34	21.92	789	25.06	902	26.78	964
35	22.72	818	25.92	933	27.67	996
40	26.78	964	30.22	1088	32.14	1157
50	35.03	1261	38.97	1403	41.17	1482
60	43.47	1565	47.86	1723	50.31	1811
70	52.03	1873	56.83	2046	59.50	2142
75	56.33	2028	61.33	2208	64.14	2309
80	60.67	2184	65.89	2372	68.81	2477
90	69.42	2499	75.00	2700	78.11	2812
100	78.22	2816	84.17	3029	87.47	3149

Appendix D

Erlang B Table

GOS	B.01		B.05		B.10	
Trunks	ERL	CCS	ERL	CCS	ERL	CCS
1	.01	.36	.05	1.8	.11	3.96
2	.15	5.40	.38	13.7	.60	21.6
3	.46	16.60	.90	32.4	1.27	45.7
4	.87	31.30	1.52	54.7	2.05	73.8
5	1.36	49	2.22	79.9	2.88	104
6	1.91	69	2.96	107	3.76	135
7	2.50	90	3.74	135	4.67	168
8	3.13	113	4.54	163	5.60	202
9	3.78	136	5.37	193	6.55	236
10	4.46	161	6.22	224	7.51	270
11	5.16	186	7.08	255	8.49	306
12	5.88	212	7.95	286	9.47	341
13	6.61	238	8.83	318	10.47	377
14	7.35	265	9.73	350	11.47	413
15	8.11	292	10.63	383	12.48	449
16	8.87	319	11.54	415	13.50	486
17	9.65	347	12.46	449	14.52	523
18	10.44	376	13.38	482	15.55	560
19	11.23	404	14.31	515	16.58	597
20	12.03	433	15.25	549	17.61	634
21	12.84	462	16.19	583	18.65	671
22	13.65	491	17.13	617	19.69	709
23	14.47	521	18.08	651	20.74	748
24	15.29	550	19.03	685	21.78	784
25	16.12	580	19.99	720	22.83	822
26	16.96	611	20.94	754	23.88	860
27	17.80	641	21.90	788	24.94	898
28	18.64	671	22.87	823	26.00	936
29	19.49	702	23.83	858	27.05	974
30	20.34	732	24.80	893	28.11	1012
31	21.19	763	25.77	928	29.17	1050
32	22.05	794	26.75	963	30.23	1088
33	22.91	825	27.72	998	31.30	1127
34	23.77	856	28.70	1033	32.26	1165
35	24.64	887	29.68	1068	33.43	1203
40	29	1044	34.6	1246	38.8	1396
50	38	1364	44.5	1603	49.5	1784
60	47	1688	54.6	1966	60.4	2174
70	56	2020	64.7	2329	71.3	2567
75	61	2185	69.7	2509	76.7	2761
80	65	2354	74.8	2693	82.2	2959
90	75	2689	85.0	3060	93.1	3352
100	84	3028	95.2	3427	107	3748

Appendix E

50% TABLE

first attm	hour con-nect	hour recal retry	hour ovfl	con nect vmr	ovfl vmr	% block GOS	lines or servr	Hours Connected Traffic on Servers						
								1	2	3	4	5	6	7
.200	.180	.020	.020	.82	1.08	18.0%	1	.180						
.200	.198	.002	.002	.97	1.07	1.7%	2	.168	.030					
.400	.323	.077	.077	.68	1.13	32.3%	1	.323						
.400	.388	.012	.012	.90	1.13	5.7%	2	.292	.097					
.400	.399	.001	.001	.98	1.11	.7%	3	.286	.093	.019				
.600	.565	.035	.036	.62	1.19	11.0%	2	.388	.177					
.600	.594	.006	.006	.95	1.17	2.0%	3	.377	.167	.050				
.800	.726	.074	.074	.74	1.24	16.9%	2	.466	.260					
.800	.783	.017	.017	.91	1.22	4.0%	3	.450	.240	.094				
.800	.797	.003	.003	.97	1.18	.8%	4	.445	.236	.090	.025			
1.00	.870	.130	.130	.66	1.27	23.1%	2	.531	.339					
1.00	.965	.035	.035	.85	1.27	6.7%	3	.509	.311	.146				
1.00	.992	.008	.008	.95	1.24	1.6%	4	.502	.302	.139	.048			
1.20	.996	.204	.204	.59	1.29	29.1%	2	.584	.412					
1.20	1.14	.062	.062	.80	1.32	9.9%	3	.558	.376	.204				
1.20	1.18	.017	.017	.92	1.29	2.7%	4	.549	.363	.192	.079			
1.20	1.2	.004	.004	.98	1.24	.6%	5	.546	.360	.189	.077	.024		
1.40	1.11	.294	.294	.52	1.31	34.8%	2	.629	.477					
1.40	1.30	.101	.101	.74	1.35	13.4%	3	.600	.435	.264				
1.40	1.37	.030	.030	.88	1.33	4.2%	4	.589	.418	.247	.116			
1.40	1.40	.008	.008	.96	1.29	1.1%	5	.585	.413	.241	.112	.041		
1.60	1.45	.151	.151	.68	1.38	17.3%	3	.637	.488	.324				
1.60	1.55	.050	.050	.84	1.38	6.1%	4	.623	.467	.300	.159			
1.60	1.59	.015	.015	.94	1.34	1.8%	5	.618	.460	.292	.152	.064		
1.80	1.60	.214	.214	.62	1.41	21.3%	3	.668	.536	.382				
1.80	1.72	.078	.076	.79	1.42	8.3%	4	.653	.512	.353	.205			
1.80	1.77	.025	.025	.91	1.38	2.8%	5	.646	.502	.341	.195	.091		
1.80	1.80	.007	.007	.97	1.33	.9%	6	.644	.499	.337	.191	.089	.034	
2.00	1.71	.289	.289	.56	1.42	25.3%	3	.696	.578	.436				
2.00	1.89	.113	.113	.75	1.45	10.7%	4	.679	.552	.402	.254			
2.00	1.96	.040	.040	.88	1.43	3.9%	5	.671	.540	.387	.239	.123		
2.00	1.99	.012	.012	.95	1.38	1.2%	6	.668	.535	.382	.233	.119	.050	
2.20	1.82	.377	.377	.51	1.43	29.3%	3	.720	.616	.487				
2.20	2.04	.158	.158	.70	1.48	13.4%	4	.702	.588	.449	.303			
2.20	2.14	.060	.060	.84	1.47	5.3%	5	.693	.574	.431	.283	.159		
2.20	2.18	.020	.020	.93	1.42	1.8%	6	.689	.568	.423	.276	.153	.071	
2.20	2.19	.006	.006	.97	1.37	.6%	7	.688	.566	.421	.273	.150	.069	.027
2.40	2.19	.212	.212	.65	1.50	16.3%	4	.723	.620	.492	.352			
2.40	2.31	.065	.085	.80	1.50	6.9%	5	.713	.605	.471	.328	.198		
2.40	2.37	.031	.031	.90	1.46	2.6%	6	.709	.598	.462	.317	.188	.095	
2.40	2.39	.010	.010	.96	1.41	.8%	7	.707	.595	.458	.313	.185	.093	.039
2.60	2.32	.277	.277	.60	1.52	19.2%	4	.742	.649	.533	.399			

(continued)

50% TABLE   *(continued)*

first attm	hour con- nect	hour recal retry	hour ovfl	con nect vmr	ovfl vmr	% block GOS	lines or servr	Hours Connected Traffic on Servers						
								1	2	3	4	5	6	7
2.60	2.48	.118	.118	.76	1.54	8.7%	5	.731	.632	.509	.371	.239		
2.60	2.55	.046	.046	.88	1.50	3.4%	6	.726	.624	.498	.358	.226	.123	
2.60	2.58	.016	.016	.95	1.45	1.2%	7	.723	.621	.493	.353	.221	.119	.055
2.80	2.50	.351	.351	.56	1.53	22.3%	4	.759	.676	.570	.445			
2.80	2.64	.157	.151	.72	1.57	10.6%	5	.747	.658	.544	.413	.281		
2.80	2.73	.064	.064	.85	1.54	4.5%	6	.741	.648	.531	.397	.265	.154	
2.80	2.77	.024	.024	.93	1.49	1.7%	7	.739	.644	.525	.390	.257	.148	.073
2.80	2.79	.008	.008	.97	1.44	.6%	8	.737	.642	.523	.387	.255	.146	.072
3.00	2.52	.483	.483	.51	1.64	27.8%	4	.755	.683	.592	.486			
3.00	2.79	.205	.205	.68	1.59	12.8%	5	.762	.681	.576	.453	.323		
3.00	2.91	.088	.088	.81	1.58	5.7%	6	.755	.670	.561	.435	.303	.187	
3.00	2.96	.035	.035	.91	1.53	2.3%	7	.752	.665	.554	.426	.294	.179	.095
3.00	2.99	.013	.013	.96	1.48	.8%	8	.751	.663	.551	.422	.290	.176	.092
3.20	2.62	.579	.579	.48	1.64	30.6%	4	.771	.705	.622	.523			
3.20	2.94	.260	.260	.63	1.61	15.1%	5	.776	.702	.606	.491	.365		
3.20	3.08	.117	.117	.78	1.61	7.1%	6	.768	.690	.590	.470	.342	.222	
3.20	3.15	.049	.049	.88	1.57	3.0%	7	.765	.684	.582	.460	.331	.212	.119
3.20	3.18	.018	.018	.94	1.52	1.1%	8	.763	.682	.578	.455	.325	.207	.115
3.40	2.78	.682	.682	.44	1.64	33.4%	4	.785	.725	.650	.558			
3.40	3.07	.325	.325	.59	1.62	17.4%	5	.788	.721	.633	.527	.406		
3.40	3.25	.153	.153	.74	1.64	8.6%	6	.780	.709	.616	.504	.380	.259	
3.40	3.33	.066	.066	.85	1.61	3.8%	7	.776	.702	.607	.492	.366	.245	.146
3.40	3.37	.026	.026	.93	1.56	1.5%	8	.774	.699	.602	.486	.360	.239	.141
3.40	3.39	.010	.010	.97	1.50	.6%	9	.773	.698	.600	.484	.357	.237	.139
3.60	3.16	.444	.444	.55	1.73	22.0%	5	.783	.723	.646	.554	.449		
3.60	3.41	.194	.194	.70	1.67	10.2%	6	.791	.725	.640	.535	.417	.296	
3.60	3.51	.088	.088	.83	1.65	4.8%	7	.787	.718	.630	.522	.401	.280	.175
3.60	3.56	.037	.037	.91	1.60	2.0%	8	.784	.715	.625	.515	.393	.272	.168
3.60	3.57	.014	.014	.96	1.54	.8%	9	.783	.713	.622	.512	.389	.268	.165
3.80	3.27	.527	.527	.52	1.73	24.4%	5	.795	.739	.669	.584	.485		
3.80	3.56	.243	.243	.67	1.69	12.0%	6	.802	.741	.662	.565	.452	.334	
3.80	3.69	.114	.114	.79	1.68	5.8%	7	.797	.733	.651	.550	.434	.315	.206
3.80	3.75	.050	.050	.89	1.63	2.6%	8	.794	.729	.645	.542	.425	.305	.197
3.80	3.78	.020	.020	.94	1.57	1.1%	9	.793	.727	.642	.539	.421	.300	.193
4.00	3.38	.618	.618	.48	1.73	26.8%	5	.806	.755	.691	.612	.519		
4.00	3.70	.299	.299	.63	1.70	13.9%	6	.811	.756	.683	.593	.487	.372	
4.00	3.85	.146	.146	.76	1.71	7.0%	7	.806	.747	.671	.577	.466	.349	.238
4.00	3.93	.066	.066	.86	1.67	3.3%	8	.803	.742	.664	.568	.456	.337	.227
4.00	3.97	.028	.028	.93	1.61	1.4%	9	.801	.740	.661	.563	.450	.332	.222
4.00	3.99	.011	.011	.97	1.55	.5%	10	.800	.739	.660	.561	.448	.329	.219
4.20	3.48	.717	.717	.45	1.73	29.2%	5	.816	.769	.710	.637	.551		
4.20	3.84	.362	.362	.59	1.72	15.9%	6	.820	.769	.702	.619	.519	.408	
4.20	4.02	.183	.183	.73	1.73	8.3%	7	.814	.760	.689	.601	.497	.383	.272
4.20	4.11	.086	.086	.84	1.71	4.0%	8	.811	.755	.682	.591	.485	.370	.258
4.20	4.16	.038	.038	.91	1.65	1.8%	9	.809	.752	.678	.586	.479	.363	.251
4.20	4.19	.015	.015	.96	1.59	.7%	10	.808	.751	.677	.584	.476	.359	.248
4.40	3.58	.823	.823	.42	1.73	31.5%	5	.825	.782	.728	.661	.580		

first attm	hour con-nect	hour recal retry	hour ovfl	con nect vmr	ovfl vmr	% block GOS	lines or servr	1	2	3	4	5	6	7
										Connected Traffic on Servers		Hours		
4.40	3.92	.480	.480	.55	1.81	19.7%	6	.815	.767	.708	.634	.547	.449	
4.40	4.17	.226	.226	.69	1.76	9.8%	7	.822	.772	.707	.625	.526	.417	.306
4.40	4.29	.110	.110	.81	1.74	4.9%	8	.819	.766	.699	.614	.513	.401	.289
4.40	4.35	.050	.050	.89	1.69	2.2%	9	.817	.763	.694	.608	.505	.393	.281
4.40	4.38	.021	.021	.95	1.63	1.0%	10	.816	.762	.692	.605	.502	.389	.277
4.60	3.66	.936	.936	.39	1.72	33.8%	5	.834	.794	.745	.683	.608		
4.60	4.04	.560	.560	.52	1.82	21.7%	6	.823	.780	.725	.656	.575	.481	
4.60	4.33	.275	.275	.66	1.78	11.3%	7	.830	.783	.723	.646	.554	.449	.340
4.60	4.46	.138	.138	.78	1.77	5.8%	8	.826	.777	.714	.635	.539	.432	.321
4.60	4.53	.065	.065	.87	1.72	2.8%	9	.824	.774	.709	.628	.531	.422	.311
4.60	4.57	.028	.028	.93	1.66	1.2%	10	.822	.772	.707	.625	.527	.417	.306
4.60	4.59	.012	.012	.97	1.60	.5%	11	.822	.771	.706	.623	.525	.415	.304
4.80	4.15	.647	.647	.49	1.82	23.8%	6	.831	.791	.740	.677	.601	.512	
4.80	4.47	.330	.330	.63	1.79	12.9%	7	.837	.794	.737	.667	.580	.481	.374
4.80	4.63	.171	.171	.75	1.80	6.9%	8	.833	.787	.728	.654	.564	.461	.353
4.80	4.71	.083	.083	.85	1.76	3.4%	9	.830	.783	.723	.647	.555	.450	.341
4.80	4.76	.037	.037	.92	1.70	1.5%	10	.829	.781	.720	.643	.550	.445	.335
4.80	4.78	.016	.016	.96	1.64	.7%	11	.828	.781	.719	.641	.548	.442	.332
5.00	4.26	.742	.742	.46	1.82	25.8%	6	.839	.802	.755	.696	.625	.542	
5.00	4.56	.437	.437	.59	1.89	16.1%	7	.831	.791	.740	.676	.600	.511	.414
5.00	4.79	.209	.209	.72	1.82	8.0%	8	.839	.797	.742	.673	.588	.490	.384
5.00	4.90	.104	.104	.83	1.79	4.1%	9	.836	.792	.736	.665	.578	.478	.370
5.00	4.95	.049	.049	.90	1.74	1.9%	10	.835	.790	.733	.660	.572	.471	.363
5.00	4.98	.021	.021	.95	1.67	.9%	11	.834	.789	.731	.658	.570	.468	.359
5.50	4.49	1.01	1.01	.39	1.80	30.9%	6	.856	.825	.786	.738	.679	.609	
5.50	4.87	.626	.626	.51	1.90	20.4%	7	.848	.814	.771	.718	.654	.577	.490
5.50	5.17	.327	.327	.64	1.87	11.2%	8	.854	.818	.772	.714	.642	.556	.460
5.50	5.33	.175	.175	.76	1.86	6.2%	9	.850	.813	.765	.704	.630	.541	.442
5.50	5.41	.088	.088	.85	1.82	3.1%	10	.848	.810	.761	.699	.622	.532	.432
5.50	5.46	.042	.042	.92	1.76	1.5%	11	.847	.808	.759	.696	.618	.527	.426
5.50	5.48	.019	.019	.96	1.70	.7%	12	.847	.808	.758	.694	.616	.525	.423
6.00	5.14	.856	.856	.44	1.89	25.0%	7	.863	.834	.798	.754	.700	.635	.559
6.00	5.47	.529	.529	.56	1.97	16.2%	8	.857	.826	.787	.739	.681	.611	.530
6.00	5.73	.274	.274	.69	1.92	8.7%	9	.863	.831	.790	.738	.674	.597	.508
6.00	5.85	.147	.147	.79	1.90	4.8%	10	.860	.827	.785	.732	.666	.586	.495
6.00	5.93	.074	.074	.87	1.85	2.4%	11	.859	.825	.782	.728	.660	.580	.487
6.00	5.96	.036	.036	.93	1.79	1.2%	12	.858	.824	.780	.726	.658	.576	.483
6.00	5.98	.016	.016	.96	1.72	.5%	13	.858	.823	.780	.725	.656	.574	.481
6.50	5.33	1.17	1.17	.38	1.95	30.5%	7	.868	.843	.813	.777	.732	.679	.617
6.50	5.77	.727	.727	.49	1.98	20.1%	8	.869	.843	.810	.770	.720	.660	.590
6.50	6.05	.447	.447	.61	2.03	12.9%	9	.864	.836	.801	.758	.705	.641	.567
6.50	6.27	.230	.230	.73	1.96	6.8%	10	.871	.842	.806	.760	.703	.634	.553
6.50	6.38	.123	.123	.82	1.93	3.7%	11	.869	.839	.802	.755	.697	.626	.543
6.50	6.44	.063	.063	.89	1.87	1.9%	12	.868	.838	.800	.752	.693	.621	.537
6.50	6.47	.030	.030	.94	1.81	.9%	13	.867	.837	.799	.751	.691	.619	.534
7.00	5.53	1.47	1.47	.33	1.92	34.7%	7	.880	.859	.834	.803	.765	.720	.667

(continued)

first attm	hour con- nect	hour recal retry	hour ovfl	con nect vmr	ovfl vmr	% block GOS	lines or servr	Hours Connected Traffic on Servers						
								1	2	3	4	5	6	7
7.00	6.04	.963	.963	.43	1.97	24.2%	8	.880	.858	.830	.796	.754	.703	.343
7.00	6.38	.618	.618	.54	2.05	16.2%	9	.875	.851	.821	.784	.739	.684	.619
7.00	6.62	.378	.378	.65	2.09	10.3%	10	.872	.846	.815	.775	.728	.670	.602
7.00	6.80	.193	.193	.76	2.00	5.4%	11	.878	.852	.820	.779	.728	.667	.593
7.00	6.90	.104	.104	.85	1.96	2.9%	12	.877	.850	.817	.776	.724	.661	.586
7.00	6.95	.053	.053	.91	1.89	1.5%	13	.876	.849	.816	.774	.721	.657	.582
7.00	6.97	.026	.026	.95	1.82	.7%	14	.875	.849	.815	.772	.720	.656	.579
8.00	6.42	1.58	1.58	.33	1.98	33.0%	8	.893	.876	.856	.831	.801	.765	.723
8.00	6.89	1.11	1.11	.42	2.10	24.4%	9	.888	.869	.847	.820	.787	.748	.701
8.00	7.30	.704	.704	.52	2.12	16.2%	10	.890	.870	.846	.817	.782	.739	.687
8.00	7.55	.446	.446	.63	2.17	10.6%	11	.887	.866	.841	.810	.773	.727	.673
8.00	7.76	.241	.241	.74	2.09	5.8%	12	.892	.871	.846	.814	.774	.726	.668
8.00	7.86	.136	.136	.82	2.05	3.3%	13	.891	.869	.843	.811	.770	.721	.661
8.00	7.93	.073	.073	.89	1.99	1.8%	14	.890	.868	.842	.809	.768	.718	.657
8.00	7.96	.038	.038	.93	1.92	.9%	15	.889	.868	.841	.808	.767	.716	.655
9.00	7.32	1.68	1.68	.32	2.04	31.5%	9	.904	.890	.873	.853	.829	.800	.765
9.00	7.80	1.20	1.20	.41	2.16	23.6%	10	.900	.884	.866	.844	.818	.786	.748
9.00	8.21	.787	.787	.50	2.19	16.1%	11	.901	.885	.866	.842	.814	.780	.739
9.00	8.49	.513	.513	.61	2.24	10.8%	12	.882	.882	..861	.837	.807	.770	.727
9.00	8.71	.290	.290	.71	2.18	6.3%	13	.886	.886	.865	.840	.809	.771	.724
9.00	8.83	.171	.171	.80	2.14	3.7%	14	.884	.884	.863	.837	.805	.766	.719
9.00	8.90	.096	.096	.87	2.09	2.1%	15	.883	.883	.862	.836	.803	.763	.715
9.00	8.95	.052	.052	.92	2.02	1.2%	16	.883	.883	.861	.834	.802	.762	.713
9.00	8.97	.027	.027	.95	1.95	.6%	17	.883	.883	.861	.834	.801	.761	.712
10	8.18	1.82	1.82	.32	2.16	30.8%	10	.909	.897	.882	.865	.845	.822	.794
10	8.70	1.30	1.30	.40	2.22	22.9%	11	.909	.896	.881	.863	.841	.815	.785
10	9.09	.910	.910	.49	2.32	16.7%	12	.906	.892	.876	.857	.833	.806	.773
10	9.42	.580	.580	.59	2.31	11.0%	13	.908	.894	.877	.857	.833	.804	.769
10	9.63	.374	.374	.68	2.32	7.2%	14	.907	.892	.875	.854	.828	.798	.761
10	9.79	.208	.208	.78	2.23	4.1%	15	.911	.896	.879	.858	.832	.801	.763
10	9.88	.122	.122	.85	2.18	2.4%	16	.910	.896	.878	.856	.830	.798	.759
10	9.93	.068	.068	.90	2.11	1.4%	17	.910	.895	.877	.855	.829	.796	.757
10	9.96	.037	.037	.94	2.04	.7%	18	.909	.895	.877	.855	.828	.795	.756
11	9.08	1.92	1.92	.32	2.21	29.7%	11	.917	.906	.894	.880	.863	.843	.820
11	9.62	1.38	1.38	.39	2.28	22.3%	12	.917	.906	.893	.878	.860	.839	.814
11	10.0	.988	.988	.48	2.37	16.5%	13	.914	.903	.889	.873	.854	.831	.804
11	10.4	.645	.645	.57	2.37	11.1%	14	.916	.904	.890	.874	.854	.830	.801
11	10.6	.426	.426	.67	2.39	7.5%	15	.915	.903	.888	.871	.850	.825	.795
11	10.8	.246	.246	.76	2.31	4.4%	16	.918	.906	.892	.874	.853	.827	.797
11	10.9	.149	.149	.83	2.26	2.7%	17	.918	.905	.891	.873	.851	.825	.794
11	10.9	.087	.087	.89	2.20	1.6%	18	.917	.905	.890	.872	.850	.824	.792
11	10.9	.049	.049	.93	2.12	.9%	19	.917	.904	.889	.871	.849	.823	.791
12	10.0	2.00	2.00	.31	2.26	28.6%	12	.923	.914	.904	.892	.877	.861	.841
12	10.5	1.51	1.51	.39	2.39	22.3%	13	.920	.911	.900	.886	.871	.853	.832
12	10.9	1.06	1.06	.47	2.43	16.3%	14	.921	.911	.900	.886	.870	.851	.828
12	11.3	.710	.710	.56	2.44	11.2%	15	.923	.913	.901	.887	.870	.850	.827
12	11.5	.478	.478	.65	2.46	7.7%	16	.922	.911	.899	.884	.867	.846	.821

first attm	hour con-nect	hour recal retry	hour ovfl	con nect vmr	ovfl vmr	% block GOS	lines or servr	Hours Connected Traffic on Servers						
								1	2	3	4	5	6	7
12	11.7	.313	.313	.73	2.45	5.1%	17	.921	.910	.897	.882	.864	.843	.818
12	11.8	.177	.177	.81	2.34	2.9%	18	.924	.914	.901	.886	.868	.847	.821
12	11.9	.106	.106	.87	2.28	1.8%	19	.924	.913	.900	.885	.867	.845	.819
12	11.9	.062	.062	.92	2.21	1.0%	20	.923	.913	.900	.885	.866	.844	.818
12	11.9	.034	.034	.95	2.13	.6%	21	.923	.913	.900	.884	.866	.844	.817
13	10.3	2.71	2.71	.26	2.23	34.5%	12	.929	.922	.913	.903	.892	.879	.863
13	10.9	2.13	2.13	.31	2.37	28.1%	13	.926	.918	.909	.898	.886	.872	.855
13	11.4	1.59	1.59	.38	2.44	21.8%	14	.926	.918	.908	.897	.884	.869	.851
13	11.9	1.14	1.14	.46	2.48	16.1%	15	.927	.919	.909	.897	.883	.867	.848
13	12.2	.812	.812	.55	2.55	11.8%	16	.926	.917	.906	.894	.879	.863	.843
13	12.5	.530	.530	.64	2.52	7.8%	17	.928	.918	.908	.895	.881	.863	.843
13	12.6	.354	.354	.72	2.52	5.3%	18	.927	.917	.906	.894	.879	.861	.840
13	12.8	.207	.207	.80	2.42	3.1%	19	.930	.920	.910	.897	.882	.864	.842
13	12.9	.127	.127	.86	2.36	1.9%	20	.929	.920	.909	.896	.881	.863	.841
13	12.9	.076	.076	.91	2.29	1.2%	21	.929	.920	.909	.896	.880	.862	.840
13	12.9	.044	.044	.94	2.21	.7%	22	.929	.919	.908	.895	.880	.861	.839
14	11.2	2.82	2.82	.26	2.33	33.6%	13							
14	11.8	2.20	2.20	.31	2.41	27.2%	14							
14	12.3	1.66	1.66	.38	2.49	21.3%	15							
14	12.8	1.21	1.21	.45	2.54	15.9%	16							
14	13.1	.874	.874	.53	2.61	11.8%	17							
14	13.4	.582	.582	.62	2.59	8.0%	18							
14	13.6	.395	.395	.70	2.58	5.5%	19							
14	13.8	.238	.238	.78	2.49	3.3%	20							
14	13.8	.150	.150	.85	2.44	2.1%	21							
14	13.9	.092	.092	.89	2.37	1.3%	22							
14	13.9	.054	.054	.93	2.29	.8%	23							
15	12.1	2.90	2.90	.26	2.37	32.4%	14							
15	12.7	2.29	2.29	.31	2.46	26.5%	15							
15	13.3	1.74	1.74	.37	2.53	20.8%	16							
15	13.7	1.32	1.32	.44	2.64	16.1%	17							
15	14.1	.935	.935	.53	2.66	11.7%	18							
15	14.4	.634	.634	.61	2.64	8.1%	19							
15	14.6	.437	.437	.69	2.65	5.7%	20							
15	14.7	.293	.293	.76	2.62	3.8%	21							
15	14.8	.173	.173	.83	2.51	2.3%	22							
15	14.8	.108	.108	.88	2.44	1.4%	23							
15	14.9	.066	.066	.92	2.37	.9%	24							
15	14.9	.039	.039	.95	2.28	.5%	25							
16	13.0	2.98	2.98	.26	2.42	31.4%	15							
16	13.6	2.36	2.36	.31	2.50	25.7%	16							
16	14.1	1.85	1.85	.37	2.63	20.8%	17							
16	14.6	1.38	1.38	.44	2.69	15.9%	18							
16	15.0	.995	.995	.52	2.71	11.7%	19							
16	15.3	.685	.685	.60	2.70	8.2%	20							

(continued)

16	15.5	.479	.479	.68	2.71	5.8%	21
16	15.7	.327	.327	.75	2.69	4.0%	22
16	15.8	.198	.198	.82	2.58	2.4%	23
16	15.9	.126	.126	.87	2.51	1.6%	24
16	15.9	.078	.078	.91	2.44	1.0%	25
16	15.9	.047	.047	.94	2.36	.6%	26

NOTE: The load on each server was purposely left off the table for 14, 15, and 16 Erlangs. Throughout the above table, the load was only shown on the first 7 servers. As the number of servers grows beyond 7, the table would need to be much larger to show the load on each and every server. In traffic engineering, we are more interested in total load, GOS desired, and number of servers required to reach that GOS. This appendix serves that function well. Should you desire to engineer to a finer degree, you will need to get a complete table.

Appendix F

JEWETT-SHRAGO FAST RETRIAL 100% TABLE

first attm	hour con- nect	hour recal retry	hour ovfl	con nect vmr	ovfl vmr	% block GOS	lines or servr	1	2	3	4	5	6	7
.200	.200	.050	.000	.80	1.09	20.0%	1	.200						
.200	.200	.003	.000	.97	1.07	1.7%	2	.169	.031					
.400	.400	.025	.000	.90	1.14	6.0%	2	.298	.102					
.400	.400	.003	.000	.98	1.11	.7%	3	.287	.019					
.600	.600	.083	.000	.81	1.20	12.2%	2	.406	.194					
.600	.600	.013	.000	.95	1.17	2.1%	3	.380	.169	.051				
.800	.800	.200	.000	.70	1.25	20.0%	2	.500	.300	.800	.036	.000	.90	1.23
.455	.247	.098												
.800	.800	.600	.000	.97	1.19	.8%	4	.446	.237	.091	.025			
1.00	1.00	.079	.000	.84	1.28	7.3%	3	.519	.324	.157				
1.00	1.00	.016	.000	.95	1.24	1.6%	4	.504	.305	.142	.049			
1.20	1.20	.151	.000	.77	1.33	11.2%	3	.575	.399	.227				
1.20	1.20	.035	.000	.92	1.29	2.8%	4	.553	.368	.197	.082			
1.20	1.20	.008	.000	.98	1.24	.6%	5	.547	.361	.190	.078	.025		
1.40	1.40	.266	.000	.70	1.37	16.0%	3	.625	.471	.304				
1.40	1.40	.066	.000	.88	1.34	4.5%	4	.595	.427	.256	.123			
1.40	1.40	.016	.000	.96	1.29	1.2%	5	.586	.415	.243	.114	.042		
1.60	1.60	.576	.000	.59	1.52	26.5%	3	.654	.539	.407				
1.60	1.60	.115	.000	.83	1.39	6.7%	4	.632	.481	.316	.172			
1.60	1.60	.031	.000	.93	1.34	1.0%	5	.620	.463	.296	.155	.066		
1.80	1.80	.889	.000	.51	1.54	33.1%	3	.702	.606	.492				
1.80	1.80	.186	.000	.77	1.43	9.4%	4	.665	.531	.376	.228			
1.80	1.80	.054	.000	.90	1.39	2.9%	5	.650	.507	.347	.200	.095		
1.80	1.80	.015	.000	.97	1.33	.8%	6	.645	.500	.339	.192	.090	.034	
2.00	2.00	.289	.000	.71	1.47	12.6%	4	.696	.578	.436	.289			
2.00	2.00	.087	.000	.87	1.43	4.2%	5	.676	.548	.397	.248	.131		
2.00	2.00	.026	.000	.95	1.38	1.3%	6	.670	.538	.384	.236	.121	.052	
2.20	2.20	.556	.000	.62	1.61	20.2%	4	.707	.614	.501	.377			
2.20	2.20	.135	.000	.83	1.48	5.8%	5	.700	.585	.455	.298	.172		
2.20	2.20	.042	.000	.93	1.42	1.9%	6	.692	.571	.427	.280	.156	.073	
2.20	2.20	.012	.000	.97	1.38	.6%	7	.689	.567	.422	.274	.151	.070	.027
2.40	2.40	.791	.000	.55	1.63	24.8%	4	.738	.658	.559	.445			
2.40	2.40	.201	.000	.78	1.52	7.7%	5	.722	.619	.491	.350	.218		
2.40	2.40	.066	.000	.90	1.47	2.7%	6	.712	.602	.468	.324	.195	.099	
2.40	2.40	.021	.000	.96	1.41	.9%	7	.708	.596	.460	.315	.187	.094	.040
2.60	2.60	1.29	.000	.47	1.73	33.1%	4	.759	.696	.618	.527			
2.60	2.60	.292	.000	.73	1.56	10.1%	5	.743	.651	.535	.402	.269		
2.60	2.60	.100	.000	.87	1.51	3.7%	6	.730	.630	.506	.368	.236	.130	
2.60	2.60	.033	.000	.94	1.45	1.3%	7	.725	.623	.496	.356	.224	.121	.056
2.80	2.80	.521	.000	.65	1.69	15.7%	5	.746	.669	.574	.464	.346		
2.80	2.80	.145	.000	.83	1.56	4.9%	6	.747	.656	.542	.411	.279	.165	
2.80	2.80	.050	.000	.92	1.50	1.8%	7	.740	.647	.529	.395	.262	.152	.076

(continued)

first attm	hour con- nect	hour recal retry	hour ovfl	con nect vmr	ovfl vmr	% block GOS	lines or servr	Hours Connected Traffic on Servers						
								1	2	3	4	5	6	7
2.80	2.80	.017	.000	.97	1.44	.6%	8	.738	.643	.524				
3.00	3.00	.706	.000	.60	1.72	19.0%	5	.767	.700	.615	.515	.403		
3.00	3.00	.205	.000	.80	1.60	6.4%	6	.762	.681	.576	.453	.323	.205	
3.00	3.00	.074	.000	.90	1.54	2.4%	7	.755	.669	.560	.432	.301	.185	.099
3.00	3.00	.026	.000	.96	1.48	.8%	8	.752	.664	.553	.424	.293	.178	.094
3.20	3.20	.948	.000	.54	1.73	22.9%	5	.788	.729	.655	.565	.463		
3.20	3.20	.284	.000	.75	1.63	8.1%	6	.777	.703	.608	.494	.369	.248	
3.20	3.20	.106	.000	.87	1.58	3.2%	7	.768	.689	.588	.468	.340	.220	.126
3.20	3.20	.038	.000	.94	1.52	1.2%	8	.764	.683	.580	.458	.329	.210	.118
3.40	3.40	1.44	.000	.46	1.82	29.8%	5	.800	.752	.692	.620	.536		
3.40	3.40	.482	.000	.68	1.76	12.4%	6	.776	.712	.632	.536	.428	.317	
3.40	3.40	.148	.000	.84	1.62	4.2%	7	.780	.708	.615	.503	.379	.258	.156
3.40	3.40	.056	.000	.92	1.56	1.6%	8	.776	.701	.605	.490	.364	.244	.144
3.40	3.40	.020	.000	.97	1.50	.6%	9	.774	.698	.601	.485	.359	.238	.140
3.60	3.60	.632	.000	.63	1.79	14.9%	6	.791	.734	.662	.574	.473	.366	
3.60	3.60	.201	.000	.81	1.66	5.3%	7	.792	.726	.641	.536	.418	.297	.190
3.60	3.60	.078	.000	.90	1.60	2.1%	8	.786	.718	.629	.521	.399	.278	.173
3.60	3.60	.029	.000	.96	1.54	.8%	9	.784	.714	.624	.514	.392	.271	.167
3.80	3.80	.822	.000	.58	1.81	17.8%	6	.806	.755	.691	.612	.519	.417	
3.80	3.80	.270	.000	.77	1.70	6.6%	7	.803	.743	.665	.568	.456	.338	.227
3.80	3.80	.108	.000	.88	1.64	2.8%	8	.796	.733	.651	.549	.433	.314	.205
3.80	3.80	.042	.000	.94	1.58	1.1%	9	.793	.729	.645	.541	.424	.304	.196
4.00	4.00	1.21	.000	.52	1.90	23.3%	6	.812	.769	.715	.649	.571	.484	
4.00	4.00	.443	.000	.71	1.83	10.0%	7	.799	.746	.678	.595	.499	.394	.289
4.00	4.00	.146	.000	.85	1.68	3.5%	8	.806	.747	.671	.577	.466	.349	.238
4.00	4.00	.058	.000	.93	1.62	1.4%	9	.802	.742	.664	.567	.455	.336	.226
4.00	4.00	.022	.000	.97	1.55	.5%	10	.801	.740	.661	.563	.450	.331	.221
4.20	4.20	1.71	.000	.45	1.97	29.0%	6	.822	.785	.740	.685	.621	.547	
4.20	4.20	.567	.000	.67	1.85	11.9%	7	.811	.762	.701	.625	.535	.435	.331
4.20	4.20	.194	.000	.82	1.72	4.4%	8	.815	.760	.690	.602	.499	.385	.273
4.20	4.20	.080	.000	.91	1.66	1.9%	9	.811	.754	.682	.591	.484	.369	.257
4.20	4.20	.031	.000	.96	1.59	.7%	10	.809	.752	.678	.586	.478	.362	.250
4.40	4.40	2.35	.000	.39	2.02	34.8%	6	.834	.803	.766	.721	.668	.607	
4.40	4.40	.721	.000	.62	1.88	14.1%	7	.822	.778	.722	.653	.571	.477	.376
4.40	4.40	.254	.000	.79	1.76	5.5%	8	.823	.773	.709	.627	.530	.421	.309
4.40	4.40	.107	.000	.89	1.70	2.4%	9	.818	.766	.698	.614	.512	.401	.289
4.40	4.40	.043	.000	.94	1.63	1.0%	10	.816	.763	.694	.607	.505	.392	.280
4.60	4.60	.912	.000	.58	1.89	16.5%	7	.833	.794	.743	.681	.606	.519	.423
4.60	4.60	.329	.000	.76	1.79	6.7%	8	.831	.785	.726	.651	.560	.456	.347
4.60	4.60	.141	.000	.86	1.74	3.0%	9	.826	.777	.714	.635	.540	.432	.322
4.60	4.60	.059	.000	.93	1.67	1.3%	10	.823	.773	.709	.628	.530	.421	.310
4.60	4.60	.024	.000	.97	1.60	.5%	11	.822	.772	.706	.624	.526	.417	.305
4.80	4.80	1.30	.000	.51	1.98	21.3%	7	.837	.802	.759	.706	.643	.568	.485
4.80	4.80	.510	.000	.70	1.91	9.6%	8	.828	.786	.733	.667	.588	.498	.399
4.80	4.80	.184	.000	.84	1.77	3.7%	9	.833	.788	.729	.655	.566	.463	.355
4.80	4.80	.079	.000	.91	1.71	1.6%	10	.830	.783	.723	.647	.555	.450	.340
4.80	4.80	.033	.000	.96	1.64	.7%	11	.829	.781	.720	.643	.550	.444	.334

first attm	hour con-nect	hour recal retry	hour ovfl	con nect vmr	ovfl vmr	% block GOS	lines or servr	1	2	3	4	5	6	7
5.00	5.00	1.78	.000	.46	2.05	26.2%	7	.843	.813	.776	.731	.678	.615	.544
5.00	5.00	.637	.000	.66	1.94	11.3%	8	.836	.798	.750	.689	.617	.532	.438
5.00	5.00	.237	.000	.81	1.81	4.5%	9	.840	.798	.743	.675	.591	.493	.388
5.00	5.00	`.104	.000	.90	1.75	2.0%	10	.836	.792	.736	.655	.578	.478	.370
5.00	5.00	.044	.000	.95	1.68	.9%	11	.835	.790	.733	.660	.572	.470	.362
5.50	5.50	1.23	.000	.54	2.05	18.2%	8	.851	.821	.784	.738	.683	.618	.544
5.50	5.50	.510	.000	.71	1.98	8.5%	9	.845	.811	.767	.712	.645	.567	.478
5.50	5.50	.196	.000	.84	1.84	3.4%	10	.851	.814	.766	.706	.631	.543	.444
5.50	5.50	.088	.000	.91	1.77	1.6%	11	.848	.810	.761	.699	.622	.532	.432
5.50	5.50	.038	.000	.96	1.70	.7%	12	.847	.808	.759	.696	.618	.527	.426
6.00	6.00	2.38	.000	.41	2.16	28.4%	8	..865	.843	.817	.785	.747	.702	.650
6.00	6.00	.849	.000	.62	2.03	12.4%	9	.863	.834	.798	.754	.700	.635	.559
6.00	6.00	.414	.000	.75	2.01	6.5%	10	.854	.823	.783	.733	.673	.601	.519
6.00	6.00	.162	.000	.86	1.86	2.6%	11	.860	.828	.786	.732	.667	.588	.497
6.00	6.00	.074	.000	.93	1.79	1.2%	12	.859	.825	.782	.728	.660	.580	.487
6.00	6.00	.033	.000	.96	1.72	.5%	13	,858	.824	.780	.725	.657	.576	.483
7.00	7.00	3.19	.000	.37	2.32	31.3%	9	.880	.863	.843	.820	.793	.761	.724
7.00	7.00	1.20	.000	.57	2.19	14.6%	10	.876	.854	.827	.794	.755	.707	.651
7.00	7.00	.544	.000	.72	2.11	7.2%	11	.874	.850	.819	.782	.736	.680	.614
7.00	7.00	.231	.000	.83	1.97	3.2%	12	.879	.853	.821	.781	.730	.669	.596
7.00	7.00	.113	.000	.90	1.90	1.6%	13	.877	.851	.817	.776	.724	.661	.587
7.00	7.00	.053	.000	.95	1.83	.8%	14	.876	.849	.816	.774	.721	.657	.582
8.00	8.00	4.03	.000	.33	2.44	33.5%	10	.892	.878	.863	.846	.825	.802	.775
8.00	8.00	1.57	.000	.52	2.32	16.4%	11	.888	.870	.849	.824	.794	.758	.716
8.00	8.00	.685	.000	.68	2.20	7.9%	12	.890	.870	.846	.817	.781	.738	.686
8.00	8.00	.309	.000	.81	2.08	3.7%	13	.893	.872	.847	.816	.777	.730	.672
8.00	8.00	.158	.000	.88	2.01	1.9%	14	.891	.870	.844	.811	.771	.722	.663
8.00	8.00	.079	.000	.93	1.93	1.0%	15	.890	.869	.842	.809	.768	.718	.658
9.00	9.00	2.00	.000	.49	2.43	18.0%	12	.897	.883	.866	.846	.823	.795	.763
9.00	9.00	.928	.000	.64	2.35	9.3%	13	.897	.881	.862	.838	.811	.777	.737
9.00	9.00	.457	.000	.77	2.25	4.8%	14	.898	.881	.861	.836	.805	.768	.724
9.00	9.00	.211	.000	.86	2.11	2.3%	15	.902	.885	.864	.838	.806	.768	.721
9.00	9.00	.110	.000	.91	2.03	1.2%	16	.901	.884	.862	.836	.803	.764	.716
9.00	9.00	.055	.000	.95	1.95	.6%	17	.901	.883	.861	.835	.802	.762	.713
10.0	10.0	2.38	.000	.46	2.54	19.3%	13	.906	.894	.880	.864	.845	.823	.798
10.0	10.0	1.09	.000	.62	2.42	9.8%	14	.908	.894	.878	.860	.837	.810	.779
10.0	10.0	.557	.000	.74	2.33	5.3%	15	.908	.894	.877	.857	.832	.803	.768
10.0	10.0	.269	.000	.84	2.20	2.6%	16	.911	.897	.880	.859	.833	.802	.765
10.0	10.0	.145	.000	.90	2.12	1.4%	17	.910	.896	.878	.857	.831	.799	.760
10.0	10.0	.076	.000	.94	2.04	.8%	18	.910	.895	.877	.855	.829	.797	.758
11.0	11.0	2.81	.000	.43	2.63	20.4%	14	.913	.903	.891	.878	.863	.845	.825
11.0	11.0	1.36	.000	.58	2.55	11.0%	15	.913	.902	.888	.873	.855	.833	.808
11.0	11.0	.663	.000	.72	2.41	5.7%	16	.916	.905	.891	.874	.854	.830	.801
11.0	11.0	.382	.000	.81	2.36	3.4%	17	.914	.902	.887	.870	.849	.824	.794
11.0	11.0	.186	.000	.88	2.21	1.7%	18	.918	.906	.891	.873	.852	.826	.795
11.0	11.0	.101	.000	.93	2.13	.9%	19	.917	.905	.890	.872	.850	.824	.792
12.0	12.0	3.25	.000	.41	2.71	21.3%	15	.919	.910	.901	.890	.877	.862	.846

(continued)

JEWETT-SHRAGO FAST RETRIAL 100% TABLE   *(continued)*

first attm	hour connect	hour recal retry	hour ovfl	con nect vmr	ovfl vmr	% block GOS	lines or servr	1 / 11 / 21	2 / 12 / 22	3 / 13 / 23	4 / 14 / 24	5 / 15 / 25	6 / 16 / 26	7 / 17 / 27	8 / 18 / 28	9 / 19 / 29	10 / 20 / 30
													Hours Connected Traffic on Servers				
12.0	12.0	1.54	.000	.56	2.60	11.4%	16				.921	.911	.900	.887	.872	.854	.833
12.0	12.0	.851	.000	.68	2.55	6.6%	17				.920	.910	.898	.884	.867	.847	.824
12.0	12.0	.456	.000	.79	2.44	3.7%	18				.922	.911	.899	.884	.866	.846	.821
12.0	12.0	.231	.000	.87	2.30	1.9%	19				.924	.914	.901	.887	.869	.848	.822
12.0	12.0	.129	.000	.91	2.22	1.1%	20				.924	.913	.901	.885	.867	.846	.820
12.0	12.0	.071	.000	.95	2.14	.6%	21				.924	.913	.900	.885	.866	.845	.818
13.0	13.0	3.85	.000	.38	2.83	22.8%	16				.923	.915	.907	.898	.887	.875	.861
13.0	13.0	1.84	.000	.53	2.71	12.4	17				.925	.916	.907	.896	.883	.868	.851
13.0	13.0	.971	.000	.67	2.62	7.0%	18				.926	.918	.907	.895	.881	.865	.846
13.0	13.0	.534	.000	.77	2.51	3.9%	19				.928	.918	.908	.895	.881	.863	.843
13.0	13.0	.318	.000	.84	2.45	2.4%	20				.926	.917	.906	.893	.878	.860	.839
13.0	13.0	.161	.000	.90	2.30	1.2%	21				.929	.920	.909	.896	.881	.863	.842
13.0	13.0	.091	.000	.94	2.22	.7%	22				.929	.920	.909	.896	.880	.862	.840
14.0	14.0	4.31	.000	.36	2.89	23.5%	17				.928	.921	.914	.906	.897	.887	.875
14.0	14.0	2.15	.000	.51	2.81	13.3%	18				.929	.921	.913	.903	.892	.880	.865
14.0	14.0	1.20	.000	.65	2.68	7.3%	19				.932	.924	.915	.905	.893	.879	.863
14.0	14.0	.616	.000	.75	2.59	4.2%	20				.933	.925	.916	.905	.892	.878	.861
14.0	14.0	.374	.000	.82	2.53	2.6%	21				.932	.924	.914	.903	.890	.875	.857
14.0	14.0	.196	.000	.89	2.38	1.4%	22				.934	.926	.917	.906	.893	.877	.859
14.0	14.0	.113	.000	.93	2.30	.8%	23				.934	.926	.916	.905	.892	.876	.858
15.00	15.00	4.923	.000	.34	.299	24.7%	18	.931	.925	.919	.912	.904	.895	.886	.875	.862	.848
								.833	.815	.796	.774	.749	.722	.692	.659		
15.00	15.00	2.462	.000	.49	2.90	14.1%	19	.932	.926	.918	.910	.901	.890	.877	.863	.847	.829
								.808	.784	.757	.726	.692	.653	.610	.563	.512	
15.00	15.00	1.312	.000	.62	2.79	8.0%	20	.934	.928	.920	.911	.900	.888	.875	.859	.840	.819
								.794	.766	.733	.696	.654	.607	.555	.499	.440	.380
15.00	15.00	.702	.000	.73	2.66	4.5%	21	.937	.930	.922	.913	.902	.890	.875	.858	.838	.815
								.788	.756	.720	.678	.630	.578	.520	.458	.394	.329
								.267									
15.00	15.00	.433	.000	.81	2.60	2.8%	22	.936	.929	.921	.911	.900	.887	.872	.854	.833	.809
								.781	.748	.710	.666	.617	.562	.502	.439	.374	.309
15.00	15.00	.234	.000	.88	2.46	1.5%	23	.938	.931	.923	.914	.902	.889	.874	.856	.834	.809
								.780	.746	.706	.660	.608	.551	.488	.422	.355	.289
								.227	.172	.125							
15.00	15.00	.138	.000	.92	2.38	.9%	24	.938	.931	.923	.913	.902	.888	.873	.854	.833	.807
								.777	.742	.702	.655	.603	.545	.481	.415	.348	.282
								.220	.166	.120	.083						
15.00	15.00	.080	.000	.95	2.29	.5%	25	.938	.931	.922	.912	.901	.888	.872	.853	.831	.806
								.776	.740	.699	.653	.600	.541	.477	.411	.343	.277
								.216	.162	.117	.080	.053					
16.00	16.00	2.67	.000	.47	2.94	14.3%	20	.937	.931	.925	.917	.909	.900	.889	.877	.864	.848
								.830	.810	.787	.761	.731	.698	.661	.620	.575	.527
16.00	16.00	1.48	.000	.60	2.85	8.3%	21	.939	.933	.926	.918	.909	.899	.887	.873	.858	.840
								.819	.795	.767	.736	.700	.660	.615	.565	.511	.455
								.396									

first attm	hour con-nect	hour recal retry	hour ovfl	con nect vmr	ovfl vmr	% block GOS	lines or servr	Hours Connected Traffic on Servers									
								1 / 11 / 21	2 / 12 / 22	3 / 13 / 23	4 / 14 / 24	5 / 15 / 25	6 / 16 / 26	7 / 17 / 27	8 / 18 / 28	9 / 19 / 29	10 / 20 / 30
16.00	16.00	.857	.000	.71	2.78	5.1%	22	.939	.933	.926	.917	.908	.897	.884	.870	.853	.833
								.811	.785	.754	.720	.680	.636	.586	.532	.475	.414
								.353	.293								
16.00	16.00	.495	.000	.79	2.67	3.0%	23	.940	.934	.927	.918	.909	.897	.884	.869	.852	.831
								.807	.779	.747	.710	.668	.620	.568	.510	.449	.386
								.323	.262	.206							
16.00	16.00	.307	.000	.85	2.60	1.9%	24	.940	.933	.926	.917	.907	.896	.882	.867	.849	.828
								.803	.774	.741	.703	.660	.611	.557	.498	.436	.373
								.310	.250	.194	.146						
16.00	16.00	.165	.000	.91	2.45	1.0%	25	.942	.935	.928	.920	.910	.898	.885	.869	.851	.829
								.804	.775	.741	.701	.656	.605	.549	.488	.424	.358
								.294	.233	.178	.131	.093					
16.00	16.00	.098	.000	.94	2.37	.6%	26	.942	.935	.928	.919	.909	.898	.884	.868	.850	.828
								.803	.773	.738	.698	.653	.602	.545	.484	.419	.354
								.289	.229	.174	.128	.090	.061				

Appendix G

JACOBSEN 70% TABLE

# TRKS	TABLE .01 OFFD CCS	TABLE .01 CRRD CCS	TABLE .03 OFFD CCS	TABLE .03 CRRD C0.7	TABLE .05 OFFD CCS	TABLE .05 CRRD CCS	TABLE .1 OFFD CCS	TABLE .1 CRRD CCS	# TRKS
1	0.2	0.2	0.7	0.7	1.2	1.2	2.5	2.4	1
2	4.7	4.7	8.4	8.4	11.2	11.0	16.8	16.3	2
3	14.8	14.8	22.7	22.5	28.0	27.5	38.0	36.7	3
4	29	29	41	40	49	48	63	60	4
5	46	46	62	61	72	71	90	86	5
6	65	65	85	84	97	95	118	114	6
7	86	85	109	108	123	121	148	143	7
8	108	107	134	133	150	148	178	172	8
9	131	130	160	159	178	185	209	202	9
10	154	154	187	185	207	204	241	233	10
11	179	178	215	212	236	232	272	263	11
12	203	203	243	240	266	261	305	295	12
13	230	229	271	269	296	291	338	326	13
14	256	255	300	297	326	321	371	359	14
15	283	282	330	326	357	351	404	391	15
16	309	308	359	356	388	382	437	423	16
17	336	335	389	385	420	413	471	456	17
18	364	362	419	415	451	444	505	488	18
19	392	391	450	445	483	475	539	521	19
20	420	418	480	476	515	507	573	555	20
21	448	447	511	506	547	538	608	588	21
22	477	475	542	537	579	570	642	621	22
23	506	504	573	568	612	602	677	655	23
24	535	533	604	599	644	634	711	688	24
25	564	562	636	630	677	666	746	722	25
26	594	592	667	661	710	699	781	755	26
27	623	621	699	693	743	731	816	789	27
28	653	651	731	724	776	764	851	823	28
29	683	680	763	756	809	796	886	857	29
30	713	710	795	788	842	829	921	891	30
31	743	740	827	828	876	862	956	925	31
32	773	770	860	852	909	895	991	959	32
33	803	801	892	884	942	927	1027	993	33
34	833	831	925	916	976	961	1062	1028	34
35	864	861	957	948	1010	994	1098	1062	35
36	895	892	990	980	1043	1027	1133	1096	36
37	929	926	1022	1013	1077	1060	1168	1131	37
38	969	957	1055	1045	1111	1093	1204	1165	38
39	991	988	1088	1078	1145	1127	1240	1199	39
40	1022	1019	1122	1112	1180	1161	1276	1253	40
41	1054	1050	1157	1146	1215	1196	1313	1270	41
42	1085	1082	1190	1179	1249	1230	1349	1305	42

	TABLE .01		TABLE .03		TABLE .05		TABLE .1		
# TRKS	OFFD CCS	CRRD CCS	OFFD CCS	CRRD C0.7	OFFD CCS	CRRD CCS	OFFD CCS	CRRD CCS	# TRKS
43	1117	1113	1223	1211	1283	1263	1384	1339	43
44	1149	1145	1256	1244	1317	1296	1420	1374	44
45	1180	1176	1288	1276	1351	1329	1455	1408	45
46	1212	1208	1321	1309	1385	1363	1491	1443	46
47	1243	1240	1355	1342	1419	1397	1527	1477	47
48	1275	1271	1388	1375	1453	1430	1563	1512	48
49	1307	1303	1421	1400	1488	1464	1599	1547	49
50	1338	1334	1455	1445	1527	1498	1635	1582	50
51	1370	1366	1489	1475	1557	1532	1671	1617	51
52	1402	1398	1522	1508	1591	1566	1707	1652	52
53	1434	1430	1566	1541	1626	1600	1743	1687	53
54	1466	1462	1590	1575	1660	1634	1779	1722	54
55	1499	1494	1623	1608	1695	1669	1815	1757	55
56	1531	1526	1657	1641	1730	1703	1851	1792	56
57	1563	1558	1691	1675	1764	1737	1887	1826	57
58	1595	1590	1724	1708	1799	1771	1924	1861	58
59	1627	1622	1758	1742	1834	1805	1960	1896	59
60	1659	1654	1792	1775	1868	1839	1996	1931	60
61	1691	1686	1826	1809	1903	1873	2032	1966	61
62	1723	1718	1860	1843	1938	1907	2068	2002	62
63	1755	1750	1894	1877	1973	1942	2105	2037	63
64	1787	1782	1928	1918	2007	1976	2141	2072	64
65	1819	1814	1963	1944	2042	2010	2177	2107	65
66	1851	1845	1997	1978	2077	2045	2214	2142	66
67	1883	1878	2031	2012	2112	2079	2250	2177	67
68	1916	1910	2065	2046	2147	2114	2286	2212	68
69	1948	1942	2099	2080	2182	2148	2323	2247	69
70	1981	1975	2134	2114	2217	2182	2359	2283	70
71	2014	2007	2168	2147	2252	2217	2395	2318	71
72	2047	2040	2202	2181	2287	2251	2432	2353	72
73	2080	2073	2236	2215	2322	2286	2468	2388	73
74	2113	2106	2270	2249	2357	2320	2505	2424	74
75	2146	2140	2304	2283	2392	2355	2541	2459	75
76	2180	2173	2338	2317	2427	2389	2578	2494	76
77	2213	2206	2372	2350	2462	2424	2614	2529	77
78	2247	2240	2407	2384	2497	2458	2650	2565	78
79	2280	2273	2441	2418	2532	2493	2687	2600	79
80	2314	2306	2475	2452	2568	2527	2723	2635	80
81	2347	2340	2509	2486	2603	2562	2760	2671	81
82	2380	2373	2544	2520	2638	2597	2796	2706	82
83	2413	2406	2578	2554	2673	2631	2833	2741	83
84	2447	2439	2612	2588	2708	2666	2870	2777	84
85	2480	2472	2647	2622	2744	2701	2906	2812	85
86	2513	2505	2681	2657	2779	2735	2943	2847	86
87	2546	2538	2716	2691	2814	2770	2979	2883	87
88	2579	2571	2751	2725	2848	2805	3016	2918	88
89	2612	2604	2785	2759	2885	2840	3052	2953	89

(continued)

# TRKS	TABLE .01		TABLE .03		TABLE .05		TABLE .1		# TRKS
	OFFD CCS	CRRD CCS	OFFD CCS	CRRD C0.7	OFFD CCS	CRRD CCS	OFFD CCS	CRRD CCS	
90	2645	2637	2820	2793	2920	2874	3089	2989	90
91	2678	2670	2854	2828	2955	2909	3126	3024	91
92	2711	2703	2889	2862	2991	2944	3162	3060	92
93	2745	2736	2923	2896	3026	2979	3199	3095	93
94	2778	2769	2958	2931	3061	3013	3235	3131	94
95	2811	2802	2993	2965	3097	3048	3272	3166	95
96	2845	2836	3027	2999	3132	3083	3309	3201	96
97	2878	2869	3062	3034	3167	3118	3345	3237	97
98	2911	2902	3097	3068	3203	3153	3382	3272	98
99	2945	2936	3131	3102	3238	3187	3419	3308	99
100	2979	2969	3166	3137	3274	3222	3455	3343	100

Appendix H

Erlang C Table Unlimited Queue

N	P	D1	D2	Q1	Q2	P8	P4	P2	P1	PP
					Offered Traffic is 0.50 Erlangs					
1	0.5000	1.00	2.00	0.50	1.00	0.47	0.44	0.39	0.30	0.18
2	0.1000	0.07	0.67	0.03	0.33	0.08	0.07	0.05	0.02	0.00
3	0.0152	0.01	0.40	0.00	0.20	0.01	0.01	0.00	0.00	0.00
4	0.0018	0.00	0.29	0.00	0.14	0.00	0.00	0.00	0.00	0.00
5	0.0002	0.00	0.22	0.00	0.11	0.00	0.00	0.00	0.00	0.00

N	P	D1	D2	Q1	Q2	P8	P4	P2	P1	PP
					Offered Traffic is 1.00 Erlangs					
2	0.3333	0.33	1.00	0.33	1.00	0.29	0.26	0.20	0.12	0.05
3	0.0909	0.05	0.50	0.05	0.50	0.07	0.06	0.03	0.01	0.00
4	0.0204	0.01	0.33	0.01	0.33	0.01	0.01	0.00	0.00	0.00
5	0.0038	0.00	0.25	0.00	0.25	0.00	0.00	0.00	0.00	0.00
6	0.0006	0.00	0.20	0.00	0.20	0.00	0.00	0.00	0.00	0.00

N	P	D1	D2	Q1	Q2	P8	P4	P2	P1	PP
					Offered Traffic is 1.50 Erlangs					
2	0.6429	1.29	2.00	1.93	3.00	0.60	0.57	0.50	0.39	0.24
3	0.2368	0.16	0.67	0.24	1.00	0.20	0.16	0.11	0.05	0.01
4	0.0746	0.03	0.40	0.04	0.60	0.05	0.04	0.02	0.01	0.00
5	0.0201	0.01	0.29	0.01	0.43	0.01	0.01	0.00	0.00	0.00
6	0.0047	0.00	0.22	0.00	0.33	0.00	0.00	0.00	0.00	0.00
7	0.0010	0.00	0.18	0.00	0.27	0.00	0.00	0.00	0.00	0.00

N	P	D1	D2	Q1	Q2	P8	P4	P2	P1	PP
					Offered Traffic is 2.00 Erlangs					
3	0.4444	0.44	1.00	0.89	2.00	0.39	0.35	0.27	0.16	0.06
4	0.1739	0.09	0.50	0.17	1.00	0.14	0.11	0.06	0.02	0.00
5	0.0597	0.02	0.33	0.04	0.67	0.04	0.03	0.01	0.00	0.00
6	0.0180	0.00	0.25	0.01	0.50	0.01	0.01	0.00	0.00	0.00
7	0.0048	0.00	0.20	0.00	0.40	0.00	0.00	0.00	0.00	0.00
8	0.0011	0.00	0.17	0.00	0.33	0.00	0.00	0.00	0.00	0.00
9	0.0002	0.00	0.14	0.00	0.29	0.00	0.00	0.00	0.00	0.00

(continued)

				Offered Traffic is 2.50 Erlangs						
N	P	D1	D2	Q1	Q2	P8	P4	P2	P1	PP
3	0.7022	1.40	2.00	3.51	5.00	0.66	0.62	0.55	0.43	0.26
4	0.3199	0.21	0.67	0.53	1.67	0.27	0.22	0.15	0.07	0.02
5	0.1304	0.05	0.40	0.13	1.00	0.10	0.07	0.04	0.01	0.00
6	0.0474	0.01	0.29	0.03	0.71	0.03	0.02	0.01	0.00	0.00
7	0.0154	0.00	0.22	0.01	0.56	0.01	0.01	0.00	0.00	0.00
8	0.0045	0.00	0.18	0.00	0.45	0.00	0.00	0.00	0.00	0.00
9	0.0012	0.00	0.15	0.00	0.38	0.00	0.00	0.00	0.00	0.00
10	0.0003	0.00	0.13	0.00	0.33	0.00	0.00	0.00	0.00	0.00

				Offered Traffic is 3.00 Erlangs						
N	P	D1	D2	Q1	Q2	P8	P4	P2	P1	PP
4	0.5094	0.51	1.00	1.53	3.00	0.45	0.40	0.31	0.19	0.07
5	0.2362	0.12	0.50	0.35	1.50	0.18	0.14	0.09	0.03	0.00
6	0.0991	0.03	0.33	0.10	1.00	0.07	0.05	0.02	0.00	0.00
7	0.0376	0.01	0.25	0.03	0.75	0.02	0.01	0.01	0.00	0.00
8	0.0129	0.00	0.20	0.01	0.60	0.01	0.00	0.00	0.00	0.00
9	0.0040	0.00	0.17	0.00	0.50	0.00	0.00	0.00	0.00	0.00
10	0.0012	0.00	0.14	0.00	0.43	0.00	0.00	0.00	0.00	0.00
11	0.0003	0.00	0.12	0.00	0.38	0.00	0.00	0.00	0.00	0.00

				Offered Traffic is 3.50 Erlangs						
N	P	D1	D2	Q1	Q2	P8	P4	P2	P1	PP
4	0.7379	1.48	2.00	5.17	7.00	0.69	0.65	0.57	0.45	0.27
5	0.3778	0.25	0.67	0.88	2.33	0.31	0.26	0.18	0.08	0.02
6	0.1775	0.07	0.40	0.25	1.40	0.13	0.09	0.05	0.01	0.00
7	0.0762	0.02	0.29	0.08	1.00	0.05	0.03	0.01	0.00	0.00
8	0.0299	0.01	0.22	0.02	0.78	0.02	0.01	0.00	0.00	0.00
9	0.0107	0.00	0.18	0.01	0.64	0.01	0.00	0.00	0.00	0.00
10	0.0035	0.00	0.15	0.00	0.54	0.00	0.00	0.00	0.00	0.00
11	0.0011	0.00	0.13	0.00	0.47	0.00	0.00	0.00	0.00	0.00
12	0.0003	0.00	0.12	0.00	0.41	0.00	0.00	0.00	0.00	0.00

				Offered Traffic is 4.00 Erlangs						
N	P	D1	D2	Q1	Q2	P8	P4	P2	P1	PP
5	0.5541	0.55	1.00	2.22	4.00	0.49	0.43	0.34	0.20	0.07
6	0.2848	0.14	0.50	0.57	2.00	0.22	0.17	0.10	0.04	0.01
7	0.1351	0.05	0.33	0.18	1.33	0.09	0.06	0.03	0.01	0.00
8	0.0590	0.01	0.25	0.06	1.00	0.04	0.02	0.01	0.00	0.00
9	0.0238	0.00	0.20	0.02	0.80	0.01	0.01	0.00	0.00	0.00
10	0.0088	0.00	0.17	0.01	0.67	0.00	0.00	0.00	0.00	0.00
11	0.0030	0.00	0.14	0.00	0.57	0.00	0.00	0.00	0.00	0.00
12	0.0010	0.00	0.12	0.00	0.50	0.00	0.00	0.00	0.00	0.00

N	P	D1	D2	Q1	Q2	P8	P4	P2	P1	PP
				Offered Traffic is 4.50 Erlangs						
5	0.7625	1.52	2.00	6.86	9.00	0.72	0.67	0.59	0.46	0.28
6	0.4217	0.28	0.67	1.26	3.00	0.35	0.29	0.20	0.09	0.02
7	0.2172	0.09	0.40	0.39	1.80	0.16	0.12	0.06	0.02	0.00
8	0.1039	0.03	0.29	0.13	1.29	0.07	0.04	0.02	0.00	0.00
9	0.0460	0.01	0.22	0.05	1.00	0.03	0.01	0.00	0.00	0.00
10	0.0189	0.00	0.18	0.02	0.82	0.01	0.00	0.00	0.00	0.00
11	0.0072	0.00	0.15	0.00	0.69	0.00	0.00	0.00	0.00	0.00
12	0.0026	0.00	0.13	0.00	0.60	0.00	0.00	0.00	0.00	0.00
13	0.0008	0.00	0.12	0.00	0.53	0.00	0.00	0.00	0.00	0.00

N	P	D1	D2	Q1	Q2	P8	P4	P2	P1	PP
				Offered Traffic is 5.00 Erlangs						
6	0.5875	0.59	1.00	2.94	5.00	0.52	0.46	0.36	0.22	0.08
7	0.3241	0.16	0.50	0.81	2.50	0.25	0.20	0.12	0.04	0.01
8	0.1673	0.06	0.33	0.28	1.67	0.11	0.08	0.04	0.01	0.00
9	0.0805	0.02	0.25	0.10	1.25	0.05	0.03	0.01	0.00	0.00
10	0.0361	0.01	0.20	0.04	1.00	0.02	0.01	0.00	0.00	0.00
11	0.0151	0.00	0.17	0.01	0.83	0.01	0.00	0.00	0.00	0.00
12	0.0059	0.00	0.14	0.00	0.71	0.00	0.00	0.00	0.00	0.00
13	0.0021	0.00	0.12	0.00	0.62	0.00	0.00	0.00	0.00	0.00
14	0.0007	0.00	0.11	0.00	0.56	0.00	0.00	0.00	0.00	0.00

N	P	D1	D2	Q1	Q2	P8	P4	P2	P1	PP
				Offered Traffic is 6.00 Erlangs						
7	0.6138	0.61	1.00	3.68	6.00	0.54	0.48	0.37	0.23	0.08
8	0.3570	0.18	0.50	1.07	3.00	0.28	0.22	0.13	0.05	0.01
9	0.1960	0.07	0.33	0.39	2.00	0.13	0.09	0.04	0.01	0.00
10	0.1013	0.03	0.25	0.15	1.50	0.06	0.04	0.00	0.00	0.00
12	0.0225	0.00	0.17	0.02	1.00	0.01	0.01	0.00	0.00	0.00
13	0.0096	0.00	0.14	0.01	0.86	0.00	0.00	0.00	0.00	0.00
14	0.0039	0.00	0.12	0.00	0.75	0.00	0.00	0.00	0.00	0.00
15	0.0015	0.00	0.11	0.00	0.67	0.00	0.00	0.00	0.00	0.00
16	0.0005	0.00	0.10	0.00	0.60	0.00	0.00	0.00	0.00	0.00

N	P	D1	D2	Q1	Q2	P8	P4	P2	P1	PP
				Offered Traffic is 7.00 Erlangs						
8	0.6353	0.64	1.00	4.45	7.00	0.56	0.49	0.39	0.23	0.09
9	0.3849	0.19	0.50	1.35	3.50	0.30	0.23	0.14	0.05	0.01
10	0.2217	0.07	0.33	0.52	2.33	0.15	0.10	0.05	0.01	0.00
11	0.1211	0.03	0.25	0.21	1.75	0.07	0.04	0.02	0.00	0.00
12	0.0626	0.01	0.20	0.09	1.40	0.03	0.02	0.01	0.00	0.00
13	0.0306	0.01	0.17	0.04	1.17	0.01	0.01	0.00	0.00	0.00
14	0.0142	0.00	0.14	0.01	1.00	0.01	0.00	0.00	0.00	0.00
15	0.0062	0.00	0.12	0.01	0.88	0.00	0.00	0.00	0.00	0.00
16	0.0026	0.00	0.11	0.00	0.78	0.00	0.00	0.00	0.00	0.00
17	0.0010	0.00	0.10	0.00	0.70	0.00	0.00	0.00	0.00	0.00
18	0.0004	0.00	0.09	0.00	0.64	0.00	0.00	0.00	0.00	0.00

(continued)

Offered Traffic is 8.00 Erlangs

N	P	D1	D2	Q1	Q2	P8	P4	P2	P1	PP
9	0.6533	0.65	1.00	5.23	8.00	0.58	0.51	0.40	0.24	0.09
10	0.4092	0.20	0.50	1.64	4.00	0.32	0.25	0.15	0.06	0.01
11	0.2450	0.08	0.33	0.65	2.67	0.17	0.12	0.05	0.01	0.00
12	0.1398	0.03	0.25	0.28	2.00	0.08	0.05	0.02	0.00	0.00
13	0.0760	0.02	0.20	0.12	1.60	0.04	0.02	0.01	0.00	0.00
14	0.0393	0.01	0.17	0.05	1.33	0.02	0.01	0.00	0.00	0.00
15	0.0193	0.00	0.14	0.02	1.14	0.01	0.00	0.00	0.00	0.00
16	0.0090	0.00	0.12	0.01	1.00	0.00	0.00	0.00	0.00	0.00
17	0.0040	0.00	0.11	0.00	0.89	0.00	0.00	0.00	0.00	0.00
18	0.0017	0.00	0.10	0.00	0.80	0.00	0.00	0.00	0.00	0.00
19	0.0007	0.00	0.09	0.00	0.73	0.00	0.00	0.00	0.00	0.00

Offered Traffic is 9.00 Erlangs

N	P	D1	D2	Q1	Q2	P8	P4	P2	P1	PP
10	0.6687	0.67	1.00	6.02	9.00	0.59	0.52	0.41	0.25	0.09
11	0.4305	0.22	0.50	1.94	4.50	0.34	0.26	0.16	0.06	0.01
12	0.2660	0.09	0.33	0.80	3.00	0.18	0.13	0.06	0.01	0.00
13	0.1575	0.04	0.25	0.35	2.25	0.10	0.06	0.02	0.00	0.00
14	0.0892	0.02	0.20	0.16	1.80	0.05	0.03	0.01	0.00	0.00
15	0.0482	0.01	0.17	0.07	1.50	0.02	0.01	0.00	0.00	0.00
16	0.0249	0.00	0.14	0.03	1.29	0.01	0.00	0.00	0.00	0.00
17	0.0123	0.00	0.12	0.01	1.12	0.00	0.00	0.00	0.00	0.00
18	0.0058	0.00	0.11	0.01	1.00	0.00	0.00	0.00	0.00	0.00
19	0.0026	0.00	0.10	0.00	0.90	0.00	0.00	0.00	0.00	0.00
20	0.0011	0.00	0.09	0.00	0.82	0.00	0.00	0.00	0.00	0.00
21	0.0005	0.00	0.08	0.00	0.75	0.00	0.00	0.00	0.00	0.00

Offered Traffic is 10.00 Erlangs

N	P	D1	D2	Q1	Q2	P8	P4	P2	P1	PP
11	0.6821	0.68	1.00	6.82	10.00	0.60	0.53	0.41	0.25	0.09
12	0.4494	0.22	0.50	2.25	5.00	0.35	0.27	0.17	0.06	0.01
13	0.2853	0.10	0.33	0.95	3.33	0.20	0.13	0.06	0.01	0.00
14	0.1741	0.04	0.25	0.44	2.50	0.11	0.06	0.02	0.00	0.00
15	0.1020	0.02	0.20	0.20	2.00	0.05	0.03	0.01	0.00	0.00
16	0.0573	0.01	0.17	0.10	1.67	0.03	0.01	0.00	0.00	0.00
17	0.0309	0.00	0.14	0.04	1.43	0.01	0.01	0.00	0.00	0.00
18	0.0159	0.00	0.12	0.02	1.25	0.01	0.00	0.00	0.00	0.00
19	0.0079	0.00	0.11	0.01	1.11	0.00	0.00	0.00	0.00	0.00
20	0.0037	0.00	0.10	0.00	1.00	0.00	0.00	0.00	0.00	0.00
21	0.0017	0.00	0.09	0.00	0.91	0.00	0.00	0.00	0.00	0.00
22	0.0007	0.00	0.08	0.00	0.83	0.00	0.00	0.00	0.00	0.00

Offered Traffic is 11.00 Erlangs

N	P	D1	D2	Q1	Q2	P8	P4	P2	P1	PP
12	0.6939	0.69	1.00	7.63	11.00	0.61	0.54	0.42	0.26	0.09
13	0.4664	0.23	0.50	2.56	5.50	0.36	0.28	0.17	0.06	0.01
14	0.3029	0.10	0.33	1.11	3.67	0.21	0.14	0.07	0.02	0.00
15	0.1898	0.05	0.25	0.52	2.75	0.12	0.07	0.03	0.00	0.00
16	0.1145	0.02	0.20	0.25	2.20	0.06	0.03	0.01	0.00	0.00
17	0.0665	0.01	0.17	0.12	1.83	0.03	0.01	0.00	0.00	0.00
18	0.0371	0.01	0.14	0.06	1.57	0.02	0.01	0.00	0.00	0.00
19	0.0199	0.00	0.12	0.03	1.38	0.01	0.00	0.00	0.00	0.00
20	0.0103	0.00	0.11	0.01	1.22	0.00	0.00	0.00	0.00	0.00
21	0.0051	0.00	0.10	0.01	1.10	0.00	0.00	0.00	0.00	0.00
22	0.0024	0.00	0.09	0.00	1.00	0.00	0.00	0.00	0.00	0.00
23	0.0011	0.00	0.08	0.00	0.92	0.00	0.00	0.00	0.00	0.00
24	0.0005	0.00	0.08	0.00	0.85	0.00	0.00	0.00	0.00	0.00

Offered Traffic is 12.00 Erlangs

N	P	D1	D2	Q1	Q2	P8	P4	P2	P1	PP
13	0.7044	0.70	1.00	8.45	12.00	0.62	0.55	0.43	0.26	0.10
14	0.4817	0.24	0.50	2.89	6.00	0.38	0.29	0.18	0.07	0.01
15	0.3192	0.11	0.33	1.28	4.00	0.22	0.15	0.07	0.02	0.00
16	0.2046	0.05	0.25	0.61	3.00	0.12	0.08	0.03	0.00	0.00
17	0.1266	0.03	0.20	0.30	2.40	0.07	0.04	0.01	0.00	0.00
18	0.0756	0.01	0.17	0.15	2.00	0.04	0.02	0.00	0.00	0.00
19	0.0435	0.01	0.14	0.07	1.71	0.02	0.01	0.00	0.00	0.00
20	0.0241	0.00	0.12	0.04	1.50	0.01	0.00	0.00	0.00	0.00
21	0.0129	0.00	0.11	0.02	1.33	0.00	0.00	0.00	0.00	0.00
22	0.0066	0.00	0.10	0.01	1.20	0.00	0.00	0.00	0.00	0.00
23	0.0033	0.00	0.09	0.00	1.09	0.00	0.00	0.00	0.00	0.00
24	0.0016	0.00	0.08	0.00	1.00	0.00	0.00	0.00	0.00	0.00
25	0.0007	0.00	0.08	0.00	0.92	0.00	0.00	0.00	0.00	0.00

Offered Traffic is 13.00 Erlangs

N	P	D1	D2	Q1	Q2	P8	P4	P2	P1	PP
14	0.7138	0.71	1.00	9.28	13.00	0.63	0.56	0.43	0.26	0.10
15	0.4957	0.25	0.50	3.22	6.50	0.39	0.30	0.18	0.07	0.01
16	0.3343	0.11	0.33	1.45	4.33	0.23	0.16	0.07	0.02	0.00
17	0.2185	0.05	0.25	0.71	3.25	0.13	0.08	0.03	0.00	0.00
18	0.1383	0.03	0.20	0.36	2.60	0.07	0.04	0.01	0.00	0.00
19	0.0847	0.01	0.17	0.18	2.17	0.04	0.02	0.00	0.00	0.00
20	0.0501	0.01	0.14	0.09	1.86	0.02	0.01	0.00	0.00	0.00
21	0.0286	0.00	0.12	0.05	1.62	0.01	0.00	0.00	0.00	0.00
22	0.0158	0.00	0.11	0.02	1.44	0.01	0.00	0.00	0.00	0.00
23	0.0084	0.00	0.10	0.01	1.30	0.00	0.00	0.00	0.00	0.00
24	0.0043	0.00	0.09	0.01	1.18	0.00	0.00	0.00	0.00	0.00
25	0.0021	0.00	0.08	0.00	1.08	0.00	0.00	0.00	0.00	0.00
26	0.0010	0.00	0.08	0.00	1.00	0.00	0.00	0.00	0.00	0.00
27	0.0005	0.00	0.07	0.00	0.93	0.00	0.00	0.00	0.00	0.00

(continued)

				Offered Traffic is 14.00 Erlangs						
N	P	D1	D2	Q1	Q2	P8	P4	P2	P1	PP
15	0.7223	0.72	1.00	10.11	14.00	0.64	0.56	0.44	0.27	0.10
16	0.5085	0.25	0.50	3.56	7.00	0.40	0.31	0.19	0.07	0.01
17	0.3483	0.12	0.33	1.63	4.67	0.24	0.16	0.08	0.02	0.00
18	0.2317	0.06	0.25	0.81	3.50	0.14	0.09	0.03	0.00	0.00
19	0.1496	0.03	0.20	0.42	2.80	0.08	0.04	0.01	0.00	0.00
20	0.0936	0.02	0.17	0.22	2.33	0.04	0.02	0.00	0.00	0.00
21	0.0567	0.01	0.14	0.11	2.00	0.02	0.01	0.00	0.00	0.00
22	0.0332	0.00	0.12	0.06	1.75	0.01	0.00	0.00	0.00	0.00
23	0.0188	0.00	0.11	0.03	1.56	0.01	0.00	0.00	0.00	0.00
24	0.0103	0.00	0.10	0.01	1.40	0.00	0.00	0.00	0.00	0.00
25	0.0055	0.00	0.09	0.01	1.27	0.00	0.00	0.00	0.00	0.00
26	0.0028	0.00	0.08	0.00	1.17	0.00	0.00	0.00	0.00	0.00
27	0.0014	0.00	0.08	0.00	1.08	0.00	0.00	0.00	0.00	0.00
28	0.0007	0.00	0.07	0.00	1.00	0.00	0.00	0.00	0.00	0.00

				Offered Traffic is 15.00 Erlangs						
N	P	D1	D2	Q1	Q2	P8	P4	P2	P1	PP
16	0.7301	0.73	1.00	10.95	15.00	0.64	0.57	0.44	0.27	0.10
17	0.5203	0.26	0.50	3.90	7.50	0.41	0.32	0.19	0.07	0.01
18	0.3613	0.12	0.33	1.81	5.00	0.25	0.17	0.08	0.02	0.00
19	0.2442	0.06	0.25	0.92	3.75	0.15	0.09	0.03	0.00	0.00
20	0.1604	0.03	0.20	0.48	3.00	0.09	0.05	0.01	0.00	0.00
21	0.1023	0.02	0.17	0.26	2.50	0.05	0.02	0.01	0.00	0.00
22	0.0633	0.01	0.14	0.14	2.14	0.03	0.01	0.00	0.00	0.00
23	0.0380	0.00	0.12	0.07	1.88	0.01	0.01	0.00	0.00	0.00

				Offered Traffic is 15.00 Erlangs						
N	P	D1	D2	Q1	Q2	P8	P4	P2	P1	PP
24	0.0221	0.00	0.11	0.04	1.67	0.01	0.00	0.00	0.00	0.00
25	0.0124	0.00	0.10	0.02	1.50	0.00	0.00	0.00	0.00	0.00
26	0.0068	0.00	0.09	0.01	1.36	0.00	0.00	0.00	0.00	0.00
27	0.0036	0.00	0.08	0.00	1.25	0.00	0.00	0.00	0.00	0.00
28	0.0018	0.00	0.08	0.00	1.15	0.00	0.00	0.00	0.00	0.00
29	0.0009	0.00	0.07	0.00	1.07	0.00	0.00	0.00	0.00	0.00

				Offered Traffic is 20.00 Erlangs						
N	P	D1	D2	Q1	Q2	P8	P4	P2	P1	PP
21	0.7606	0.76	1.00	15.21	20.00	0.67	0.59	0.46	0.28	0.10
22	0.5679	0.28	0.50	5.68	10.00	0.44	0.34	0.21	0.08	0.01
23	0.4157	0.14	0.33	2.77	6.67	0.29	0.20	0.09	0.02	0.00
24	0.2981	0.07	0.25	1.49	5.00	0.18	0.11	0.04	0.01	0.00
25	0.2091	0.04	0.20	0.84	4.00	0.11	0.06	0.02	0.00	0.00
26	0.1434	0.02	0.17	0.48	3.33	0.07	0.03	0.01	0.00	0.00
27	0.0961	0.01	0.14	0.27	2.86	0.04	0.02	0.00	0.00	0.00

N	P	D1	D2	Q1	Q2	P8	P4	P2	P1	PP
					Offered Traffic is 20.00 Erlangs *(Continued)*					
28	0.0628	0.01	0.12	0.16	2.50	0.02	0.01	0.00	0.00	0.00
29	0.0401	0.00	0.11	0.09	2.22	0.01	0.00	0.00	0.00	0.00
30	0.0250	0.00	0.10	0.05	2.00	0.01	0.00	0.00	0.00	0.00
31	0.0151	0.00	0.09	0.03	1.82	0.00	0.00	0.00	0.00	0.00
32	0.0090	0.00	0.08	0.01	1.67	0.00	0.00	0.00	0.00	0.00
33	0.0052	0.00	0.08	0.01	1.54	0.00	0.00	0.00	0.00	0.00
34	0.0029	0.00	0.07	0.00	1.43	0.00	0.00	0.00	0.00	0.00
35	0.0016	0.00	0.07	0.00	1.33	0.00	0.00	0.00	0.00	0.00
36	0.0009	0.00	0.06	0.00	1.25	0.00	0.00	0.00	0.00	0.00

N	P	D1	D2	Q1	Q2	P8	P4	P2	P1	PP
					Offered Traffic is 25.00 Erlangs					
26	0.7824	0.78	1.00	19.56	25.00	0.69	0.61	0.47	0.29	0.11
27	0.6030	0.30	0.50	7.54	12.50	0.47	0.37	0.22	0.08	0.01
28	0.4573	0.15	0.33	3.81	8.33	0.31	0.22	0.10	0.02	0.00
29	0.3410	0.09	0.25	2.13	6.25	0.21	0.13	0.05	0.01	0.00
30	0.2499	0.05	0.20	1.25	5.00	0.13	0.07	0.02	0.00	0.00
31	0.1798	0.03	0.17	0.75	4.17	0.08	0.04	0.01	0.00	0.00
32	0.1269	0.02	0.14	0.45	3.57	0.05	0.02	0.00	0.00	0.00
33	0.0878	0.01	0.12	0.27	3.12	0.03	0.01	0.00	0.00	0.00
34	0.0596	0.01	0.11	0.17	2.78	0.02	0.01	0.00	0.00	0.00
35	0.0396	0.00	0.10	0.10	2.50	0.01	0.00	0.00	0.00	0.00
36	0.0258	0.00	0.09	0.06	2.27	0.01	0.00	0.00	0.00	0.00
37	0.0164	0.00	0.08	0.03	2.08	0.00	0.00	0.00	0.00	0.00
38	0.0103	0.00	0.08	0.02	1.92	0.00	0.00	0.00	0.00	0.00
39	0.0063	0.00	0.07	0.01	1.79	0.00	0.00	0.00	0.00	0.00

N	P	D1	D2	Q1	Q2	P8	P4	P2	P1	PP
					Offered Traffic is 25.00 Erlangs					
40	0.0038	0.00	0.07	0.01	1.67	0.00	0.00	0.00	0.00	0.00
41	0.0022	0.00	0.06	0.00	1.56	0.00	0.00	0.00	0.00	0.00
42	0.0013	0.00	0.06	0.00	1.47	0.00	0.00	0.00	0.00	0.00
43	0.0007	0.00	0.06	0.00	1.39	0.00	0.00	0.00	0.00	0.00

N	P	D1	D2	Q1	Q2	P8	P4	P2	P1	PP
					Offered Traffic is 30.00 Erlangs					
31	0.7989	0.80	1.00	23.97	30.00	0.71	0.62	0.48	0.29	0.11
32	0.6302	0.32	0.50	9.45	15.00	0.49	0.38	0.23	0.09	0.01
33	0.4905	0.16	0.33	4.90	10.00	0.34	0.23	0.11	0.02	0.00
34	0.3764	0.09	0.25	2.82	7.50	0.23	0.14	0.05	0.01	0.00
35	0.2846	0.06	0.20	1.71	6.00	0.15	0.08	0.02	0.00	0.00
36	0.2119	0.04	0.17	1.06	5.00	0.10	0.05	0.01	0.00	0.00
37	0.1553	0.02	0.14	0.67	4.29	0.06	0.03	0.00	0.00	0.00
38	0.1119	0.01	0.12	0.42	3.75	0.04	0.02	0.00	0.00	0.00

(continued)

N	P	D1	D2	Q1	Q2	P8	P4	P2	P1	PP
				Offered Traffic is 30.00 Erlangs *(Continued)*						
39	0.0793	0.01	0.11	0.26	3.33	0.03	0.01	0.00	0.00	0.00
40	0.0552	0.01	0.10	0.17	3.00	0.02	0.00	0.00	0.00	0.00
41	0.0378	0.00	0.09	0.10	2.73	0.01	0.00	0.00	0.00	0.00
42	0.0254	0.00	0.08	0.06	2.50	0.01	0.00	0.00	0.00	0.00
43	0.0168	0.00	0.08	0.04	2.31	0.00	0.00	0.00	0.00	0.00
44	0.0109	0.00	0.07	0.02	2.14	0.00	0.00	0.00	0.00	0.00
45	0.0069	0.00	0.07	0.01	2.00	0.00	0.00	0.00	0.00	0.00
46	0.0043	0.00	0.06	0.01	1.88	0.00	0.00	0.00	0.00	0.00
47	0.0027	0.00	0.06	0.00	1.76	0.00	0.00	0.00	0.00	0.00
48	0.0016	0.00	0.06	0.00	1.67	0.00	0.00	0.00	0.00	0.00
49	0.0009	0.00	0.05	0.00	1.58	0.00	0.00	0.00	0.00	0.00

N	P	D1	D2	Q1	Q2	P8	P4	P2	P1	PP
				Offered Traffic is 40.00 Erlangs						
41	0.8229	0.82	1.00	32.92	40.00	0.73	0.64	0.50	0.30	0.11
42	0.6706	0.34	0.50	13.41	20.00	0.52	0.41	0.25	0.09	0.01
43	0.5409	0.18	0.33	7.21	13.33	0.37	0.26	0.12	0.03	0.00
44	0.4317	0.11	0.25	4.32	10.00	0.26	0.16	0.06	0.01	0.00
45	0.3407	0.07	0.20	2.73	8.00	0.18	0.10	0.03	0.00	0.00
46	0.2658	0.04	0.17	1.77	6.67	0.13	0.06	0.01	0.00	0.00
47	0.2049	0.03	0.14	1.17	5.71	0.09	0.04	0.01	0.00	0.00
48	0.1560	0.02	0.12	0.78	5.00	0.06	0.02	0.00	0.00	0.00
49	0.1172	0.01	0.11	0.52	4.44	0.04	0.01	0.00	0.00	0.00
50	0.0870	0.01	0.10	0.35	4.00	0.02	0.01	0.00	0.00	0.00
51	0.0636	0.01	0.09	0.23	3.64	0.02	0.00	0.00	0.00	0.00
52	0.0459	0.00	0.08	0.15	3.33	0.01	0.00	0.00	0.00	0.00
53	0.0327	0.00	0.08	0.10	3.08	0.01	0.00	0.00	0.00	0.00

N	P	D1	D2	Q1	Q2	P8	P4	P2	P1	PP
				Offered Traffic is 40.00 Erlangs						
54	0.0230	0.00	0.07	0.07	2.86	0.00	0.00	0.00	0.00	0.00
55	0.0159	0.00	0.07	0.04	2.67	0.00	0.00	0.00	0.00	0.00
56	0.0108	0.00	0.06	0.03	2.50	0.00	0.00	0.00	0.00	0.00
57	0.0073	0.00	0.06	0.02	2.35	0.00	0.00	0.00	0.00	0.00
58	0.0048	0.00	0.06	0.01	2.22	0.00	0.00	0.00	0.00	0.00
59	0.0032	0.00	0.05	0.01	2.11	0.00	0.00	0.00	0.00	0.00
60	0.0020	0.00	0.05	0.00	2.00	0.00	0.00	0.00	0.00	0.00
61	0.0013	0.00	0.05	0.00	1.90	0.00	0.00	0.00	0.00	0.00
62	0.0008	0.00	0.05	0.00	1.82	0.00	0.00	0.00	0.00	0.00

N	P	D1	D2	Q1	Q2	P8	P4	P2	P1	PP
				Offered Traffic is 50.00 Erlangs						
51	0.8397	0.84	1.00	41.99	50.00	0.74	0.65	0.51	0.31	0.11
52	0.6996	0.35	0.50	17.49	25.00	0.54	0.42	0.26	0.09	0.01
53	0.5781	0.19	0.33	9.64	16.67	0.40	0.27	0.13	0.03	0.00

N	P	D1	D2	Q1	Q2	P8	P4	P2	P1	PP
				Offered Traffic is 50.00 Erlangs *(Continued)*						
54	0.4736	0.12	0.25	5.92	12.50	0.29	0.17	0.06	0.01	0.00
55	0.3845	0.08	0.20	3.85	10.00	0.21	0.11	0.03	0.00	0.00
56	0.3093	0.05	0.17	2.58	8.33	0.15	0.07	0.02	0.00	0.00
57	0.2465	0.04	0.14	1.76	7.14	0.10	0.04	0.01	0.00	0.00
58	0.1944	0.02	0.12	1.22	6.25	0.07	0.03	0.00	0.00	0.00
59	0.1518	0.02	0.11	0.84	5.56	0.05	0.02	0.00	0.00	0.00
60	0.1173	0.01	0.10	0.59	5.00	0.03	0.01	0.00	0.00	0.00
61	0.0897	0.01	0.09	0.41	4.55	0.02	0.01	0.00	0.00	0.00
62	0.0678	0.01	0.08	0.28	4.17	0.02	0.00	0.00	0.00	0.00
63	0.0507	0.00	0.08	0.19	3.85	0.01	0.00	0.00	0.00	0.00
64	0.0375	0.00	0.07	0.13	3.57	0.01	0.00	0.00	0.00	0.00
65	0.0274	0.00	0.07	0.09	3.33	0.00	0.00	0.00	0.00	0.00
66	0.0198	0.00	0.06	0.06	3.12	0.00	0.00	0.00	0.00	0.00
67	0.0141	0.00	0.06	0.04	2.94	0.00	0.00	0.00	0.00	0.00
68	0.0099	0.00	0.06	0.03	2.78	0.00	0.00	0.00	0.00	0.00
69	0.0069	0.00	0.05	0.02	2.63	0.00	0.00	0.00	0.00	0.00
70	0.0048	0.00	0.05	0.01	2.50	0.00	0.00	0.00	0.00	0.00
71	0.0032	0.00	0.05	0.01	2.38	0.00	0.00	0.00	0.00	0.00
72	0.0022	0.00	0.05	0.00	2.27	0.00	0.00	0.00	0.00	0.00
73	0.0014	0.00	0.04	0.00	2.17	0.00	0.00	0.00	0.00	0.00
74	0.0010	0.00	0.04	0.00	2.08	0.00	0.00	0.00	0.00	0.00

N	P	D1	D2	Q1	Q2	P8	P4	P2	P1	PP
				Offered Traffic is 60.00 Erlangs						
61	0.8524	0.85	1.00	51.15	60.00	0.75	0.66	0.52	0.31	0.12
62	0.7218	0.36	0.50	21.65	30.00	0.56	0.44	0.27	0.10	0.01
63	0.6070	0.20	0.33	12.14	20.00	0.42	0.29	0.14	0.03	0.00

N	P	D1	D2	Q1	Q2	P8	P4	P2	P1	PP
				Offered Traffic is 60.00 Erlangs						
64	0.5069	0.13	0.25	7.60	15.00	0.31	0.19	0.07	0.01	0.00
65	0.4201	0.08	0.20	5.04	12.00	0.22	0.12	0.03	0.00	0.00
66	0.3455	0.06	0.17	3.45	10.00	0.16	0.08	0.02	0.00	0.00
67	0.2818	0.04	0.14	2.42	8.57	0.12	0.05	0.01	0.00	0.00
68	0.2280	0.03	0.12	1.71	7.50	0.08	0.03	0.00	0.00	0.00
69	0.1829	0.02	0.11	1.22	6.67	0.06	0.02	0.00	0.00	0.00
70	0.1455	0.01	0.10	0.87	6.00	0.04	0.01	0.00	0.00	0.00
71	0.1147	0.01	0.09	0.63	5.45	0.03	0.01	0.00	0.00	0.00
72	0.0895	0.01	0.08	0.45	5.00	0.02	0.00	0.00	0.00	0.00
73	0.0693	0.01	0.08	0.32	4.62	0.01	0.00	0.00	0.00	0.00
74	0.0531	0.00	0.07	0.23	4.29	0.01	0.00	0.00	0.00	0.00
75	0.0403	0.00	0.07	0.16	4.00	0.01	0.00	0.00	0.00	0.00
76	0.0303	0.00	0.06	0.11	3.75	0.00	0.00	0.00	0.00	0.00
77	0.0225	0.00	0.06	0.08	3.53	0.00	0.00	0.00	0.00	0.00
78	0.0166	0.00	0.06	0.06	3.33	0.00	0.00	0.00	0.00	0.00
79	0.0121	0.00	0.05	0.04	3.16	0.00	0.00	0.00	0.00	0.00
80	0.0087	0.00	0.05	0.03	3.00	0.00	0.00	0.00	0.00	0.00

(continued)

				Offered Traffic is 60.00 Erlangs *(Continued)*						
N	P	D1	D2	Q1	Q2	P8	P4	P2	P1	PP
81	0.0062	0.00	0.05	0.02	2.86	0.00	0.00	0.00	0.00	0.00
82	0.0044	0.00	0.05	0.01	2.73	0.00	0.00	0.00	0.00	0.00
83	0.0031	0.00	0.04	0.01	2.61	0.00	0.00	0.00	0.00	0.00
84	0.0021	0.00	0.04	0.01	2.50	0.00	0.00	0.00	0.00	0.00
85	0.0015	0.00	0.04	0.00	2.40	0.00	0.00	0.00	0.00	0.00
86	0.0010	0.00	0.04	0.00	2.31	0.00	0.00	0.00	0.00	0.00

				Offered Traffic is 70.00 Erlangs						
N	P	D1	D2	Q1	Q2	P8	P4	P2	P1	PP
71	0.8624	0.86	1.00	60.37	70.00	0.76	0.67	0.52	0.32	0.12
72	0.7396	0.37	0.50	25.89	35.00	0.58	0.45	0.27	0.10	0.01
73	0.6305	0.21	0.33	14.71	23.33	0.43	0.30	0.14	0.03	0.00
74	0.5341	0.13	0.25	9.35	17.50	0.32	0.20	0.07	0.01	0.00
75	0.4497	0.09	0.20	6.30	14.00	0.24	0.13	0.04	0.00	0.00
76	0.3760	0.06	0.17	4.39	11.67	0.18	0.08	0.02	0.00	0.00
77	0.3123	0.04	0.14	3.12	10.00	0.13	0.05	0.01	0.00	0.00
78	0.2576	0.03	0.12	2.25	8.75	0.09	0.03	0.00	0.00	0.00
79	0.2109	0.02	0.11	1.64	7.78	0.07	0.02	0.00	0.00	0.00
80	0.1714	0.02	0.10	1.20	7.00	0.05	0.01	0.00	0.00	0.00
81	0.1382	0.01	0.09	0.88	6.36	0.03	0.01	0.00	0.00	0.00
82	0.1106	0.01	0.08	0.65	5.83	0.02	0.01	0.00	0.00	0.00
83	0.0878	0.01	0.08	0.47	5.38	0.02	0.00	0.00	0.00	0.00
84	0.0691	0.00	0.07	0.35	5.00	0.01	0.00	0.00	0.00	0.00
85	0.0540	0.00	0.07	0.25	4.67	0.01	0.00	0.00	0.00	
86	0.0418	0.00	0.06	0.18	4.38	0.01	0.00	0.00		
87	0.0321	0.00	0.06	0.13	4.12	0.00	0.00			
88	0.0244	0.00	0.06	0.09	3.89	0.00				

				Offered Traffic is 70.00 Erlangs						
N	P	D1	D2	Q1	Q2	P8	P4	P2	P1	PP
89	0.0184	0.00	0.05	0.07	3.68					
90	0.0138	0.00	0.05	0.05	3.50					
91	0.0102	0.00	0.05	0.03	3.33					
92	0.0075	0.00	0.05	0.02	3.18					
93	0.0055	0.00	0.04	0.02	3.04					
94	0.0039	0.00	0.04	0.01	2.92					
95	0.0028	0.00	0.04	0.01	2.80					
96	0.0020	0.00	0.04	0.01	2.69					
97	0.0014	0.00	0.04	0.00	2.59					
98	0.0010	0.00	0.04	0.00	2.50					

				Offered Traffic is 80.00 Erlangs						
N	P	D1	D2	Q1	Q2	P8	P4	P2	P1	PP
81	0.8706	0.87	1.00	69.65	80.00	0.77	0.68	0.53	0.32	0.12
82	0.7542	0.38	0.50	30.17	40.00	0.59	0.46	0.28	0.10	0.01
83	0.6499	0.22	0.33	17.33	26.67	0.45	0.31	0.15	0.03	0.00

				Offered Traffic is 80.00 Erlangs *(Continued)*						
N	P	D1	D2	Q1	Q2	P8	P4	P2	P1	PP
84	0.5571	0.14	0.25	11.14	20.00	0.34	0.20	0.08	0.01	0.00
85	0.4748	0.09	0.20	7.60	16.00	0.25	0.14	0.04	0.00	0.00
86	0.4024	0.07	0.17	5.37	13.33	0.19	0.09	0.02	0.00	0.00
87	0.3390	0.05	0.14	3.87	11.43	0.14	0.06	0.01	0.00	0.00
88	0.2838	0.04	0.12	2.84	10.00	0.10	0.04	0.01	0.00	0.00
89	0.2361	0.03	0.11	2.10	8.89	0.08	0.02	0.00	0.00	0.00
90	0.1951	0.02	0.10	1.56	8.00	0.06	0.02	0.00	0.00	0.00
91	0.1602	0.01	0.09	1.17	7.27	0.04	0.01	0.00	0.00	0.00
92	0.1306	0.01	0.08	0.87	6.67	0.03	0.01	0.00	0.00	
93	0.1058	0.01	0.08	0.65	6.15	0.02	0.00	0.00	0.00	
94	0.0851	0.01	0.07	0.49	5.71	0.01	0.00	0.00		
95	0.0679	0.00	0.07	0.36	5.33	0.01	0.00	0.00		
96	0.0538	0.00	0.06	0.27	5.00	0.01	0.00			
97	0.0423	0.00	0.06	0.20	4.71	0.01	0.00			
98	0.0330	0.00	0.06	0.15	4.44	0.00				
99	0.0256	0.00	0.05	0.11	4.21	0.00				
100	0.0196	0.00	0.05	0.08	4.00					
101	0.0150	0.00	0.05	0.06	3.81					
102	0.0113	0.00	0.05	0.04	3.64					
103	0.0085	0.00	0.04	0.03	3.48					
104	0.0063	0.00	0.04	0.02	3.33					
105	0.0047	0.00	0.04	0.01	3.20					
106	0.0034	0.00	0.04	0.01	3.08					
107	0.0025	0.00	0.04	0.01	2.96					
108	0.0018	0.00	0.04	0.01	2.86					
109	0.0013	0.00	0.03	0.00	2.76					
110	0.0009	0.00	0.03	0.00	2.67					

				Offered Traffic is 90.00 Erlangs						
N	P	D1	D2	Q1	Q2	P8	P4	P2	P1	PP
91	0.8775	0.88	1.00	78.97	90.00	0.77	0.68	0.53	0.32	0.12
92	0.7665	0.38	0.50	34.49	45.00	0.60	0.46	0.28	0.10	0.01
93	0.6665	0.22	0.33	19.99	30.00	0.46	0.31	0.15	0.03	0.00
94	0.5767	0.14	0.25	12.98	22.50	0.35	0.21	0.08	0.01	0.00
95	0.4966	0.10	0.20	8.94	18.00	0.27	0.14	0.04	0.00	
96	0.4254	0.07	0.17	6.38	15.00	0.20	0.09	0.02	0.00	
97	0.3625	0.05	0.14	4.66	12.86	0.15	0.06	0.01	0.00	
98	0.3072	0.04	0.12	3.46	11.25	0.11	0.04	0.01		
99	0.2589	0.03	0.11	2.59	10.00	0.08	0.03	0.00		
100	0.2169	0.02	0.10	1.95	9.00	0.06	0.02	0.00		
101	0.1807	0.02	0.09	1.48	8.18	0.05	0.01	0.00		
102	0.1496	0.01	0.08	1.12	7.50	0.03	0.01			
103	0.1231	0.01	0.08	0.85	6.92	0.02	0.00			
104	0.1007	0.01	0.07	0.65	6.43	0.02	0.00			
105	0.0818	0.01	0.07	0.49	6.00	0.01	0.00			
106	0.0660	0.00	0.06	0.37	5.62	0.01				
107	0.0529	0.00	0.06	0.28	5.29	0.01				
108	0.0421	0.00	0.06	0.21	5.00	0.00				

(continued)

N	P	D1	D2	Q1	Q2	P8	P4	P2	P1	PP
					Offered Traffic is 90.00 Erlangs *(Continued)*					
109	0.0333	0.00	0.05	0.16	4.74	0.00				
110	0.0262	0.00	0.05	0.12	4.50					
111	0.0204	0.00	0.05	0.09	4.29					
112	0.0158	0.00	0.05	0.06	4.09					
113	0.0122	0.00	0.04	0.05	3.91					
114	0.0093	0.00	0.04	0.03	3.75					
115	0.0070	0.00	0.04	0.03	3.60					
116	0.0053	0.00	0.04	0.02	3.46					
117	0.0040	0.00	0.04	0.01	3.33					
118	0.0029	0.00	0.04	0.01	3.21					
119	0.0022	0.00	0.03	0.01	3.10					
120	0.0016	0.00	0.03	0.00	3.00					
121	0.0011	0.00	0.03	0.00	2.90					
122	0.0008	0.00	0.03	0.00	2.81					

N	P	D1	D2	Q1	Q2	P8	P4	P2	P1	PP
					Offered Traffic is 100.00 Erlangs					
101	0.8833	0.88	1.00	88.33	100.00	0.78	0.69	0.54	0.32	0.12
102	0.7771	0.39	0.50	38.85	50.00	0.61	0.47	0.29	0.11	0.01
103	0.6808	0.23	0.33	22.69	33.33	0.47	0.32	0.15	0.03	0.00
104	0.5939	0.15	0.25	14.85	25.00	0.36	0.22	0.08	0.01	0.00
105	0.5157	0.10	0.20	10.31	20.00	0.28	0.15	0.04	0.00	0.00
106	0.4458	0.07	0.17	7.43	16.67	0.21	0.10	0.02	0.00	
107	0.3835	0.05	0.14	5.48	14.29	0.16	0.07	0.01	0.00	
108	0.3283	0.04	0.12	4.10	12.50	0.12	0.04	0.01	0.00	
109	0.2797	0.03	0.11	3.11	11.11	0.09	0.03	0.00		

N	P	D1	D2	Q1	Q2	P8	P4	P2	P1	PP
					Offered Traffic is 100.00 Erlangs					
110	0.2370	0.02	0.10	2.37	10.00	0.07	0.02	0.00		
111	0.1998	0.02	0.09	1.82	9.09	0.05	0.01	0.00		
112	0.1675	0.01	0.08	1.40	8.33	0.04	0.01	0.00		
113	0.1397	0.01	0.08	1.07	7.69	0.03	0.01			
114	0.1158	0.01	0.07	0.83	7.14	0.02	0.00			
115	0.0954	0.01	0.07	0.64	6.67	0.01	0.00			
116	0.0782	0.00	0.06	0.49	6.25	0.01				
117	0.0637	0.00	0.06	0.37	5.88	0.01				
118	0.0516	0.00	0.06	0.29	5.56	0.01				
119	0.0415	0.00	0.05	0.22	5.26	0.00				
120	0.0332	0.00	0.05	0.17	5.00	0.00				
121	0.0264	0.00	0.05	0.13	4.76					
122	0.0208	0.00	0.05	0.09	4.55					
123	0.0163	0.00	0.04	0.07	4.35					
124	0.0127	0.00	0.04	0.05	4.17					
125	0.0099	0.00	0.04	0.04	4.00					
126	0.0076	0.00	0.04	0.03	3.85					
127	0.0058	0.00	0.04	0.02	3.70					

| | | | | Offered Traffic is 100.00 Erlangs *(Continued)* | | | | | | |
N	P	D1	D2	Q1	Q2	P8	P4	P2	P1	PP
128	0.0044	0.00	0.04	0.02	3.57					
129	0.0033	0.00	0.03	0.01	3.45					
130	0.0025	0.00	0.03	0.01	3.33					
131	0.0019	0.00	0.03	0.01	3.23					
132	0.0014	0.00	0.03	0.00	3.12					
133	0.0010	0.00	0.03	0.00	3.03					
134	0.0007	0.00	0.03	0.00	2.94					

Appendix I

EQEEB Table

1 Minute Maximum Wait Queue

hour first attm	hour con- nect	hour ovfl	delayed portion	Delay Length Call Factor	lines or servr	Hours Connected Traffic on Servers									
						1	2	3	4	5	6	7	8	9	10
.20	.180	.020	.190	.141	1	.180					.097	.097	.097	.097	
.40	.321	.079	.355	.252	1	.321									
.40	.388	.012	.058	.044	2	.291	.097								
.60	.564	.036	.114	.086	2	.388	.176								
.80	.724	.076	.178	.132	2	.465	.258								
.80	.783	.017	.041	.032	3	.449	.240	.093							
1.0	.866	.134	.246	.180	2	.529	.337								
1.0	.964	.036	.069	.052	3	.508	.310	.146							
1.2	.991	.209	.315	.226	2	.582	.409								
1.2	1.136	.064	.102	.077	3	.557	.375	.203							
1.2	1.183	.017	.028	.021	4	.549	.363	.192	.079						
1.4	1.100	.300	.383	.270	2	.626	.473								
1.4	1.296	.104	.140	.105	3	.599	.434	.263							
1.4	1.369	.031	.043	.033	4	.588	.418	.246	.116						
1.6	1.444	.156	.181	.135	3	.636	.486	.322							
1.6	1.548	.052	.062	.048	4	.622	.467	.300	.159						
1.8	1.580	.220	.225	.165	3	.667	.534	.380							
1.8	1.720	.080	.085	.065	4	.652	.511	.352	.205						
1.8	1.774	.026	.028	.022	5	.646	.502	.341	.194	.091					
2.0	1.704	.296	.271	.196	3	.695	.576	.433							
2.0	1.883	.117	.111	.084	4	.678	.551	.401	.253						
2.0	1.959	.041	.040	.030	5	.671	.540	.387	.238	.123					
2.2	1.815	.385	.317	.227	3	.719	.613	.483							
2.2	2.037	.163	.139	.104	4	.702	.587	.448	.301						
2.2	2.138	.062	.054	.041	5	.693	.574	.430	.283	.159					

hour first attm	hour con-nect	hour ovfl	delayed portion	Delay Length Call Factor	lines or servr	Hours Connected Traffic on Servers									
						1	2	3	4	5	6	7	8	9	10
2.4	1.915	.485	.363	.257	3	.740	.646	.528							
2.4	2.182	.218	.170	.127	4	.722	.619	.491	.350						
2.4	2.312	.088	.070	.054	5	.713	.604	.471	.327	.197					
2.6	2.004	.596	.407	.285	3	.759	.676	.569							
2.6	2.316	.284	.203	.150	4	.741	.648	.531	.397						
2.6	2.478	.122	.089	.068	5	.731	.632	.508	.370	.238					
2.6	2.553	.047	.035	.027	6	.726	.624	.497	.358	.226	.123				
2.8	2.441	.359	.236	.173	4	.758	.674	.567	.442						
2.8	2.683	.162	.110	.083	5	.747	.657	.543	.412	.279					
2.8	2.773	.067	.046	.035	6	.741	.648	.530	.397	.264					
3.0	2.555	.445	.271	.197	4	.773	.698	.600	.484						
3.0	2.789	.211	.132	.100	5	.762	.680	.575	.451	.321					
3.0	2.909	.091	.058	.045	6	.755	.670	.561	.434	.303	.186				
3.2	2.660	.540	.306	.220	4	.787	.719	.631	.523						
3.2	2.932	.268	.157	.117	5	.775	.701	.604	.489	.363					
3.2	3.079	.121	.072	.055	6	.768	.690	.589	.469	.341	.221				
3.2	3.150	.050	.030	.023	7	.765	.684	.581	.459	.330	.211	.119			
3.4	2.756	.644	.341	.243	4	.800	.738	.658	.559						
3.4	3.067	.333	.183	.135	5	.788	.720	.632	.524	.404					
3.4	3.242	.158	.088	.067	6	.780	.708	.615	.503	.379	.258				
3.4	3.331	.069	.039	.030	7	.776	.702	.606	.491	.366	.245	.145			
3.6	2.843	.757	.376	.266	4	.811	.756	.683	.593						
3.6	3.193	.407	.210	.154	5	.799	.737	.657	.557	.443					
3.6	3.400	.200	.106	.080	6	.791	.725	.639	.534	.415	.295				
3.6	3.509	.091	.049	.037	7	.787	.718	.629	.521	.400	.279	.174			
3.8	2.922	.878	.411	.287	4	.822	.771	.706	.623						
3.8	3.331	.489	.237	.174	5	.810	.753	.679	.588	.480					
3.8	3.550	.250	.124	.094	6	.801	.740	.661	.564	.451	.332				
3.8	3.682	.118	.060	.046	7	.796	.733	.651	.549	.434	.314	.205			
3.8	3.748	.052	.026	.020	8	.794	.729	.645	.542	.425	.305	.197	.113		
4.0	3.420	.580	.266	.193	5	.819	.768	.700	.616	.516					
4.0	3.693	.307	.144	.108	6	.811	.755	.682	.591	.485	.369				
4.0	3.849	.151	.072	.055	7	.805	.747	.670	.576	.466	.348	.237			
4.0	3.931	.069	.033	.025	8	.803	.742	.664	.567	.455	.337	.226	.136		
4.2	3.521	.679	.294	.212	5	.828	.781	.720	.642	.549					
4.2	3.828	.372	.166	.123	6	.820	.768	.701	.617	.517	.406				
4.2	4.011	.189	.086	.065	7	.814	.759	.689	.600	.496	.382	.270			
4.2	4.111	.089	.041	.031	8	.811	.755	.682	.591	.484	.369	.257	.162		
4.4	3.614	.786	.323	.231	5	.837	.793	.737	.666	.580					
4.4	3.956	.444	.188	.139	6	.828	.780	.719	.641	.547	.441				
4.4	4.167	.233	.100	.076	7	.822	.771	.706	.624	.525	.415	.304			
4.4	4.286	.114	.050	.038	8	.818	.766	.698	.613	.512	.400	.289	.189		
4.6	3.700	.900	.352	.250	5	.845	.805	.753	.688	.609					
4.6	4.076	.524	.211	.155	6	.836	.792	.735	.663	.576	.475				
4.6	4.317	.283	.116	.088	7	.829	.782	.722	.645	.553	.448	.338			

(continued)

EQEEB TABLE *(continued)*

hour first attm	hour con-nect	hour ovfl	delayed portion	Delay Length Call Factor	lines or servr	1	2	3	4	5	6	7	8	9	10
4.6	4.458	.142	.059	.045	8	.826	.777	.714	.634	.539	.431	.320	.218		
4.6	4.453	.067	.028	.022	9	.823	.774	.709	.628	.531	.422	.311	.209	.13	
4.8	3.779	1.02	.380	.268	5	.852	.815	.768	.709	.635					
4.8	4.189	.611	.235	.172	6	.843	.802	.750	.684	.603	.508				
4.8	4.461	.339	.133	.100	7	.836	.793	.737	.665	.579	.479	.372			
4.8	4.624	.176	.070	.054	8	.832	.787	.728	.654	.564	.460	.352	.247		
4.8	4.714	.086	.034	.026	9	.830	.783	.723	.647	.555	.450	.340	.237	.15	
5.0	3.851	1.15	.408	.286	5	.858	.825	.781	.727	.659					
5.0	4.295	.705	.259	.188	6	.850	.812	.764	.703	.628	.539				
5.0	4.597	.403	.151	.113	7	.843	.803	.750	.684	.603	.509	.405			
5.0	4.785	.215	.082	.062	8	.839	.796	.741	.672	.587	.489	.383	.278		
5.0	4.892	.108	.041	.032	9	.836	.792	.736	.664	.577	.477	.370	.265	.17	
5.5	4.528	.972	.320	.229	6	.865	.834	.794	.744	.683	.608				
5.5	4.909	.591	.200	.147	7	.858	.824	.781	.726	.658	.577	.484			
5.5	5.163	.337	.116	.087	8	.853	.817	.771	.713	.641	.555	.458	.356		
5.5	5.139	.181	.063	.048	9	.850	.813	.765	.704	.629	.540	.441	.338	.240	
5.5	5.409	.091	.032	.025	10	.848	.810	.761	.699	.622	.532	.431	.327	.230	.149
6.0	4.722	1.28	.381	.269	6	.878	.852	.819	.779	.728	.666				
6.0	5.179	.821	.251	.183	7	.871	.843	.807	.761	.705	.636	.556			
6.0	5.504	.496	.155	.116	8	.866	.835	.796	.747	.687	.613	.527	.432		
6.0	5.718	.282	.090	.068	9	.862	.830	.789	.738	.673	.596	.507	.410	.312	
6.0	5.848	.152	.049	.037	10	.860	.827	.785	.731	.665	.586	.494	.396	.297	.207
6.5	5.408	1.09	.305	.219	7	.882	.858	.828	.791	.744	.687	.618			
6.5	5.806	.694	.198	.147	8	.877	.851	.818	.777	.726	.663	.589	.504		
6.5	6.084	.416	.121	.091	9	.873	.846	.811	.767	.712	.646	.567	.479	.384	
6.5	6.263	.237	.070	.053	10	.871	.842	.805	.760	.703	.633	.552	.461	.365	.272
6.5	6.372	.128	.038	.029	11	.869	.839	.802	.755	.696	.626	.543	.450	.354	.260
						.178									
7.0	5.602	1.40	.359	.254	7	.892	.872	.847	.815	.776	.729	.671			
7.0	6.070	.930	.244	.179	8	.887	.865	.837	.802	.759	.707	.643	.569		
7.0	6.414	.586	.157	.117	9	.883	.859	.829	.792	.745	.689	.621	.542	.455	
7.0	6.650	.350	.095	.072	10	.880	.855	.823	.784	.735	.675	.604	.522	.432	.340
7.0	6.601	.199	.055	.042	11	.878	.852	.819	.779	.728	.666	.593	.509	.417	.324
						.236									
7.0	6.893	.107	.030	.023	12	.877	.850	.817	.775	.724	.661	.586	.500	.408	.315
						.227	.153								
7.5	5.675	1.73	.411	.288	7	.901	.884	.862	.836	.803	.764	.715			
7.5	6.297	1.20	.292	.211	8	.896	.877	.853	.824	.788	.743	.690	.626		
7.5	6.708	.792	.196	.145	9	.892	.871	.845	.813	.774	.726	.667	.599	.521	
7.5	7.005	.495	.125	.094	10	.888	.867	.839	.805	.763	.712	.650	.577	.495	.407
7.5	7.205	.295	.075	.057	11	.886	.863	.835	.799	.756	.702	.637	.562	.478	.388
						.299									
7.5	7.333	.167	.043	.033	12	.884	.861	.832	.796	.750	.695	.629	.552	.466	.376
						.286	.205								
8.0	6.491	1.51	.340	.242	8	.904	.887	.867	.842	.812	.774	.729	.675		

hour first attm	hour con-nect	hour ovfl	delayed portion	Delay Length Call Factor	lines or servr	Hours Connected Traffic on Servers									
						1	2	3	4	5	6	7	8	9	10
8.0	6.967	1.03	.238	.174	9	.900	.882	.859	.832	.799	.758	.708	.649	.580	
8.0	7.326	.674	.158	.118	10	.896	.877	.853	.824	.788	.744	.690	.627	.554	.473
8.0	7.581	.419	.099	.075	11	.894	.873	.849	.818	.780	.733	.677	.610	.534	.450
						.363									
8.0	7.751	.249	.060	.046	12	.892	.871	.845	.813	.774	.726	.667	.599	.520	.435
						.347	.262								
8.0	7.859	.141	.034	.026	13	.890	.869	.843	.811	.770	.721	.661	.591	.511	.425
						.337	.252	.177							
8.5	6.656	1.84	.387	.273	8	.911	.896	.879	.858	.832	.800	.762	.717		
8.5	7.192	1.31	.281	.203	9	.907	.891	.872	.848	.820	.785	.743	.692	.633	
8.5	7.614	.886	.194	.143	10	.903	.886	.866	.840	.809	.772	.726	.671	.607	.534
8.5	7.927	.573	.127	.096	11	.900	.883	.861	.834	.801	.761	.712	.654	.586	.510
						.427									
8.5	8.146	.354	.079	.061	12	.898	.880	.857	.829	.795	.753	.701	.641	.570	.492
						.407	.322								
8.5	8.290	.210	.047	.036	13	.897	.878	.855	.826	.790	.747	.694	.632	.560	.480
						.394	.309	.229							
8.5	8.381	.119	.027	.021	14	.896	.877	.853	.824	.788	.743	.690	.626	.553	.472
						.386	.300	.221	.153						
9.0	7.386	1.61	.324	.232	9	.913	.899	.883	.863	.838	.809	.773	.730	.679	
9.0	7.869	1.13	.232	.170	10	.910	.895	.877	.855	.828	.796	.756	.709	.654	.589
9.0	8.242	.758	.158	.118	11	.907	.891	.872	.848	.820	.785	.743	.692	.632	.564
						.488									
9.0	8.513	.487	.103	.078	12	.904	.888	.868	.843	.813	.776	.732	.678	.616	.545
						.466	.384								
9.0	8.700	.300	.064	.049	13	.903	.886	.865	.840	.808	.770	.724	.669	.604	.531
						.450	.367	.285							
9.0	8.823	.177	.038	.029	14	.902	.884	.863	.837	.805	.766	.718	.662	.596	.521
						.440	.356	.274	.200						
9.5	7.553	1.95	.367	.260	9	.919	.907	.892	.875	.854	.829	.798	.762	.718	
9.5	8.092	1.41	.271	.197	10	.915	.902	.887	.868	.844	.817	.783	.743	.695	.639
9.5	8.524	.976	.191	.142	11	.912	.898	.882	.861	.836	.806	.770	.726	.674	.614
						.545									
9.5	8.851	.649	.129	.097	12	.910	.895	.877	.856	.829	.797	.759	.712	.657	.593
						.522	.444								
9.5	9.086	.414	.083	.063	13	.908	.893	.874	.852	.824	.791	.750	.701	.644	.578
						.504	.424	.343							
9.5	9.246	.254	.051	.039	14	.907	.891	.872	.849	.821	.786	.744	.694	.635	.567
						.491	.410	.329	.251						
9.5	9.351	.149	.030	.023	15	.906	.890	.871	.847	.818	.783	.740	.689	.628	.559
						.483	.401	.320	.242	.173					
10	7.696	2.30	.409	.287	9	.924	.913	.901	.885	.867	.846	.820	.789	.751	
10	8.287	1.71	.311	.223	10	.921	.909	.895	.879	.859	.835	.806	.771	.730	.682
10	8.776	1.22	.226	.166	11	.918	.905	.890	.872	.851	.825	.793	.755	.711	.658
						.597									

(continued)

hour first attm	hour con-nect	hour ovfl	delayed portion	Delay Length Call Factor	lines or servr	Hours Connected Traffic on Servers									
						1	2	3	4	5	6	7	8	9	10
10	9.160	.840	.158	.118	12	.915	.902	.886	.867	.844	.816	.782	.742	.694	.637
						.573	.501								
10	9.446	.554	.105	.080	13	.913	.900	.883	.863	.839	.809	.773	.731	.680	.621
						.554	.480	.401							
10	9.648	.352	.067	.051	14	.912	.898	.881	.860	.835	.804	.767	.722	.670	.609
						.539	.464	.384	.305						
10	9.785	.215	.041	.032	15	.911	.896	.879	.858	.832	.800	.762	.717	.663	.600
						.529	.453	.373	.293	.220					
11	8.601	2.40	.389	.74	10	.930	.921	.910	.897	.882	.864	.843	.817	.787	.751
11	9.192	1.81	.299	.215	11	.927	.917	.905	.891	.875	.855	.832	.804	.770	.731
11	9.686	1.31	.221	.162	12	.924	.914	.901	.886	.869	.847	.822	.791	.755	.715
						.661	.604								
11	10.08	.920	.157	.117	13	.922	.911	.898	.882	.863	.840	.813	.780	.741	.695
						.641	.580	.513							
11	10.38	.621	.107	.081	14	.920	.909	.895	.878	.859	.835	.806	.771	.730	.681
						.625	.562	.491	.417						
11	10.60	.404	.070	.054	15	.919	.907	.893	.876	.855	.830	.800	.764	.721	.671
						.613	.547	.475	.400	.323					
11	10.75	.254	.044	.034	16	.918	.906	.891	.874	.853	.827	.796	.759	.715	.664
						.604	.537	.464	.387	.311	.239				
11	10.85	.154	.027	.021	17	.918	.905	.891	.873	.851	.825	.794	.756	.711	.659
						.598	.530	.456	.379	.303	.231	.168			
12	9.51	2.49	.372	.263	11	.935	.927	.917	.907	.894	.879	.861	.840	.815	.785
12	10.01	1.90	.288	.208	12	.932	.924	.914	.902	.888	.872	.852	.829	.802	.769
12	10.60	1.40	.216	.159	13	.930	.921	.910	.898	.883	.865	.844	.819	.789	.754
						.713	.665	.610							
12	10.00	.997	.156	.116	14	.928	.919	.907	.894	.878	.859	.837	.810	.778	.740
						.696	.645	.587	.523						
12	11.31	.687	.108	.082	15	.927	.917	.905	.891	.874	.855	.831	.803	.769	.729
						.683	.629	.569	.502	.430					
12	11.54	.457	.073	.056	16	.926	.915	.903	.889	.872	.851	.826	.797	.762	.721
						.672	.617	.554	.486	.413	.340				
12	11.71	.294	.047	.036	17	.925	.914	.902	.887	.869	.848	.823	.793	.757	.714
						.665	.608	.544	.474	.401	.327	.256			
12	11.82	.183	.029	.023	18	.924	.913	.901	.886	.868	.846	.821	.790	.753	.710
						.659	.601	.536	.466	.392	.318	.248	.184		
13	10.42	2.58	.356	.253	12	.939	.932	.924	.915	.904	.891	.876	.858	.837	.813
						.784	.751								
13	11.01	1.99	.279	.202	13	.937	.929	.921	.911	.899	.885	.869	.849	.827	.800
						.768	.731	.689							
13	11.52	1.48	.211	.156	14	.935	.927	.918	.907	.894	.880	.862	.841	.816	.787
						.753	.713	.667	.615						
13	11.93	1.07	.155	.116	15	.933	.925	.915	.904	.891	.875	.856	.834	.808	.776
						.740	.697	.648	.593	.531					
13	12.25	.752	.110	.083	16	.932	.923	.913	.901	.887	.871	.851	.828	.800	.767
						.729	.684	.633	.575	.511	.443				

hour first attm	hour con-nect	hour ovfl	delayed portion	Delay Length Call Factor	lines or servr	Hours Connected Traffic on Servers									
						1	2	3	4	5	6	7	8	9	10
13	12.49	.510	.075	.057	17	.931	.922	.912	.899	.885	.868	.847	.823	.794	.760
						.720	.673	.620	.560	.495	.425	.354			
13	12.66	.336	.050	.038	18	.930	.921	.910	.898	.883	.865	.844	.819	.790	.754
						.713	.665	.611	.550	.483	.413	.341	.272		
13	12.79	.214	.032	.024	19	.930	.920	.910	.897	.882	.864	.842	.817	.786	.751
						.709	.660	.604	.542	.475	.404	.332	.263	.200	
14	10.67	3.33	.422	.294	12	.945	.939	.932	.925	.916	.906	.894	.880	.864	.846
						.824	.799								
14	11.39	2.66	.343	.244	13	.943	.937	.929	.921	.912	.901	.888	.873	.855	.835
						.811	.783	.750							
14	11.93	2.07	.270	.196	14	.941	.934	.927	.918	.908	.896	.882	.866	.847	.824
						.798	.767	.732	.691						
14	12.44	1.56	.207	.153	15	.939	.932	.924	.915	.904	.891	.877	.859	.838	.814
						.786	.753	.714	.670	.620					
14	12.85	1.15	.154	.115	16	.938	.931	.922	.912	.901	.887	.872	.853	.831	.805
						.775	.740	.699	.651	.598	.539				
14	13.18	.815	.110	.083	17	.937	.929	.920	.910	.898	.884	.867	.848	.825	.798
						.766	.728	.685	.636	.580	.519	.454			
14	13.44	.563	.077	.059	18	.936	.928	.919	.908	.896	.881	.864	.844	.820	.792
						.758	.719	.675	.623	.566	.503	.437	.368		
14	13.62	.378	.052	.040	19	.935	.927	.918	.907	.894	.879	.862	.841	.816	.787
						.753	.713	.666	.614	.555	.491	.424	.355	.287	
14	13.75	.246	.034	.026	20	.934	.926	.917	.906	.893	.878	.860	.838	.813	.783
						.748	.708	.660	.607	.547	.483	.414	.345	.277	.215
15	11.59	3.41	.404	.283	13	.948	.943	.937	.930	.922	.914	.903	.892	.878	.862
						.844	.822	.797							
15	12.26	2.74	.330	.236	14	.946	.941	.934	.927	.919	.909	.898	.885	.870	.853
						.833	.809	.782	.750						
15	12.85	2.15	.263	.191	15	.944	.939	.932	.924	.915	.905	.893	.879	.863	.844
						.822	.797	.767	.732	.692					
15	13.36	1.64	.203	.150	16	.943	.937	.930	.922	.912	.901	.889	.874	.856	.836
						.812	.784	.752	.715	.672	.624				
15	13.78	1.22	.152	.114	17	.942	.935	.928	.919	.909	.898	.884	.869	.850	.829
						.803	.774	.739	.700	.654	.603	.547			
15	14.12	.878	.111	.084	18	.941	.934	.926	.917	.907	.895	.881	.864	.845	.822
						.796	.764	.728	.686	.639	.585	.527	.464		
15	14.38	.616	.078	.060	19	.940	.933	.925	.916	.905	.893	.878	.861	.841	.817
						.789	.757	.719	.676	.626	.571	.511	.447	.380	
15	14.58	.420	.054	.041	20	.939	.932	.924	.915	.904	.891	.876	.858	.837	.813
						.784	.751	.712	.667	.617	.560	.499	.434	.367	.300
15	14.72	.279	.036	.028	21	.939	.931	.923	.914	.903	.889	.874	.856	.835	.810
						.781	.746	.707	.661	.610	.552	.490	.424	.357	.291
						.229									
1.0	.984	.016	.295	.447	2	.579	.405								
1.0	.997	.003	.074	.113	3	.518	.323	.156							

(continued)

hour first attm	hour con- nect	hour ovfl	delayed portion	Delay Length Call Factor	lines or servr	Hours Connected Traffic on Servers									
						1	2	3	4	5	6	7	8	9	10
1.0	.999	.001	.016	.025	4	.504	.305	.141	.049						
1.6	1.511	.089	.640	.935	2	.789	.722								
1.6	1.583	.017	.218	.332	3	.668	.535	.381							
1.6	1.595	.005	.068	.104	4	.631	.480	.314	.171						
1.6	1.599	.001	.019	.030	5	.620	.463	.296	.155	.065					
2.0	1.754	.246	.886	1.22	2	.888	.866								
2.0	1.959	.041	.359	.541	3	.749	.661	.548							
2.0	1.989	.011	.128	.195	4	.694	.576	.433	.286						
2.0	1.997	.003	.042	.065	5	.676	.547	.396	.248	.130					
2.6	2.459	.141	.632	.925	3	.875	.823	.779							
2.6	2.565	.035	.262	.398	4	.775	.700	.603	.487						
2.6	2.589	.011	.102	.156	5	.742	.649	.532	.399	.266					
2.6	2.596	.004	.037	.058	6	.729	.630	.206	.367	.235	.130				
3.0	2.692	.308	.829	1.16	3	.912	.898	.881							
3.0	2.931	.069	.385	.580	4	.823	.773	.708	.627						
3.0	2.978	.022	.161	.246	5	.779	.707	.614	.501	.377					
3.0	2.992	.008	.065	.099	6	.762	.680	.575	.451	.321	.203				
3.0	2.997	.003	.025	.038	7	.754	.668	.559	.432	.300	.184	.099			
3.6	3.412	.188	.620	.909	4	.890	.869	.843	.810						
3.6	3.456	.054	.287	.434	5	.831	.784	.724	.649	.557					
3.6	3.580	.020	.126	.193	6	.803	.743	.666	.570	.458	.340				
3.6	3.592	.008	.054	.082	7	.791	.725	.640	.535	.416	.296	.189			
3.6	3.597	.003	.022	.033	8	.786	.717	.628	.520	.399	.278	.173	.095		
4.0	3.640	.360	.788	1.12	4	.927	.917	.905	.891						
4.0	3.904	.096	.398	.599	5	.863	.832	.791	.740	.677					
4.0	3.965	.035	.184	.281	6	.829	.781	.720	.643	.549	.444				
4.0	3.986	.014	.083	.127	7	.813	.757	.686	.597	.491	.377	.265			
4.0	3.994	.006	.036	.055	8	.805	.747	.670	.576	.466	.348	.237	.145		
4.0	3.998	.002	.015	.022	9	.802	.742	.664	.566	.454	.336	.225	.136	.073	
4.6	4.370	.230	.607	.891	5	.910	.896	.878	.856	.830					
4.6	4.525	.075	.302	.457	6	.865	.834	.794	.744	.682	.607				
4.6	4.570	.030	.144	.221	7	.842	.801	.748	.681	.598	.503	.398			
4.6	4.587	.013	.067	.103	8	.831	.785	.725	.649	.558	.454	.345	.241		
4.6	4.594	.006	.030	.046	9	.826	.777	.714	.634	.539	.431	.321	.218	.135	
5.0	4.594	.406	.756	1.08	5	.937	.929	.920	.910	.898					
5.0	4.876	.124	.405	.608	6	.889	.867	.839	.806	.764	.712				
5.0	4.952	.048	.200	.305	7	.860	.827	.785	.732	.666	.587	.495			
5.0	4.979	.021	.097	.149	8	.846	.807	.757	.693	.615	.523	.421	.317		
5.0	4.991	.009	.046	.070	9	.839	.797	.743	.674	.590	.492	.386	.281	.188	
5.0	4.996	.004	.021	.032	10	.836	.792	.736	.664	.577	.477	.370	.265	.174	.104
5.5	4.758	.742	.915	1.248	5	.959	.956	.952	.948	.943					
5.5	5.264	.236	.561	.829	6	.918	.905	.891	.873	.851	.825				
5.5	5.415	.085	.291	.441	7	.883	.859	.829	.791	.745	.688	.620			
5.5	5.464	.036	.147	.224	8	.864	.833	.794	.743	.981	.606	.519	.423		
5.5	5.483	.017	.073	.111	9	.855	.820	.775	.718	.648	.564	.468	.367	.269	

hour first attm	hour con-nect	hour ovfl	delayed portion	Delay Length Call Factor	lines or servr	Hours Connected Traffic on Servers									
						1	2	3	4	5	6	7	8	9	10
5.5	5.492	.008	.035	.053	10	.850	.813	.765	.705	.631	.542	.443	.340	.243	.159
5.5	5.496	.004	.016	.025	11	.848	.810	.761	.699	.622	.532	.431	.328	.230	.149
						.088									
6.5	5.723	.777	.877	1.212	6	.962	.959	.956	.952	.949	.944				
6.5	6.229	.271	.552	.817	7	.929	.919	.908	.895	.879	.861	.838			
6.5	6.396	.104	.299	.453	8	.900	.882	.860	.833	.800	.759	.710	.651		
6.5	6.453	.047	.158	.241	9	.884	.861	.831	.795	.749	.694	.627	.549	.463	
6.5	6.477	.023	.082	.126	10	.876	.849	.815	.773	.721	.657	.581	.495	.402	.308
6.5	6.489	.011	.042	.064	11	.871	.843	.807	.762	.705	.637	.557	.466	.371	.277
						.193									
6.5	6.495	.005	.021	.032	12	.869	.840	.802	.755	.697	.626	.544	.451	.355	.262
						.179	.114								
7.5	6.689	.811	.845	1.180	7	.964	.962	.959	.956	.953	.949	.945			
7.5	7.196	.304	.544	.805	8	.937	.929	.921	.910	.898	.884	.868	.848		
7.5	7.377	.123	.305	.461	9	.913	.899	.882	.862	.837	.807	.771	.728	.677	
7.5	7.442	.058	.167	.255	10	.899	.881	.858	.831	.797	.755	.705	.645	.575	.497
7.5	7.471	.029	.091	.139	11	.891	.871	.845	.813	.773	.724	.665	.596	.518	.432
						.344									
7.5	7.485	.015	.048	.074	12	.887	.865	.837	.802	.759	.707	.643	.569	.486	.397
						.308	.225								
7.5	7.492	.008	.025	.039	13	.885	.862	.833	.797	.752	.697	.631	.554	.469	.378
						.289	.208	.139							
8.5	7.675	.843	.818	1.151	8	.966	.964	.962	.959	.956	.953	.950	.946		
8.5	8.165	.335	.536	.794	9	.943	.937	.930	.922	.912	.901	.889	.874	.857	
8.5	8.358	.142	.309	.468	10	.923	.912	.898	.883	.864	.841	.814	.782	.743	.698
8.5	8.431	.069	.175	.267	11	.910	.896	.878	.857	.831	.799	.761	.715	.660	.597
						.526									
8.5	8.464	.036	.098	.150	12	.904	.887	.866	.841	.811	.773	.728	.673	.610	.538
						.458	.375								
8.5	8.481	.019	.054	.084	13	.900	.882	.860	.832	.799	.758	.708	.649	.581	.503
						.420	.335	.254							
8.5	8.490	.010	.029	.045	14	.897	.897	.856	.827	.792	.749	.697	.635	.564	.484
						.399	.314	.234	.164						
8.5	8.495	.005	.015	.024	15	.896	.877	.853	.824	.788	.744	.691	.627	.554	.473
						.388	.302	.233	.154	.100					

Glossary

ACK A positive acknowledgment (ACK) that the data was good. The receiving modem sends an ACK to the transmitting modem to indicate that the data is being received okay.

Addressing The process of identifying which telephone has been called. The telephone issues an address as dialed digits using dial pulses or DTMF tones.

Advanced Mobile Phone System (AMPS) Also called *cellular radio.* The serving area for AMPS is broken up into cells. Low-powered transmitters are used for AMPS. Radio frequencies can be reused in nonadjacent cells. Cellular radio uses frequencies from 825 to 845 MHz and from 870 to 890 MHz.

Alerters A ring detector circuit that results in a chirping sound from a speaker.

Alerting Signals Let someone know a call is waiting. Ringing signals alert a called party that the phone should be answered. Alerting is closely tied to supervisory signals.

All Trunks Busy (ATB) Refers to a condition where all circuits in a group are busy. In SPC switching systems, a memory variable serves as an ATB meter and keeps track of how many call attempts were made to a group of circuits when all the circuits in the group were busy. An ATB meter exists for each circuit group. ATB is a count of the number of blocked calls in a SPC switching system.

Alternating Current (AC) A signal that is continually alternating the direction of current flow due to changes in the polarity of the voltage of the signal. In other words, it is a signal that starts at zero voltage, rises to a maximum positive potential, declines to zero, and continues the decline until it reaches a maximum negative potential, then returns to zero to complete one cycle of a signal. The number of cycles the signal completes in one second is the frequency of the signal. This is the type of electricity supplied by the local power company to our homes and businesses.

American Standard Code for Information Interchange (ASCII) The use of a seven-level binary code to represent letters of the alphabet and the numerals 0 to 9.

Ampere-Hours Units in which a cell or battery is rated, indicating how much electrical storage it has. A 1600 ampere-hour battery will deliver 1600 amps for 1hr, 1 amp for 1600hr, 400 amps for 4 hr, 200 amps for 8hr, and so on.

Amplitude Modulation (AM) Used by AM broadcast stations. In amplitude modulation, the amplitude of the carrier frequency coming out of the mixer will vary according to the changing frequency of the input voice signal.

Analog Signal An electrical signal that is analogous (similar) to a voice signal. The signal continuously varies in amplitude and frequency. An analog signal has an infinite number of values for voltage, current, and frequency.

Architecture Refers to the way components of a switching system are physically and logically organized and how these components are connected. Each manufacturer uses a proprietary design to interconnect the various components of a switching system. Thus, the architecture for CBXs will be different for each manufacturer and will even vary between the different models of switches sold by the same manufacturer. The difference in architecture makes the models different.

Asymmetric Digital Subscriber Line (ADSL) A digital subscriber line technology that can be used to deliver the speed of a T1 line over the local loop. An Asymmetric Digital Subscriber Line has a high bit rate in one direction and a low bit rate in the opposite direction.

Asynchronous Transmission also called *start-and-stop transmission*. The transmitting device sends a start bit prior to each character and sends a stop bit after each character. The receiving device will synchronize from the received start bits. Thus, synchronization occurs at the beginning of each character. Data is sent between two devices as a serial bit stream.

Automatic Call Distributor (ACD) A switching system that has more trunks than lines attached to it. ACDs can serve as inbound systems and will automatically connect an incoming call to an agent. They can also serve as outbound systems and will automatically dial telephone numbers. When the called party answers, an agent is connected to the call. ACDs distribute calls to agents according to which agent has been most idle (least busy).

Automatic Number Identification (ANI) Usually refers to the transmission of the originating telephone number (used for billing purposes) from the LEC's class 5 exchange to the IEC's intertoll switch. Billing numbers are not always the same as the telephone number for a business, but they are the same for a residence. ANI and caller ID are two different processes. The IECs will not substitute ANI for Caller ID. ANI takes place between two switches. Caller ID is an end-to-end process using SS7.

Automatic Retransmission Request (ARQ) The method of error detection used between high-speed modems. ARQ is either discrete ARQ or continuous ARQ. The receiving device returns a positive acknowledgment (ACK) when it receives a good block of data, and returns a negative acknowledgment (NAK) when the block of data received contains an error.

Average Bouncing Busy Hour (ABBH) An average of various busy hour statistics such as PC, CCS, and ATB when the busy hour does not occur at the same time.

Average Busy Hour (ABH) If the busy hour for every day included in a study occurs at the same time (for example, if the busy hour for a five-day study occurred at 10:00 AM every day), the average for all hours is ABH.

Average Busy Hour Peg Count (ABHPC) An average of the number of calls placed over a group of circuits during the hours when most calls occur for several days.

Average Hold Time (AHT) The average length of time a circuit is in use. It is found by dividing CCS by PC. For example, if the ABBH CCS is 500 CCS and ABBH PC is 100 PC for a five-day study, we can conclude that the average length of every call will be 5 CCS based on our sample (study).

Bandwidth Refers to the width of a signal. The width of a signal is determined by subtracting the highest frequency of the signal from its lowest frequency.

Base Station Controller (BSC) Connects to all the base transceiver station (BTS) sites serving a BTA. The BTS provides an interface between the mobile phone and the BSC. The BTS is controlled by the BSC. Communication is constantly taking place between the BSC and the BTS. The BSC tells the BTS which radio channel to use for communication with a particular mobile station.

Base Transceiver Station (BTS) The BTS site for a PCS 1900 system will consist of a small building to house the base station transmitters and receivers and an antenna tower approximately 150 ft high. The antenna tower will contain 9 antennas arranged in a triangle array. Each side of the triangle will contain one transmitting antenna and two receiving antennas.

Basic Rate Interface (BRI) The basic building block of ISDN. A BRI consists of two 64-Kbps bearer channels and one 16-Kbps delta channel (2B+D). The delta (D) channel is used for signaling and sets up calls for the B-channels. The B-channels are used to carry customer voice or data.

Basic Trading Area (BTA) Each of the 51 major trading areas (MTAs) was subdivided into a BTA. There are 492 BTAs. PCS licenses were sold for each BTA.

Battery The connection of two or more cells in series to form a power supply. A typical car battery has six 2.17-V cells connected in series to form what is commonly referred to as a 12-V battery. In reality it is a 13.02-V battery.

Baud The number of times a signal changes its state. If the amplitude of a signal changes 2400 times a second, the signal changes states 2400 times per second or at 2400 baud. If a signal changes back and forth between a frequency of 1800 Hz and 2200 Hz 3200 times per second, the baud rate is 3200. If a signal changes phase 2400 times a second, the baud rate is 2400 baud. Many people confuse baud rate and bit rate. Even some terminal emulation programs will provide an option to change Baud Rates from 9200 to 19,200. *This is incorrect.* They should state that you can change the *bit rate* from 9200 to 19,200 bps. The selection of baud rate is done automatically by modems.

B-Channel The channel that carries the customer's information. This information may be voice, data, or video. Each B-channel is 64 Kbps.

Bell Operating Companies (BOCs) Often referred to after deregulation as the "Baby Bells." These were the local Bell Telephone companies. Before the 1984 Modified Final Judgment, 23 BOCs existed as subsidiaries of AT&T.

Blocked Calls Calls that receive a fast busy signal or a recording stating all circuits are busy. When there are insufficient circuits to handle the volume of calls being placed through a switching system or over trunk circuits to a particular location, the system will block some calls.

Blocked Calls Cleared (BCC) When insufficient circuit paths exist and blockage occurs, the blocked calls are cleared from the system by attaching them to a fast busy tone or all circuits busy recording.

Blocked Calls Delayed (BCD) When insufficient circuit paths exist and blockage occurs, the blocked calls are held in a waiting queue by the system until a circuit becomes available to serve the caller.

Blocked Calls Held (BCH) When insufficient circuit paths exist and blockage occurs, the blocked calls are not cleared or put in a queue. If callers wait, they will be served. This methodology occurs for dial registers. If no dial registers are available, callers must wait or hold until one is available. This is why we are supposed to listen for dial tone. The presence of dial tone tells us our telephone has been attached to a dial register and we can dial a number.

Blocked Load The amount of offered load that does not get carried but is blocked due to insufficient circuits. Blocked load is estimated by multiplying the number of attempts blocked (ATB) times AHT.

Blocking A term used to describe a situation where a call cannot be completed because all circuit paths are in use.

Branch Feeder Cables Cables that connect distribution cables to main feeder cables.

Bridged Ringer A term indicating that the ringer is wired between the Tip and Ring leads. Bridged ringers were used on private lines as well as on two- and four-party lines.

Broadband ISDN Broadband ISDN provides speeds faster than 100 Mbps and can be used to establish wide area networks. Broadband ISDN is asynchronous transfer mode transmission over fiber optic cables.

Busy Hour The hour of a day when the most telephone usage occurs.

Caller Identification One of the Custom Local Area Signaling Services (CLASS) sold by the LECs. This feature requires SS7 for the transfer of the originating telephone number from the originating exchange to the terminating exchange. If the originating office is not a SPC switch, it cannot connect to the SS7 network, and the message "Not Available" will be received on the caller ID display.

Calling Capacity The number of simultaneous calls that can be supported by a switching system. Sometimes this is stated as the number of calls that can be handled in one hour. Since these two approaches are not the same, I would have the salesperson state calling capacity in both ways.

Call Sequencer A system that uses keysystems and beehive lamp fields to indicate the presence of an incoming call to a call center. The agents answer the calls by pushing a button on the keysystem.

Call Store A part of memory set aside to store information that is temporary in nature. The call store will keep track of each call that is in progress or that is currently up on a conversation. Call store is also used to load temporary programs used by maintenance and administration personnel.

Capacity A measure of a switching system or circuit group's ability to handle telephone calls. Capacity can be stated as either ABBH PC or ABBH CCS.

Carried Load The load carried by a circuit, group of circuits, or switching system. It is measurable. If the switch carries traffic, it can measure the traffic or load carried. Carried load equals the ABBH PC times the ABBH AHT.

Carrier Serving Area (CSA) The establishment of DS0 capabilities to a remote area of the exchange by using *subscriber carrier* to establish a remote wire center.

Carterphone Decision of 1968 A ruling by the Federal Communications Commission forcing AT&T to allow the attachment of a Carterphone to telephone lines at the customer's residence. This ruling was the beginning of deregulation of station equipment.

CCS Centium call seconds or hundred call seconds. One CCS equals 100s; 1hr is 3600s or 36 CCS.

Cell A device that generates electricity by a chemical reaction

Cellular Radio *See AMPS*

Central Office The central wire center or central exchange. All telephones in a small geographic area are wired to a central exchange, which serves all telephones in that area.

Central Office (CO) Trunk Circuit A trunk circuit in a PBX/CBX that connects to the local central exchange. Most of the trunk circuits in a PBX/CBX are central office trunks, and most connect to the central office using a twisted-pair local loop.

Channel Service Unit (CSU)/Data Service Unit (DSU) Also called *Customer Service Unit/Data Service Unit* The CSU/DSU is a DCE used to interface a computer to a digital leased line. It is a digital-to-digital interface. It can connect a low-speed digital device to a high-speed digital highway.

Channel Unit Performs the function of interfacing the multiplexer to a specific type of input signal. The channel unit makes changes to the input signal so it can be combined (multiplexed) with other signals for transmission over the transmission medium.

Circuit Switching The process of connecting one circuit to another.

Class 5 Exchange Also called the end office. The lowest-level switch in the PSTN hierarchy; the local exchange.

Code Division Multiple Access (CDMA) Has been adopted as IS-95 standard for PCS networks. IS-95 Q-CDMA was codified as a standard in 1993. CDMA can provide ten times the capacity of DAMPS. CDMA is not backward compatible with analogue AMPS in the same way that D-AMPS is. CDMA cannot reuse DAMP technology. To implement CDMA, the service provider must install a new network. CDMA is spread spectrum technology. Many conversations are multiplexed over one frequency. Each conversation is assigned a code. The receiver strips out conversations individually by using the code.

Code Excited Linear Predictive Coding (CLIP)
Also known as *vector-sum excited linear predictive (VSELP)* coding. This speech coding algorithm is EIA Standard IS-54. This is the speech coding technique recommended for TDMA cellular radio systems. The bit rate is 7.95 Kbps. VSELP is also available in a 16-Kbps chip. AT&T makes a digital signal processor using this technology; it is called a DSP-1616.

Common Carrier A company that offers telecommunications services to the general public as part of the Public Switched Telephone Network.

Communication A process that allows information to pass between a sender (transmitter) and a receiver over some medium.

Communications Act of 1934 Legislation passed by Congress and signed into law by the president in 1934 to regulate interstate telecommunications and radio broadcasts. This law created the Federal Communications Commission to administer the act.

Computer Inquiry Any of a series of rulings by the Federal Communications Commission. Issued in 1971 Computer Inquiry I; stated that the Federal Communications Commission would not regulate computer services.

Computer Inquiry II Issued in 1981; mandated that station equipment was to be deregulated and could not be provided by a local exchange carrier.

Computer Inquiry III In 1986, detailed the extent to which AT&T and the Bell Operating Companies could compete in the nonregulated enhanced services arena.

Computerized Branch Exchange (CBX) A SPC switching system used at a private business location.

Concentration The use of fewer outlets than inlets.

Conductor A device or transmisson medium that readily conducts or carries an electrical signal. Copper wire is the most commonly used transmission medium in telecommunications and is a good conductor because it has a low resistance to electric current flow.

Configuration Refers to the types of interfaces, the number of interfaces, dialing patterns, station numbers, system features, station features, and other database entries that uniquely define a CBX.

Continuous ARQ Also known as *Sliding Window ARQ*. Continuous ARQ eliminates the need for a transmitting device to wait for ACKs after each block of data. The device continuously transmits blocks of data and sends a block number with each block. The receiving device checks the blocks for errors and continuously returns positive ACKs to the transmitter. If an error in data occurs, the receiving device returns a NAK along with the block number affected. On receipt of a NAK, the transmitting device will retransmit the bad block of data.

Controller A microprocessor designed to control a specific function. The controller is directed by programming in a PROM chip. The PROM chip resides on the controller circuitboard. The type of programming contained in the PROM dictates whether the controller is a line, trunk, or network controller. Controllers also receive instructions from the main processor over messaging links.

Crosstalk A condition where a circuit is picking up signals being carried by another circuit. An undesirable condition. For example, a customer hears other conversations or noise on their private line. This is usually caused by a breakdown in the insulation between wires of two different circuits.

Custom Local Area Signaling Services (CLASS) Features of a SPC switch, which also require the SPC to have Signaling System 7. Some Class Features are: Caller ID, Caller ID Blocking, Automatic Callback, Automatic Recall, Selective Call Forwarding, Selective Call Rejection, and Ring Again.

Customer-Provided Equipment Also called *customer-premise equipment*. This is station equipment, at a customer's premise, which according to Computer Inquiry II must be provided by someone other than the local exchange carrier. The

1996 Telecommunications Reform Act overrides Computer Inquiry II and allows LECs to reenter the CPE market.

Cyclic Redundancy Check (CRC) A form of error checking used between modems. A transmitting modem treats the block of data transmitted as representing a large binary number. This number is divided by a 17-bit divisor (CRC-16) or a 33-bit divisor (CRC-32). The remainder is attached in the trailer behind the block of data. The receiving modem performs the same division on the block of data received and compares its calculated remainder to the remainder sent in the trailer.

Data Raw facts, characters, numbers, and so on that have little or no meaning in themselves. When data is processed, it becomes information. In telecommunications when we speak of data, we are referring to information represented by digital codes.

Data Communications Equipment (DCE) Also called *Data Circuit Termination Equipment*. A device that interfaces *Data Terminal Equipment (DTE)* to the PSTN. A modem is a DCE used to interface a DTE to an analog line circuit on the PSTN. A CSU/DSU is a DCE used to interface a DTE to a leased digital line in the PSTN.

Data Store That part of RAM set aside to contain the database for the switch. The database will consist of translations software (line and trunk tables) and parameters software.

Data Switching Exchange (DSE) Also known as *Packet Switching Exchange*. These are the switches of a packet switched network. The DSE routes packets of information based on the packet address information found in each packet header.

Data Terminal Equipment (DTE) A DTE is used to transmit and receive data in the form of digital signals. The personal computer is the most common form of DTE.

dBm Stands for "decibels referenced to 1mW." A unit of measure used in telecommunications to measure voice signals against a reference signal level of 1 mW. Also see *decibel*.

D-Channel The channel that carries signaling and control information. The D-channel is used to set up calls for the B-channels.

DC Voltage The voltage produced by a chemical reaction. DC stands for "direct current." Devices that produce electricity via a chemical reaction are called *batteries* or *cells*.

Decibel (dB) The unit of measure used to compare two power levels. A decibel is not an *absolute* measurement; it is a *relative* measurement. The decibel tells the relationship of one power level to another power level. The formula for the decibel is as follows: dB = 10 log (power ratio). Also see *dBm*.

Decimonic Ringing A ringing system used by Independent telephone companies. The system could send out five different ringing signals (10-, 20-, 30-, 40-, and 50-cycle signals).

Delay Time Refers to the length of time that a call for a customer is held in queue before being served.

Demand Offered load, or the amount of traffic that users are trying to place over a circuit, group of circuits, or switching system.

Deregulation The change of the telecommunications industry from a regulated monopoly to a competitive nonregulated market. One of the major provisions of the 1984 Modified Final Judgment was to deregulate long distance services.

Digital Advanced Mobile Telephone System (DAMPS) A technique that places multiple calls over one radio frequency using pulse code modulation (PCM) and time division multiplexing (TDM). The PCM signal is not the standard PCM signal found in the PSTN. The PSTN converts an analog signal into 64,000 bps. DAMPS technology converts the analog signal into 16,000 bps. DAMPS was introduced in 1992 for the existing advanced mobile telephone system. DAMPS allowed cellular operators to carry four times as many calls as a regular AMPS system. DAMPS is backward compatible with analog AMPS.

Digital Multiplexed Switching System A Stored Program Control switching system that uses a PCM/TDM switching matrix. All voice paths inside the switch are digital paths, and the SPC system switches these digital paths.

Digital Signal An electrical signal that has two states. The two states may be represented by voltage or current. For example, the presence of voltage could represent a digital logic of 1, while the absence of voltage could represent a digital logic of 0.

Digital Subscriber Line Circuit (DSL) Usually refers to a local-loop cable pair that is handling a digital signal using 2B1Q line coding.

Digital Trunk Controller (DTC) A controller card that interfaces a digital trunk facility, such as a T1 carrier system line, directly to the time division multiplex loop of the SPC switch.

Direct Current (DC) The electric current flow is at a constant rate and flows continuously in one direction. The direction of electric current flow (outside the battery) is from the negative terminal of the battery, through a device attached to the battery, to the positive terminal of the battery. The amount of electric current flowing is measured in amperes.

Direct Inward Dial (DID) Central office trunks allowing incoming calls to bypass the switchboard operator. CBX station numbers are part of the central exchange dialing plan, and CBX stations can be reached by dialing a regular telephone number.

Direct Outward Dial (DOD) Central office trunk circuits accessed by the station user dialing a 9. The CBX operator is not needed to connect to a central office trunk.

Discrete ARQ Also called *stop-and-wait ARQ*. An error control protocol that requires an acknowledgment from the receiver after each block of data sent. The transmitting modem sends a block of data and then waits for an acknowledgment before sending the next block of data.

Distributed Processing A SPC switch architecture that uses multiple processors. Each line shelf, trunk shelf, and network shelf contains a processor to control the activities of equipment in its shelf. These processors are called *controllers*.

Distribution Cable A cable that is fed by a *branch feeder cable* from the central exchange and that connects to telephones via drop wires. The distribution cable is usually a small cable of 25 to 400 pairs.

Divestiture A major provision of the 1984 Modified Final Judgment. This provision forced AT&T to divest itself of the 23 Bell Operating Companies.

Drop Wire A pair of wires connecting the telephone to the distribution cable. One end of the drop wire connects to the protector on the side of a house or business, and the other end connects to a terminal device on the distribution cable.

Dry Cell A chemical cell that generates electricity via chemical means between an electrolyticsolution and two dissimilar metals. The electrolytic solution is in the form of a paste.

Dual Simultaneous Voice and Data (DSVD) A modem that can handle voice and data at the same time.

Dual-Tone Multifrequency (DTMF) Dial The dial on a 2500 set that sends out a combination of two tones when a digit on the keypad is depressed.

EIA-232 Interface Also called *RS-232*. The most common interface standard for data communications. CCITT standard V.24 is the same as EIA-232. EIA-232 defines the voltage levels needed for the various signal leads of EIA-232.

Electrically Programmable Read Only Memory (EPROM) Chip Used on controller cards. The EPROM is programmed to perform a specific function. The programming makes the controller a line, trunk, or network controller.

Equal Access A key provision of the 1984 Modified Final Judgment. This provision forced the Bell Operating Companies to provide 1+ access to toll for all interexchange carriers.

Equipped for An installed switching system has X number of printed circuit cards installed to support X number of lines, trunks, register, and so on.

Erlang A traffic usage measurement that is equal to 1hr or 36 CSS.

Exchange Boundary Pertains to the limits of the serving area. The outermost customers being served by a particular exchange establish the boundary of the exchange.

Exchange Code The first three digits of a seven-digit telephone number.

Expansion Capacity The ability to add additional equipment to a switching system. This allows your system to support your company's future communication needs.

Extended Area Service (EAS) Tandem A class 4 switching system used in a metropolitan area to connect class 5 switching systems together on EAS calls. Extends the rate base of a local exchange so that calls to neighboring exchanges are not toll calls but are local calls.

Facility A term typically used to refer to the transmission medium.

Feature Group A series of services established by the Bell Operating Companies to provide access to interexchange carriers (IECs). Feature Group A gave customers access to IECs by having them dial a seven- or ten-digit telephone number plus a personal identification number. Feature Group B provided access via a seven-digit code of the form 950-10XX. This access code was later changed to 10XXX. Feature Group D is the 1+ access to IECs that was mandated by the 1984 Modified Final Judgment.

Feature Phone A telephone containing electronics that allow it to provide features. When additional features have been added to the basic single-line telephone such as last number redial and speed dialing lists, the phone is called a feature phone.

Federal Communications Commission (FCC) The federal agency created by the Communications Act of 1934 to oversee enforcement of the act by regulating interstate telecommunications and broadcast communications.

Feeder Cable Includes various types. Cables leaving the central exchange are called main feeder cables. The main feeder cables connect to branch feeder cables. The branch feeder cables are used to connect the main feeder cables to distribution cables.

Final Circuit Group A group of circuits for which there is no alternate route. A final group is usually the alternate route for a high-usage group. Final groups are engineered for lower levels of efficiency to provide low levels of blockage.

Finite Queue A queue that has a definite waiting period.

First-Attempt Traffic The offered load in retrial traffic engineering methodology.

Flow Control Controlling the flow of data from one device to another usually via hardware flow control (RTS/CTS) or by software flow control (XON/XOFF). A modem contains a memory buffer to allow it to compress data before transmitting and to allow it to convert asynchronous data to synchronous data. It must be able to stop the transmitting PC when this memory buffer approaches a near-full condition. Some modems contain a very large memory buffer, which negates the need for flow control.

Frequency Division Multiplexing (FDM) The process of converting each speech path to different frequency signals and then combining the different frequencies so they may be sent over one transmission medium.

Full-Duplex Transmission If a transmission system allows signals to be transmitted in both directions at the same time, the system is called a full duplex transmission system.

Generic Program The program that controls call processing in a SPC switch. The generic program contains software for all the features of the switch. To add features to a switch, the generic program must be replaced with a later version containing the desired features. The generic program is loaded into program store.

Glare Condition The simultaneous seizure of a trunk from both ends. This condition occurs in electromechanical switching systems.

Grade of Service (GOS) A probability measure of blockage occurring. Using statistics (traffic tables), we can predict the probability that blockage will occur for a specific load placed on a specific number of servers.

Graham Act of 1921 Legislation passed by Congress and signed by the president in 1921 to exempt telecommunications from antitrust legislation.

Grounded Ringer A term indicating that the ringer is wired between the Tip lead and ground or between the Ring lead and ground. Grounded Ringing was used on eight- and ten-party lines.

Ground Start Line When a line circuit has been modified for ground start, -50-V DC battery is present on the Ring lead and the Tip lead has no connection. The trunk circuit at the PBX cannot use loop start to seize the line relay because no ground is present on the Tip lead. The PBX trunk must place a ground on the Ring lead to seize the central office line circuit.

Ground Start Signaling A type of signaling used with PABX trunk circuits. The line circuit of the central office is operated by placing ground on the Ring lead.

Half Duplex A type of transmission where transmitters on each end of a medium take turns sending over the same medium.

Harmonic Ringing A ringing system used by Independent telephone companies. The system could send out five different ringing signals (162/3-, 25-, 332/3-, 50-, and 662/3-cycle signals).

Header A term describing the placement of control information in front of a block of data transmitted using synchronous transmission. The header contains a beginning flag (or sync signal), the destination address, and a control field.

High-Usage Trunk Group A group of trunk circuits purposely engineered to high levels of efficiency with a high probability of blockage. Traffic blocked is routed over an alternate route.

Hookswitch Often called a *switch-hook*. This is the device in the telephone that closes an electrical path between the central office and the telephone, by closing contacts together, when the receiver is lifted out of its cradle.

Hybrid Keysystem A keysystem designed to handle more than 24 central office lines and 40 telephone sets. Central office lines are accessed by dialing the digit 9 for access.

Hybrid Network A network that consists of a transformer and an impedance matching circuit. The transformer performs a two- to four-wire conversion and vice versa. In a telephone, the hybrid network connects the two-wire local loop to the four-wire transmitter/receiver.

Hypertext Markup Language (HTML) A software language used to create documents, such as home pages, for use on the Internet.

Improved Mobile Telephone System (IMTS) IMTS was introduced around 1964. The major benefit of IMTS was that it used several different frequencies and could support many different conversations at the same time. IMTS was also connected to the local class 5 office instead of the toll office. It was connected to regular telephone numbers and line circuits. An IMTS phone was assigned a regular PSTN telephone number. Anyone could reach an IMTS phone by dialing the PSTN number assigned to the mobile phone. This eliminated the need for operators to handle mobile phone calls.

In-Band Signaling Involves sending signals over the same circuit used for the accompanying voice signal.

Integrated Services Digital Network (ISDN) The use of digital line circuits to provide end-to-end digital service. The Basic Rate Interface (BRI) provides the user with two DS0 channels. The Primary Rate Interface (PRI) provides the user with 23 DS0 channels.

Interexchange Carriers (IECs or IXCs) Common carriers that provide long distance telephone service. Major IECs are AT&T, MCI, Sprint, LDI, and so on.

Internet A network of computers that can be accessed by other members of the Internet community. Most individuals become members of the community by purchasing access to the Internet from a local Internet Service Provider or Commercial on-line Information Access Provider such as Prodigy or America-On-Line (AOL).

Internet Service Provider (ISP) A company that provides individuals with access to the Internet.

Interoffice Calls Calls completed between telephones served by two different central exchanges.

Intraoffice Calls Calls completed between telephones served by the same central exchange.

Jumper Wire A short length of twisted-pair wire used to connect wire pairs of two different cables. A jumper wire is a very small gauge of wire (usually 24 or 26 gauge). Except for its small size and much longer length, a jumper wire is similar to the jumper cable used to start a car with a dead battery.

Keysystem A telephone system used by small businesses to allow several central office lines to terminate at each telephone attached to the system.

Kingsbury Commitment of 1913 An agreement signed by the vice president of AT&T in 1913. This agreement stated that AT&T would allow other telephone companies access to its network, would sell its Western Union stock, and would not buy any more Independent telephone companies without getting permission from the government.

Kirchhoff's Current Law States that at any given point in a circuit, the total current leaving a point equals the total current entering the point.

Kirchhoff's Voltage Law States that the sum of all voltage drops in an enclosed loop equals the total voltage supplied to the loop.

Line Card An electronic printed circuitboard that contains line circuits. The line cards of most SPC central exchanges contain one line circuit per circuit card. The first AT&T SPC switch had line cards containing 8 line circuits. Many CBXs have line cards containing from 8 to 16 line circuits.

Line Circuit The electronic printed circuit card in a SPC switch that interfaces a local-loop telephone line to the SPC switching system. The line circuit has many functions. One function is to interface the analog signal from a telephone to the digital switching matrix of the switch.

Line Controller A circuit card in the line shelf that contains a microprocessor and EPROM to control the activities of line circuits in that shelf.

Line Drawer A shelf in a digital switching system that contains line circuits. A typical line drawer contains 32 line circuits.

Line Relay An electromechanical device attached to a telephone line at the central exchange. When a subscriber took the handset of the telephone off-hook, electric current flowing in the relay caused it to operate and signal the operator.

Link Access Procedure on the D-Channel LAPD The software protocol used at the data link layer (layer 2) of the D-channel. LAPD uses Carrier Sense Multiple Access/with Collision Resolution (CSMA/CR) to allow multiple TEs on the S/T bus. LAPD is a multipoint protocol.

Load Balance The physical distribution of residential and business telephones across all line circuits in such a manner that the same percentage of business to residential occurs in all line groups. This is done to prevent a heavy concentration of business lines in one group.

Load Coils The devices that introduce additional inductance into a circuit. Load coils are added to cable pairs over 18,000 ft to improve the ability of the pair to carry voice signals.

Loading Purposely adding inductance to a cable pair over 18,000 ft long to offset the mutual capacitance of the cable pair.

Local Access Transport Area (LATA) A key provision of the 1984 Modified Final Judgment (MFJ). This provision established 184 geographic regions that conform to the Standard Metropolitan Statistical Index used by marketing organizations. The LATA was defined as an area within which calls must be carried by a local exchange carrier (LEC) and could not be carried by an interexchange carrier (IEC or IXC). These calls within the LATA are called *intra-LATA calls*. The MFJ further stated that calls between LATAs (*Inter-LATA calls*) must be carried by an IEC, not by a LEC. The 1996 Telecommunications

Reform Act replaced the 1984 MFJ and opened the door to competition in this area. It allows either type of call to be carried by either a LEC or an IEC.

Local Exchange The switching system that telephones are connected to.

Local Exchange Carrier (LEC) The provider of local telephone services. Prior to the 1996 Telecommunications Reform Act, the LEC was your local telephone company. With the passage of that act, the local telephone company is now called the *incumbent local exchange carrier (ILEC)* and its competitors are called *competitive local exchange carriers (CLECs)*.

Local Loop A term used to describe the facilities that connect a telephone to the central exchange. These facilities usually consist of a twisted pair of wires.

Loop Extender A hardware device. Can be connected between the line circuit of the switch, and the cable pair serving a remote area of the exchange, in order to extend the range of a central exchange. A loop extender adds a voltage boost to the circuit. By doubling the voltage applied to a line, the distance served can be doubled.

Loop Signaling Activation of a signal by opening or closing the loop to complete an electrical circuit and causing current to flow.

Loop Treatment A term indicating that a local-loop cable pair has been specially conditioned. The most prevalent use of this term is to describe the addition of extra equipment to a local-loop cable pair. To extend the range of a central exchange, a device called a *loop extender* can be connected between the line circuit of the switch and the cable pair serving a remote area of the exchange. A loop extender adds a voltage boost to the circuit. By doubling the voltage applied to a line, the distance served can be doubled. Another device called a *voice frequency repeater (VFR)* is also added to long local loops. The VFR will amplify voice signals in both directions to compensate for the extra decibel loss of long loops.

Magneto A device that generates an AC voltage by turning a coil of wire inside a magnetic field. The old hand-crank telephone had a magneto attached to the hand crank.

Main Distributing Frame (MDF) In a central exchange, the place where all cables that connect to the switch are terminated. Outside plant cables that connect to telephones and to other central offices are also terminated at the MDF. Short lengths of a twisted-wire pair (called a *jumper wire*) are used to connect wires from the switch to wires in the outside plant cable. In a PBX environment, the MDF is where all cables from the switch, IDFs, and cables from the central exchange are terminated and jumpered together.

Main Feeder Cable Cable that leaves the central exchange and connects to branch feeder cables.

Main Processor The CPU that controls all activities in a SPC switch.

Major Trading Area (MTA) The United States is divided into 51 areas for mobile telephone licensing purposes. Each of these segments is called a major trading area (MTA). Each MTA was subdivided into a basic trading area (BTA). There are 492 BTAs.

MCI Decision of 1976 A ruling by the federal court in MCI's favor against AT&T; it allowed MCI and other common carriers to handle long distance services for the general public.

MCI Ruling of 1969 The ruling in 1969, by the Federal Communications Commission, that forced AT&T to allow private-line customers of Microwave Communications Inc. (MCI) to use local telephone lines for access to MCI's private-line network.

Medium The device used to transport information or a message between the transmitter and receiver.

Mixer (Modulator) An electronic device that has two inputs. One input is a low-frequency signal that contains intelligence. The other signal is a pure sine wave at a high frequency. This high-frequency signal is called the *carrier signal* because it carries our intelligent signal after modulation. The mixer combines the two signals so that we end up with a high-frequency signal that has the intelligence of the low-frequency signal imposed on it.

Mobile Telephone Serving Office (MTSO) The central controller and switching system for cellular radio. The MTSO can be a stand-alone switch owned by a private cellular company, or it can be integrated into a local switch if the LEC is the cellular service provider.

Mobile Telephone System (MTS) Mobile telephone service began in 1946 and was called MTS. These systems used the radio frequencies between 35 and 45 MHz. Although MTS stands for "mobile telephone service," it also meant "manual telephone system." All calls had to be handled by an operator.

Modem Acronym for Modulator/Demodulator A DCE device that interfaces the digital signal from a DTE to the analog local loop and line circuit of the PSTN, by converting the digital signals into modulations of an analog signal.

Modified Final Judgment (MFJ) The 1984 agreement—reached on August 24, 1982, between AT&T and the Department of Justice—to settle an antitrust suit brought against AT&T by the Justice Department on November 20, 1974. This judgment modified the 1956 Final Judgment. The major provisions of the 1984 MFJ were *deregulation* of long distance services and *divestiture* of the Bell Operating Companies by AT&T.

Modified Long Route Design (MLRD) Outside plant design that involves adding loop treatment to cable pairs serving remote areas of an exchange. Local loops longer than 24,000 ft are designed using MLRD.

Monopoly A form of market where one company is the sole provider of goods and/or services. There is no competition. Market demand does not set the price of goods or services. In a nonregulated monopoly, price is set by the monopoly service provider. In a *regulated monopoly*, price is set by the government agency charged with regulating the monopoly service provider.

Moves, Adds, and Changes (MACs) The changes made in the database for a SPC switch when a telephone is moved or added or has its features or number changed.

Multiplexer A device that can combine many different signals or calls (data or voice channels) so they can be placed over one facility. A multiplexer also contains a demultiplexer so that it can demultiplex a received multichannel signal into separate voice channels.

Multiplexing In telecommunications, a process combining many individual signals (voice or data) so they can be sent over one transmission medium.

NAK A negative acknowledgment transmitted by the receiving modem when it detects that an error has occurred in a block of transmitted data. If a NAK is received, the modem retransmits the last block of data from its memory.

Narrowband ISDN Designates BRI and PRI, which use synchronous transmission.

Negative Exponential Distribution Assumes the probability of long service times in a queuing environment is low and the probability of short service times is high.

Network An overused term. In telecommunications, the word *network* has many meanings, usually determined by the context. The Public Switched Telephone *Network* (PSTN) refers to the interconnection of switching systems in the PSTN. Switching *network* refers to the component inside a switching system that switches one circuit to another circuit. The *network* in a telephone refers to the hybrid network, which performs a two- to four-wire conversion process.

Network Termination 1 (NT1) An interface device provided by a customer to terminate an ISDN line to the customer's premises. The NT1 interfaces the customer's ISDN equipment to the Digital Subscriber Line of the LEC.

Network Termination 2 (NT2) An interface device that interfaces devices to a PRI NT1. A NT2 is usually used in a PBX environment. NT2s will only be found in a PRI environment.

Network Termination 12 (NT12) A device combining a NT1 and a NT2.

Null Modem No modem. A null modem cable is used to connect two PCs via their serial ports, when they are connected directly without using a modem.

Numbering Plan Area (NPA) An area represented by an area code.

Offered Load The demand or amount of traffic that callers are trying to place over a system. It is equal to carried load plus blocked load.

Ohm's Law A formula showing the relationship between the voltage applied to a circuit, the resistance in the circuit, and the resulting current flow in the circuit. $I = V/R$.

Out-of Band Signaling Signaling that occurs over a different circuit than the circuit used for the voice signal. Many times the signals for several different voice circuits are multiplexed over one facility and called *Common Channel Signaling*.

Outside Plant Cables, telephone poles, pedestals, and anything that is part of the telecommunications infrastructure and not inside a building.

Outside Plant Engineer The person charged with properly designing the facilities that connect telephones to central exchanges and the facilities that connect central exchanges together (outside plant designer).

Overlay Programs The term used to describe maintenance and administrative programs that do not permanently reside in memory but are contained on a secondary storage device. These programs are only loaded into memory as needed; when loaded, they overwrite the last overlay program contained in memory.

Packet Assembler/Disassembler (PAD) The device that interfaces data to a packet network. The PAD accepts data from a user and arranges the data into packets that can be processed by the packet switching network.

Packet Network A data network that transmits packets of data

Parallel Transmission The use of several transmission leads to allow the simultaneous transmission of several bits at one time. A parallel data bus with 8 leads can process data 8 bits at a time. Or 64 leads on the data bus allows information to be processed 64 bits at a time. Parallel transmission is much faster than serial transmission but requires many more transmission leads.

Peg Count (PC) A term used to indicate how many calls are handled by a circuit, a group of circuits, or a switching system. This term is a holdover from electromechanical traffic registers. When the relay of the register operated, it was said to "peg the meter." Each peg of a meter made it turn a numbered dial one step.

Peripheral Equipment Equipment that exists on the periphery of a switching system matrix. Peripheral equipment interfaces external devices to the computer of a SPC switch to provide capabilities and features to these devices. Most peripheral equipment consists of line circuits to interface a telephone to the switch and trunk circuits to interface interoffice circuits to a switch. Also included in this category are dial registers, modems, and voice mail.

Personal Communications System (PCS) PCS is not so much a technology as a concept. The concept of PCS is to assign someone a personal telephone number (PTN). This PTN is stored in a database on the SS7 network. That database keeps track of where a person can be reached. When a call is placed for that person, the artificial intelligence network (AIN) of the SS7 determines where the call should be directed.

Personal Communications System 1900 (PCS 1900) PCS is provided by radio frequencies in the 1900-MHz range. PCS 1900 is the latest evolution in mobile communication. PCS 1900 can be provided by the time division multiple access (TDMA) technology of DAMPS or can use code division multiple access (CDMA) technology.

Plastic Insulated Cable (PIC) Cable that contains wires electrically isolated from each other by plastic insulation around each wire. PIC can also mean preferred interexchange carrier.

Point of Presence (POP) The point at which the local exchange carrier and interexchange carrier facilities meet each other.

Port The interface point between a switching system matrix and external devices. Each switch will be equipped with a certain number of ports to handle the number of devices that will be attached to the system. If a line circuit card is inserted in the port, it becomes a line port. If a trunk circuit card is inserted into a port, it becomes a trunk port. If a dial register card is inserted in a port, it becomes a dial register port. Line circuits usually use one port per circuit. Some trunk circuits, dial registers, and special circuits may use two or more ports for each circuit.

Port Capacity The number of different devices that can be connected to a CBX. This definition can be misleading because some devices require more than one port. Line circuits require one port. Trunk circuits *may* require two ports for each circuit. DTMF register circuits *may* require four ports per circuit.

Preferred Interexchange Carrier (PIC) Determined as follows: A customer tells the local exchange carrier who it wants to use as its long distance service provider. The LEC makes an entry in the database of the switching system serving this customer. That database entry will inform the switch which IEC should be used on long distance calls placed by that customer. With the 1996 Telecommunications Reform act, customers must be able to choose a carrier for toll calls within a LATA as well as the same or a separate carrier for long distance service.

Primary Rate Interface (PRI) Consists of twenty-three 64-Kbps B-channels and one 64-Kbps D-channel.

Printed CircuitBoard (PCB) A plastic card containing electronic circuitry, integrated circuit chips, and solder tracks connecting the circuitry.

Private Automated Branch Exchange (PABX) A term used to describe electromechanical switching systems at a private business location.

Parameters Software Defines a particular SPC, by indicating the number of lines, trunks, registers and so on.

Private Branch Exchange (PBX) The name given to manual switchboards in a private business location. A PBX may also be small SPC switching system used by large businesses.

Private-Line Services Services provided by a common carrier to a private organization to help that organization establish its own private network. A private network cannot be accessed or utilized by the public. A private network is exactly that—it is private and can only be accessed and used by the private organization.

Program Store The section of memory reserved for the generic program.

Progress Signals Indicate the progress of a call to the calling party; examples include dial tone, 60-ipm busy tone, 120-ipm busy tone, ring-back tone, and recordings.

Proprietary Telephone A telephone designed by its manufacturer to only work with certain keysystems or PABXs made by that same manufacturer.

Protocol The rules of communication. Each protocol defines a formal procedure for how data is to be transmitted and received using that protocol.

Public Data Network (PDN) The packet data network accessible to the general public for the transmission of packet data.

Public Switched Telephone Network (PSTN) A network of switching systems connected together to allow anyone to call any telephone located in the United States. A public network is accessible to any members of the general public. The PSTN is composed of many nodes (switching systems) and many transmission links connecting these nodes. Stations are the end points of the network and connect to the PSTN via their local exchange.

Public Utilities Commission (PUC) The state government agency that regulates telecommunications within the state.

Pulse Amplitude Modulation (PAM) The process used to take samples of analog voice signals so they can be multiplexed using TDM. The samples appear as pulses in the TDM signal.

Pulse Code Modulation(PCM) The industry standard method used to convert an analog signal into a digital 64,000-bps signal. PCM takes a PAM signal and converts each sample (pulse) into an 8-bit code.

Quadrature Amplitude Modulation (QAM) The transmission of 4 bits per baud. A V.29 modem transmits at 2400 baud with 4 bits per baud to achieve 9600 bps transmission. QAM uses 2 different amplitudes for each of 8 different phases of a 1700-Hz signal to achieve the 16 detectable events necessary to code, and decode 4 bits at a time.

Queuing Theory Assumes that people are willing to wait in the queue or holding line for service and that based on the length of the queue, the behavior of people in the queue, and how the queue handles new arrivals, the number of servers needed can be forecast.

Random Access Memory (RAM) Highly volatile memory that will lose its contents when power to the RAM is lost. RAM is divided into three stores: *program store, data store,* and *call store*. The main processor accesses RAM to direct switching operations.

Receiver The device responsible for decoding or converting received information into an intelligible message.

Reference Points The points at which various devices are connected in an ISDN circuit. The U interface serves as a demarc. It is where the cable pair from the LEC attaches to a NT1. The S/T reference is the interface where a TE1 or TA connects to a NT1. The R reference is the interface where a non-ISDN device attaches to a TA. The V reference is the interface point between the LE and ET in the central exchange.

Register (Dial Register) The device in a common control switching system that receives dialed digits.

Resistance The amount of opposition to electric current flow that a device has.

Regulated Monopoly A sole provider of goods or services; regulated by a government agency. A regulated monopoly must seek approval from the government agency for anthing it wishes to do. The government agency sets the price that a monopoly can charge for its goods and services.

Resistance Design Designing a local loop, which uses twisted-pair copper wire, so that the resistance of the wire serving a telephone does not exceed the resistance design limitations of the central exchange line circuit. Each switching system manufacturer will state how much resistance its switch is designed to support.

Retrial Methodology Assumes people will redial their call upon being blocked. It makes an adjustment to blocked load before adding it to carried load to get offered load. It makes this adjustment because this methodology assumes some of the blocked calls were eventually carried due to retry by the caller. Blocked load that was carried should not be added to carried load because it will make offered load seem higher than it actually was.

Revenue Sharing The sharing of toll revenue between the Independent Telephone Companies and AT&T prior to the 1984 MFJ.

Reverse Battery Signaling In a SXS central office, the connector switch would signal that the called party had answered by reversing the battery feed to the calling party.

Revised Resistance Design (RRD) A design approach in which the outside plant engineer uses the length of the local loop, as well as the design limitations of the central exchange, to determine the local-loop design criteria. Loops up to 18,000 ft are designed to use nonloaded cable pairs with a maximum loop resistance of 1300 Ω. Loops between 18,000 ft and 24,000 ft are designed to use loaded cable pairs with a maximum loop resistance of 1500 Ω. Loops longer than 24,000 ft are designed according to *Modified Long Route Design (MLRD)*.

Ring Detector An electronic solid-state transistorized device designed to detect the presence of a ringing signal (a 90+V AC signal).

Ringer An electromechanical device (relay) that vibrated when a 90+V AC signal was received by the telephone. The vibrating device would strike metal gongs to create a ringing sound.

Roaming The mobile telephone of AMPS or DAMPS is continuously reporting its location to the closest cell site. As you drive across country and pass from one service provider to another, the central controller of the MTSO owned by other service providers will report your location over the SS7 network to your home base.

RS-232 See *EIA-232*

Rural Electrification Administration (REA) Department of the federal government established during the Depression under President Franklin D. Roosevelt by the Rural Electrification Act of 1936. This act put people to work bringing electricity to rural America. The "Act" was amended in 1949 to bring telephone service to rural America. This department was renamed to the Rural Utility Services Department.

Selective ARQ With selective ARQ, only frames that have errors are retransmitted. When a transmitting modem is using selective ARQ and receives a NAK, only the frame for which the NAK was received is retransmitted.

Serial Transmission The transmission of bits, one behind the other, over one transmission medium.

Servers Circuits available to handle the offered load.

Service Profile Identifier (SPID) Contained in a database maintained by the LEC of ISDN line equipment and the services available to each line. When a terminal is activated, the SPID is used to determine which services a terminal can have access to.

Sidebands The two additional signals generated by mixing two signals. The upper sideband represents the sum of the carrier and modulating frequencies. The lower sideband will have a frequency that is the difference between the carrier frequency and the modulating frequency. Both sidebands contain the intelligence of the modulating signal.

Sidetone In a telephone set, some of the transmitted signal is purposely coupled by the *hybrid network* to the receiver so you can hear yourself in the ear covered by the telephone receiver. This signal is called sidetone.

Signaling Anything that serves to direct, command, monitor, or inform. In telecommunications, signals are needed to inform a local switching system that someone wishes to place a call, provide directions to the PSTN on how the call is to be connected, and inform the called party that a call is waiting. Signaling is divided into four basic categories: (1) supervisory, (2) addressing, (3) alerting, and (4) progress.

Signaling System 7 (SS7) A signaling network that can only interface to computer-controlled (SPC) switching systems. CPUs on the signaling network are called signal transfer points (STPs) and pass messages between the service signaling points (SSPs) of switching systems to provide switching instructions. The CPU of the SPC switch is programmed to take specific actions for each individual message received. This Common Channel Signaling technique, using a separate network of computers to send and receive messages as signals, has been adopted by the International community as CCIS #7 and is referred to as Signaling System 7 (SS7).

Simplex Transmission Transmission of signals in one direction only is called simplex communications.

Sine Wave A graphical representation of an AC signal along a time line.

Sliding Window ARQ See *Continuous ARQ*

Space Division Multiplexing A process in which multiple communications are placed over many different wire pairs inside one cable. Each communication channel (voice or data) occupies its own space (occupies its own set of wires).

Space Division Switching A form of switching (SXS and XBAR) where each conversation occupies its own distinct and separate wire path through the switching system.

SPC Central Exchange An automated switching system which is controlled by a microprocessor and Stored Program Control. Digital switching systems

Specialized Common Carrier Ruling of 1971 An extension of the 1969 MCI Decision by the FCC. This ruling allowed any Common Carrier to handle Private Line Networks.

Standby Processor A central processor that runs in parallel with the main processor. If the main processor fails, the standby processor takes over.

Station Equipment The largest segment of station equipment is the telephone but the term *station equipment* has been broadened to include anything a customer attaches to a telephone line. The most common piece of station equipment is the telephone. A modem, CSU/DSU, personal computer, and keysystem are all referred to as station equipment.

Straight-Line Ringer A ringer that will operate on any ringing signal. This type of ringer is used on single-party private lines.

Stored Program Control (SPC) Digital switching systems controlled by a computer using stored programs.

Subscriber Carrier A device used to multiplex more than one customer onto one facility. Older systems were analog devices using FDM to multiplex up to eight customers over one cable pair. The most prevalent system in use today is *Subscriber Line Carrier–96 (SLC-96)*.

Subscriber Line Carrier–96 (SLC-96) SLC-96uses TDM to multiplex 96 lines over two cable pairs. Multiple SLC-96 systems are often multiplexed onto a fiber facility.

Superimposed Ringer The type of ringer used by Bell on four-party lines. The ringer included a diode. Ringing signals were superimposed on top of a DC voltage. The diode only allowed the signal with the correct polarity of DC for that station to pass to the ringer.

Supervisory Signals Used to monitor the status of a line or trunk circuit. A switching system is informed whether a line or trunk is idle or busy by supervisory signals.

Switchboard The manual switchboard was used by operators to establish calls between two telephones. The manual switchboard was replaced by an automated switch.

Synchronous Transmission The transmission of data as blocks of bytes. Synchronization of the receiver occurs from a special bit pattern called a *sync signal* that is placed in front of the block of data information.

Tandem Exchange Class 1, 2, 3, and 4 exchanges that are part of the PSTN.

Tariff The document filed by a Common Carrier which defines in detail any service proposed by the Carrier and the charge proposed for the service. For Interstate Service, the Tariff is filed with the FCC. For Intrastate services, the Tariff is filed with the PUC.

Telecommunications The communications of voice or data over long distances using the Public Switched Telephone Network or privately owned networks.

Telecommunications Reform Act of 1996 Effectively replaces the 1984 MFJ. This act is designed to accelerate competition for providing services in the local and long distance markets. It eliminates the franchised local territory concept. The LEC is no longer the only service provider in town. Other common carriers can come into an *incumbent local exchange carrier (ILEC)* territory and provide local telephone service. These new local service providers are called *competitive local exchange carriers (CLECs)*. Prior to this act, the older incumbent LECs (ILECs) were restricted to providing local services only. Under this act, if the ILEC can prove that it has not hindered competition in its local exchange territories, it can now offer long distance services.

Terminal Adapter (TA) A device that interfaces a non-ISDN device to a NT1 in BRI and to a NT2 in a PRI environment.

Terminal Equipment 1 (TE1) Equipment designed to ISDN standards and for direct interface to a NT1.

Terminal Equipment 2 (TE2) Equipment that is non-ISDN compliant. TE2 equipment must be attached to a terminal adapter.

Terminal Equipment Identifier (TEI) Specific number assigned to each TE1 and TA. The TEI is usually assigned automatically the first time the terminal device is used. LAPD will invoke a query of the SPID database to obtain the TEI. The TEI is used by the LE and ET to determine what service to provide for a B-channel.

Time Division Multiple Access (TDMA) TDMA technology is used by DAMPS. With TDMA, each conversation occupies a time slot on a common transmission medium. In DAMPS, each conversation occupies a time slot on a particular radio frequency.

Time Division Multiplexing (TDM) The process of converting each speech path into samples, then combining the different samples by transmitting each sample at a different time, so they may be sent over one transmission medium.

Time Division Multiplex (TDM) Loop The time division switching component of a SPC switch. The early TDM loops were 30-channel loops. TDM loops are now 512-channel or higher.

Time Division Switching A form of switching (Stored Program Control or SPC) where each conversation occupies its own distinct and separate time slot on a common wire path through the switching system.

Time Slot Interchange (TSI) The heart of the switching matrix in a digital switching system. The TSI is a space division switching component. One TSI is needed for every channel on a TDM loop. The TSI connects a time slot on the TDM's transmit bus to a time slot on the TDM's receive bus.

Traffic A term used for the volume of calls placed over a circuit, group of circuits, or a switching system. If a system is handling 10,000 calls and hour, we say the traffic is 10,000 calls an hour.

Trailer A trailer is used in synchronous communications. A trailer is data placed behind the block of information transmitted. The trailer cnotains parity checking information and the address of the sender.

Transducer A device that converts energy from one form into another. The transmitter converts sound energy into electrical energy, and the receiver converts electrical energy into sound energy.

Translations Software A database counting line translations tables and trunk translations tables. A database of how ports are assigned and what features are assigned to each port.

Translator In a Stored Program Control switching system, a database that converts dialed digits into switching instructions.

Transmitter The device responsible for sending information or a message in a form that the receiver and medium can handle. The device in a telephone that converts the air pressure of a voice signal into an electrical signal that represents the voice.

Trellis Coded Modulation (TCM) TCM is a forward error-correction technique. An extra bit is added to the bits of data transmitted to help the receiver decode the data more reliably. QAM transmits 4 bits at a time. TCM added a 5th bit to the 4-bit code for error correction purposes.

Trunk Controller A circuit card in the trunk shelf that contains a microprocessor and EPROM, to control the activities of trunk circuits in that shelf.

Tuned Ringer A ringer that will operate on only one ringing signal. These ringers were used on party lines. They were tuned to only ring on the signal assigned to that party.

2500 Telephone Set The standard single-line telephone, which contains a DTMF dialpad.

2B1Q 2 bits = 1 quat. The coding technique used for a digital subscriber line. This coding technique allows 2 bits to be transmitted at one time by using four distinct voltage levels. Each voltage level can represent a particular 2-bit code.

Universal Asynchronous Receiver/Transmitter (UART) A piece of hardware (an integrated circuit chip) whose purpose is to interface a device using parallel transmission to a device using serial transmission. Every serial port contains a UART between the parallel data bus of the PC (or modem) and the serial port.

Universal Call Distributor (UCD) The first automated system for distributing calls to agents. Calls were distributed to agents according to their position in a queue. The first agent received more calls than the second agent, and so on.

Unlimited Queue Assumes customers in a queue are willing to wait forever to get service.

Varistor A specially designed resistor. As the voltage across the resistor increases, the resistance increases.

Voice Frequency Repeater (VFR) A device that will amplify voice signals in both directions to compensate for the extra decibel loss of long loops.

Voltage A measure of how much potential difference or electrical pressure exists between the two terminals of a battery (or some other device that generates electricity). This electrical pressure is called electromotive force (EMF) and is measured in volts.

Wave Division Multiplexer The hardware device attached to each end of a fiber pair that multiplexes 16 OC-48 systems onto the fiber pair.

Wave Division Multiplexing (WDM) Technology that splits the light spectrum into many different frequencies. A *time division multiplexer* is then assigned to each of the different light wave frequencies. WDM is used to multiplex multiple OC-48 (or OC-192) systems over one fiber pair.

Wet Cell A chemical cell that generates electricity via chemical means between an electrolytic solution and two dissimilar metals. The electrolytic solution is a liquid. The battery in a Stored Program Control central exchange is composed of 24 wet cells.

Wired for The wiring is in place to support X number of lines. A system is usually installed with "wired for" greater than "equipped for." This allows additional lines, trunks, registers, and so on to be added to a system simply by plugging the appropriate *printed circuitboard (PCB)* in a vacant card slot and adding information for the PCB in the database.

Word In a PC, the number of adjacent bits that can be manipulated or processed. This depends on the number of leads comprising the data bus, which is in turn dependent on the number of registers attached to the data bus. A word can be 8 bits or 1 byte in an 8088 microprocessor environment. A word can be 32 bits or 4 bytes in a 80486-based PC, or it can be 64 bits (8 bytes) in a Pentium-based PC.

X.25 The interface standard to a packet data network.

References

American Telephone and Telegraph Company Bell Laboratories, 1983. *Engineering and Operations in the Bell System.* Murray Hill, NJ: AT&T Bell Laboratories.

American Telephone and Telegraph Company, 1974. *Telecommunications Transmission Engineering, vol. 1.* Murray Hill, NJ: AT&T Bell Laboratories.

American Telephone and Telegraph Company, 1975. *Telecommunications Transmission Engineering, vol. 3.* Murray Hill, NJ: AT&T Bell Laboratories.

American Telephone and Telegraph Company, 1977. *Telecommunications Transmission Engineering, vol. 1.* Murray Hill, NJ: AT&T Bell Laboratories.

Bell Communications Research, 1990. *Telecommunications Transmission Engineering, vol. 1 (Principles), 3rd ed.* Moristown, NJ: BellCore.

Bell Communications Research, 1990. *Telecommunications Transmission Engineering, vol. 2 (Facilities), 3rd ed.* Moristown, NJ: BellCore.

Bell Communications Research, 1990. *Telecommunications Transmission Engineering, vol. 3 (Net-works and Services), 3rd ed.* Moristown, NJ: BellCore.

Comer, Douglas, 1988. *Internetworking with TCP/IP: Principles, Protocols, and Architectures.* Englewood Cliffs, NJ: Prentice Hall.

Feher, Kamilo, 1995. *Wireless Digital Communications.* Upper Saddle River, NJ: Prentice Hall.

Floyd, Thomas L., 1995. *Electronic Circuit Fundamen-tals, 3rd ed.* Englewood Cliffs, NJ: Prentice Hall.

Freeman, Roger L., 1981. *Telecommunications Trans-mission Handbook, 2nd ed.* New York, NY: John Wiley & Sons.

Freeman, Roger L., 1989. *Telecommunications System Engineering, 2nd ed.* New York, NY: John Wiley & Sons.

Gary, Vijay K., and Patrick J. O'Connor, 1986. *Voice/ Data Telecommunications Systems.* Englewood Cliffs, NJ: Prentice Hall.

Green, James H., 1992. *The Irwin Handbook of Telecommunications, 2nd ed.* Burr Ridge, IL: Irwin Professional Publishing.

Gurrie, Michael L., and Patrick J. O'Connor, 1986. *Voice/Data Telecommunications Systems.* Englewood Cliffs, NJ: Prentice Hall.

Hioki, Warren, 1995. *Telecommunications, 2nd ed.* Upper Saddle River, NJ: Prentice Hall.

Kessler, Gary C., and Peter Southwick, 1997. *ISDN, Third Edition.* New York, NY: McGraw Hill.

Maritime, Roberta R., 1994. *Basic Traffic Analysis.* Upper Saddle River, NJ: Prentice Hall.

Ramteke, Timothy, 1994. *Networks.* Englewood Cliffs, NJ: Prentice Hall.

Scott, Stan, and Steven Fox, 1990. *Voice/Data Telecom-munications for Business.* Englewood Cliffs, NJ: Prentice Hall.

Sherman, Kenneth, 1990. *Data Communications: A User's Guide, 3rd ed.* Englewood Cliffs, NJ: Prentice Hall.

Stallings, William, 1996. *Data and Computer Communications, 5th ed.* Upper Saddle River, NJ: Prentice Hall.

World Wide Web Site References:

1. http://www.adsl.com
2. http://www.analysys.com
3. http://www.att.com
4. http://www.gte.com
5. http://www.mci.com
6. http://www.pactel.com
7. http://www.sbc.com
8. http://www.sprint.com
9. http://thomas.loc.gov/cgibin/bdquery/z?d104:sn00652:

Index

armature, 85
ARQ (Automatic Retransmission Request), 245
arrival rate, 405
    of calls, 392
arrivals, 420
ASCII coding, 238
Asymmetric Digital Subscriber Line (ADSL), 232, 314
asynchronous transfer mode (ATM), 312
asynchronous transmission, 241
AT&T, 3, 12, 21, 50
    wireless, 450
ATB (All Trunks Busy), 385, 386, 388, 401
    meter, 386, 398
ATM (asynchronous transfer mode), 312
atom, 56
AUC (authentication center), 452
audiotext, 375
authentication center (AUC), 452
automated
    attendant, 167, 360, 371, 376
    call distribution, 280, 284, 344, 355, 435
    switching system, 174
Automated Number Identification (ANI), 13, 277
automatic call distributor (ACD), 280, 284, 344, 355, 435
automatic retransmission request (ARQ), 245
average bouncing busy hour (ABBH), 387, 389
average busy hour
    data, 387
    peg count, 324
average carried load, 388
average conversation time (ACT), 436
average hold time (AHT), 324, 388–389, 392, 398, 403, 427
average number of customers waiting, 425
average time delay, 425

B

B8ZS (Binary 8 Zero Substitution), 111
B bit, 275
B-channels, 295
B-ISDN (Broadband ISDN), 312
Baby Bells, 11
backplane, 330
balancing network, 155
bandwidth, 97
base station
    controller, 452
    transmitter, 444

base transceiver station (BTS), 449, 452, 453
basic rate interface, 294
basic rate ISDN (BRI ISDN), 294, 313, 316
basic trading area (BTA), 450
battery, 33, 59, 186
baud rate, 248
BCC (Blocked Calls Cleared), 390, 414
BCD (Blocked Calls Delayed), 390
BCH (Blocked Calls Held), 390, 414
Bearer (B) channel, 165, 294
Bell LECs, 16, 20
Bell T1 Standard, 107
BH (Busy Hour), 324, 386, 401
BIC (Binary Interface Card), 221
Binary 8 Zero Substitution (B8ZS), 111
binary signal, 232
bipolar signal, 232
bipolar violation, 111
bit, 243, 275
bit-robbed signaling, 110, 273
blockage, 399
blocked call(s), 183, 324, 386
    blocked calls cleared (BCC), 390
    blocked calls held (BCH), 390, 414
blocked load, 389, 403, 416
blocking, 181, 329
    network, 183
blue box, 200, 276
BOCs, 15, 24
BPV, 111
branch feeder cable, 140
BRI ISDN (Basic Rate ISDN), 294, 313, 316
bridged ringer, 159
broadband ISDN (B-ISDN), 312
BSC (Base Station Controller), 452
BTA (Basic Trading Area), 450
BTS (Base Transceiver Station), 449, 452, 453
bus interface card, 221
busy hour, 324, 386
busy signal, 267
byte, 241

C

C bit, 275
C conditioning, 290
C message

multiplexing, 47, 91, 143
mutual capacitance, 134

N

n (nano), 108
NAK (negative acknowledgement), 246
nano (n), 108
narrowband ISDN, 312
negative exponential probability distribution, 421
network, 184, 215
    design, 383–384, 390, 407, 413
    engineering department, 383
    hierarchy, 18
    links, 221
    termination (1, 2, and 12), 298, 300, 311
No. 4 ESS, 219
No. 5 ESS, 219
noise on line, 139
noise on local loop, 139
nonblocking, 181
    matrix, 399
    network, 328–329, 399
North America Numbering Plan, 185
North American Telephone Numbering Plan, 44
NPA (number plan area), 185, 199
NT1 (network terminator 1), 299
NT12 (network terminator 12), 311
NT2 (network terminator 2), 300
Null Modem, 234
numbering plan area (NPA), 185, 199
Nyquist, 105

O

OC-192, 147
OC-48, 113, 147
OC-96, 147
offered load, 389–392, 394, 398, 403, 407, 415
off-premise extension (OPX), 341
Ohm's Law, 62, 126
OML (outgoing matching loss), 400
open systems interconnection, 282
open wire, 47, 121
Operator Switchboard, 39
Optical Network, 148

OPX (off-premise extension), 341
OSI, 282
out-of-band signaling, 275, 293
outside plant, 125
    engineers, 121
overhead channel, 295
overlay programs, 219, 326

P

P.011 GOS, 400
p (pico), 108
PABX (Private Automated Branch Exchange), 46, 181, 321
packet network, 254
packet switched network, 254
PAD (packet assembler-disassembler), 254
PAM (pulse amplitude modulation), 102–103
parallel resistance formula, 66
parallel transmission, 241
parameters software, 218
patch cord, 38
PBX (private branch exchange), 7, 46, 169, 181, 321
PC (peg count), 324, 385, 388, 398, 401
    meter, 386–87, 398
PCM (pulse code modulation), 103–104, 449
PCM/TDM (Pulse Code Modulated/Time Division Multiplex), 206
PCS (personal communication system), 15, 16, 442, 449
    1900, 442
    Technology, 16, 50
PDN (public data network), 174, 254, 311
PE (peripheral equipment), 323
pedestals, 142
peg count (PC), 324, 385, 388, 398, 401
percent occupancy, 405
peripheral equipment (PE), 323
peripherals, 322
Personal Communication System (PCS), 15, 16, 442, 449
personal telephone number (PTN), 442
phase modulation (PM), 249
PIC (preferred interexchange carrier), 14, 176
pico (p), 108
plastic insulated cable, 47, 121
PM (phase modulation), 249
PN (pseudorandom number), 451

touchtone telephone, 192
traffic, 324, 383
    design tables, 390
    engineering, 384
    on the system, 383
    reports, 393
    study, 401
    usage recording, 388
    usage register (TUR), 385, 388
transceiver, 33, 84, 154
transducer, 33, 84
translations, 218
translator, 192, 197, 199, 396
transmitter, 31, 33, 153, 159
trellis coded modulation (TCM), 251, 316
trunk, 42
    controllers, 205
    side access, 13
    translation table, 218
TSI (time slot interchange), 212, 220
Tuned Circuits, 78
tuned ringer, 82, 157
TUR (traffic usage register), 385, 388
twisted pair, 119

U

U interface point, 299
U reference point, 299
UART (Universal Asynchronous Receiver Transmitter),
        106, 239, 312
UCD (uniform call distributor), 365
unbalanced line, 139
uniform call distributor (UCD), 365
uniform dialing plan, 341
United Telecommunications, 14
Universal Asynchronous Receiver Transmitter
        (UART), 106, 239, 312
Universal Service, 21
unlimited-length queue, 423

V

V reference point, 311
VANs (Value Added Networks), 12
varistor, 159
VDU (video display unit), 165
vertical cables, 332
VFR (voice frequency repeater), 144
virtual private network (VPN), 342
virtually non-blocking network, 329
visitor location register database (VLR), 452, 454
visual display unit (VDU), 165
VLR (visitor location registry), 452, 454
voice
    mail, 163,369
    path, 398
    signals, 35
voice frequency repeater (VFR), 144
VPN (virtual private network), 342

W

waiting-line theory, 420
WATS (wide area toll service), 283, 340
wave division multiplexer, 114
wave division multiplexing (WDM), 114, 147
wet cell, 59
wide area toll service (WATS), 283, 340
wire centers, 128
wired for, 325
word, 241
wrap-up, 362
    time, 436
WUT, 436

X

XBAR, 187, 191–192